The United States and Canada:

The Land and the People

Second Edition

The
United States
and
Canada:

The Land and the People

EDITORS:

Arthur Getis
San Diego State University

Judith Getis

I. E. Quastler
San Diego State University

CONTRIBUTORS:

Edward Aguado
San Diego State University

Jerome D. Fellmann
*University of Illinois at
Urbana-Champaign*

Larry R. Ford
San Diego State University

Arthur Getis
San Diego State University

Judith Getis

David H. Kaplan
Kent State University

Bob R. O'Brien
San Diego State University

Philip R. Pryde
San Diego State University

Gerard Rushton
University of Iowa

John R. Weeks
San Diego State University

Boston   Burr Ridge, IL   Dubuque, IA   Madison, WI   New York   San Francisco   St. Louis
Bangkok   Bogotá   Caracas   Lisbon   London   Madrid
Mexico City   Milan   New Delhi   Seoul   Singapore   Sydney   Taipei   Toronto

McGraw-Hill Higher Education ℘

A Division of The **McGraw-Hill** *Companies*

THE UNITED STATES AND CANADA: THE LAND AND THE PEOPLE
SECOND EDITION

Published by McGraw-Hill, an imprint of The McGraw-Hill Companies, Inc., 1221 Avenue of the Americas, New York, NY 10020. Copyright © 2001, 1995 by The McGraw-Hill Companies, Inc. All rights reserved. No part of this publication may be reproduced or distributed in any form or by any means, or stored in a database or retrieval system, without the prior written consent of The McGraw-Hill Companies, Inc., including, but not limited to, in any network or other electronic storage or transmission, or broadcast for distance learning.

Some ancillaries, including electronic and print components, may not be available to customers outside the United States.

♻ This book is printed on recycled, acid-free paper containing 10% postconsumer waste.

1 2 3 4 5 6 7 8 9 0 QPH/QPH 0 9 8 7 6 5 4 3 2 1 0

ISBN 0–07–235677–4

Vice president and editor-in-chief: *Kevin T. Kane*
Publisher: *JP Lenney*
Sponsoring editor: *Robert Smith*
Developmental editor: *Renee Russian*
Editorial assistant: *Jenni Lang*
Senior marketing manager: *Lisa L. Gottschalk*
Senior project manager: *Kay J. Brimeyer*
Associate producer: *Judi David*
Senior production supervisor: *Sandra Hahn*
Design manager: *Stuart D. Paterson*
Cover designer: *Annis Leung*
Cover image: Canada, Ontario, Toronto, Skyline—*Tony Stone Images,* top photo and three at base—*PhotoDisc*
Senior photo research coordinator: *Lori Hancock*
Photo research: *Alexandra Truitt & Jerry Marshall*—www.pictureresearching.com
Compositor: *Shepherd, Inc.*
Typeface: *10/12 Times Roman*
Printer: *Quebecor Printing Book Group/Hawkins, TN*

The credits section for this book begins on page 421 and is considered an extension of the copyright page.

Library of Congress Cataloging-in Publication Data

The United States and Canada : the land and the people / editors, Arthur Getis, Judith Getis,
Imre E. Quastler. — 2nd ed.
    p. cm.
   Includes bibliographical references and index.
   ISBN 0–07–235677–4 (acid-free paper)
    1. United States—Geography. 2. Canada–Geography. I. Getis, Arthur, 1934– .
II. Getis, Judith, 1938– . III. Quastler, I.E., 1940–.

E161.3 .U715 2001
971—dc21   00–029207
                                   CIP

www.mhhe.com

BRIEF CONTENTS

CONTENTS

5

Political Geography   115

6

Agriculture, Gathering, and Extractive Industries   141

7

Industrial and Commercial Organization   175

8

Modern Transportation and Communication Systems   201

9

Cities   227

10

Neighborhoods   257

11

Recreational Resources   283

12

Human Impact on the Environment   315

PREFACE

This second edition of *The United States and Canada: The Land and the People* continues in the tradition of the first edition of 1995. In it we have endeavored to update the chapters to the most recent data available, and to make certain changes suggested by reviewers that we thought desirable. However, the fundamental philosophy and organization of the original edition remain in place.

Most geography textbooks about the United States and Canada are organized within a regional framework. They divide the two countries into major regions and then discuss selected topics, such as the physical environment, agriculture, and economy, as they pertain to those regions. In this book, the process has been reversed. The editors believe that such identification of major regions tends to be somewhat arbitrary and overly generalized. A region is a portion of the earth's surface that differs from other areas because of some distinctive feature or features. There are as many regions as there are features, and they do not all coincide in any single area. Thus, the physiographic provinces of the United States and Canada differ from regions distinguished on the basis of agriculture, which in turn differ from regions defined by ethnic composition, religious affiliation, or a host of other factors.

Our approach is to let the systematic topics that geographers study provide the organizational framework for the book. We present a systematic view of the interaction of people and the land. Each factor, whether it is inches of rainfall or the location of iron ore deposits, contributes to the continental mosaic and helps us better understand how Americans and Canadians affect and are affected by the landscape.

Far from being ignored, regions are given their appropriate place by being defined on the basis of the topic under consideration. Thus, Chapter 2 identifies the landform and climatic regions of the United States and Canada, while culture regions are discussed in Chapter 13.

One area, Canada, does receive single-chapter treatment in Chapter 14, "The Canadian Difference." Although Canadian examples are included in earlier chapters where relevant, we believe that Canada also merits a special chapter. Instructors of courses that consider only the geography of the United States can omit this chapter.

A main feature of the text is its emphasis on the changes that have occurred in the last half-century. We attempt to explain the present state of the people-land interface by emphasizing the important trends and events of the last fifty years.

Learning Aids

Most chapters begin with a vignette designed to arouse students' curiosity and sustain their interest in the subject matter that follows. Each chapter contains several boxed inserts that amplify ideas in the text or introduce related examples. The Summary that appears at the end of each chapter reiterates the main points.

New terms and special usages of common words and phrases appear in boldface or italic type. Those in boldface are in the Key Words list at the conclusion of each chapter and are defined in the Glossary at the end of the text.

Gaining Insights questions focus on the important concepts developed in the chapter. They are designed to help students check their understanding of the chapter material. The Selected References listing includes relatively widely available books and journal articles that expand on topics presented in the chapter and provide the basis for further study.

Acknowledgments

Although eleven authors contributed to this book, our hope is that it does not read like an edited work. It was conceived of as a single piece, and all authors followed the same guidelines in writing their chapters. Many of the authors have years of experience teaching courses about the geography of the United States and Canada.

As editors, we owe a special debt of gratitude to our colleagues for their interest in this project, the quality of their work, and their willingness to abide by publisher's deadlines. The authors of individual chapters are:

Chapter 1: Judith M. Getis
Chapter 2: Edward Aguado, Philip R. Pryde, and I. E. Quastler, *San Diego State University*
Chapter 3: I. E. Quastler, *San Diego State University*
Chapter 4: John R. Weeks, and I. E. Quastler, *San Diego State University*
Chapter 5: I. E. Quastler, *San Diego State University*
Chapter 6: Jerome D. Fellmann, *University of Illinois at Urbana-Champaign*

Chapter 7: Gerard Rushton, *University of Iowa*
Chapter 8: I. E. Quastler, *San Diego State University*
Chapter 9: Arthur Getis and Larry R. Ford, S*an Diego State University*
Chapter 10: Arthur Getis, *San Diego State University*
Chapter 11: Bob R. O'Brien, *San Diego State University*
Chapter 12: Judith M. Getis
Chapter 13: Larry R. Ford, *San Diego State University*
Chapter 14: David H. Kaplan, *Kent State University*

A number of colleagues from around the country responded to a publisher's survey and made thoughtful suggestions' about the first edition's organization and topic coverage. With pleasure, we acknowledge the contributions of:

Michael Albert
 University of Wisconsin-River Falls
James G. Ashbaugh
 Portland State University
D. Gordon Bennett
 University of North Carolina-Greensboro
Kathryn Black
 William Paterson College
Elaine F. Bosowski
 Villanova University
Marshall E. Bowen
 Mary Washington College
Ayse Can
 Syracuse University
Robert E. Clarke
 University of Northern Iowa
Roman Cybriwsky
 Temple University
Leslie Dean
 Riverside Community College
Lee Dexter
 Northern Arizona University
Joseph Enedy
 James Madison University
Gerald Fish
 Winthrop University
Michael M. Folsom
 Eastern Washington University
Jack Ford
 Shippensburg University of Pennsylvania
Gary S. Freedom
 McNeese State University
Jerry Gerlach
 Winona State University
Martin Glassner
 Southern Connecticut State University
Anthony F. Grande
 Hunter College-City University of New York
Roger Henrie
 Central Michigan University
Virgil Holder
 University of Wisconsin-LaCrosse

Deryck W. Holdsworth
 Pennsylvania State University
William M. Holmes
 University of North Texas
John F. Jakubs
 University of South Carolina
Robert L. Janiskee
 University of South Carolina
Ronald A. Janke
 Valparaiso University
David T. Javersak
 West Liberty State College
John V. Jezierski
 Saginaw Valley State University
Eric S. Johnson
 Illinois State University
Les A. Joslin
 Central Oregon Community College
Larry King
 Portland Community College
Maria S. Koshel
 Hofstra University
Charles F. Kovacik
 University of South Carolina
Steve LaDochy
 California State University-Los Angeles
Joseph S. Leeper
 Humboldt State University
Ann M. Legreid
 Central Missouri State University
Alan A. Lew
 Northern Arizona University
Ralph E. Lewis
 Eastern Oregon State College
Francis E. Lindsay
 Keene State College
Joseph T. Manzo
 Concord College
Harold M. Mayer
 University of Wisconsin-Milwaukee
Harold McConnell
 Florida State University
Harold Meeks
 University of Vermont
Cynthia A. Miller
 Mankato State University
Douglas C. Munski
 University of North Dakota
Stanley F. Norsworthy
 California State University-Fresno
Richard S. Palm
 University of Wisconsin-Eau Claire
Bob Phillips
 University of Wisconsin-Platteville
Leon S. Pitman
 California State University-Stanislaus

William D. Puzo
California State University-Fullerton

Nicholas Polizzi
Cypress College

Karl Raitz
University of Kentucky

Roger Reede
Southwest State University

Timothy F. Reilly
University of Southwestern Louisiana

Roger L. Richman
Moorehead State University

Charles Roberts
Florida Atlantic University

Gregory S. Rose
Ohio State University at Marion

Thomas A. Rumney
State University of New York-Plattsburgh

Richard Santer
Ferris State University

Steven Scott
University of Northern Colorado

Daniel A. Selwa
Coastal Carolina College

Paul Shott
Plymouth State College

Rodman E. Snead
University of New Mexico

Dennis Spetz
University of Louisville

George A. Van Otten
Northern Arizona University

C. Edwin Williams
Southeast Missouri State University

Robert Wingate
University of Wisconsin-LaCrosse

Perry Wood
Mankato State University

Reviews of the 1st edition manuscript helped shape and improve its content. We are indebted to the following reviewers for their advice, suggestions, and corrections:

Roger W. Stump
State University of New York-Albany

Daniel Jacobson
Michigan State University

Marvin W. Baker, Jr.
University of Oklahoma

Barton M. Hayward
Lander University

Samuel R. Sheldon
St. Bonaventure University

James Bingham
Western Kentucky University

Albert J. Larson
University of Illinois at Chicago

Robert B. Gould
Moorehead State University

As the person primarily responsible for updating the second edition, I would like to thank certain individuals for their help. At McGraw-Hill, I owe special thanks to Sponsoring Editors Daryl Bruflodt and Bob Smith, to Developmental Editors Renee Russian and Kristine Fisher, and to Senior Project Manager Kay J. Brimeyer, all of whom were very helpful whenever I needed answers to my many questions. In addition, Judith Getis provided numerous useful suggestions in the early phases of the work, Dr. John O'Leary of San Diego State University gave me some much-appreciated feedback on the chapter on physical geography, and Dr. Kenneth C. Martis of West Virginia University was most helpful in reviewing the chapter on political geography. Most of all, I would like to thank my wife, Reba Wright-Quastler, for many hours of surfing the Internet to help me find the most recent data on a whole host of topics. Similarly, many thanks are due to those people who conceived of and developed the Internet. Without their pioneering work, the task of updating this book would have been far more difficult and time-consuming.

I. E. Quastler
San Diego State University

CHAPTER
1

Introduction

On December 19, 1606, three small ships set sail from England. The *Discovery, Goodspeed,* and *Sarah Constant* carried 118 men and two women, would-be settlers in a distant land. By the end of the voyage, 16 of the men had died. The remainder disembarked on May 24, 1607, and in a malarial swamp on the James River near the mouth of Chesapeake Bay founded what would become the first successful English settlement in North America (Figure 1.1). The cluster of small thatched shelters that was called Jamestown was a fragile foothold on the edge of a vast continent sparsely settled by perhaps 2 to 4 million American Indians.

For months, it appeared as though Jamestown might perish, just as three earlier attempts by the English to settle America had failed. By the end of the year, 76 of the 104 original settlers were dead, victims of disease, malnutrition, and starvation. One of the survivors, George Percy, described the situation:

> *Our men are destroyed with cruel diseases, as swelling, flixes [dysentery], burning fevers, and by wars, and some departed suddenly, but for the most part they died of mere famine. There were never Englishmen left in a foreign country in such misery as we were in this new discovered Virginia. Our food was but a small can of barley sod in water, our drink cold water taken out of the river, which was at flood very salty, at low tide full of slime and filth, which was the destruction of many of our men. Our men night and day groaned in every corner of the fort most pitiful to hear. Some departed out of the world, many times three or four a night; in the morning their bodies trailed out of their cabins like dogs to be buried.*

("Observations by Master George Percy" in Lyon G. Tyler, ed., *Narratives of Early Virginia. 1606–1625,* pp. 21–22. Barnes & Noble, New York, 1907.)

Despite these inauspicious beginnings, the colony endured. Within a decade, nearly 2000 English immigrants—among them hundreds of women—joined their compatriots in Virginia. Farmers and artisans from Germany, Italy, and Poland took their places alongside the English. Not all immigration was voluntary; the arrival in 1619 of 20 Africans in chains marked the beginning of slavery in what would become the United States.

Less than 400 years have elapsed since the founding of Jamestown. In that time, two of the countries that comprise North America and that are the subject of this book—the United States and Canada—have become prosperous industrial nations. Both are among the Group of Seven industrialized countries. Their citizens rank high on such

Figure 1.1

Arrival at Jamestown, 1607. The colony was funded by the London Company, which was chartered by King James I to establish settlements on the American coast between the parallels of 34°and 41° North latitude. After a series of disasters, the capital of Virginia was moved to Williamsburg in 1699, and Jamestown fell into decay. Québec, founded in New France a year after Jamestown, has endured.

indices of material and social well-being as gross national product per person, energy consumption, caloric intake, life expectancy, and adult literacy. Before we begin our examination of the transformation of the United States and Canada from American Indian lands and colonial experiments to world powers, it will be helpful to review some basic geographic facts that put the continent in a global perspective.

Size and Location

Both Canada and the United States are extremely large, the second and fourth largest countries, respectively, in area in the world (Figure 1.2). Russia is first, and China is third largest. Canada covers almost 10 million square kilometers (3.8 million sq mi), the United States about 9.4 million square kilometers (3.6 million sq mi). Together, the two countries equal about one-eighth of the land area of the earth, or about twice the area of Europe.

Both countries are continental in dimension, extending from the Atlantic to the Pacific Ocean, a distance of more than 4800 kilometers (3000 mi). Canada, however, has a significantly greater portion of its territory in the northern latitudes than does the United States. It stretches over 40° of latitude, from the Great Lakes to the ice-covered islands of the Arctic Ocean, a distance of some 4670 kilometers (2900 mi). Its northernmost point, Cape Columbia on Ellesmere Island (83° North), lies just 765 kilometers (475 mi) from the North Pole. In contrast, the greatest north-south distance of the United States, excluding Alaska, is 2573 kilometers (1598 mi). Sparsely populated Alaska is located to the west of the northern part of Canada. As we shall see, the more northerly location of Canada with respect to the conterminous United States has significant implications for its climate and people. The greatest portion of the United States is generally at the same latitude as the Mediterranean Sea, while Canada's position is much more like that of Russia (Figure 1.3).

Climate

The physiography and broad latitudinal and longitudinal extent of North America ensure that nearly every type of climatic region exists somewhere on the continent (Figure 1.4). The United States seems to have it all, from the extreme, continuous cold of the polar region to the hot deserts of the Southwest. The hot, dry summers and mild winters of the coastal Mediterranean climate zone of California nurture such crops as oranges, grapes, and avocados, while the grasslands of the prairies support livestock and grain production. Only the tropical rain forest climate is missing on the mainland, but it can be found in parts of Hawaii. Not only is there a variety of climates in the United States, but compared to Europe, the daily fluctuations in weather are more pronounced.

As Figure 1.4 makes clear, Canada's northern location limits the types of climates found there. Much of the

---

AGREEMENT ON TERMS

The names given to geographic areas often are confusing. One standard dictionary gives four definitions of *America:* It is either the United States; North America; South America; or North, Central, and South America taken together. Likewise, *North America* can be taken to mean that portion of the Western Hemisphere that contains the countries of Central America, Mexico, Canada, the United States, Greenland, and the Caribbean islands. Frequently, however, it is considered to include just Mexico, the United States, and Canada, and some sources confine it to simply the latter two countries.

For many years, the name "Anglo-America" was applied to that part of the continent consisting of the United States and Canada, to distinguish it from the rest of the Western Hemisphere known as "Latin America." It was a convenient way to draw attention to major differences in the hemisphere, and its use is still common. In recent years, however, the appellation has fallen into disfavor. The critics of the term suggest it ignores major influences in the culture of both the United States and Canada that are not British in origin, and this rankles those who are proud of their own distinctive cultural heritage.

Faced with the dilemma of what to call the area under study in this book, we have chosen to use "the United States and Canada." We use it because no good shorthand term for the region consisting solely of those two countries is acceptable to all. "American" is used to describe things pertaining to or characteristic of the United States, such as people, cities, and culture; its counterpart is "Canadian."

Finally, we should note that a number of islands and island groups are possessions of the United States. These include, among others, Guam and the Northern Mariana Islands in the Pacific Ocean and Puerto Rico and the U.S. Virgin Islands in the Caribbean. A discussion of these possessions is outside the scope of this book, in which the "United States," denotes only the 50 states.

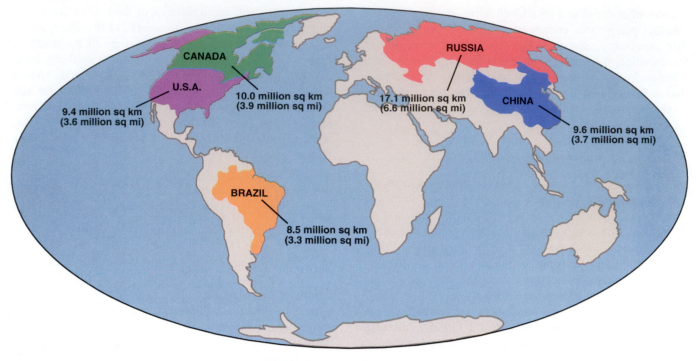

Figure 1.2
The five largest countries in the world in terms of area. These five occupy more than one-third of the land area of the earth.

Figure 1.3
The relative locations of the United States, Canada, and Europe.

country lies north of 50° latitude and is characterized by arctic and subarctic climates. Summers are short and cool, winters are long and very cold, and there is little precipitation in any season. Only in Canada's southern portion is the growing season long, warm, and wet enough for agriculture, and it is there that almost all Canadians live.

Physiography

Both Canada and the United States have a diversified physical environment. As Chapter 2 indicates, each has a variety of landform features, climates, soils, and vegetation. At the grossest level of generalization, one can say that the United States consists mainly of a rugged western highland area, a vast interior plain, and a lesser eastern upland characterized by rolling hills and low mountains (Figure 2.1). The western highlands extend parallel to the Pacific Coast in north–south trending ranges. The Rocky Mountains, the largest and easternmost of these ranges, extend from the Yukon basin in the north almost to the Rio Grande in the south. Along the western part of the Rocky Mountains is the Wasatch Range. Closer to the Pacific Ocean are the Sierra Nevada, Coastal, and Cascade Ranges.

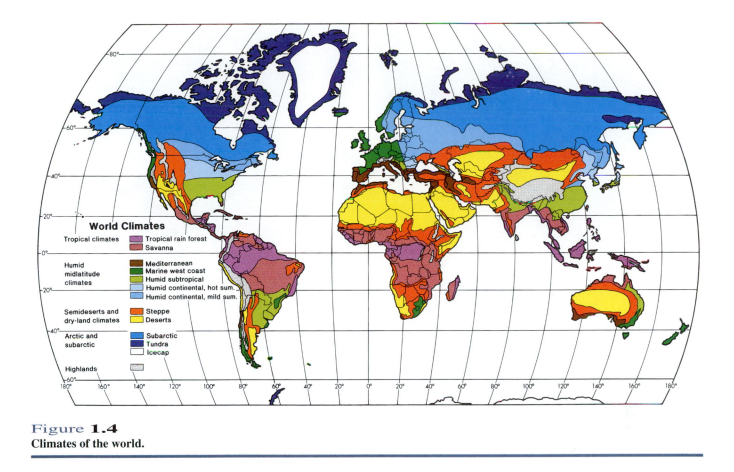

Figure 1.4
Climates of the world.

The great central plain of North America reaches from the Gulf of Mexico to the Arctic Ocean. It slopes eastward to the Appalachian Mountains and to the Laurentian Highland of Canada. Between the Appalachians and the Atlantic Ocean lies the long Atlantic coastal plain, which is widest in Georgia and becomes increasingly narrow as it extends northeastward to Long Island.

One of the largest river basins in the world drains the central plain. The Mississippi River is like the main vein of a huge leaf, with its many tributaries (such as the Missouri, Ohio, Tennessee, and Arkansas Rivers) the auxiliary veins. Together, the rivers drain more than 2.6 million square kilometers (1 million sq mi). Two centuries ago, the rivers gave explorers access to the continental interior and were a source of fresh water for settlements; today, these inland waterways, linked by canals, are important for waterborne transportation.

Natural Resources

Societies depend for their economic and material well-being on a variety of naturally occurring, exploitable materials. These include land, minerals, and most forms of energy. The availability of natural resources is a function of two things: the physical characteristics of the resources and human economic and technological conditions. The processes that govern the formation, distribution, and occurrence of natural resources are determined by physical laws over which people have no direct control. We take what nature gives us. What is considered a resource, however, is a cultural, not purely a physical, circumstance. American Indians may have viewed the resource base of Virginia as its forests, which provided shelter, fuel, and game animals on which they depended for food. European settlers viewed the forests as the unwanted covering of the resource they perceived to be of value: soil for agriculture.

Because the distribution of natural resources is the result of long-term geologic processes, it follows that the larger the country, the more likely it is to contain such resources. Canada and the United States possess abundant and diverse resources. Great size is no guarantee of natural resources, however. Other large countries, such as China and Brazil, are not as well endowed with as many natural resources as are the United States and Canada.

Soils

People's own raw energy resource is food, and securing an adequate supply of food is a paramount daily concern. The three most important determinants of the location of potentially productive agricultural resources are sunshine, water, and soil. Plants must receive an adequate amount of solar

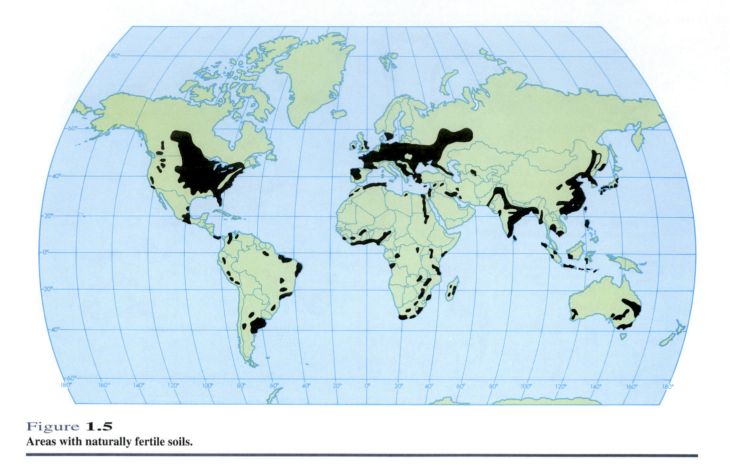

Figure 1.5
Areas with naturally fertile soils.

energy and fresh water for photosynthesis to take place. They also need soils rich in such nutrients as phosphorus, potassium, and nitrogen. As Figure 1.5 indicates, soils that are naturally fertile occupy a relatively small portion of the earth, and they are unevenly distributed among the continents.

North America is fortunate in having a large area with fertile soils. In general, the portion of the continent that contains productive farming soils and that receives adequate precipitation lies south of 50° latitude and east of the 98th meridian (Figure 1.6). That means that compared to Canada, the United States has an abundance of fertile land. Particularly well endowed are the interior plains and prairies and the lower Mississippi Valley. Areas with soils least suitable for agriculture include the unirrigated deserts of the southwestern states, the western highlands, the Laurentian Highland, and the boreal forests north of 50° latitude. When irrigated, some desert areas (such as the Imperial Valley of California) are unusually productive.

Minerals

Agricultural societies depend on such resources as fertile soil, domesticated animals, wood for cooking and heating, and wind and water to power windmills and waterwheels. However, it was the shift from such renewable resources to those derived from nonrenewable mineral resources that sparked the Industrial Revolution and gave population-supporting capacity to areas far in excess of what would be possible without these mineral energy sources. The enormous increase in individual and national wealth in industrialized countries has been built in large measure on an economic base of coal, oil, and natural gas. These *fossil fuels* provide heat, generate electricity, and run engines (Figure 1.7). Equally important are a variety of nonfuel minerals, both metallic and nonmetallic. They can be processed into steel, aluminum, and other metals, and into glass, cement, and other products. Our buildings, tools, and weapons are chiefly mineral in origin.

Mineral resources are unevenly distributed throughout the world. Some areas possess substantial quantities of the economically important mineral resources, while others have almost none. No country contains all of the important minerals, but, in general, the United States and Canada were bountifully provisioned by nature. They had ample supplies of the fossil fuels; of iron ore and some of the iron alloys, such as nickel; base metals, such as copper, lead, and zinc; precious metals (gold, silver, uranium); and nonmetallic minerals, such as phosphates and nitrates. The United States and Canada are among the top ten producers of cadmium, copper, lead, and other minerals.

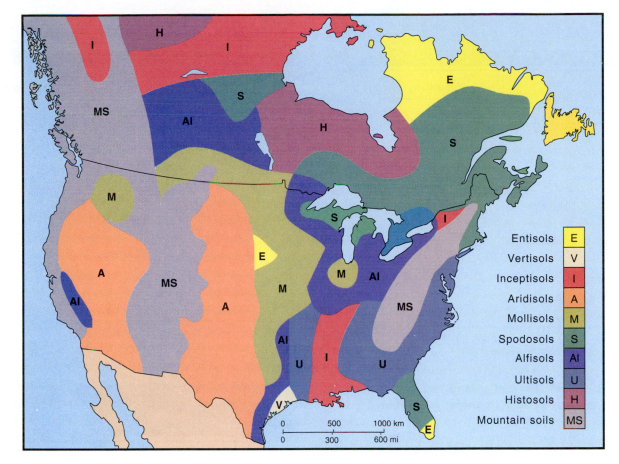

Figure 1.6
Soil types. The most highly productive agricultural soils are the alfisols and mollisols.

Figure 1.7
The American petroleum industry got its start in 1859 in Titusville, located along Oil Creek in northwestern Pennsylvania, when Edwin L. Drake and Billy Smith sank a well that yielded 400 gallons of oil a day. The ensuing oil boom was likened to the California gold rush as miners, operators, and speculators flooded into the area, seen here in 1865. Refined petroleum was used chiefly for burning in lamps and lubricating machinery until 1911, when gasoline for automobiles became the largest-selling petroleum product.

Population

The world's population, which now exceeds 6 billion, is unevenly distributed over the earth's surface. Indeed, one of the most striking features of a world population distribution map is the unevenness of the pattern. Some land areas are nearly uninhabited, others are sparsely settled, and still others contain dense agglomerations of people. More than 3.5 billion people, or almost 60% of the world total, live in Asia. Canada and the United States together have about 306 million people (2000), or just 5% of the world total.

Table 1.1 shows the world's ten leading countries for population in 2000. Note that all but three of the ten are in Asia. Its population of Table 1.1 shows 275 million makes the United States the third most populous country in the world. Canada's population (31 million) is significantly smaller than that of the United States—smaller, in fact, than that of the state of California. Population projections for 2025 show 335 million people in the U.S. and 38 million in Canada.

In world terms, both Canada and the United States have low population densities (Table 1.2). Although easy to calculate, measures of crude population density—the

Introduction

Table 1.1		
Ten Leading Countries with a Population Greater than 100 Million, 2000 and 2025 (Estimates)		
	Population (in millions)	
	2000	**2025**
China	1256	1408
India	1018	1415
United States	275	335
Indonesia	219	288
Brazil	174	210
Russia	146	139
Pakistan	141	212
Bangladesh	129	179
Japan	126	120
Nigeria	117	203

Source: U.S. Census Bureau, International Data Base.

Table 1.2	
Crude Population Densities of the Ten Most Populous Countries, Plus Canada, 1995	
	People per Square Kilometer
Bangladesh	925
Japan	332
India	315
Pakistan	182
China	131
Nigeria	123
Indonesia	109
United States	27
Brazil	19
Russia	9
Canada	3

Source: From World Resources 1996–1997.

number of people per unit area of land—conceal variations in the way people are actually distributed over a large area such as a country. As Figure 1.8 reveals, population densities vary significantly over the United States and Canada, a topic discussed in more detail in Chapter 4. There is a significant concentration of people in the northeastern United States and the adjacent portion of Canada, while very few people live poleward of 60° North or at very high altitudes. As Gertrude Stein noted, "In the United States there is more room where nobody is than where everybody is."

Both Canada and the United States are classified as urban countries; most people live in cities or their suburbs (Table 1.3), and an even higher percentage are expected to be urban in 2025. The ratio between urban and rural dwellers varies widely from country to country, from 100% urban in a city-state such as Singapore to less than 10% urban in Burundi and Rwanda in Africa. In general, urban residents greatly outnumber their rural counterparts in highly industrialized countries and in agricultural countries (for example, Australia and Argentina) where mechanization permits high yields with comparatively little farm labor.

Table 1.3		
Urban Population as a Percentage of Total Population, 2000 and 2025 (Estimates)		
	2000	**2025**
Brazil	78.3	88.9
Japan	77.6	84.7
Canada	76.7	83.7
United States	76.2	84.9
Russia	76.0	85.7
Nigeria	39.3	61.6
Indonesia	35.4	60.7
Pakistan	34.7	56.7
China	30.3	54.5
India	26.8	45.2
Bangladesh	18.3	40.4

Source: From World Resources 1996–1997.

What Is Geography?

Many academic disciplines study Canada and the United States, each bringing its own distinctive focus and questions. The economist, for example, might study differences and similarities in the foreign trade of the two economies, or how the labor markets differ. Sociologists might be concerned with family formation and divorce rates, or antisocial behavior. Geographers have two tradi-

tions that they use when studying these countries. One is the *spatial tradition,* concerned with explaining the location and distribution of things. The second is the *environmental tradition.*

Under the spatial approach, geographers try to make sense out of the location of a wide variety of phenomena.

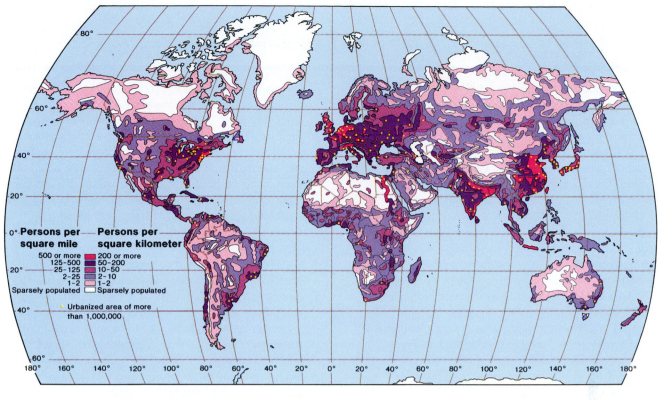

Figure 1.8

World population density. Latitude, temperature, aridity, soil type, and length of growing season all affect the distribution of population. Two-thirds of all people dwell in the midlatitudes, between 20° and 60° North.

Some might investigate where computer manufacturers or software firms locate, and why they choose such places. Others may want to locate the areas with the highest percentages of radio stations specializing in country music and to explain this geographic (spatial) pattern. Geographers in this tradition focus especially on two questions, (1) *where are these things located?* and (2) *why there?*

Maps can help to answer the "where" question. Geography has a long history of mapmaking (cartography), and students will see that this book contains numerous maps. Just answering the "where" question is never enough; the geographer must also convincingly explain "why there," i.e., why things are located as they are. She/he can study a wide range of subjects under this approach, including such physical geography topics as climates, vegetation, and soils, and cultural geography topics like population, city location, transportation, recreation, and manufacturing.

Under the environmental tradition (sometimes called the people–land tradition), geographers focus on the mutual relationship between people and their environments ("land" in the broadest sense). Environmental geographers study human-caused changes to the environment, such as changes to air and water quality from emissions and chemicals, and losses of open space, especially near cities. Their main objective is to identify the causes of these problems and, more importantly, to seek possible solutions to them. Other geographers in this tradition are interested in natural resources, addressing such questions as the likely future rates of oil and natural gas consumption, energy alternatives, the depletion of forests or fisheries, or the national and local policies that protect and conserve natural resources. Still others emphasize recreation geography, perhaps studying ecotourism or the environmental impacts of increasing numbers of recreational vehicles.

Organization of the Book

We begin our study by examining physical geography. Chapter 2, "Physical Features of the United States and Canada," is concerned with the broad patterns of landforms, climates, soils, and vegetation. It starts with a description of the major plains, mountains, and plateaus. This is followed by an examination of the main causes of climates and climate regions. The chapter ends with brief discussions of the geographies of soils and natural vegetation, whose locations are influenced by landforms and climate.

The following three chapters focus on the people who inhabit this portion of the world. The European exploration and settlement of the continent, the main subject of

One topic that interests most students is job prospects for majors in a particular discipline. There are many options for geography majors, the answer depending on what they emphasize within the major.

Students who follow the environmental (people-land) tradition can find jobs with a wide variety of government agencies (Figure 1.9). Some may work for the federal government, perhaps with an office concerned with environmental protection. Geography graduates have found jobs with such federal agencies as the National Park Service, the Bureau of Land Management, the Army Corps of Engineers, and the National Forest Service. Many others work for lower levels of government, such as states or provinces, regional planning agencies, counties, and cities. The largest number of jobs are in environmental planning, such as in the environmental section of a county or regional planning department. In the broadest sense, such geographers are concerned with trying to protect the quality of the environment, perhaps by monitoring air quality or by trying to assure that wildlife has enough space to reproduce.

Many other graduates work for private environmental consultants. These companies are hired by both private industry and government to help resolve environmental issues. For example, if a city is concerned about the environmental impacts of a proposed power generating plant, it may hire environmental consultants. Their job may be to propose alternative locations for the facility and to predict the likely environmental impacts of each alternative. The consultants may also be asked to recommend measures, such as alternative fuels or the use of air scrubbers, to minimize negative impacts. Physical geographers who specialize in soils, climatology, or natural vegetation often find jobs with environmental agencies or consultants.

Those who follow the spatial tradition, perhaps emphasizing urban or economic geography, usually get quite different jobs. A large number work in urban planning. In most states, such planning is either required or recommended, and there are thousands of planning departments scattered across Canada and the United States. These geographers perform a wide variety of jobs, such as checking plans for new urban developments (for example, proposed new housing tracts or freeways), making sure that they meet the quality standards set by the city or county. Others may help to create plans to guide future urban growth, in line with community preferences. There are also many positions with local, state, provincial, and national transportation agencies

A geographer at work at an environmental agency.

Chapter 3, occurred virtually simultaneously in what would become Canada and the United States, and their early populations came largely from the same stock. Chapter 4 focuses on the changing locations of Canadians and Americans, especially in the period of 1950–2000. It also discusses the evolving composition and location of the ethnic populations. In Chapter 5 we look at people's political behavior from a geographic perspective.

Chapters 6, 7, and 8 examine geographic aspects of the economies of the United States and Canada. Again, the where geographers may help develop new bus or rail systems, design future freeway routes, or participate in developing transportation policies. On the private side, graduates may work for major home-builders, locate new shopping malls, or conduct market research about consumers' geographic behavior.

Recently, advances in computer mapping have brought many new job opportunities. This is especially true in what is called GIS, or Geographic Information Systems, wherein huge amounts of data can be stored in computers and retrieved quickly to produce a wide variety of highly useful maps (Figure 1.10). GIS specialists are hired by many organizations, ranging from fire and police departments (which can use them to track the patterns of crime or accidents) to environmental agencies (which can, for example, quickly get updated maps via satellite data to track the spread of tree diseases or forest fires), and urban or rural planning departments. Real estate firms also can use GIS to respond quickly to clients' requests for information. One customer might be interested in knowing the location of all land parcels within a metro area that are larger than ten acres, zoned for industry, within one mile of a freeway, and currently for sale. A GIS system, with the appropriate data, can quickly produce a custom map showing all qualifying parcels, thereby saving time and effort for both buyer and seller.

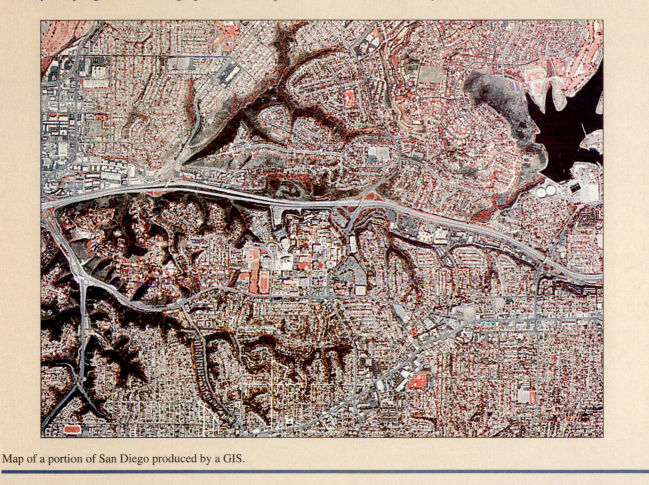

Map of a portion of San Diego produced by a GIS.

focus is on developments since 1950. Chapter 6, "Agriculture, Gathering and Extractive Industries," describes the countries' large-scale, mechanized, commercial, and specialized farming, as well as fishing, forestry, and mining activities, while Chapter 7 focuses on manufacturing and information technology. Both Canada and the United States made the transition from agricultural to industrial societies some decades ago. Now they have advanced to *postindustrial societies* that specialize in providing personal and professional services, often using computers and telecommunications to collect, process, and exchange information. The extensive transportation and communication networks that link the various parts of the countries and promote economic development are described in Chapter 8.

In Chapters 9 and 10, we view cities in two different ways. First, in Chapter 9, we consider the major factors responsible for their size and location. Then we describe the nature and location of land uses within those cities. In the following chapter we discuss recent research into how large metropolitan areas are segmented into many socioeconomic and lifestyle areas; such research is widely used not only in geography, but also in sociology and marketing.

North Americans spend a significant portion of their lives pursuing recreation, and a surprising amount of land is devoted to it. In Chapter 11 we explore some implications of this fact. A related theme, the often negative impacts of people on the environment, is the subject of Chapter 12.

The two concluding chapters are devoted to regions. The introduction to Chapter 13 discusses the nature of culture regions in the United States; the body of the chapter consists of a series of regional vignettes. As its title, "The Canadian Difference," suggests, Chapter 14 is devoted solely to a discussion of that country, examining its regions as well as ways in which Canada differs from its neighbor to the south.

2

Physical Features of the United States and Canada

As people of European descent first began to cross North America from east to west in what would become the central United States, they traversed a long stretch of gently rolling land between the Appalachian Mountains and current day central Colorado. Here they encountered the front range of the Rocky Mountains, which rises abruptly as the Great Plains end. This was a formidable barrier, and it must have discouraged many from venturing much farther.

Eventually the pioneers "discovered" what they called **South Pass,** a significant break in the mountains in southern Wyoming. This route, which is really a westward extension of the Great Plains called the **Wyoming Basin,** offered an easy crossing of the Rocky Mountains barrier. It had long been known by Native Americans, who passed their knowledge on to the "mountain men" (early fur trappers). In 1842, when the first substantial numbers of Americans began to settle on the West Coast, in "Oregon Country," they took this route, as did the Mormons heading for Utah shortly thereafter. In 1849 large numbers of miners heading for the gold country of California passed this way. When the first transcontinental railroad linked the Pacific and Atlantic shores in the 1860s, the engineers chose the Wyoming Basin as the path of least resistance. More recently, one of the busier long-distance freeways, Interstate 80, was built along this alignment. Today, when small airplanes, many of which are not pressurized, cross the continent, they logically select this low route through the mountains.

Throughout history, prominent physical features like mountains, plains, and rivers have influenced the movement of peoples and their ideas. Note that we are saying *influenced,* and not *determined,* for the latter would be to espouse the long-rejected doctrine of environmental determinism. The great passes of the Alps were (and remain) important links between northern and southern Europe. In the United States and Canada, great rivers like the Mississippi and St. Lawrence were the main routes in and out of the continent for more than two centuries. New York City benefited enormously from its easy, natural route to the west through the Appalachian Mountains, the Mohawk Gap west of Albany, an advantage unmatched by its main rivals, Boston and Philadelphia. For many reasons, people have shown a long-standing preference for some climates at the expense of others: few would choose to live under arctic conditions all year long, and before air conditioning, only a handful were willing to live in Phoenix. The facts of physical geography also have practical ramifications, for insurance rates fluctuate with an area's potential for earthquakes, hurricanes, tornadoes, and flood damage. For such reasons, and because physical geography can be seen as an important part of the broader setting in which our lives unfold, we study the physical features of the two countries.

This chapter is concerned with the landforms, climates, soils and vegetation of the United States and Canada. In such a large area, almost necessarily there is a great variety of such features. After all, the national territories encompass 17.4 million square kilometers (6.7 million sq mi) that extend over about 60 degrees of latitude and 136 degrees of longitude. The main goal of this chapter is to present a highly generalized, but useful, look at the physical geography of these countries.

We start the discussion with landforms, focusing on the broad patterns of plains, mountains, and plateaus. Before beginning, it is necessary to define a term that is used throughout the text, **region.** A region is an area of relative uniformity, such as a zone with a certain type of climate or a particular type of natural vegetation. No area on earth is completely uniform, so we are satisfied with relative uniformity. For example, places five kilometers apart do not have exactly the same climate, but we are willing to ignore small differences and focus on overall similarity.

Initially, landforms are described at the most generalized level, with only three major regions identified for all the area north of Mexico. This is followed by a more detailed breakdown, illustrating some of the variety found even within these relatively uniform areas. The three major regions are: (1) the Coastal and Interior Lowlands, (2) the Eastern Mountains, and (3) the Western Plateaus, Mountains, and Valleys. The reader should constantly refer to Figure 2.1, which not only shows these regions but is also keyed to the place names used in the chapter.

Coastal and Interior Lowlands

This is by far the largest landform region of Canada and the United States, comprising well over half the total area. As shown in Figure 2.1, it extends from the northernmost extremes of Canada and Alaska, where it is widest, and gradually tapers southward. The lowlands reach their minimum east-west extent of 2400 km (1500 mi) between the southern end of the Eastern (Appalachian) Mountains and the Rocky Mountains, and then widen again to include the Gulf and Atlantic coastal zones. In the north, this region has very few people, but this changes substantially near the international border, especially around the southern Great Lakes, where its largest cities include Chicago, Toronto, Detroit, Cleveland, and Buffalo. Farther south, some particularly large metropolitan areas in the Lowlands are Dallas–Fort Worth, Houston, and Atlanta.

The Coastal and Interior Lowlands region encompasses the entire east coast of the U.S. south of New York City, including all of the peninsula of Florida, and then extends westward along the Gulf Coast to West Texas. Here surface rocks are rather soft, and it is one of the flatter parts of the larger region. The area slopes gently to the sea, a slope that continues for some distance under the water as the continental shelf. The region's flatness is illustrated by the fact that Florida's highest point is a mere 100 meters (325 ft) above sea level, or only about one-quarter of the

Figure 2.1
The three major landform regions of the United States and Canada. Also shown are the place names used in this chapter.

Physical Features of the United States and Canada

height of New York's Empire State Building. In recent geologic times, the outer edges of the strip have been submerged somewhat, drowning the mouths of many rivers, thereby creating such excellent natural harbors as Chesapeake Bay and New York harbor. Off most of the Atlantic and Gulf coasts are semi-continuous barrier islands, consisting mostly of sand, which are widely used for recreation. A few contain famous resorts such as Atlantic City and Miami Beach.

Most of the rest of this vast region is gently rolling, in contrast to the extreme flatness that many people living outside the area seem to assume. There are also some fairly important features within the Lowlands that do not fit the general pattern, especially the higher and more rugged areas called: (1) the Ozarks (maximum elevation 800 meters, or 2700 ft), which can be considered an outlier of the Eastern Mountains, (2) the Black Hills (which rise about 1000 meters, or 3250 ft above the surrounding plains), where ancient rocks break through the younger surface materials, and (3) the higher parts of the Canadian Shield (see below). But most of the region is rather flat, and it includes some of the most productive agricultural lands in the world. In the United States, the Corn Belt and parts of the Great Plains are excellent for agriculture; the same applies to the Canadian Prairies.

The Canadian Shield

The Coastal and Interior Lowlands can be divided into many subparts. One of these is the **Canadian Shield,** a vast area of more than 2.6 million square kilometers (1 million sq mi) that covers a large portion of eastern Canada, with two small extensions into the northern United States (Figure 2.2). Shields are stable areas of ancient rock that usually have been weathered and eroded down to an uneven plain. The Canadian Shield does contain a few substantial hills, particularly in eastern Quebec province and neighboring Labrador. It has the oldest rocks in North America, and is considered to be the landform around which the rest of the continent developed. While much of it occurs at the surface, the Shield also underlies large parts of central Canada and the United States.

Wherever the Canadian Shield appears at the surface, there are few people. As described below, the shield was eroded by glaciers, and much of its soil was transported away, leaving a rocky surface with many low spots, the latter now occupied by tens of thousands of lakes. Further, virtually all of this area has too harsh a climate to attract large numbers of people. All over the world, such ancient shields have been subjected over eons of time to severe forces of compression, leading to mineralization. North America's shield contains such important resources as copper, nickel, and iron ore, often in substantial deposits. At these widely scattered sites, a moderate number of people have settled in mining towns.

Glacial Features

Until about ten thousand years ago, but just yesterday in geological time, much of the northern lowland was covered with glaciers. These great, thick sheets of ice left many marks of their passing, including some low hills (moraines) and a chaotic drainage system compared to preglacial times. Toward the north, where the glaciers mostly eroded the landscape, they left the areas affected poorer for their passing, but in the south, where they mostly deposited materials, their impacts made the areas considerably more suited to agriculture and settlement than in preglacial times.

The great **continental glaciers** that repeatedly covered much of Canada and the northeastern United States developed initially over Hudson Bay, well north of the Great Lakes, and then spread out in all directions. At one time over three-quarters of the area north of Mexico was covered by them and the related **mountain (or alpine) glaciers** to the west (Figure 2.3). As the continental glaciers spread, they picked up and transported large quantities of materials, such as soil and rocks. When they began to melt, much of this material (called **glacial drift**) was deposited in what is now the northeastern quarter of the United States and the southern parts of the province of Ontario, Canada. Eventually the ice sheets extended about as far south as the present courses of the Ohio and Missouri rivers, the New York City area, and northern Pennsylvania. As they retreated, the drift was deposited in a sheet from less than one meter to 50 meters (150 ft) thick. Today, knowledgeable observers can easily identify many signs of the glaciers' presence.

To someone looking at a map of North America, probably the most obvious feature formed by the glaciers is the Great Lakes (Figure 2.4). In fact, they are but remnants of much larger lakes that were formed by the deepening of preexisting lowlands and the blocking effect of glacial deposits that impounded the water as the glaciers melted. Lake Erie, for example, once extended about 160 km (100 mi) westward of its present location; the former lake bed, now exposed in northern Ohio and Indiana, forms some of the flattest lands in North America. Historically, the Great Lakes provided an easy transportation route for both Native Americans (First Nations people in Canada) and Europeans into and out of the interior of the continent. Today they are important to both national economies, offering a cheap way to move vast quantities of bulk materials, such as coal, iron ore, and wheat, within the region. In conjunction with the St. Lawrence River, which is their natural outlet to the Atlantic Ocean, they also allow ships from all over the world to sail to and dock at such far inland cities as Toronto and Chicago.

Whenever the glaciers' outer edges were stationary for some time (melting about as fast as they were advancing), they left great deposits of heterogeneous materials called **terminal moraines.** Some of these stand out today as conspicuous hills in an otherwise rather flat area. Along the coasts, a few terminal moraines formed islands. Long

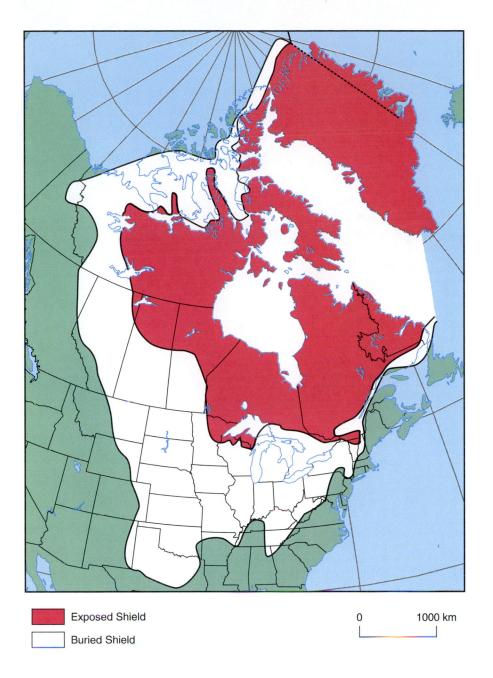

Figure 2.2

The Canadian Shield, the oldest rock area of North America, covers much of eastern Canada, and underlies large parts of Canada and the central United States.

Island, the largest island in the coterminous United States, is mostly the work of such a terminal moraine. Similar features include the resort islands of Martha's Vineyard and Nantucket off the Massachusetts mainland. The scenic eleven Finger Lakes of upper New York state were formed when the glaciers advanced down some preexisting north-south valleys and deepened them (Figure 2.5). When the glaciers retreated, they left moraines that acted like dams to block the north ends of the valleys, producing the lakes.

In general, however, the glaciers left areas they affected significantly flatter than they were in preglacial times. Acting somewhat like enormous bulldozers, they sheared off the tops of hills when they advanced, and when retreating they deposited drift that filled lower parts of former valleys. The net result was a landscape much gentler in its features than what had prevailed before them. This made many areas much more useful to people, especially for agriculture, than they would be in the absence of glaciers.

Figure 2.3

Farthest extent of glaciation in the Northern Hemisphere. Separate centers of snow accumulation and ice formation developed. Large lakes were created between the western mountains of the United States and Canada and the advancing ice front. To the south, huge rivers carried away glacial meltwaters. Since large volumes of moisture were trapped as ice on the land, sea levels were lowered and continental margins were extended.

One way to confirm these impacts is to examine the **Driftless Area** of southwestern Wisconsin. Because of the blocking effects of some higher ground to the north, this area was missed by all advances of the glaciers and became an island of land in the middle of a glacial sea. Today southwestern Wisconsin is significantly more hilly than adjacent areas that were covered by ice, sometimes strikingly so, and its soils often are poorer (Figure 2.6). Much of it has forest-covered hills, interspersed by relatively deep (for this part of the country) valleys. Overall, a significantly smaller percentage of the area is in productive agriculture

compared to nearby places that were glaciated. Without the glacial era, much of the Midwest probably would look like the Driftless Area, and national farm output would be substantially lower.

Great Plains and Prairies

Another well-known feature of these lowlands is called the Great Plains in the United States and the Prairies in Canada. These areas have a reputation for being among the flattest, most uniform, parts of both countries, but they too consist mostly of gently rolling terrain. Still, of all the populated

Figure 2.4

The Great Lakes today are remnants of much larger lakes formed as the glaciers retreated. They include over 11,000 miles of shoreline, much of it scenic. The lakes are an enormous source of fresh water, and they are important for transporting bulk materials like iron ore, limestone, and coal.

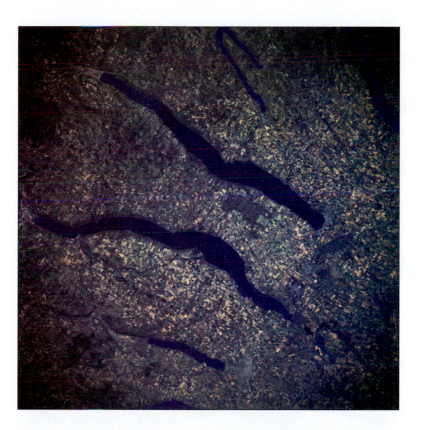

Figure 2.5

The Finger Lakes of northern New York state were formed as advancing glaciers deepened valleys and retreating glaciers then deposited moraines that acted like dams. This high altitude view shows several of the lakes.

Physical Features of the United States and Canada

Figure 2.6
The Driftless Area of southwestern Wisconsin was missed by all advances of the glaciers. As a result, its terrain is substantially rougher than nearby areas that were covered by ice.

parts of the two countries, they may come closest to forming a featureless plain (Figure 2.7).

The elevations of the Great Plains and Prairies increase slowly from east to west. The Great Plains rise at about 3 meters (10 ft) per mile from around 600 meters (2000 ft) on their eastern edge to about 1500 meters (5000 ft) in the west (as at Denver), where they end at the Rocky Mountains. Calgary, in Canada, lies near the western edge of the Prairies at about 1100 meters (3500 ft). The Great Plains and Prairies are rather dry, receiving less than 500 mm (20 in) of rainfall per year. When Columbus first came to the Americas, they were cloaked with a natural grassland, in contrast to the forested areas found farther east.

Mississippi Alluvial Valley

Another noteworthy part of the Lowlands is the Mississippi Alluvial Valley. It occupies a large area south of the

Figure 2.7
Seemingly endless wheat fields cover the Great Plains near Glasgow, Montana.

confluence of the Ohio and Mississippi Rivers, which was once under the sea. The Mississippi Alluvial Valley was formed over millions of years through the deposition of huge quantities of **alluvium** (stream-borne materials), such as silt and clay, by the Mississippi River. As it entered the sea, the river's sedimentary load was deposited on the seafloor. Eventually these deposits were extensive enough to form new land, and thereby many thousands of square kilometers of flat terrain were added to North America. This process continues today in the Mississippi delta, about 160 km (100 mi) south of New Orleans, where the river dumps huge quantities of materials into the Gulf of Mexico (Figure 2.8).

All over the world, such extensive alluvial valleys are among the most productive lands for agriculture, and historically they have supported some of the world's largest population concentrations (such as in India and China). The Mississippi Alluvial Valley is not known for its teeming millions, but it does form one of the best agricultural areas of the United States, producing vast quantities of cotton, rice, soybeans, and other valuable crops.

Eastern Mountains

In the eastern United States and the Atlantic provinces of Canada is a system of low, old, worn down mountains, remnants of a much higher range that once rose to over 3000 meters (10,000 ft). They trend in a southwest-northeast direction, stretching from central Alabama to the Gulf of St. Lawrence and the island of Newfoundland; there are some related ranges as far north as Quebec and Labrador. Most peaks in this system today do not reach 1200 meters (4000 ft), but there are exceptions. In some areas, they rise to over 1800 meters (6000 ft), forming the highest points in the eastern half of the continent. The loftiest peak is Mount Mitchell in the Great Smoky Mountains in North Carolina, at 2037 meters (6684 ft).

Figure 2.8

LANDSAT image of the delta of the Mississippi River. Notice the ongoing deposition of silt and the effect that both river and gulf currents have on the movement of the silt.

Figure 2.9
A view along the Blue Ridge west of Mount Mitchell in the Great Smoky Mountains. Northeast of Mount Mitchell, the Blue Ridge forms the watershed divide between the Atlantic Coast and Mississippi River drainages.

In the United States, most of this system is known as the Appalachian Mountains, which form a continuous chain from central Alabama to southern New York state. Although they are not especially high, in some areas they have many steep slopes, which made transportation across them very difficult before the modern era. Dense forests compounded this problem, discouraging travel on anything other than foot or horseback. As a result, for the first 150 years of European settlement, these mountains effectively blocked large-scale westward movement, restricting farmers essentially to the Atlantic lowlands. Not until after the Revolutionary War (1783) did large numbers of people begin to move over the mountains into what was then called the west (Figure 2.9).

The westernmost part of this system is known as the Appalachian Plateau. A **plateau** is a relatively flat, elevated area. In this case, the former flat plateau surface has been deeply eroded by hundreds of streams, so that today most people would call it a hill country. Here are found the vast quantities of high quality coal, easily reached in thick, layered beds that constitute perhaps the most valuable resource of Appalachia. Much of this coal is well-suited to steel re-

fining, and it played a major role in the industrialization of the United States and (through export) Canada.

In New England (the six northeasternmost states of Maine, New Hampshire, Vermont, Massachusetts, Rhode Island, and Connecticut), the Eastern Mountains were heavily glaciated, and much of the detail of the landscape today is the result of the actions of ice (glacial processes). Their peaks, however, were never reached by the ice, which advanced down the valleys along the path of least resistance. The highest elevation in the northeastern United States is at Mount Washington in New Hampshire, 1917 meters (6288 ft) high. These mountains are a major element of scenic northern New England, and the whole region sees many tourists during most of the year (Figure 2.10).

From south of the Potomac River, near Washington, D.C., to northern Georgia, the eastern edge of this highland forms a remarkably continuous wall of folded mountains called the Blue Ridge. This feature is broken through by only two rivers, and it formed a particularly effective barrier to western movement before the Revolution. Near the southern extremity of the Blue Ridge are the Great Smoky Mountains, with one of the country's most visited national parks.

Figure 2.10
The mountains of northern New England are scenic, and in winter they draw large numbers of tourists for skiing and other snow-related recreation. This view, from Maine, also shows a coniferous forest.

In Canada's Atlantic Provinces, these eastern mountains are low, seldom exceeding 600 meters (2000 ft). In some areas, such as central New Brunswick and parts of Nova Scotia, they consist mostly of volcanic materials, called igneous intrusions. These have been severely weathered and eroded over millions of years, and today they present scant evidence of their origins. Most of these low mountains were also covered by glaciers, and a good part of the original soil was moved south into what became the United States. Partly as a result, much of the area is not particularly valuable for agriculture.

Western Plateaus, Mountains, and Valleys

Going westward over the Great Plains and Prairies, the lowlands stop abruptly with the beginnings of the Western Plateaus, Mountains, and Valleys, and a region that continues to the Pacific shores. In the United States this region

consists mostly of plateaus of varying heights, but three mountain chains are also conspicuous, and much of the plateau country incorporates numerous short, north-south oriented ranges.

Plateaus

The plateau zone lies between the Rocky Mountains on the east and a twin row of mountains that parallel the Pacific Coast. It is broadest (about 1300 kilometers, or 800 mi) near the southern end along the Mexican border and narrows to the north. Several distinct plateaus are located in this general area, with widely varying elevations and landscapes. Perhaps the best known is the Colorado Plateau, centered in northern Arizona, much of which lies at around 2100 meters (7000 ft). Into it the Colorado River has cut a gorge more than a mile deep, the Grand Canyon (Figure 2.11). Another plateau, sometimes called the Basin and Range landform province, is mostly 300–600 meters (1000 to 2000 ft) above sea level and contains many short mountain ranges. In Canada and Alaska, the plateau is narrow, and much of it

Figure 2.11
The Grand Canyon was formed when the Colorado River cut deeply into a plateau surface 2100 meters (7000 ft) above sea level.

has been so severely eroded that it resembles hill country or even the nearby mountains.

Mountains

The Rocky Mountains, a chain of young and rugged highlands, extend in a southeast-northwest direction from central New Mexico into central Alaska, a distance of roughly 4800 kilometers (3000 mi). They rise abruptly from the adjacent flatlands, sometimes to considerable heights (Figure 2.12); the state of Colorado, for example, has 54 mountains that reach over 4300 meters (14,000 ft). As mentioned earlier, in the United States, the range is broken at only one point, in southern Wyoming, where the

Wyoming Basin forms a rather easy route across this otherwise considerable barrier. This natural break divides the mountain system into the northern and southern Rockies. In Canada the range continues to the northwest into the far north in Yukon Territory, where it merges with the Coast Mountains. The Canadian Rockies also have one particularly easy crossing west of Edmonton, where Yellowhead Pass (Figure 2.13) is at only 1133 meters (3717 ft).

Along the U.S. West Coast there are two parallel mountain ranges, with some discontinuous but important valleys between. Farthest west are the Coast Ranges, which extend from Los Angeles in the south all the way to northernmost Washington state. Most of them do not rise over

Figure 2.12
The change from the Great Plains to the Rocky Mountains is often abrupt and dramatic. Here the Choteau Prairie meets the Rocky Mountains in Montana.

1200 meters (4000 ft), although they reach 2424 meters (almost 8000 ft) in the Olympic Peninsula of Washington. Despite its generally moderate height, this range has a considerable impact on the climates of some of the most populated parts of the Far West. In California it helps to define a coastal zone with a considerably milder climate than areas farther east. North of the state of Washington to Alaska, these coastal mountains continue as a series of islands, notably as Canada's Vancouver Island, off the mainland of British Columbia.

These coastal mountains are in the contact zone between two **earth plates.** The latter are major portions of the earth's crust that slowly move over semi-molten materials at greater depths. Here the Pacific Plate is moving northward against the North American Plate. This causes tremendous pressure, which eventually is released in the form of earthquakes. The differential movement of earth along a linear break in its surface is known as a fault, and the coastal ranges are full of **fault lines.** The most famous is the San Andreas Fault, which occasionally produces major earthquakes, such as the Loma Prieta (or the World Series) earthquake of 1989. Figure 2.14 shows the major fault lines in California, not all of which are related to the Coast Ranges.

California is not the only state with a potential for earthquakes. In fact, most western states, southwest Canada, and Alaska are in "earthquake country." Further, possibly the largest earthquake in U.S. and Canadian history occurred in 1812 in the Mississippi Valley near the confluence of the Mississippi and Missouri rivers. It was so powerful it caused the Mississippi River to be relocated. Major earthquakes also have occurred near Boston, New York City, and Charleston, South Carolina.

Figure 2.13
Yellowhead Pass in Canada forms an easy crossing of the Rocky Mountain Chain west of Edmonton. It is used by a major railroad and an important highway.

One hundred sixty to 320 km (100 to 200 mi) inland are the much higher Sierra Nevada and Cascade Mountains of California, Oregon, and Washington. California's Sierra Nevada range has a steep rise on the east, but a much gentler one on the west (Figure 2.15). The former was a major barrier to 19th century immigrants, more so than the Rocky Mountains. The Sierra includes Mount Whitney, which at 4418 meters (14,494 ft) is the highest point in the 48-state part of the United States.

In central California the Sierra gradually integrates northward into the Cascade Mountains, which are distinguished by their volcanic origins. The highest of these is Mount Rainier near Seattle, which at 4392 meters (14,410 ft) is only 26 meters (84 ft) lower than Mount Whitney (Figure 2.16). Both ranges receive a great deal of precipitation, but the areas to the east of them are deserts or near deserts. In California much of this precipitation falls in the winter as snow, and the spring and summer meltwater from the resulting snowpack is critical to the state's water supply in the dry summer months.

In southern Canada, the Cascades chain becomes the Coast Mountains, which are separated from the Canadian Rockies by a narrow plateau. Farther north, as said earlier, these two ranges merge. Near the Alaskan border they reach Canada's highest point at Mount Logan, 5951 meters (19,520 ft). In Alaska this range splits in two, separated by a broad lowland, and the mountains change directions to east-west. The southern of these branches, the Alaska Range, includes Mount McKinley (Mount Denali), which at 6194 meters (20,320 ft) is the highest point in North America (Figure 2.17). South of the Alaskan border only three rivers cut through these mountains, the Skeena and Fraser in Canada and the Columbia in the United States. All form important routes for travel, especially the Fraser and the Columbia.

The western ranges were once largely covered by alpine glaciers, which formed at the time when continental glaciers were developing farther east. These glaciers greatly modified the mountains, sharpening peaks, lowering many valleys, and creating various distinctive landform types.

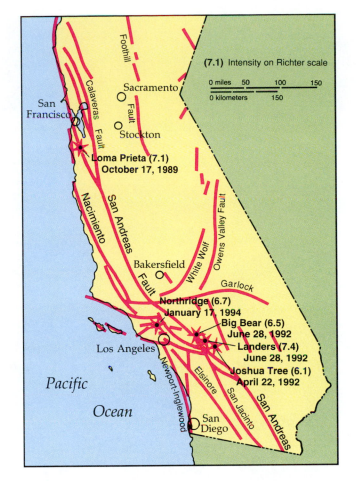

Figure 2.14
The major fault systems of California, showing the epicenters of some of the larger recent earthquakes.

Figure 2.15
On their east side, the Sierra Nevada mountains of California and Nevada rise steeply, forming a major barrier to east-west travel. This view shows that barrier, as seen from the Owens Valley in eastern California.

Physical Features of the United States and Canada

Figure 2.16

Mount Rainier is a dormant volcano 161 kilometers (100 mi) southeast of Seattle. Numerous glaciers descend from its permanent snow fields. It is the largest individual mountain in the contiguous 48 states.

California's Yosemite Valley is a world-famous example of how such alpine glaciers sculpted the area.

The Hawaiian Islands are of volcanic origin. Essentially, they are the tops of a submarine mountain range that trends in a northwest-southeast direction. The only active volcanoes today are on the largest island, Hawaii, where two peaks reach over 4100 meters (13,600 ft). The Kilauea Crater in Hawaii Volcanoes National Park is one of the world's most active, with major eruptions every few years.

Magma flowing from it sometimes reaches the sea, adding to the island's land area (Figure 2.18).

Valleys

Between the Coast Ranges and the Sierra Nevada-Cascades are several important valleys and lowlands. The largest is the Central Valley of California, an area once below sea level but which has since been filled with a deep layer of sediments (alluvium) from the surrounding mountains.

Figure 2.17
Located approximately 350 kilometers (220 mi) north of Anchorage, Alaska, Mount McKinley, at 6194 meters (20,320 ft), is the highest mountain in the United States and Canada. Cloud-covered most of the time, it is the centerpiece of Denali National Park.

Figure 2.18
An eruption at the Kilauea Crater in Hawaii Volcanoes National Park on the island of Hawaii. This is one of the world's most active volcanoes, with major eruptions every few years.

When the first Europeans came it was also largely a grassland, and this combination of alluvium and grasslands produced some of the best agricultural lands in the United States. Its counterpart in Oregon is the Willamette Valley, where the city of Portland is located. Washington's Puget Sound Lowland, around Seattle, once covered by coniferous forests rather than grasslands, has lower agricultural potential. A similar lowland in southwest British Columbia contains Vancouver, the largest Canadian city west of Toronto.

Major Climates of the United States and Canada

Weather refers to the conditions of the atmosphere in an area at a particular point in time, in terms of such meas-ures as wind, cloud cover, temperature, and precipitation. **Climate,** in contrast, refers to the *typical* conditions of the atmosphere in a place or region, based on long-term measurements, and considers seasonal variations in those conditions. In this section we take up the major climates of Canada and the United States. We begin with a discussion of the major causes of climates and end with an explanation of the resulting climatic regions. First, the two main elements of climate, temperature and precipitation, are discussed.

Temperature
Temperature is a measure of the energy present at any given time, and it is mainly a result of the amount of energy that reaches the earth's surface from the sun. That amount, in turn, is importantly a function of the directness of the angle

Before 1980, most Washingtonians thought of Mount St. Helens as one of the most symmetrical and picturesque of all American volcanoes, a delightful place to hike, picnic, boat, climb, fish, and relax. American Indians, however, knew better; in 1843, the mountain had vented steam, ash, and lava in a major, but nondestructive, eruption. It was on the basis of their legends that nearby Spirit Lake received its American name.

On May 18, 1980, the peaceful, symmetrical cone became the violent agent of its own destruction, when it produced one of the two largest volcanic explosions ever witnessed in the United States (the Katmai eruption in Alaska was the other). Days later, when the ash plume had died down, the entire top of the mountain had been blown away, and an estimated four cubic kilometers (1 cubic mi) of cinder and ash material that used to comprise the mountain were scattered over Washington and Idaho. Mudflows tore down the drainages on the northwest side of the mountain and blocked the Toutle and Cowlitz Rivers, destroying roads and bridges and flooding low-lying lands. Four hundred square kilometers (150 sq mi) of forests were destroyed (most instantly stripped and blown over by the force of the initial blast). Spirit Lake was buried in mud and uprooted trees, and several people lost their lives.

Mount St. Helens in Washington as it appeared before the big eruption in 1980. Spirit Lake in the foreground.

If the first lesson was that dormant volcanoes should be treated with great respect, the second was that nature's recuperative powers are impressive. Today, a new paved road takes sightseers into the "devastated area" of the new Mount St. Helens National Volcanic Monument and on to an overlook above the reborn and much larger Spirit Lake. Wildflowers and other types of opportunistic vegetation are appearing on the bare slopes, and birds and mammals are exploring the possibilities of settlement. Slowly, the barren landscape, denuded by a blast equal to a large hydrogen bomb, will be transformed into a new, and continually evolving, montane ecosystem, eagerly studied by botanists and zoologists to document the remarkable ways in which nature heals itself.

Mount St. Helens as it appeared following the 1980 eruption.

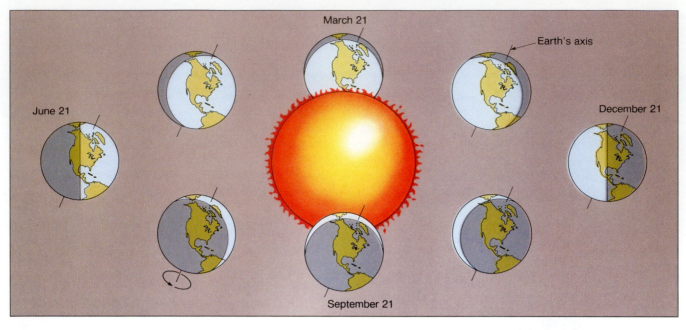

Figure 2.19

The earth's axis is titled in the same direction throughout its annual orbit around the sun.

of the sun's rays with respect to the earth's surface; the closer to 90 degrees those rays are, the more solar radiation is received. One major factor that accounts for seasonal temperature differences between places is latitude, for the angle of the sun's rays varies greatly with latitude and seasons. In summer, the northern hemisphere is tilted toward the sun, and it receives a good deal of solar radiation, which is converted to heat (Figure 2.19). Not only are the sun's rays more direct then, but days are long and nights short (and increasingly so the farther north one goes). At this time of the year even places far north of the equator can have warm to hot days. In winter, the northern hemisphere faces away from the sun, leading to a low angle of the sun's rays and short days. Near the equator the sun's rays are always fairly direct and the number of hours of daylight varies little, leading (at low altitudes) to warm to hot conditions all year long.

Another important factor in explaining climates is differences in the way land surfaces and water bodies heat and cool. Land surfaces tend to warm up and cool rapidly, while the opposite is true of water bodies. Thus, large bodies of water (especially oceans) have temperatures that vary little during the year, despite considerable seasonal variations in the amount of energy received from the sun. To illustrate, the average ocean temperature in August off the coast of the state of Washington is about 13.8° C (57° F), which is only 5.4° C (10° F) more than the average there in January, the coldest winter month.

This fact helps to explain the moderate climates along the West Coast. When winds from the west (the prevailing direction) sweep over the Pacific Ocean they tend toward the same temperature as the water. In summer, when adjacent land masses otherwise would heat up rapidly, these winds sweep inland and keep temperatures mild. In winter, when the land tends to cool rapidly, the ocean stays relatively warm, also keeping adjacent land temperatures mild. Therefore, areas along the West Coast as far north as the panhandle of Alaska and as far south as the Mexican border have both mild summers and mild winters. These ocean influences decrease rapidly as one moves inland; the farther east one goes, the warmer the summers and the colder the winters. Because of the prevailing west-to-east flow of winds in the U.S. and Canada. the Atlantic Ocean has much less of an influence of this type.

In sharp contrast, mid-latitude areas distant from oceans have marked **continentality.** Here, far from the ameliorating effects of the oceans, summers are often very hot and winters extremely cold. Continentality is especially pronounced north of about latitude 40° North. The eastern Midwest of the United States and southernmost Ontario, Canada, are good examples of places with continental climates that have large populations (Figures 2.20 and 2.21).

Precipitation

The other major aspect of climate is precipitation. For it to occur, air has to be rising. As it gains in altitude, it expands and cools, and cooler air can hold less moisture than

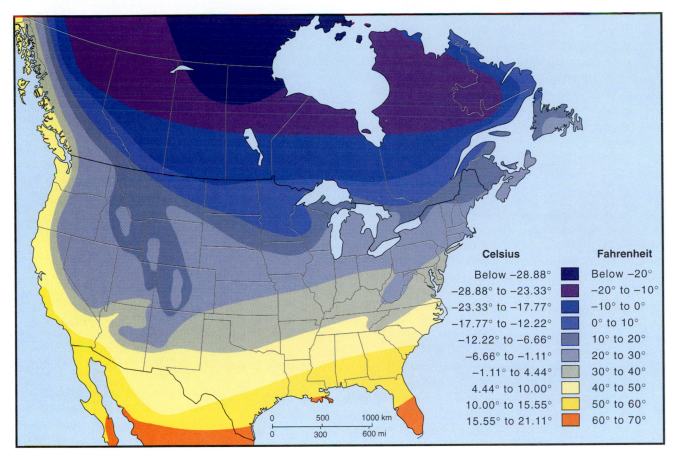

Celsius	Fahrenheit
Below −28.88°	Below −20°
−28.88° to −23.33°	−20° to −10°
−23.33° to −17.77°	−10° to 0°
−17.77° to −12.22°	0° to 10°
−12.22° to −6.66°	10° to 20°
−6.66° to −1.11°	20° to 30°
−1.11° to 4.44°	30° to 40°
4.44° to 10.00°	40° to 50°
10.00° to 15.55°	50° to 60°
15.55° to 21.11°	60° to 70°

Figure 2.20
Average daily temperature in January (° C).

Source: Data from Richard A. Anthes et al., *The Atmosphere,* 3d edition. Merrill, 1981.

warmer air. If it rises enough, the moisture in the air condenses and may be released in the form of rain, snow, or some other form of precipitation (Figure 2.22).

Three different mechanisms cause air to rise, leading to precipitation. When moving air encounters a mountain, it is forced up the slope. When this process continues long enough, the result is called **orographic precipitation** (Figure 2.23). Much of the precipitation along the west coasts of the two countries, and in Hawaii, is orographic. When frequent storm systems are forced over mountains, this can lead to very high precipitation totals. This happens especially in the coastal zone between Alaska's panhandle and Washington state, where some west-facing mountain slopes receive more than 5100 mm (200 inches) per year (Figure 2.24). These are some of the highest totals anywhere in the world. Orographic lift also contributes to the high precipitation of the southern Appalachians, as warm and moist tropical air masses moving north encounter the mountains.

Another major cause of precipitation is the lifting of air along weather fronts in association with cyclonic storms, or low pressure systems (Figure 2.25). In central North America, **cyclonic, or frontal, precipitation** occurs when cold, dry polar air masses moving south meet warm, moist air coming north from the Gulf of Mexico. When these masses collide, the warmer, lighter air is forced to rise over the denser, colder air, often leading to precipitation along *fronts*. The northeastern United States and southeastern Canada see a constant parade of such cyclonic storms followed by dry, cooler high pressure systems. This leads to regular changes in weather throughout the year.

The final cause of precipitation is differential surface heating. The result, called **convectional precipitation,** occurs primarily in the southern half of the United States (Figure 2.26). When adjacent areas have markedly different properties for heating, such as when a large plowed field sits next to a substantial woodlot, the sun heats the field much faster than the woodlot. This heats the air, which begins to rise, and if that process continues long enough it will lead to condensation, cloud formation, and precipitation, often in the form of thunderstorms. During the warmer months, such thunderstorms are nearly a daily occurrence in the southeastern part of the country. They are fairly common as far west as Arizona.

Physical Features of the United States and Canada

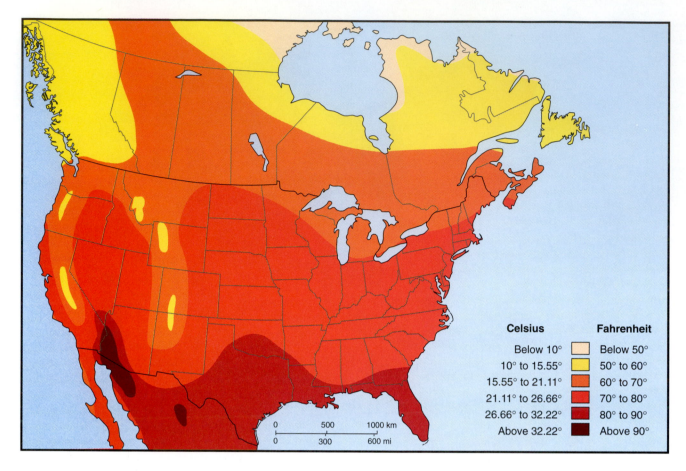

Celsius		Fahrenheit
Below 10°		Below 50°
10° to 15.55°		50° to 60°
15.55° to 21.11°		60° to 70°
21.11° to 26.66°		70° to 80°
26.66° to 32.22°		80° to 90°
Above 32.22°		Above 90°

0   500   1000 km
0   300   600 mi

Figure 2.21
Average daily temperature in July (° C).

Source: Data from Richard A. Anthes et al., *The Atmosphere,* 3d edition. Merrill, 1981.

Climate Regions of Canada and the United States

Humid Continental Climates

The distributions of temperature and precipitation go a long way towards explaining the geography of climates. As shown in Figure 2.27, there are many different climates in the two countries. The one with the largest number of people is centered on the southern Great Lakes and is labeled the "Humid Continental, Hot Summer" climate. In summer this zone has long days and fairly direct rays of the sun, and it has a classic continental location. These facts combine to produce a climate where much of the summer ranges between warm and hot. Winters are cold because of short days, low angle of the sun's rays, and the great distance from moderating oceans (continentality).

Precipitation falls year-round, mostly from the serial passage of cyclonic storms. During the summer, the area is in the zone of intense interaction between warm tropical air masses from the Gulf of Mexico and colder polar continental air masses from the north (Figure 2.28). Where these air masses meet there is a zone of conflict, and cyclonic storms are generated all along the contact zone and then move from west to east. Every few summer days, on the average, one of these storms passes over the area, bringing rain. The Great Lakes influence this pattern, as they seem to attract such storms. Cyclonic storms also pass over the area in winter, usually bringing snow, but precipitation totals are distinctly higher in summer. Such large cities as Boston, New York, Pittsburgh, Chicago, and Kansas City are located in this climate region.

The area just to the north has a similar climate, but with cooler summers and more severe winters. In fact, here winters are the dominant season. This is the "Humid Continental, Mild Summer" climate region. Partly because of greater temperature extremes, it attracted fewer people than the adjoining climate area to the south. Still, such major cities as Montreal, Toronto, Minneapolis, and Winnipeg fall into this region.

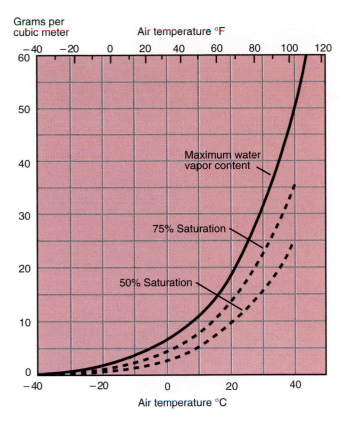

Grams per cubic meter

Air temperature °F

Figure 2.22
The amount of water vapor that can exist in the atmosphere increases with temperature. The existing percentage of this maximum is the relative humidity.

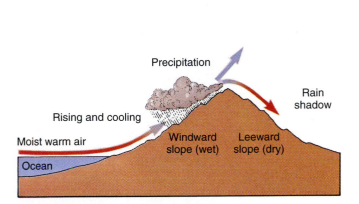

Precipitation

Rain shadow

Rising and cooling

Moist warm air

Windward slope (wet)

Leeward slope (dry)

Ocean

Figure 2.23
Orographic precipitation occurs when air is forced to rise when it encounters a mountain. As it rises it cools and if it is cooled sufficiently, precipitation occurs. As it then descends on the leeward side it warms and its capacity to hold moisture increases, and precipitation is unlikely.

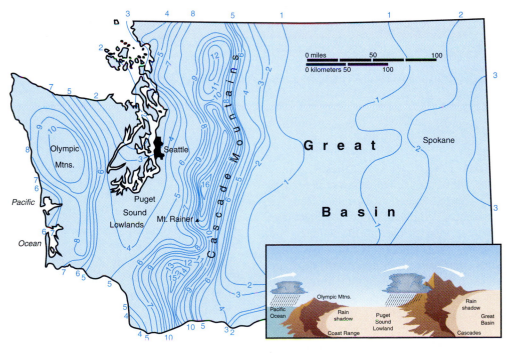

Figure 2.24
November 1985 precipitation for Washington. West-to-east flowing air is forced to rise as it approaches the Olympic and Cascade Ranges, leading to orographically enhanced precipitation.

Physical Features of the United States and Canada

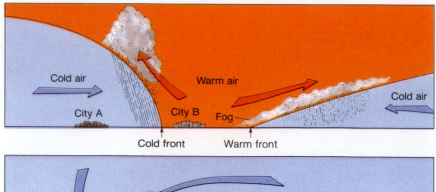

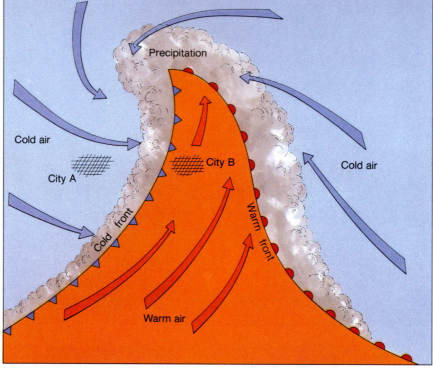

Figure 2.25
Side view (top) and surface pattern (bottom) of a typical midlatitude cyclone. Lifting along the fronts causes cloud formation and precipitation.

Subtropical and Tropical Climates

In the southeast quadrant of the United States summer is a time when moist tropical air masses dominate. Here we have a "Humid Subtropical" climate. With more direct rays of the sun all year than areas further north, summers are even hotter. Winters are much milder, for not only are the sun's rays more direct but days are longer. Still, there are occasional outbreaks of arctic air masses from the north that can extend as far as southern Florida, sometimes causing considerable frost damage to plants. Precipitation occurs all year, with differential heating (convectional precipitation) prominent in summer and cyclonic storms traversing the area in winter, as the main contact zone between polar and tropical air masses shifts south to this re-

gion. The abundant moisture is derived from the Gulf of Mexico, and precipitation decreases slowly with increasing distance from that body of water.

In southernmost Florida we have an area with a "Tropical Savannah" climate. Here summers are also hot and rainy, but winters are mild and relatively dry. Key West, a small city on an island off the south coast of Florida, has the only weather station in the coterminous United States that has never recorded a temperature below freezing.

Except near the tops of its highest mountains, the climates of the Hawaiian Islands are tropical. All the larger islands have a wet and a dry side. As moisture-laden winds off the Pacific Ocean strike the volcanic mountains from the north and east, they produce a great deal of orographic precipitation on the **windward** side. The rainfall totals here are among the highest anywhere in the world; combined with the warm to hot temperatures, they produce a "Tropical Rain Forest" climate. Mount Waialeale, on the island of Kauai, gets more than 10,000 millimeters (400 in) of rainfall per year, and is advertised as the world's wettest spot.

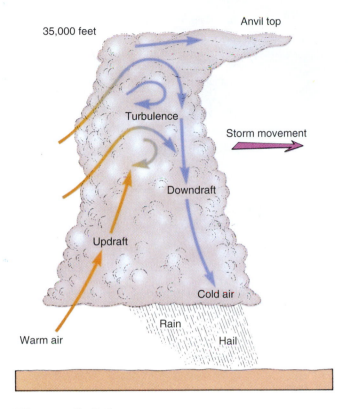

Figure 2.26

When warm air laden with moisture rises, a cumulonimbus cloud may develop and convectional precipitation occur. The turbulence within the system creates a downdraft of cold upper-altitude air.

With so much rain, precipitation is almost continuous. In contrast, the **leeward** sides of the islands are much drier, sometimes so dry that they have only a scrubby, drought-resistant vegetation. The largest city, Honolulu, is on the leeward side of some mountains, and therefore has only about 600 millimeters (25 in) of rainfall per year. Some of its windward suburbs, however, see several times that much rain.

Steppe Climates

West of about the 98th meridian, the climate changes as one reaches the Great Plains and Prairies. This area also gets its moisture from the Gulf of Mexico, but it receives rather little because most of the humid air masses from that body of water are deflected away from the region toward the northeast. This deflection is the result of the earth's rotation and of a high pressure system over the Atlantic Ocean that moves air clockwise (toward the northeast). Only occasionally do the moist air masses make it to the Great Plains (Figure 2.29). The eastern parts of the Plains and Prairies are more likely to experience such air masses than those farther west, where aridity increases rapidly.

This whole area has a "Steppe" climate. There are two main variations of this climate, according to latitude. In the north the area has warm to hot, dry summers, and cold, dry winters. Perhaps to the maximum extent in the more populated parts of North America, this area (the northern Great Plains and the Canadian Prairies) shows the impacts of continentality. It gets much of its summer precipitation from cyclonic storms, and a little snowfall in winter from the same mechanism. In the southern Great Plains, the population experiences hot, dry summers and mild, dry winters. In the summer the main mechanism leading to precipitation here is differential (convectional) heating. Both the north and south have a good deal of intense, dramatic weather, including hailstorms, blizzards, and tornadoes (Figure 2.30).

TORNADO WATCHES AND WARNINGS

When conditions are likely to lead to a tornado, the National Severe Storms Forecast Center issues **tornado watches** for the affected area, which may cover several counties. Tornado watches are quite common, and about 60% of the time no tornadoes actually hit the watch area. **Tornado warnings** are issued by the local offices of the National Weather Service when a trained weather spotter sights a tornado or local radar indicates the presence of one. When a warning is issued, civil defense sirens sound, and local television and radio stations advise their listeners to take cover immediately.

It takes about four minutes on average between when a tornado touches down and a warning is disseminated. This amount of time is just about equal to the average life span of a tornado, so warnings are often issued just as the danger is over. On the other hand, since fatalities are usually due to larger, longer-lived tornadoes, the local warning system has been very effective in reducing the number of deaths.

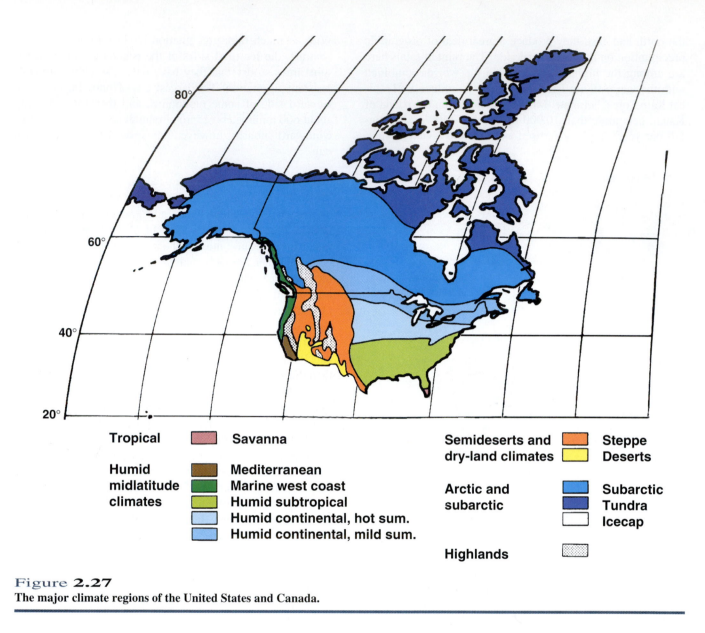

Figure 2.27
The major climate regions of the United States and Canada.

Legend:

Tropical
- Savanna

Humid midlatitude climates
- Mediterranean
- Marine west coast
- Humid subtropical
- Humid continental, hot sum.
- Humid continental, mild sum.

Semideserts and dry-land climates
- Steppe
- Deserts

Arctic and subarctic
- Subarctic
- Tundra
- Icecap

Highlands

West Coast Climates

There are two main climates on the West Coast, according to latitude. Because of their proximity to the Pacific Ocean, coastal areas throughout have mild summers and winters. Orographic precipitation is common, as sequential cyclonic storms coming onshore from the northwest Pacific Ocean encounter the two rows of mountains. The major cities of the U.S. northwest, Portland and Seattle, as well as Vancouver, are in the **rainshadow** (dry side) of the Coast Range, which means they have rather low precipitation totals for the region. Were it not so, far fewer people would be willing to live in them than the millions who now do. The coasts of northernmost California, Oregon, Washington, British Columbia, and the Alaska Panhandle have a "Marine West Coast" climate. Winters are long and very wet, while summers are short and rather drier.

From just north of San Francisco to the Mexican border, we find the "Mediterranean" climate, with its wet winters and dry summers. The summer drought is caused by a stable high pressure system (characterized by dry, descending air) that dominates the area, effectively blocking cyclonic storms moving toward the area from the north Pacific. This system weakens and moves southward in winter, allowing some storms to enter the area. Much of the resulting precipitation falls as snow in the Sierra Nevada mountains. Annual precipitation decreases as one goes southward along the coast, from about 500 millimeters (20 in) annually at San Francisco to about 250 millimeters (10 in) at the Mexican border.

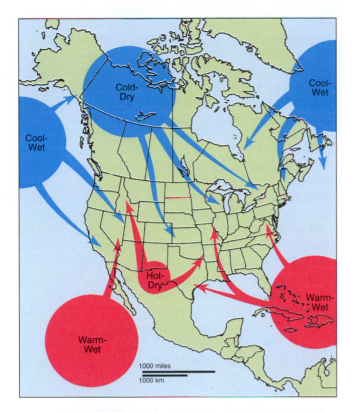

Figure 2.28

Source regions for air masses in North America. The United States and Canada, lying between major contrasting air-mass source regions, are subject to numerous storms and weather changes.

From T. McKnight, *Physical Geography: A Landscape Appreciation,* 4e, © 1993. Adapted by permission of Prentice-Hall, Englewood Cliffs, New Jersey.

Climates of the Intermontaine Area and the North

In the intermontaine area, between the Sierra Nevada–Cascades–Coast Mountains and the Rockies, is another zone, largely in the United States, that is generally dry. This area gets most of its moisture from the Pacific Ocean, but the high mountains to the west wring out the great majority of the water vapor before it gets here. One positive consequence is that this interior region is noted for its

Figure 2.29

Location and air flow of the Bermuda high pressure system.

Figure 2.30

Tornado. Among the most violent of all storms, tornadoes occur more frequently in the United States and Canada than in any other part of the world.

abundant sunshine. Its northern half has dry, hot summers and cold winters, and is also classified as having a "Steppe" climate. Toward the south the summers are hotter but the winters are mild. The south is also drier, for not only do the mountains block the rainfall, but so does the summer high pressure system over California. Therefore, the result is a "Desert" climate. Historically, these two climate areas attracted few people, but more recently population growth has been rapid, especially since the 1980s.

The higher mountain areas of southwest Canada, and the Rockies, Cascades, and Sierra Nevada in the United States, are labeled as having a "Highlands" climate. This term refers to a variety of microclimates, with great variations within small areas, based mostly on elevation. As one goes up a high mountain, for example, conditions can change from hot and dry at the base, to moderate temperatures and precipitation halfway up, to year-round cold, dry conditions near the summit. Especially toward the north, a few larger areas are high enough to experience significant orographic precipitation and therefore can support large forests.

Cold, generally dry climates cover large parts of northern Canada and Alaska, and few people live there. These areas, with a "Subarctic" climate, are bitterly cold in the winter, and have short and generally cool summers. For short periods some inland locations, because of their continentality, get surprisingly hot at the height of summers, a time of feverish activity for plants and animals. For example, temperatures of 32° C (90° F) and above have been recorded at Fairbanks, in the interior of Alaska. The northernmost continental areas of Canada and Alaska, as well as Canada's arctic islands, have a "Tundra" climate, with very cold conditions all year, except for a very short, cool summer.

Severe Weather

Sometimes climate is marked by notable departures from the norm. The United States and Canada are subject to a wide variety of extreme weather events, such as flash floods, large hail, tornadoes, hurricanes, and blizzards. These uncommon events can have a major impact on people.

One of the more potentially destructive weather events that can occur is hail. Like other forms of precipitation, hail requires updrafts of air to provide the necessary cooling for condensation. In the case of hail, a precipitating ice particle is lifted repeatedly by updrafts that carry it above the freezing level (Figure 2.31). As an ice crystal falls through a cloud, it gets a veneer of water as it collides with water droplets. When these crystals are again lifted above the freezing level by an updraft, the coating of water freezes. When this happens repeatedly, hailstones can grow to several centimeters in diameter, even reaching baseball or softball size. Eventually they fall to the ground, where they can cause considerable damage, especially to agriculture. A single hailstorm can completely destroy a field of wheat or corn, for example. Fortunately, human fatalities directly attributable to hail are extremely rare. The average

number of days with hail is greatest over the Rocky Mountains and western Great Plains, as shown in Figure 2.32.

Nowhere in the world is the incidence of tornadoes as great as in Canada and, especially, the United States. In the U.S., about 800 tornadoes a year touch down, killing about 100 people. Ever since Dorothy was transported to the Land of Oz, a popular myth has existed that Kansas has the most tornadoes. As can be seen from Figure 2.33, the greatest number per unit area actually occur in Oklahoma. However, the entire region extending from the southern Great Plains to the southern Great Lakes region sees many tornadoes and is popularly called **Tornado Alley.** Most such storms have diameters between 100 and 600 meters (about 300 to 2000 ft), but they can extend to more than 1.6 kilometers (1 mi) in diameter. Horizontal wind speeds are normally less than 225 kilometers per hour (140 MPH), but they can be as high as 500 kilometers per hour (about 300 MPH).

Tornadoes are most likely to form when strong temperature and moisture contrasts exist over fairly short distances, such as where especially cold polar air masses from the north meet warm, moist, tropical air masses. For this reason, the majority of tornado events occur in spring

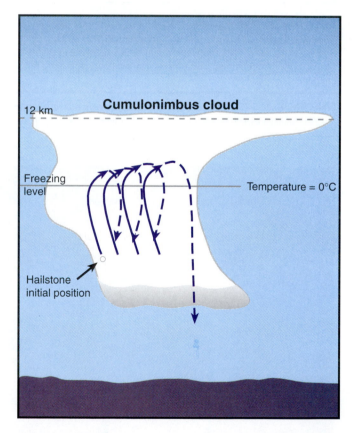

Figure 2.31
Formation of hail. Hailstones result from repeatedly being carried above the freezing level in clouds by updrafts (solid arrows). In this example, a single hailstone has been carried above the freezing level four times before falling from the cloud.

and early summer, when warm air masses start to move northward and when the temperature contrasts with the polar air masses are strongest. In May, there are five occurrences on the average day. Further, there is a latitudinal shift in the zone of greatest tornado occurrence through the season, so that tornadoes are more probable in winter in the Deep South and in May in the southern Great Plains. By June, the Great Lakes region is the most vulnerable.

Hurricanes begin as tropical disturbances over the eastern Atlantic Ocean, and then migrate westward at about 15 kilometers per hour (9 MPH). Most such disturbances die out, but a few each year develop into hurricanes, which are defined as having winds in excess of 119 kilometers per hour (74 MPH). Most hurricanes occur from August through October, with a September peak.

The energy involved in a typical hurricane is comparable to about one-half the electrical energy used in the U.S. in an entire year. This is due not only to the tremendous wind speeds involved but also the great size of hurricanes, which average about 500 kilometers (310 mi) in diameter. Average sustained speeds near the eye of the hurricane are about 160 kilometers per hour (100 MPH); within the eye there is near calm, but just outside this area the storm is at its most powerful. When such storms hit land, they not only bring high winds but also elevate sea level, bringing damaging storm surges. The entire East Coast of the U.S. and Canada, as far north as Labrador, as well as the Gulf Coast, can be hit by these hurricanes, sometimes bringing enormous damage (Figure 2.34).

One of the worst occurred on August 24, 1992, when Hurricane Andrew, with peak winds of 264 kilometers per hour (164 MPH) passed through southern Florida. The town of Homestead, just south of Miami, was virtually destroyed. After leaving Florida, Andrew curved through the Gulf of Mexico and hit the coast of Louisiana west of New Orleans. All together, 59 people died as a result of this storm, and 61,000 homes were completely destroyed.

A blizzard combines heavy snow with high winds. Although considered a winter phenomenon, many severe blizzards occur in late fall or early spring, when higher temperatures allow more water vapor to exist in the atmosphere. Reduced visibility from blowing snow and accumulation of snow on the ground can disrupt and even paralyze transportation, while the weight of heavy snow can collapse roofs. Blizzards are especially common on the Canadian Prairies and Great Plains (they have been called the "Grizzlies of the Plains"), but they can also occur over most of the eastern United States and southeast Canada.

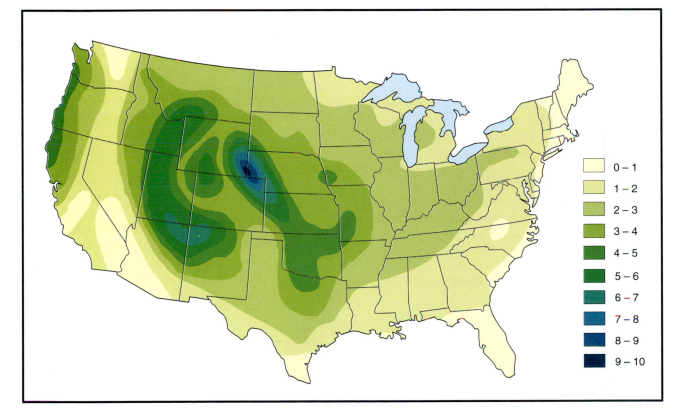

Figure 2.32
Mean annual number of days of hail.

	0 – 1
	1 – 2
	2 – 3
	3 – 4
	4 – 5
	5 – 6
	6 – 7
	7 – 8
	8 – 9
	9 – 10

Source: Data from C. Donald Ahrens, *Meteorology Today,* 4th edition. West, 1991.

Physical Features of the United States and Canada

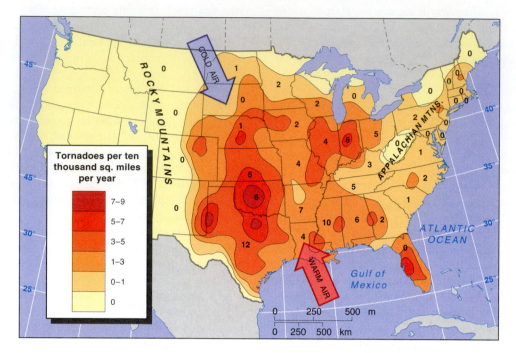

Figure 2.33

Mean annual number of tornadoes per 26,000 square kilometers (10,000 sq mi), 1950–1985. The numbers in the states indicate the average annual state death toll from tornadoes.

Vegetation Regions and Soils of Canada and the United States

To a considerable degree, an area's natural vegetation is a reflection of its climate. Therefore the discussion that follows will focus on the various climate zones, and describe the natural vegetation (before disturbance by people) that typically occurs in these areas. In reality, vegetation in the populated parts of Canada and the United States has been greatly modified by people, but the great zones of natural vegetation can still be reconstructed.

This chapter also discusses the major soil types. Soils are highly complex phenomena, made up of a combination of mineral and organic materials. They occur in very complicated geographic patterns, but climate and vegetation (along with the underlying rock, or parent material) have been major influences in the development of most soils (Figure 2.35).

Soils and Vegetation of the Far North, Great Plains, and Prairies

The harsh climate of the arctic north allows a rather limited range of vegetation, referred to as **tundra,** typically consisting of low grasses, sedges, mosses, and lichens. In these areas, soil development is minimal. Southward is a huge area of **boreal** (northern) **forest,** which covers a large part of Canada and Alaska and extends a short distance south of the border into northern Minnesota, Wisconsin, and Michi-

gan (Figure 2.36). Because of the short growing season and difficult climate, trees in the northern part of the boreal forest are stunted and of little practical value to people. Toward the south, however, they are much larger and more useful; most of Canada's tree cut comes from this area. The forest consists mostly of a few species of pines, spruce, and firs, which are coniferous softwoods, but there are some nonconifers like birch and aspen. These forests produce poor soils that are shallow, acidic, heavily leached, and quite low in nutrients.

Since most trees require substantial moisture, they are rare in the Steppe climate zone of the Great Plains and Prairies. Here they usually are found only along the few permanent streams. Instead of trees, this is the domain of natural grasslands, the particular type of grass varying considerably with location (Figure 2.37). In the wetter, more easterly areas, tall grasses dominate, and some of the world's finest soils developed here. Being large perennials, such grasses regenerate early in spring from established roots. When these plants die back at the end of each growing season, a large amount of organic material starts to become a part of the soil. This high organic content, along with the deep penetration of the roots (leading to deep soils), helps to explain these soils' great natural fertility. Farther west, in the drier parts of the Plains and Prairies, shorter grasses also produce fairly good soils, but they are not as useful as those that develop under taller grasses. In drier portions of the south, such as in central and south Texas, grasses are present but the shrubby mesquite is the dominant species.

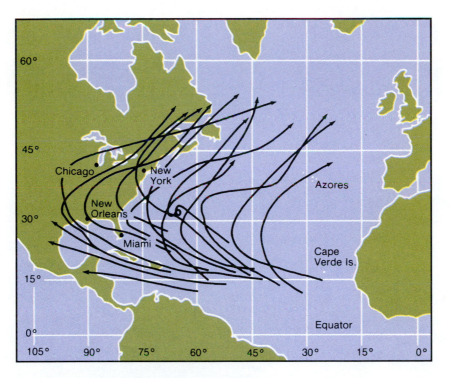

Figure 2.34
Typical hurricane paths.

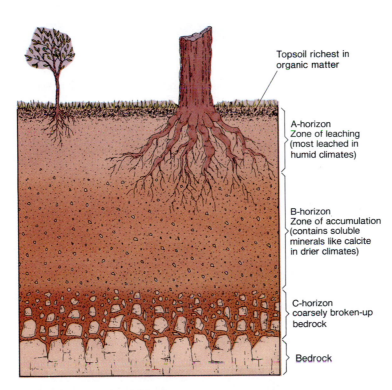

Topsoil richest in organic matter

A-horizon
Zone of leaching
(most leached in
humid climates)

B-horizon
Zone of accumulation
(contains soluble
minerals like calcite
in drier climates)

C-horizon
coarsely broken-up
bedrock

Bedrock

Figure 2.35
A representative soil profile, showing its component soil horizons and their generalized structure.

Physical Features of the United States and Canada

Figure 2.36
Boreal forest east of Fairbanks, Alaska. Due to the harsh climate, these forests replace themselves very slowly. Mount Hayes in the Alaska Range is in the distance.

Soils and Vegetation of the Humid Continental and Subtropical Climates

The highly populated areas of the northeast United States and southeast Canada originally were covered by a dense tree cover. Toward the north these were largely **mixed forests** of needle leaf **conifers** (trees that bear seed cones) and **broadleaved deciduous** (those that lose their leaves each fall) **trees.** At the time of Christopher Columbus, the main deciduous trees found here included oaks, maples, and beech, with a considerable variety of other types as well. Mixed forest areas produced soils that were moderately fertile, easily worked, and extensively used. The higher the percentage of conifers, the more acidic and less fertile the soils tend to be; therefore, those that were found farther south are generally better than more northerly soils.

South of the mixed forests, in an area that includes most of the southern Great Lakes region, southern New Eng-

land, New Jersey, and a strip of land from southern Pennsylvania to Tennessee, the natural vegetation in 1492 consisted of broadleaved deciduous forests, which produced moderately fertile and deep soils that were eagerly sought by pioneer farmers (Figure 2.38). The general north-south trend of vegetation is nicely illustrated in the Eastern Mountains, where the northern areas had almost pure stands of conifers (the boreal forest), the middle range had a mixed forest, and the southern end supported almost only deciduous trees.

In the southeastern United States, with its hot summers and mild winters, there was also enough moisture year-round to support large forests. In areas with sandy soils, such as large tracts along the Atlantic and Gulf coasts, vegetation consisted mostly of coniferous trees (especially pines), a type of vegetation that can survive the many short periods when the sandy soil (which allows rainwater easily through it) can become quite dry. The resulting softwood forests proved valuable for construction and nu-

Figure 2.38
When Columbus landed in 1492, there was a large swathe of broadleaved deciduous forest that extended from Pennsylvania to Tennessee. This is fall at a state park in eastern Pennsylvania.

Figure 2.37
The natural vegetation of the Great Plains and Prairies consists of various types of grasses. This view is of the Black Hills Grasslands in Wind Cave National Park, South Dakota.

merous other uses, and they are highly exploited to this day (Figure 2.39). Sandy areas tend to make poor soils for most kinds of farming, however. Farther inland, there were large areas of mixed and broadleaf deciduous forests. In both areas, the subtropical environment subjected the soils to heavy leaching, and they are not very fertile, though some of them are widely used. When weathered long enough, these forest soils turn red or yellow.

The Savannah climate area of southern Florida supported a tropical grassland, with widely scattered trees. Much of the area was swampy. When drained, these "black muck" soils, with a high organic content, can be very productive.

West Coast Soils and Vegetation
Along the West Coast, the area within California north of San Francisco originally had magnificent stands of conifers dominated by the Giant Redwood (Figure 2.40). With variations in species, similar coniferous forests with huge individual specimens stretched far to the north to include the panhandle of Alaska. In this zone, such forests

occur from sea level to 1800–2400 meters (6000–8000 ft); above that altitude conditions are too difficult for tree growth. In Washington, Oregon, and parts of British Columbia, the Douglas Fir is the dominant species, while in coastal Alaska it is the Sitka Spruce. These softwoods are eminently useful to people, both as recreation areas and as a vast source of the type of wood that is ideal for construction. Despite the size of the trees, such forests do not produce especially fertile or deep soils.

South of San Francisco, the Mediterranean climate of central and southern California is characterized by a specialized shrubland vegetation that is well adapted to the long, dry summers (Figure 2.41). Up to about 1200 meters (4000 ft), the natural vegetation consists largely of dense evergreen shrubs called **chaparral,** which does not produce particularly deep or fertile soils. However, some valleys here, notably large parts of the state's Central Valley, were originally cloaked with grasses. This huge depression, about 650 kilometers (400 mi) long by perhaps 150 kilometers (90 mi) wide, with alluvial soils once covered by grasses, forms a highly productive agricultural area.

Intermontaine Area
In the largely dry zone between the Rocky Mountains and the Sierra Nevada–Cascades, the vegetation patterns are highly complex, with greater variations according to altitude than with latitude. Much of the lower land, both north and south, has a combination of sagebrush and grasses

Figure 2.39

In the sandy Atlantic Coastal Plain of the southeastern United States, the natural vegetation consisted largely of coniferous trees that proved highly exploitable. This view shows Loblolly Pine in Virginia.

Figure 2.40

The humid coast of northwestern California is the home of the famous Giant Redwoods, which include some of the world's largest trees. Shown here is an old growth stand in Humboldt Redwoods State Park near Weott.

(Figure 2.42); this type of vegetation covers about 650,000 square km (250,000 million sq mi). In the middle elevations of the mountains there is significant precipitation, and large coniferous trees (especially spruce, larch, pines, and firs) grow here. These kinds of forests are extensive in the wetter north, particularly in British Columbia, Idaho, and Montana. In some drier areas, such as the lower plains of the extensive Columbia and Snake River valleys, the natural vegetation consisted of grasslands, and these areas proved well-suited to agriculture.

In the intermontaine south, popularly called the Southwest, conditions are often so dry that desert vegetation dominates. In the relatively wetter areas short grasses are common, along with scattered mesquite and creosote bushes, whereas cacti are more prevalent in drier situations (Figure 2.43). The rather alkaline soils formed here are often conducive to agriculture, so long as alkalinity can be neutralized and enough water can be brought here.

Summary

The physical landforms of the United States and Canada can be generalized into just three regions. The largest, the Coastal and Interior Lowlands, extends from the Arctic Ocean south to the Gulf of Mexico. Such regions are areas of *relative,* not absolute, uniformity; thus, there is some variety within these Lowlands. One portion is called the Canadian Shield, an area of ancient rocks that is considered the core around which North America developed. The Eastern Mountains are geologically old and worn down, usually not reaching over 1200 meters (4000 ft). The Western Plateaus, Mountains, and Valleys region is geologically much younger, and mountains here are both higher and more rugged.

The energy responsible for all weather and climate originates with the sun. The main variables of climate are temperature and precipitation. Temperature is a measure of the energy present at a given time, and it varies especially with the directness of the angle of the sun's rays and the number of hours of daylight. For precipitation to occur, air has to be rising, and in North America three main mechanisms cause air to rise. The net result of the geographies of temperature and rainfall is climate regions, or large areas with a relatively uniform climate. In turn, climate is a major influence on natural vegetation. Soils are extremely complex, but climate also exerts an important influence on the geography of soils.

Figure 2.41
Chaparral vegetation consists largely of dense evergreen shrubs. This is chaparral in the Santa Ana mountains of southern California.

Figure 2.42
This view shows sagebrush in the foreground, narrowleaf cottonwood trees in fall colors in the middle distance, and the Grand Teton Range of the Rocky Mountains of Wyoming in the background.

Physical Features of the United States and Canada

Figure 2.43

The stately saguaro cactus is the indicator species of the Sonoran Desert. This typical landscape is in Saguaro National Monument, west of Tucson.

Key Words

alluvium	leeward
boreal forest	mixed forest
broadleaf deciduous trees	mountain (or alpine) glaciers
Canadian Shield	orographic precipitation
chaparral	plateau
climate	rainshadow
conifers	region
continental glaciers	South Pass
continentality	terminal moraine
convectional precipitation	tornado Alley
cyclonic (or frontal) precipitation	tornado warnings
deciduous trees	tornado watches
driftless area	tundra
earth plates	weather
fault line	windward
glacial drift	wyoming Basin

Gaining Insights

1. Define the term *region.* What are some examples of the lack of complete uniformity in the three main landform regions, or the major climate regions, of the U.S. and Canada?

2. Where is the Canadian Shield, and what is it? Why is it useful to the economies of the U.S. and Canada? Why does it have so few people?

3. What were some impacts, both pro and con, of the glaciers?

4. Where is the Mississippi Alluvial Valley, and how was it formed? Why do you think it has a low population density, compared to similar areas in Asia?

5. What's the difference between weather and climate? Apply the concepts to conditions in your area.

6. Explain seasonal variations in the amounts of solar radiation at your location.

7. Why do large bodies of water moderate seasonal temperature differences over nearby land masses?

8. Explain the causes and impacts of continentality.

9. Discuss the three main lift mechanisms that can lead to precipitation. In which parts of the U.S. and Canada are each of these mechanisms important?

10. Explain the differences in the two main West Coast climates of Canada and the U.S.

11. Explain the main causes of the differences between the Humid Continental, Hot Summer, and the Humid Subtropical climates.
12. What areas of the U.S. and Canada are most susceptible to hail, tornadoes, hurricanes, and blizzards?
13. Why do high grass areas tend to produce such excellent soils?
14. List and explain the changes in vegetation as one moves from south to north within the eastern U.S. and southeastern Canada. Do the same for soils.

Selected References

Ahrens, C. Donald. *Meteorology Today.* St. Paul: West Publishing Co., 1991.

Barbour, M. G., and W. D. Billings, eds. *North American Terrestrial Vegetation,* Cambridge: Cambridge University Press, 1988.

Barry, R. G. *Atmosphere, Weather and Climate,* 5th ed. New York: Methuen, 1987.

Bird, J. B. *The Natural Landscapes of Canada.* Toronto: Wiley Publishers, 1978.

Eagleman, Joe R. *Meteorology: The Atmosphere in Action,* 2nd ed. Belmont, CA: Wadsworth, 1985.

Gersmehl, P. J. "Soil Taxonomy and Mapping," *Annals, Association of American Geographers* 67 (1977): 419–28.

Henderson-Sellers, Ann, and Peter J. Robinson. *Contemporary Climatology.* New York: John Wiley & Sons, 1986.

Hidore, John J., and John E. Oliver. *Climatology: An Atmospheric Science.* New York: Macmillan, 1993.

Kuchler, A. W. *Potential Natural Vegetation of the Conterminous United States.* New York: American Geographical Society, Special Publication 36, 1964.

Linacre, Edward. *Climate Data and Resources.* New York: Routledge, 1992.

Lydolph, Paul E. *The Climate of the Earth.* Totowa, N.J.: Rowman and Allanheld, 1985.

Pielke, Roger A. *The Hurricane.* New York: Routledge, 1990.

Putnam, D. F. *Canadian Regions: A Geography of Canada,* 8th ed. Toronto: Dent Co., 1968.

Scott, R. C. *Essentials of Physical Geography.* St. Paul: West Publishing Co., 1991

Strahler, A. N., and A. H. Strahler. *Elements of Physical Geography,* 4th ed. New York: John Wiley & Sons, 1990.

Vale, T. R. *Plants and People: Vegetation Change in North America.* Washington, D.C.: Association of American Geographers, 1982.

Wallen, R. N. *Introduction to Physical Geography.* Dubuque, Iowa: Wm. C. Brown, 1992.

Weatherwise. Issued six times a year by Weatherwise, Inc., Princeton, N.J.

Physical Features of the United States and Canada

Settlement Patterns before 1950

In 1620, when the Pilgrims landed at what is now Plymouth, Massachusetts, they found an area with many signs of native occupation but few people. Extensive "Indian Olde Lands," clearings in the forest, had once been farmed, but were now abandoned. What had happened?

Recent studies show that eastern North America in 1492 had a far larger native population than once thought. These people, long isolated from the Old World, had no immunity to European diseases. With the arrival of European explorers and fishermen shortly after 1492, epidemic diseases such as influenza, measles, and smallpox spread quickly and took a terrible toll.

The American Indians of Massachusetts, being directly in the path of European interests, were early victims of this deadly invasion. Infected tribes spread the diseases to more inland groups, often leading to catastrophic losses of lives. One author estimates that largely because of these epidemics, the native population of North America north of Mexico dropped 74% between 1492 and 1800. The Pilgrims and Puritans thought that Providence had intervened to remove most of the "heathen" and to open up the land to their "divinely inspired occupance." What the surviving American Indians thought was not recorded, but it can easily be imagined.

Status of American Indians before 1607

It is now estimated that at the time of Columbus's first voyage, there were somewhere between 2 million and 4 million American Indians in what became the United States and Canada, instead of the 500,000 to 1 million accepted by most academics until quite recently. These people lived in a wide variety of environments, from the Arctic to the semi-tropical, and were highly differentiated in language and culture, with hundreds of distinct societies (Figure 3.1).

In 1492, American Indians occupied virtually all of North America, although the intensity of that occupation varied greatly according to environment and livelihood. Some were hunters and gatherers, such as the early Californians and the Inuit (Eskimos) of northern Alaska and Canada. East of the Mississippi River and the Great Lakes, natives practiced a fairly intensive agriculture using simple tools, such as digging sticks and fire. In the Southwest, some tribes had developed sophisticated irrigation systems, an innovation that had been adopted from groups living in what now is Mexico. Irrigation led to larger and more dependable food supplies, and populations grew quickly to

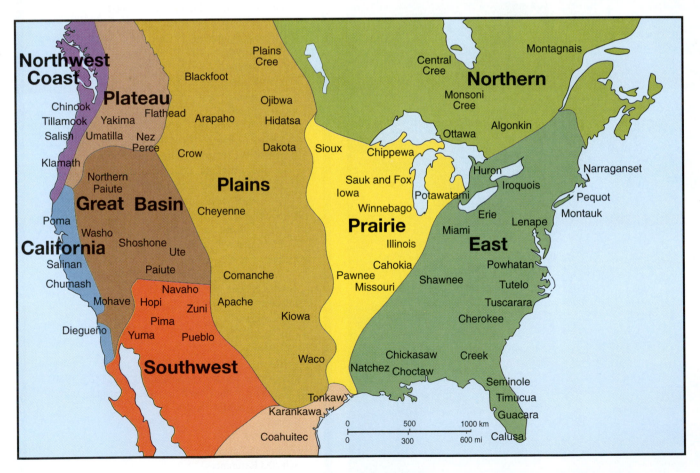

Figure 3.1
Location of major native groups, 1492. Many of the tribes later moved (see also Figure 14.3).

match the enhanced output. On the Great Plains, a few hunters and gatherers relied largely on the buffalo for sustenance. When these natives acquired horses in the 1600s, this way of life became highly successful, and several nearby tribes were induced to move onto the Great Plains.

Contrary to popular impressions, wherever they lived, American Indians substantially modified their environments. Many early explorers of eastern North America wrote of seeing large areas of intensive agriculture (Figure 3.2). Both the open eastern forests and the prairies of Indiana and Illinois, where forests had once stood, give strong evidence of native impacts on the environment by the use of fire. Eastern North America was not an idyllic "forest primeval," as both plant and animal life had been highly modified by humans. The notion that natives had developed a life-style that used the land totally in harmony with nature is largely a myth.

Wherever there was sustained contact with Europeans after 1492, local livelihoods changed quickly.

Figure 3.2

Intensive American Indian agriculture, 1585. John White's sketch of agriculture in a village shows several fields of corn, in various stages of ripeness, next to the road. This village was near the coast of what is now North Carolina.

Settlement Patterns before 1950

Natives soon developed a strong demand for many European goods, such as tools, blankets, weapons, and decorations. In return, Europeans wanted some goods natives could provide, particularly animal skins and furs. A vigorous trade developed well before extensive European settlement, an exchange with far-reaching consequences. In the eastern United States, for example, the tribes of the Iroquoian confederation of central New York soon gained great economic and political power and spread their influence over other groups far inland. With European weapons, they also wreaked havoc on traditional enemies. Tribes tended to become allied with particular European nations, and sometimes they were pulled into conflicts between them.

When the British began systematically colonizing North America, starting at Jamestown, Virginia, in 1607, the American Indians they found were only remnants of much larger tribes reduced by warfare, disease, and migration west. The English found little fault in taking land from these survivors, for they believed strongly that they could make much more effective use of the land and that God would therefore approve. This was reinforced by the English settlers' low opinion of the people they called heathens, who were ignorant of the Christian god and salvation.

Spanish, French, and Dutch Settlement to 1700

Spain, France, and the Netherlands all established early, important colonies in the area that became the United States and Canada (Figure 3.3). Although some of their efforts were temporarily successful, all had serious weaknesses and eventually were taken over by the British or the Americans. Still, some of their settlements proved to be permanent, and evidence of early Spanish and French efforts is still clearly discernible in the landscape.

Spanish Settlement

Early Spanish settlements in North America were widely scattered near the southern edge of today's United States. In 1565, Spaniards established the town of St. Augustine on Florida's east coast, which is recognized today as the oldest permanent European settlement in the United States. Here, they built a fort, which has been reconstructed as a national monument. This location was chosen to protect Spanish shipping, which took advantage of the offshore North Atlantic drift to return to Spain. Spain's occupation of Florida was always weak, attracting only 3000–4000 settlers, and

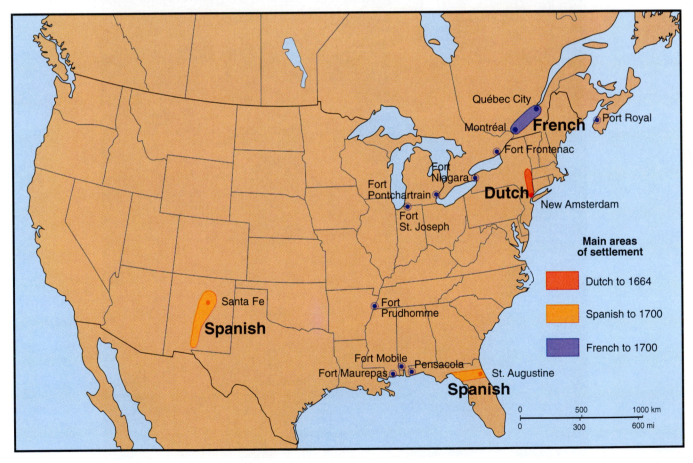

Figure 3.3

Main areas of Spanish, French, and Dutch settlement. The map shows the approximate area, somewhat exaggerated in size for visual clarity, settled by the Dutch before 1664 and by the Spanish and French in 1700.

the colony was a drain on its treasury. Some missions to convert the natives were established, another fort was built on Florida's west coast, and a little agriculture was practiced, but the Spanish hold was always tenuous, and in 1819, the area was acquired by the United States.

Far more lasting and influential were the Spanish settlements in what is today northern New Mexico, along the upper reaches of the Rio Grande. With the permission of the Spanish crown, this area was colonized in 1598 as a private venture by Juan de Oñate and 210 other Spanish colonists, as well as an unknown number of native followers. The area had a considerable population of Pueblo natives, living in adobe-and-stone structures and practicing irrigation, who were prospects for missionary work. By 1610, Santa Fe had been founded, and settlers had begun to receive large grants of land (haciendas). A ranching economy was established in this area, which was very isolated from other European settlements. Although the Spaniards were expelled by a native rebellion in 1680, they returned permanently in 1693. By 1821, when the area became a part of Mexico, about 30,000 Spaniards were scattered over a large area. This Hispanic community successfully withstood the change from Mexican to U.S. sovereignty in 1848 and even continued to expand, such as into southern Colorado.

French Role

The French were the first Europeans to establish permanent settlements in northeastern North America. Their initial enterprise was in 1606 at Port Royal in what is now the Canadian province of Nova Scotia. The second, and more important, permanent French settlement was in 1608 at Québec City, a defensible site on the St. Lawrence River, the outlet for the Great Lakes. Later, Montréal was founded farther up the St. Lawrence, at a location favorable for river transportation in several directions. All their early important bases were in eastern Canada, but from there, the French spread their influence far inland.

For many years, French-Canadians could make their best living in the fur trade, leading to many water-borne expeditions ever farther inland to explore and obtain furs. These French-Canadian **voyageurs** were the first Europeans to see large parts of the Great Lakes area and the Mississippi Valley. Such place names as Detroit, Mobile, New Orleans, and St. Louis reflect the widespread activities of this small but influential group of explorers, traders, missionaries, and settlers. As early as 1682, they established a trading-farming community at Kaskaskia, on the Mississippi River in what is now southern Illinois. The water-oriented French divided their lands into **long lots,** properties with narrow river fronts that extended perhaps ten times as far inland away from these streams (Figure 3.4). Even today, these long lots can easily be seen from aircraft flying over early French settlements, such as in Québec and along the Mississippi River north of New Orleans.

Since the heart of New France, the Québec City–Montréal area, had severe winters and mostly mediocre farmland, it did not have great appeal for potential French settlers. Despite the government's efforts, only slightly more than 10,000 French eventually moved to this area. Most settled on farms under the feudal **seigneurial system,** whereby large tracts of lands (seigneuries) were granted to favored individuals who were then charged with bringing in tenant farmers. These seigneuries, laid out in the long lot pattern, were concentrated along the St. Lawrence River between Montréal and Québec City. Although French immigration to Canada was light, birth rates were exceptionally high; by 1763, the population had risen to about 65,000. That was the year, following the military defeat of the French, that New France became a part of the British Empire. The French remained, however, and grew rapidly because of continued high birth rates. They were the largest ethnic group in Canada until 1834, and clinging tenaciously to their language and Catholicism, they continue to exert a major influence on Canadian affairs. French-speakers now constitute about one quarter of the Canadian population.

The Dutch

Another European country to found an early northeastern settlement was the Netherlands, which established a fur-trading post at Fort Nassau, near present-day Albany, New York, in 1614. This post was about as far upstream on the Hudson River as ocean-going vessels of the day could go. Here, the Dutch traded with the Iroquois, who had direct and easy access to the west via the *Mohawk Gap* through the Appalachian Mountains. In 1625, the Dutch founded New Amsterdam at the southern edge of Manhattan Island, near the mouth of the Hudson River. New Amsterdam grew into a walled city of about 800 people by 1664, when a hostile English fleet sailed into the harbor and the overwhelmed Dutch surrendered. Soon both the city and the colony were renamed New York.

The Dutch had tried to establish a feudalistic settlement pattern in New Netherlands, a feature that attracted few Dutch settlers. This encouraged the colony to adopt a liberal immigration policy, so that when the English took over, only 5000 of the 8000 colonists were of Dutch descent. Partly because of this feudal legacy, New York's growth was long retarded.

English Settlement to 1783

The 13 highly varied colonies that England established along the eastern seaboard proved very successful in attracting European settlers. Their physical environments ranged from the semitropical coastal lowlands of Georgia and South Carolina to the midlatitude shores of New England.

From the beginning, cities played an important role in the settlement of North America. All the colonies began with the establishment of coastal towns, from

Figure 3.4
L'Isletville, Québec, reflects the pattern of early land concessions along the St. Lawrence River. Note the distinctive landscape of long, narrow fields and linear settlement along the range roads (see also Figure 14.7).

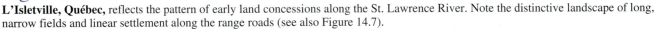

which subsequent settlement could be organized. As settlement spread into the interior, smaller towns often acted as vanguards of this expansion. Because the early colonial economies were overwhelmingly rural-based, however, only a small percentage of the population lived in cities. The largest of these cities were all on waterways, acting as points of commercial and political contact with, especially, the home country.

The 13 colonies are often classified into three broad groups, based on their similarities of economies and what might today be called life-styles. These economic-cultural regions began to take shape early in American history in areas called **culture hearths,** then spread outward to influence the rest of the country (Figure 3.5). The three culture hearths, which had been formed by 1700, were (1) the *Chesapeake Bay* area, focused originally on southeast Virginia; (2) *southern New England,* focused on the Boston area; and (3) the *Midland,* or Middle Colonies, based in Philadelphia and vicinity.

Chesapeake Bay Culture Hearth
In 1607, an area on southern Chesapeake Bay in the colony of Virginia was chosen for what turned out to be the first successful British settlement in North America (Figure 3.6). The British had decided on this location because it is at about the same latitude as Europe's Mediterranean lands. They assumed that the Virginia–North Carolina area could produce such Mediterranean products as grapes, citrus fruits, and silk, which were in demand in England. Experience soon proved otherwise.

As noted in Chapter 1, the first settlement, Jamestown, was poorly located in a swampy area with an unhealthy water supply, which led to a high mortality rate, and the Virginia Company's colony almost did not survive. The venture was saved by moving, after a long delay, to a more healthy location and especially by the discovery that the "stinking weed," tobacco, could be grown successfully. Tobacco proved highly profitable, ensuring not only survival but growth.

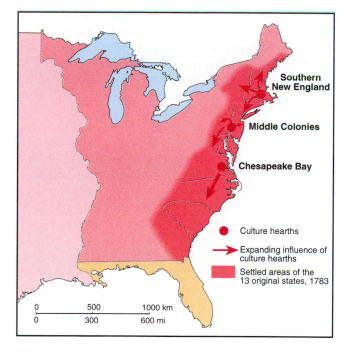

Figure 3.5
Territorial limits of the United States and the settled area in 1783. The map also shows the culture hearths and the directions of their growth. At this time, there were already some unofficial settlers west of the Appalachians, an area the British had reserved for American Indians.

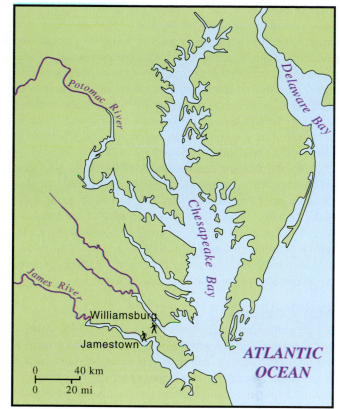

Figure 3.6
Sites of Jamestown and Williamsburg, Virginia. The capital of the Virginia colony was moved 11 kilometers (7 mi) from Jamestown to Williamsburg in 1699. The westward movement of population in the 18th century made a more central location desirable, and in 1779, the capital was transferred to Richmond. Williamsburg has been recreated—"better" than it ever was, for tourist comfort—and is today a major tourist attraction.

Tobacco production required much labor, the bulk of which for most of the 17th century was supplied by poor young Englishmen who came as indentured servants. In return for their passage to the New World, they worked for their sponsor for four to seven years, after which they received help in becoming established on their own. This system had its problems, producing a great sex imbalance and social cleavages.

Although African slaves were brought into the colony as early as 1619, they were not widely used at first, and in 1660, fewer than 1000 were living in the Chesapeake region. They did not become important to the plantation economy until the late 1670s, when, because of reduced costs, large numbers were brought in. Thereafter, blacks became a major portion of the population in all plantation areas. Many came directly from West Africa, but others had previously worked on Caribbean islands. Initially, most planters could not afford slaves, and as late as 1700, fewer than 20% of plantations had any. Because slaves were concentrated on the few largest plantations, however, their labor accounted for far more than 20% of plantation output.

Maryland, also on Chesapeake Bay, was founded by Lord Baltimore in 1634. He wanted to establish a semifeudal society, but economic reality soon led to the adoption of the Virginia pattern of tobacco plantations. In the 1650s, these two colonies together were exporting 4 million to 5 million pounds of tobacco per year, a figure that grew to 35 million pounds in 1700. By the 1680s, Maryland was also buying large numbers of slaves. All together, more than 12,000 slaves were brought into the Chesapeake region between 1670 and 1700.

The plantations were relatively self-sufficient, producing most of the food, fuel, and other items they needed. Nearby areas also had skilled rural workmen, such as blacksmiths and coopers. For these reasons, and because many plantations had their own docks for overseas shipping, there was little need for the services normally provided by towns. Therefore, by 1700, not a single place in either Virginia or Maryland had more than 500 people. No wonder this culture hearth has been called "overwhelmingly rural, agrarian, dispersed, and decentralized."

South Carolina was influenced by the Chesapeake culture hearth, but different crops were raised here, and the organizers were especially interested in attracting a wealthy elite. In the 1680s, the colony began to experiment with rice cultivation near Charleston, and within ten years, this experiment was considered a success. Exports of rice to the Caribbean islands and Southern Europe soon began. Since growing rice was even more labor-intensive than tobacco, slaves greatly outnumbered Europeans in the plantation areas. The dye indigo was also grown successfully on plantations in South Carolina. In contrast to the culture hearth, South Carolina did develop an important city, Charleston, which had a population of about 2000 in 1700.

By the time of the American Revolution, southern plantation crops dominated the overseas trade of the 13 colonies. Tobacco, still primarily grown in the Chesapeake area, accounted for almost one-third of the value of all colonial exports. Although some cotton was grown in coastal South Carolina and Georgia, it was still a minor crop in the 1770s. Overall, the value of southern exports exceeded those of the North by almost two to one.

New England

The area around Massachusetts Bay was first settled in 1620 by the Pilgrims of Plymouth Colony, people who generally had little education or means. A few years later and farther to the north, a second group began to arrive. This hearth is unusual in that the second group, the Puritans of the Massachusetts Bay Colony, played a far more important role than the original group in shaping the region's economy and culture. Puritans were quite well educated and from a higher social class than their predecessors. The first of them arrived in the Boston area in 1629, followed by a large influx of 16,000 more by 1640. The Puritans soon were both far more numerous and influential than the Pilgrims.

At first, New Englanders settled in loosely organized villages, making the area one of the few places in the future United States with *farm villages.* However, farmers gradually dispersed into *individual homesteads,* the pattern that was virtually universal elsewhere in the English colonies. These people had an extremely high birth rate, and the rapid population growth forced settlement farther into the interior. As settlement expanded, New Englanders created agricultural townships, with centrally located churches, a meeting house, and small towns. They practiced a near-*subsistence agriculture:* that is, each family grew about as much food as it consumed, with little surplus for sale. For almost 200 years, until the early 19th century, New Englanders were reluctant to leave their region, to which they were closely tied by sentiment.

Although most Puritans were farmers, many had urban backgrounds, and they helped establish several important towns. Boston, in particular, was centrally located with a good harbor. The area tributary to Boston (its *hinterland*) lacked good agricultural lands, so the city began to concentrate, with considerable success, on handling other people's goods throughout the Atlantic economy. In 1700, Boston had close to 7000 inhabitants, and by 1740, it had grown to 16,000 and was the largest British colonial city in North America. Other early towns included Salem, Providence, and Newport. These places generated a strong demand for food, which encouraged some nearby farmers to practice commercial rather than subsistence farming.

Midland Colonies

When William Penn founded his colony, Pennsylvania, in 1681, the area had few American Indians, for most had died of disease or war. From the start, Pennsylvania had liberal policies on religious freedom and immigration, and it attracted a more varied European population than most other colonies. As the colony actively recruited overseas and made it easy for settlers to obtain land, its population grew rapidly. Only 19 years after its founding, the colony had almost 18,000 people and a diversified economy.

This culture hearth began to assume its distinctive form in southeast Pennsylvania, west of Philadelphia (Figure 3.7), an area of excellent soils. Farmers specialized successfully in growing grain (especially wheat and corn). With the growth of population, this agriculture soon was carried west to the edge of the Appalachian Mountains and even beyond into some fertile valleys. Pennsylvanians soon were a significant proportion of the pioneer farmers who were pushing the colonial frontier farther to the south as well. Overall, as one author has put it, Pennsylvania "came to symbolize the heterogeneous, yeoman-farming, free labor world of the colonies north of the Chesapeake."

In the early 1700s, trading of the grain surplus began, first with the West Indies sugar colonies and the southern colonies, and then with Southern Europe. As a result, a successful agricultural society soon became a commercial one as well, with Philadelphia as its center. By 1700, Philadelphia had 3000 inhabitants and was the third largest city (after Boston and New York) in the 13 colonies. By the time of the American Revolution, Philadelphia was the largest colonial city, with about 30,000 people.

The nearby colonies of New York and New Jersey also belong to this Midland region, and together, the three are sometimes called the "bread colonies." After a slow start as New Amsterdam under the Dutch, British New York City gained considerably as a commercial city, taking full advantage of its magnificent harbor at the mouth of the Hudson River and the meat and grain production of its back country. New York's trade pattern was similar to Philadelphia's, with grain as its leading export. In 1700, the city had a population of 4500. At the start of the Revolution, New York was the second largest colonial city after Philadelphia.

Developments, 1700–1783

Between 1700 and 1776, the 13 English colonies grew rapidly, their combined population expanding tenfold, to almost 2.5 million. Most of this increase was by natural growth, but about 370,000 Europeans and 250,000 Africans

Figure 3.7
Colonial Philadelphia, about 1750. This view was painted on the New Jersey shore, across the Delaware River from the city.

George Heap, *East Prospect of Philadelphia from the Jersey Shore.*

also arrived from overseas. With this great growth, farms and plantations filled most good agricultural land east of the Appalachian Mountains. Settlement west of the Appalachians was prohibited by the British proclamation of 1763, which reserved that area for American Indians. This ruling was difficult to enforce, however, and was evaded by some.

Despite some liberal immigration policies that led to an increasing influx from other countries, most colonists were of British descent at the time of the American Revolution. However, about 20% of the population had an African background, and 10% had come from the German states. The Scots-Irish (Scots who had lived for a while in Ireland) were also well represented.

In 1776, the area south of the *Mason-Dixon line* (the Maryland-Pennsylvania border) contained almost half the population of the 13 colonies. This helps to explain why Washington, located just south of this line, was chosen for the capital of the new nation in 1791.

Urbanization, Industrialization, and Settlement Frontiers, 1783–1865

When the United States gained its independence in 1783, the country consisted of most of the territory east of the Mississippi River, excluding only a strip of land (an expanded Florida) along the Gulf of Mexico. Between that year and the eve of the Civil War, the area of the United States tripled, population grew by 15 times, the economy by 20 times, and urban population 30-fold. The four most significant additions of territory were the Louisiana Purchase of 1803, the admittance of Texas as a state in 1846, the settling of the "Oregon Country" border in 1846, and the southwestern lands acquired after the Mexican War in 1848. By 1853, with the purchase of land from Mexico in southern Arizona and New Mexico (the Gadsden Purchase), the conterminous United States had reached its ultimate size. Alaska, which had been the site of Russian coastal settlements since 1784, was purchased in 1867, and Hawaii became a U.S. territory in 1898 (Figure 3.8).

Because of areal and economic expansion, the United States had the third largest economy in the world by 1865, behind only the United Kingdom and France. Also by 1865, three major *specialized economic regions* had evolved. These were the Northeast, which was growing particularly in manufacturing and urbanization; the South, which specialized in plantation agriculture; and the Midwest (the *Old Northwest*), where small farmers concentrated on selling grains and animals for meat.

Urbanization and Industrialization of the Northeast

At the time of the first U.S. census in 1790, only about 5% of the population was urban, and as late as 1850, only 15% was living in cities. Still, absolute urban growth was impressive. For example, between 1800 and 1840, the urban population doubled, and it almost doubled again during the next twenty years: by 1860, almost 20% of the population was urban. Early American cities necessarily were very compact, for these were pedestrian-oriented places with no public transportation. Because transportation by water was far cheaper at this time than any other mode, all of the larger cities were located on oceans, lakes, rivers, or canals.

Perhaps the major urban development in the United States between independence and the mid-1800s was the consolidation of the power of the northeastern port cities of Boston, New York, Philadelphia, and Baltimore. In 1830, for example, these were the only cities with more than

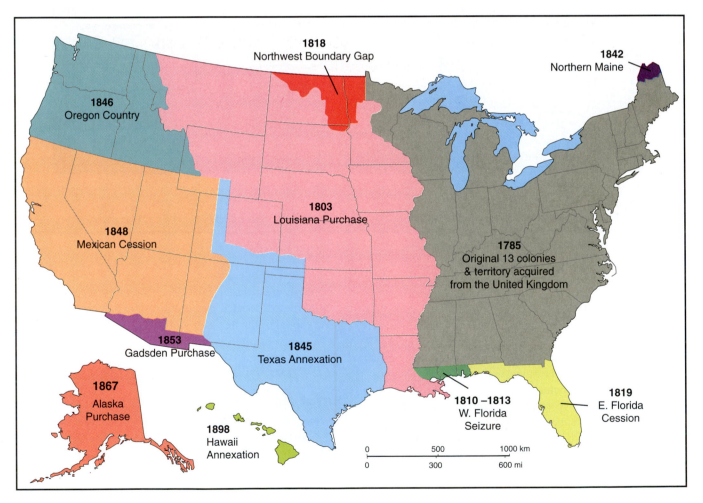

Figure 3.8
Territorial acquisitions of the United States, 1783–1898.

50,000 people. This period was also noteworthy for the emergence of New York as the country's dominant (*primate*) city, far outstripping its rivals. At the time of the 1790 census, New York had a population of 33,000 and had recently passed Philadelphia to become the country's largest city; ten years later, its population exceeded 60,000. By 1830, New York had more than 200,000 people, making it two and a half times larger than its nearest rival, Philadelphia. Critical to its rise to primate city was New York's control of the external trade of the South, especially cotton. Because of this control, southern urban growth lagged. In 1830, the largest southern cities were New Orleans and Charleston, which were the fifth and sixth largest U.S. cities, respectively.

As early as the 1790s, American urban growth was affected by industrialization (Figure 3.9). America's first important manufacturing industry produced cotton textiles, and the first cotton spinning machine was set up in Providence, Rhode Island, in 1790. In 1822, Lowell, Massachusetts, often called the first true American industrial city,

was established. As the new factory system of production required a large number of workers to be concentrated in one place, industrialization usually led to rapid urban growth. By 1825, both Lowell and Providence were among the country's 25 largest cities.

Southern New England (Massachusetts, Connecticut and Rhode Island), with its abundant water power, was the first industrial region in America, and from an early date, it specialized in the production of cotton textiles, woolens, shoes, and boots. In New England, much of the industry was located in medium-sized industrial cities, but elsewhere, it tended to concentrate in the largest cities. Still, Boston was the dominant wholesaling and trading center for all these industries, as well as a leading industrial city.

Although southern New England was America's first industrial region, the focus of development soon shifted southward to the cities of New York, Philadelphia, and Baltimore and their vicinities. These places had better access to coal, the important new industrial fuel, than New England. They were also on excellent harbors and had good access to

Figure 3.9

Whitneyville, Connecticut, an early American industrial town. The town grew up around a factory established by Eli Whitney in 1798 to make rifles using interchangeable parts.

Settlement Patterns before 1950

cheap immigrant labor. New York had by far the best route to the interior markets via the Erie Canal (described on page 65), but railroads eventually helped Philadelphia and Baltimore overcome somewhat their transportation disadvantages.

Overall, manufacturing grew especially rapidly in the older states of the northeast United States, concentrating in the country's four largest cities and some nearby areas, such as southern New England, upstate New York, and southeast Pennsylvania. By 1860, the northeast region had emerged as the clear-cut *economic core* of the expanding United States economy. In that year, it had 72% of the nation's manufacturing jobs; of the 23 cities in 1860 with more than 5000 manufacturing employees, only six were outside the northeast. In 1860, New York and Philadelphia were the largest industrial cities, accounting for about 15% of national manufacturing employment. Manufacturing's contribution to total national commercial output about doubled during the last two decades before 1860 and accounted for close to one-third of the country's total in that year.

Settling Kentucky, Tennessee, and the Old Northwest

Despite the British Proclamation line of 1763, which prohibited European expansion west of the Appalachians, many settlers were already there when the Revolutionary War ended in 1783. Thereafter, land seekers began to pour across the mountains into Kentucky and Tennessee, the only large trans-Appalachian areas open to settlement at the time (Figure 3.10). By 1790, more than 100,000 people had moved west, most traveling through the mountains via the famous **Cumberland Gap,** which was actually a series of gaps in the mountains near the westernmost tip of Virginia. The largest number came from Virginia, but many others were from Pennsylvania, Maryland, and North Carolina.

In 1783, American Indians still held rights to most of the land both north and south of Kentucky and Tennessee, and U.S. policy allowed settlement only after title to the land had been transferred by formal treaty. In the Old Northwest, which eventually produced the midwestern states of Ohio, Michigan, Indiana, Illinois, Wisconsin, and a part of Minnesota, these American Indian lands were ceded in a series of treaties. Perhaps the most important was the **Treaty of Greenville** of 1795, which opened the southern half of Ohio and a part of Indiana to settlement.

Although this area was cleared of American Indian land rights, federal policy required surveying before settlers could claim land. In the Old Northwest, this was done using the **township and range system,** under which land was subdivided into a series of approximate squares, with boundary lines running north-south and east-west (Figure 3.11). A township consisted of a square 6 miles (9.7 km) on a side that was divided into 36 "sections" of 1 square mile (640 acres or 259 hectares) each. A quarter of a section, or 160 acres (65 hectares), was considered the standard size for an American farm. Since roads and field boundaries

usually follow the straight lines of this surveying system, it is easily recognized from an airplane by its checkerboard patterns. It replaced the older **metes and bounds system,** which used natural features (e.g., rivers, rocks, trees) to demarcate property lines; that system dominates along the East Coast and in the Southeast.

Even before 1800, many settlers crossed the Appalachian Mountains via several roads to Pittsburgh. Here, many built rafts on which they floated down the Ohio River until, perhaps at Cincinnati, they disembarked to travel overland to choose land for a farm. The easier route west through the mountains via the Mohawk Gap became available later.

The earliest important western cities were Pittsburgh, Cincinnati, and Louisville, all of which occupied strategic points along the Ohio River. Pittsburgh was at the point where two rivers joined to form the Ohio (today, the site of Three Rivers Stadium); Cincinnati was where the river changed direction; and Louisville was where rapids often necessitated transferring freight and passengers between upper- and lower-river boats.

In the Midwest, early farmers concentrated on wheat, corn, and livestock. The southern portion near the Ohio River was the first to develop, and farmers there specialized in hogs and corn. Located near the center of this area, Cincinnati became an important meatpacking center.

Because water transport was much cheaper than overland shipment by wagon, for many years the Midwest's

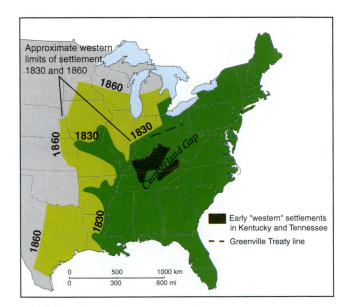

Figure **3.10**

The push westward. Soon after 1783, Americans began to pour across the Appalachians, commonly via the famous Cumberland Gap, to settle the first American "west" in Kentucky and Tennessee. Also shown are the Greenville Treaty line and the approximate limits of the frontier in 1830 (when major river routes allowed tongues of settlement westward and northward) and 1860.

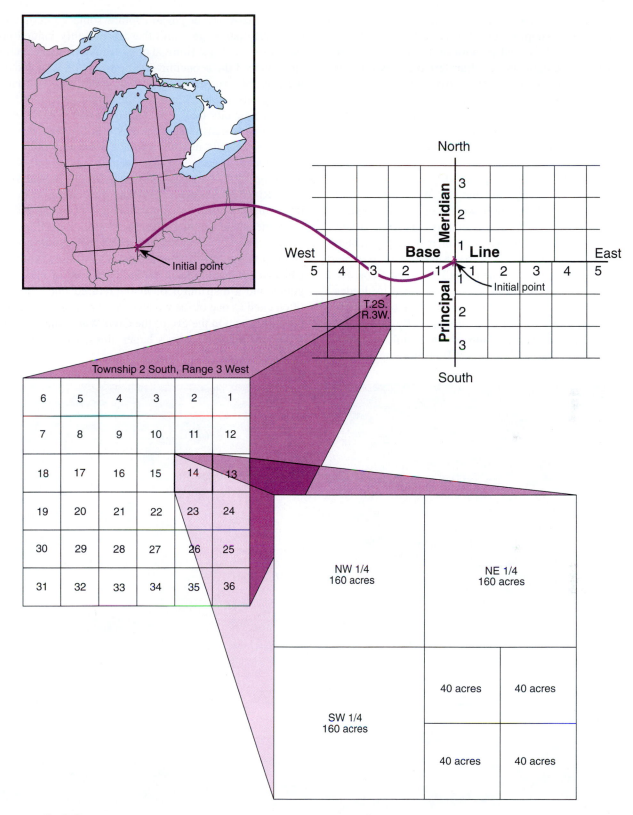

North

West   Base   Line   East

South

3
2
1
Meridian

5 4 3 2 1 | 1 2 3 4 5
1
Initial point
2
3
Principal

Initial point

T.2S.
R.3W.

Township 2 South, Range 3 West

6	5	4	3	2	1
7	8	9	10	11	12
18	17	16	15	14	13
19	20	21	22	23	24
30	29	28	27	26	25
31	32	33	34	35	36

NW 1/4
160 acres

NE 1/4
160 acres

SW 1/4
160 acres

40 acres | 40 acres
40 acres | 40 acres

Figure 3.11

After independence was achieved, the U.S. government decided that the public domain should be systematically surveyed and subdivided before being opened for settlement. The resulting township and range survey system, adopted in the Land Ordinance of 1785, established survey lines oriented in the cardinal directions and divided the land into townships 6 miles (9.7 km) square that were further subdivided into sections 1 mile (1.6 km) on a side. For a long time, the standard American farm consisted of one quarter of a section, or 160 acres (65 hectares). The American grid system of land division was first used in eastern Ohio.

agricultural surplus largely went to market downstream via the Ohio and Mississippi Rivers. The largest river port was New Orleans, with its excellent position near the mouth of the Mississippi. Here, freight was transferred to ocean-going vessels for shipment to eastern or overseas consumers. Pittsburgh and Louisville, and especially Cincinnati, were the first inland cities to benefit from an effective export link downriver to New Orleans, and they grew accordingly. Later, when steamboats also made upriver transportation feasible, they benefited as distribution points for imports brought from New Orleans (Figure 3.12). In 1830, Cincinnati was the country's largest inland city, with about 30,000 people; Louisville and Pittsburgh were about half as large.

New Orleans' dominance of this "western" trade lasted until the late 1830s, when an increasing percentage of the traffic began to use the shorter route to and from the major northeastern cities via the Great Lakes and the newly opened **Erie Canal.** By 1850, this route, aided by feeder railroads and canals, had already captured 50% of the western trade, and it grew even more dramatically during the 1850s. In that decade, several railroads were built westward from New York, Philadelphia, and Baltimore across the Appalachians and began to capture some western traffic, especially passengers and the more highly valued commodities, such as flour. These transportation changes strengthened the economic link between the East and Midwest on the eve of the Civil War, a link that was largely missing between the South and the other two regions.

A new city that participated heavily in these emerging east-west transportation routes was Chicago, a port located strategically near the extreme southwest point of the Great Lakes. It grew especially rapidly after 1850, following completion of a canal connecting it to the whole Mississippi River transportation system, and began to develop into a major rail center. With excellent water and rail links to the newly developing wheat belt of northern Illinois and Wisconsin and to the Corn Belt (with its emphasis on livestock production) of central Indiana, Illinois, and Iowa, Chicago grew remarkably as an assembler of agricultural goods and a distributor of manufactured items, and was often cited as one of the wonders of 19th-century American development. On the eve of the Civil War, Chicago was already one of the country's larger cities, with a population of 109,000.

Figure 3.12
Western side-wheeler steamboats on the Mississippi River. Such flat-bottomed, shallow-draft boats were well suited to the western rivers, where sandbars and other obstacles were a constant source of danger. Steamboats made upriver travel economically feasible and, before the coming of the railroads, they were the major form of heavy transportation in the West.

F. F. Palmer, *A Midnight Race on the Mississippi*, 1860.

Plantation System Expands Westward

Before the 19th century, plantation agriculture was confined largely to tobacco, which was concentrated in the states of Virginia, Maryland, and North Carolina. By 1800, production had started to move inland from its original coastal location to the Virginia-Carolina Piedmont, primarily because the coastal soils were depleted. At the turn of the century, cotton was still a minor crop, as only the sea-island variety, grown on islands along the South Carolina and Georgia coasts, was profitable. The upland variety could be grown inland, but the cost of separating cotton fiber from the seed by hand made it unprofitable. Then came the invention of the **cotton gin,** a machine for removing the seed that revolutionized the cotton industry.

Also revolutionized was the geography of plantation agriculture. As cotton production grew rapidly after 1800,

---

ERIE CANAL

The Erie Canal was a "ditch" 584 kilometers (363 mi) long, 12 meters (40 ft) wide, and 1.2 meters (4 ft) deep that was built from Albany to Buffalo, New York, via the Mohawk Gap, between 1815 and 1825. Since Albany was connected to New York City via the navigable Hudson River and Buffalo was a port on Lake Erie, the Erie Canal offered a low-cost, all-water route from the Atlantic Ocean at New York to the Great Lakes. This route led to a 90% reduction of eastbound freight charges along the canal from $100 a ton by land to $10 a ton by water. Agricultural products originating in Chicago could be shipped cheaply on Great Lakes vessels to Buffalo, where they were transferred to canal barges for movement to Albany and then downriver to New York.

Manufactured products could flow just as cheaply in the opposite direction, and many westbound settlers rode the canal boats.

With the canal's completion in 1825, New York City had by far the cheapest route to the interior, putting its main rivals (Boston, Philadelphia, and Baltimore) at a tremendous disadvantage. As these cities had no equivalents of the natural advantage of the Mohawk Gap, they could not overcome New York's lead. The completion of the Erie Canal was a major factor in New York's continued rapid growth after it had become the country's primate city. Buffalo also benefited greatly and for a time was one of the ten largest cities in the United States.

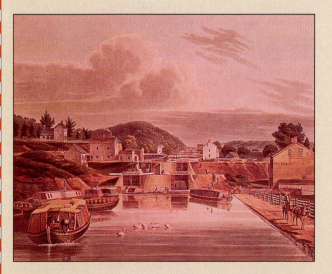

Scene on the Erie Canal, West Troy, New York. This painting shows various kinds of canal boats at the junction of the Erie and Northern Canals. Between Albany and New York City, such canal boats were towed by steamboats; on the Erie Canal, they were pulled by horses walking on towpaths, as shown on the right.

Route of the Erie Canal. The canal extended from Albany on the Hudson River to Buffalo on Lake Erie, using the gap through the Appalachian Mountains formed by the Mohawk River. In combination with the Hudson River, it offered an inexpensive route between the Northeast seaboard and the Great Lakes region.

Settlement Patterns before 1950

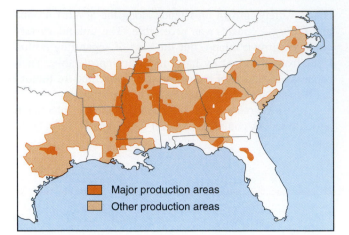

Figure 3.13
Main cotton producing areas, 1860. Cotton production was possible in much of the South, but it was concentrated in several fertile areas particularly well-suited to the crop. The slave population also became concentrated in these areas.

Major production areas
Other production areas

and in central Missouri, both hemp and tobacco were grown. Contrary to popular impressions, not all plantations could afford slaves, and in 1860, almost half the southern plantations had none at all. Nevertheless, as slaves were expensive, much of the South's wealth was tied up in this human form.

Canadian Expansion

Settlement expanded more slowly in Canada than in the United States. Still, by 1865, most of the better agricultural lands within 160 kilometers (100 mi) of the U.S. border east of Lake Huron had been settled. These formed a virtually continuous band along the border, except for the agriculturally unpromising land adjacent to northernmost Maine (Figure 3.14). Farther east, settlers had pushed about as far as they could go, establishing themselves along the shores of the Gaspé Peninsula, the northeast coast of New Brunswick, and in most of Nova Scotia. Both north and west of this settled zone, expansion was virtually blocked by lands unsuited to agriculture; in the west, the Canadian Shield formed a 2100-kilometer (1300-mi) barrier between the agricultural lands of the East and the potential farmlands of the Canadian Prairies. Still, the barrier was penetrable (if only barely, at first), for by the late 1860s, there were nascent settlements in the vicinity of Winnipeg, Manitoba, as well as on southern Vancouver Island in the Far West.

Western Settlement Frontiers, 1841–1865

Nineteenth-century frontier Americans always seemed to be restless and eager for new opportunities, pressing ever farther west (Figure 3.15). By the 1840s, some frontiersmen were already approaching the edge of land still in native reserves or otherwise not yet open for settlement. As early as 1841, a few pioneers bypassed the intermediate areas and jumped across the continent to settle in Oregon.

The motivations for going to the "Oregon Country" (today, the states of Washington, Oregon, and Idaho) ranged from patriotism to a search for fortune, with a good deal of variety in between. For some time, the area had been claimed by both the United States and the United Kingdom, with few occupants from either side. The first American settlers chose to locate in the fertile Willamette Valley, south of today's Portland, Oregon. These settlers had come overland, traversing hundreds of miles of Great Plains, several mountain ranges, and the interior basins. About 8000 Americans had arrived by 1846, the year in which the Oregon Country border between the United States and Canada was established at 49° North latitude. Earlier, it had been widely assumed that the Columbia River, farther south, would serve as the boundary, but the presence of so many American settlers gave the United States a decided advantage in the negotiations.

the area under cultivation expanded impressively in such old slave states as Georgia and South Carolina, but its most noteworthy expansion was on virgin lands farther west (Figure 3.13). Initially, American Indians were in the way, but most of them were forcibly removed, many to the newly established *Indian Territory* (later renamed Oklahoma). These western lands, which had a better climate for the crop and fewer cotton diseases, were in the Gulf Coast states of Alabama, Mississippi, Louisiana, and Texas. Although upland cotton was produced in a long arc-like belt from south-central Virginia to east Texas, by 1860, 75% was grown west of Georgia. Many planters from the Atlantic coastal states saw the opportunities and moved west with their slaves.

Geographically, this expansion was highly selective, as only the best lands went into cotton. Large areas were unsuited to the crop, and here, poor farmers of European descent were often found. The more productive western areas were well served by New Orleans and Mobile, and by 1835, these two cities accounted for about 50% of all cotton shipments.

Despite the growth of these ports, the South overall remained slow to urbanize. This was partly because New York City controlled much of the cotton trade, but the South also had been highly rural from the start. In 1870, New Orleans, with 191,000 people, was the sole large city in the Deep South. Only four other cities—Charleston, Memphis, Mobile, and Richmond—had more than 25,000 people, and of those, only Richmond had more than 50,000.

Plantation agriculture also expanded in the upper South, where tobacco and hemp were the main slave crops. Parts of Kentucky, for example, became noted for tobacco,

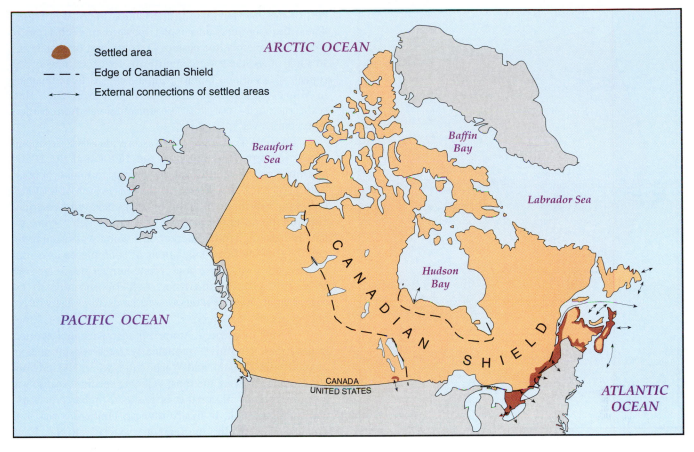

Figure 3.14
Settled areas of Canada, 1865. Settlement north of the main population areas was largely limited by environmental factors, while settlement of the western prairies was hindered by a large intervening area with no agricultural potential, the Canadian Shield.

Another western settlement frontier was occupied by the Mormons, a religious minority that had been persecuted for its beliefs in various states, including Ohio, Missouri, and Illinois. In 1846 and 1847, under the leadership of Brigham Young, Mormons left the Missouri River frontier to find a new home, which they did in a seemingly unpromising environment near the Great Salt Lake. The Mormons lived under a theocracy, were highly motivated and well organized, and soon turned their new Utah homeland into a productive agricultural area. They did this by building irrigation works that brought water from the nearby Wasatch range of the Rocky Mountains; they were the first Americans of European descent to practice irrigation successfully.

Agricultural villages spread rapidly from the Mormon core around Salt Lake City. The Mormons were the only large group of Northern European descent, other than the early New Englanders, to settle extensively in villages. Today, Mormons make up a sizable element of the population not only in Utah, which is about 75% Mormon, but also in southern Idaho and northern Arizona. The Salt Lake City area is about 50% Mormon.

In 1854, Congress passed the *Kansas-Nebraska Act,* opening the lands of these two future states for settlement. Many settlers were attracted, but as farmers spread westward into the Great Plains, they found the land became ever drier and less promising. The former wavelike expansion of the frontier came to a stop, as the environmental limits had, temporarily, been met near the centers of these states. The western parts of the Great Plains, both in the United States and Canada, would not be settled by farmers for several decades.

Following the Mexican War, a huge area covering nearly all the Southwest was ceded in 1848 to the United States. Almost simultaneously, gold was discovered in California, and one of the great gold rushes of all time took place. Americans poured into California, coming both overland from the settlement frontier and by several ocean routes. Other gold seekers came from England, Germany, Australia, China, Mexico, and many other countries. Less than three years after the discovery, California had a large enough population to be admitted as a state, and the miners' supply point of San Francisco had grown into an important city of about 35,000 people. Nothing but the lure of gold

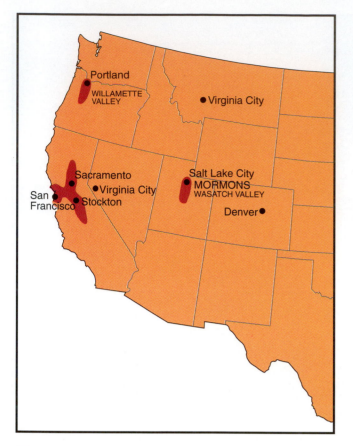

Figure 3.15

Some western U.S. settlement frontiers, 1841–1865. California attracted, by far, the largest number of settlers. The California goldfields were supplied through San Francisco, which grew spectacularly and became a major city (complete with a stock market) within just a few years. There were two Sierra Nevada goldfields; the northern one was served by Sacramento, while Stockton served the southern one.

could have produced such radical changes so quickly in what had been a thinly populated Mexican frontier zone.

The West had many other scattered mining frontiers. Among the states where mining towns were the first substantial settlements were Colorado, Montana, Idaho, and Nevada. In Colorado, the gold discovery of 1858 led to the establishment of many mining towns, of which the first, Denver, soon grew into an important supply center. In Montana, another gold rush also led to the founding of several towns in the 1860s, including Virginia City. Nevada was the child of the discovery in 1859 of silver, the famous Comstock Lode at another Virginia City; five years later, Nevada became a state, undoubtedly many decades before it would have without this spectacular find.

Frontier Settlement, 1865–1931

Between the end of the Civil War and 1890, much of the rest of the farming country in the relatively humid areas east of the 98th meridian in the United States was settled. Thus, the frontier was pushed forward in such areas as western Minnesota, the eastern Dakotas, and central Texas. In some areas, there were significant concentrations of European national and ethnic groups, such as Scandinavians in Minnesota and Germans in parts of Texas. On a finer scale, such as in townships, group settlement from particular countries was common.

From 1865 to 1931, there were also two prominent lumbering frontiers. Between the 1870s and the 1910s, Michigan and Wisconsin led the country in lumber production. They had valuable trees within easy reach of the Great Lakes, on which wood could be sent cheaply to emerging cities on the lower lakes. From Chicago, which became the nation's leading lumber market, this wood was taken to the rapidly growing farm populations of such places as central Illinois and Iowa, where appropriate lumber was scarce. By 1920, most of the northern forests had been ruthlessly cut over, leaving behind a devastated landscape that still has not completely recovered.

During this period, lumbering also became important in the Pacific Northwest, particularly in the coastal sections of California, Oregon, and Washington, where the redwood and Douglas fir were the most valuable species. Seattle was one of the many places that began as a timber processing and shipping point. In the Northwest, lumbering was long oriented to the California market, especially to San Francisco. In the 1890s the Pacific Coast became nationally prominent for lumber and remains so today.

After 1890, only one large farming frontier remained, the semiarid Great Plains of the United States and the Canadian Prairies (Figure 3.16). Early in the 19th century, two American explorers, Zebulon Pike and Stephen Long, declared this area a virtual desert, unfit for settlement. The Great Plains and Prairies grasslands were the ideal environment for millions of bison, the main source of sustenance for many American Indian tribes. By the late 1880s, however, bisons had nearly disappeared, most of them slaughtered and often merely for sport. This made it impossible for the natives, who were excellent hunters and fierce warriors, to continue their old ways; most became essentially wards of the government. At about this time, farmers made their first tentative thrusts into the region.

The first American use of the Great Plains had occurred even earlier, however. **Open-range** (without fencing) **cattle ranching** started in Texas shortly after the Civil War. In the famous cattle drives, hopelessly romanticized in movies, near-wild longhorn cattle were taken north from Texas to the nearest railroad, from where they were shipped to market (Figure 3.17). These drives gave rise to such infamous Kansas cattle towns as Abilene, Wichita, and Dodge City, which were the main railheads at different times.

Eventually, open-range ranching spread to the northern Great Plains, in Nebraska, Wyoming, and Dakota Territory. In the late 1880s, the whole industry virtually disappeared,

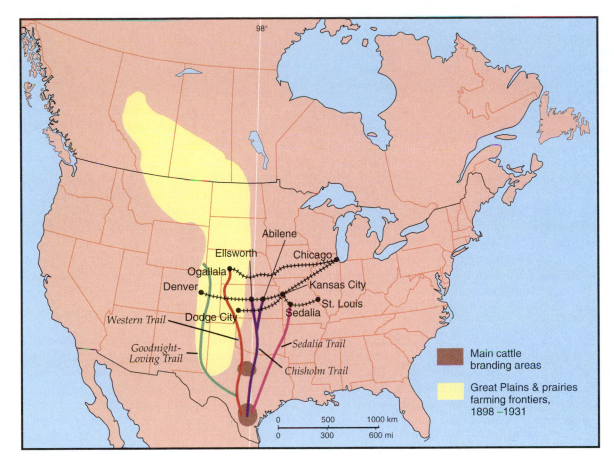

Figure 3.16

Great Plains and Canadian Prairies: main cattle trails and final settlement frontiers. Starting in 1865, semiwild Texas cattle were driven north to the railheads, for movement to the meat-packing cities. The cattle trails were pushed ever farther west, in part because of the westward spread of farming and the railroads. Dodge City, Kansas, was one of the last and (thanks to Hollywood) is the most famous of the cattle-shipping towns. The final frontiers were settled by wheat farmers in the 1898–1931 period.

Map labels: Abilene, Ellsworth, Chicago, Ogallala, Denver, Kansas City, Dodge City, St. Louis, Sedalia, Western Trail, Goodnight-Loving Trail, Sedalia Trail, Chisholm Trail, 98°

Legend:
- Main cattle branding areas
- Great Plains & prairies farming frontiers, 1898–1931

Scale: 0 500 1000 km / 0 300 600 mi

Figure 3.17

A 19th-century cattle drive. Tough longhorn cattle were well-suited to the long drives, but they did not make for the tenderest of steaks.

W. H. D. Koerner, *Trail Herd to Wyoming*, 1923. Buffalo Bill Historical Society.

Settlement Patterns before 1950

destroyed by such factors as Kansas quarantine laws against Texas cattle, the westward spread of farmers, and two major blizzards that greatly reduced the northern herds. Open-range animals were replaced by higher-grade cattle from enclosed ranches and midwestern farms, where breeding could be controlled.

Farmers settling the western Great Plains faced two noteworthy environmental problems: the unpredictability of rainfall and great distances from markets and supplies. The second was largely overcome with the construction of railroads, but the problems of climate have never been completely solved.

While farmers began to settle the eastern third of the Great Plains by the 1880s, the drier western parts of the Plains were mostly an early 20th-century frontier, with settlement largely taking place between 1898 and 1917. During this period, millions of acres were first plowed in central North Dakota and South Dakota, eastern Montana, and western Nebraska, Kansas, Oklahoma, and Texas.

Large-scale settlement of the Canadian Prairies started in about 1900 and lasted until the outbreak of World War I in 1914. British settlers were sought and many came, joined by a mass movement of land seekers from the United States. Large numbers also came from Eastern Europe, especially from the Ukraine, where physical conditions for agriculture were roughly comparable. High wheat prices during World War I encouraged expansion of the area under cultivation, sometimes into places usually too dry for satisfactory yields, with the inevitable later retreat.

The last part of the Great Plains to be settled was the dry, shortgrass country in southwest Kansas, western Texas and Oklahoma, and eastern Colorado and New Mexico. Active settlement started around the turn of the century and did not end until 1931, when the *Dust Bowl* drought devastated the area. Thousands of farmers were ruined; many migrated to California, a saga chronicled in the classic John Steinbeck novel *Grapes of Wrath.* Still, by 1950, most of the abandoned farmland had been returned to crops, and many additional acres were plowed for the first time. Today, crop failures continue to be common, despite such modern advantages as better strains of crops and methods of cultivation.

A recent work has argued that, given the dry and unpredictable weather, crop failures, and other environmental problems, the western Great Plains should never have been plowed. The authors point out that most parts of the Plains have a long history of population decline. They recommend that agriculture be abandoned and that the area become a *Buffalo Commons,* or a type of national grassland for the American bison. Most people who live in the area disagree sharply, and it is unlikely that such a radical idea will ever become national policy. Still, it has sparked lively discussion about the future of this area.

Migration, 1865–1950

After a pause during the American Civil War, the flow of immigrants to the United States resumed. Through the 1870s, earlier patterns persisted, with most newcomers coming from Northwest Europe to both farms and cities. The other immigrants included Canadians and the vanguards from Eastern and Southern Europe, as well as a few Asians and Hispanics.

Europeans

By the 1880s, a fundamental shift in the origin and main destinations of European immigrants to the United States began, a change that was firmly established by the next decade. By then, population growth in Northwest Europe, the traditional immigrant source area, had begun to slow down, and economic development greatly decreased the need to leave. Before 1880, about 80% of the immigrants were from Northwest Europe; after 1900, only about 20% were from there.

In contrast, by the 1880s, the disruptive effects of the Industrial Revolution, such as in rural crafts and farming, had begun to affect Eastern and Southern Europe, making emigration virtually necessary for many. In that decade, about 15% of immigrants came from these regions; in the 1890s, the figure was more than 50%. After 1900, the new source areas (primarily Italy, the Austro-Hungarian Empire, and the Russian Empire) dominated, accounting for more than two-thirds of the total until immigration virtually stopped with the outbreak of war in 1914.

This change led Americans to distinguish between old immigrants, who had arrived before 1880, and new ones. The old immigrants were viewed more favorably as part of the broader Northwest European culture from which American values had been derived. Most were quickly and easily assimilated; they tended to disperse to interior farms and small towns, and many already spoke English. Almost all Southern and Eastern Europeans moved directly to the expanding industrial cities of the Northeast, where job opportunities were brightest. By 1910, almost half of the immigrant population was in the northeastern cities, and another third in cities of the north-central states, such as Chicago, Cleveland, and Detroit. Immigrants usually lived in ethnic ghettos near the evolving central business district, which offered diverse employment opportunities for unskilled workers. In these ghettos, many had hardly any contact with native-born Americans. To house them, former single-family homes often were subdivided into multiple units, and cheap shacks were erected in the yards. In a few areas, notably in New York, tenements were built (Figure 3.18).

Old-line Americans often saw these newcomers as less well prepared than their predecessors for American city life, society, and politics, and they were concerned by the immensity of the flow. Between 1880 and 1920, an average of 6 million immigrants per decade entered the country,

Figure 3.18
Immigrant tenement in Boston, about 1900. High-density housing like this was associated with high infant mortality rates, high crime rates, and other signs of social disorganization, although more so with some ethnic groups than with others. At this time, such tenements were especially occupied by new immigrants from Eastern and Southern Europe.

with a staggering 8 million between 1901 and 1910. Given these numbers in a country with fewer than 100 million inhabitants, by the early 1900s, most American writers found the concentration of foreigners in northeastern cities alarming and called for immigration restrictions. The first quota system, which discriminated against Eastern and Southern Europeans, was established in 1921 and tightened in 1924. The immigration law passed in 1929 let in only 150,000 newcomers per year, apportioned according to the national origins of the American population in 1920. This system strongly favored Northwest Europeans, who had little reason to migrate, and it effectively stopped large-scale European migration for good. These restrictions did not apply to people from Western Hemisphere countries, such as Canada and Mexico.

For a long time after Canada became a British possession, immigration was dominated by people from the British Isles, who were favored by government policy. In the late 19th century, a policy encouraging immigration from continental Europe was adopted. Beginning about the turn of the century, therefore, new immigrants started to arrive from Scandinavia and Eastern Europe. In the 1910s, the Russian Empire became a major source of immigrants, most notably Ukrainians, many of whom migrated to the grasslands of the Prairie Provinces. Legislation that restricted general immigration was passed in the 1930s, about a decade after similar changes in the United States.

African-Americans

With their emancipation during the Civil War, African-Americans hypothetically became free to move wherever they chose. Between 1870 and 1910, a small number, about 250,000, left the South. Most went to the northeastern cities, but a few moved to the farming frontier. The great majority stayed in the South, however, for they had few skills and resources and prejudice against them was

Settlement Patterns before 1950

widespread wherever they moved. Most had known only farming, but as they owned no land, they usually became **sharecroppers** on former plantation lands. Under this system, a sharecropper agreed to work a certain number of acres for a landlord, paying 50% of the crop as rent. This system kept most of the African-American population (as well as many white sharecroppers) in poverty, and if they owed money to the landowner, they were not free to move. As late as 1910, about 90% of the African-American population lived in the South.

This situation began to change during World War I, when there was a tremendous demand for labor in northern industrial cities after European immigration was cut off by the start of hostilities. Labor recruiters went south to encourage African-Americans to migrate. Soon the trickle northward became a huge flow that continued until the Great Depression of the 1930s. This northward flow was highly zonal, with people from the south Atlantic states going to the northeastern coastal cities, such as Philadelphia and New York, and those from the more westerly states, such as Mississippi and Louisiana, going to more inland places, such as St. Louis and Chicago (Figure 3.19).

The job opportunities of World War II caused a renewed massive outmigration from the South. Most African-

Figure 3.19
Zonal pattern of African-American migration, 1914–1930.
Those who lived in the east south-central states tended to move to the cities of the Northeast Coast, while those who lived farther west tended to follow the Mississippi Valley to such destinations as Chicago and Detroit.

Americans still went to northern cities, but now a westward flow was added, as West Coast industries were booming. Substantial African-American communities were established in such cities as Los Angeles, Oakland, and San Francisco. In 1950, about two-thirds of the African-American population was still living in the South, but by 1970, this had been reduced to one-half.

Hispanics

There was only a minuscule immigration of Mexicans to the United States in the second half of the 19th century, but that flow increased substantially thereafter. Between 1900 and 1950, about 1.9 million Mexicans legally migrated northward, mostly to such border states as Texas and California. After 1920, some more distant large cities began to receive immigrants, and Chicago eventually developed a fair-sized Mexican-American community.

Substantial numbers of immigrants from other parts of the Americas came to the United States in the years before 1950. The largest number emigrated from Puerto Rico, starting largely during the job boom of the 1940s. Since Puerto Rico is politically associated with the United States, its people had free access to the country. Most moved to the northeastern cities, particularly to New York. Immigrants also came from other Latin countries, such as Cuba, as well as from such nearby non-Hispanic places as Jamaica, Trinidad, and the Bahamas.

Asians

Asian migration to the United States began with the California Gold Rush, when Chinese miners began to arrive. The early Chinese faced formidable prejudices. Starting in the 1860s, Chinese laborers were brought in to help build railroads, where they proved to be first-class workers. The Chinese were widely resented, partly because of racial prejudice and partly because they were willing to accept low wages, which depressed salaries. Many returned home after their work, but others decided to remain in the United States, where they tended to move to cities and to do a variety of service jobs. Between 1854 and 1882, an average of about 10,000 Chinese immigrants per year entered the United States.

In response to pressures from West Coast voters, Congress in 1882 enacted the **Chinese Exclusion Act,** which prohibited Chinese immigration for ten years. This law was renewed in 1892, and in 1902 the prohibition was made indefinite. Although the law did not totally halt Chinese immigration (from 1920 to 1929, for example, there was an average of about 3000 immigrants per year), it is noteworthy that the country's Chinese-American population actually declined between 1890 and 1930. In 1950, there were 118,000 Chinese-Americans, including about 55,000 who had been born in China.

The other noteworthy pre-1950 Asian immigrant group was the Japanese, who moved in moderate numbers to Hawaii starting in 1868, when those islands were not yet a part of the United States. Most worked in the sugar cane fields, and soon they became Hawaii's largest ethnic group (which is still the case). When the islands became U.S. territory in 1898, about 25,000 people of Japanese descent lived there.

After Hawaii became a U.S. possession, some Japanese-Americans moved to the mainland, often to work in agriculture. In the early part of the 20th century, many unskilled workers began to emigrate to California directly from Japan, and between 1900 and 1907, about 15,000 arrived each year. Driven by racial prejudice and a fear that they would lower wages, political pressure soon mounted in the Pacific Coast states to halt this flow. Ensuing negotiations with the Japanese government resulted in the **Gentlemen's Agreement** of 1907, whereby the Japanese government (to avoid the humiliation of a discriminatory U.S. immigration law) agreed to tightly regulate emigration to the United States. During World War II, most Japanese-Americans (but not those living in Hawaii) were forcibly moved to internment camps because of questions about their loyalty. Eventually, most Japanese-Americans left rural areas to settle in cities, where they tended to be economically successful. In 1950, there were about 142,000 Japanese-Americans.

Development of the Manufacturing Belt, 1865–1950

Between the end of the Civil War and 1880, the outlines of the *Manufacturing Belt* began to take shape (Figure 3.20). Before 1860, and especially before 1840, manufacturing usually served restricted regional markets. After the Civil War, with more effective corporate organization and better transportation available, many companies began to increase the size of their factories and to serve national or near-national markets. In the United States, such markets were usually most effectively served by locating in the zone between Boston and Baltimore, which had such advantages as easy access to cheap fuel and immigrant labor, the nation's most concentrated market, and excellent transportation to the rest of the country and overseas. As a result, this zone (including some nearby smaller industrial cities) became the early Manufacturing Belt.

After 1870, this original small economic core expanded westward into areas with locations and resources appropriate to the growth of manufacturing. In particular, heavy industry began to grow rapidly in a broad zone west of Pittsburgh and Buffalo, an area that became the western part of the Manufacturing Belt. Eventually, this belt consisted of the area enclosed by lines connecting Portland (Maine), Baltimore, St. Louis, and Milwaukee, including the most populous part of the province of Ontario. By 1910, this expanded Manufacturing Belt had 34 of the 50 U.S.

cities with more than 100,000 people. In that year, heavy industry was especially concentrated in the western two-thirds of this area, enclosed by Buffalo, Milwaukee, St. Louis, and Pittsburgh, as well as adjacent parts of Canada (especially in the Toronto-Hamilton area).

The expansion of industrialization westward resulted in the rapid growth of many cities, particularly those along the lower Great Lakes. This urban growth was correlated especially with the development of the steel industry and the growth of related steel-using industries. When cheap steel became available after about 1870, it was widely used by an array of industries for producing nails and stoves to rails and locomotives (Figure 3.21). So important was steel that, for many decades, it was considered as the most basic of American industries, and its geography helped to explain the location of many other industries.

After 1870, the low-cost area for producing steel was on and near the lower Great Lakes. At such lake ports as Buffalo, Cleveland, and Detroit, iron ore could be received economically via lake vessels from the new ore fields of northern Minnesota, Michigan, and Wisconsin. From these ports, it was but a short distance by rail to coal, the other critical raw material for making steel. Some of the world's best coal for steel production was

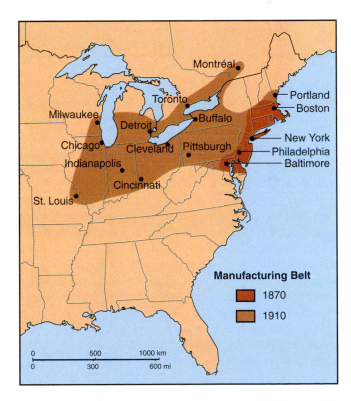

Figure 3.20

Development of the Manufacturing Belt, 1870–1910. In 1870, the belt was concentrated in the Northeast between Baltimore and greater Boston; by 1910, it had expanded westward as far as St. Louis and Milwaukee.

Figure 3.21
Homestead Iron and Steel Works in Pittsburgh, about 1925. This area along the Allegheny River, known as the Strip, contained Pittsburgh's largest concentration of foundries between the 1850s and the end of World War II. There are no iron and steel mills in this area today.

found in western Pennsylvania and West Virginia. Soon, dozens of steel mills were built at the lower lake ports, on the coal fields, and at various locations in between. So many industries located here that, by 1890, the Great Lakes area already exceeded the U.S. national average in the percent of its labor force in manufacturing.

Pittsburgh, which for many decades was the leading steel city, was located virtually atop the coal fields. Its rise to leading steel city can best be explained, however, by the chance fact that Andrew Carnegie, an important innovator who adopted the large-scale Bessemer process for making inexpensive steel, lived there. His company was highly successful and came to dominate the industry, both because of its early adoption of the Bessemer process and because of rigorous cost controls. For some years, until the supply was exhausted, Pittsburgh also had the advantage of a nearby supply of iron ore. Once achieved, the city's leading position was maintained, in part, by a system of freight charges on steel that favored Pittsburgh. It remained the leading steel city in the United States until recent decades, when it was passed by Chicago.

As mentioned earlier, since steel was literally basic to 19th-century industrialization, large steel users were attracted to the region. Therefore, when the auto industry be-

came successful early in the next century, it also began to concentrate in the western part of the Manufacturing Belt, including its Canadian section. For many decades, auto and steel were the two most important heavy industries in the United States. Until at least mid-century, it remained the undisputed heavy industry zone with a major concentration of large cities.

Geography of the United States and Canada at Mid-Century

A simplified, but useful, representation of the geography of both the United States and Canada in 1950 shows it to consist of two major regions, the **core** and the **periphery** (Figure 3.22). In the United States, the core consisted of a fully developed Manufacturing Belt, still in the same location as in 1910 but with greatly amplified urban and industrial growth. In Canada, the core consisted of the strip of land between Montréal on the northeast and Windsor on the southwest, with Toronto roughly in the middle. This region housed about 45% of the United States' 152 million people; in both countries, the core produced about two-thirds of the industrial output. In the United States, the

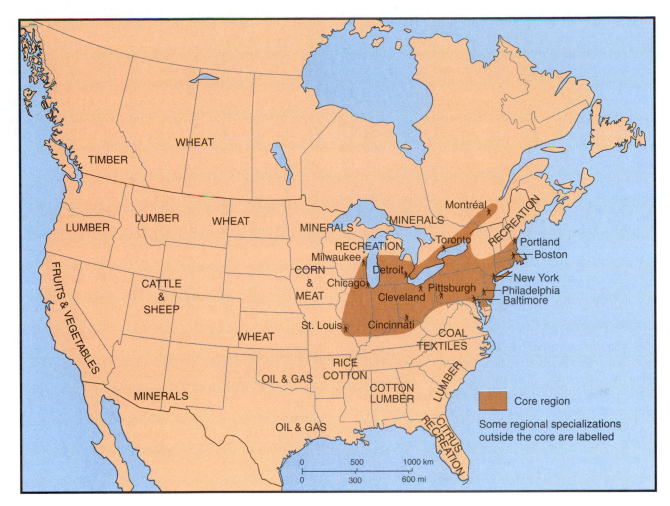

Figure 3.22
Core and periphery, 1950. The U.S. and Canadian core region in 1950 was the Manufacturing Belt, while the periphery tended to produce various goods and services demanded by the population and industries of the core.

area also contained eight of the nation's ten leading industrial centers. Not only did the core supply most of the industrial products required by an increasingly complex economy, but it also exercised control over that economy through the many corporate headquarters located in such cities as New York, Detroit, Chicago, Montréal, and Toronto. This centralized control was made possible by advanced transportation and communications systems. Because of their wealth and population, the cores also had tremendous political power.

In both countries in 1950, the rest of the national space could be labeled the peripheries, areas highly dependent on the core for their prosperity. Here were produced the mineral, agricultural, and other nonmanufactured products demanded in the core; any decrease in that demand had severe negative repercussions in these outlying areas. In a political and economic context, the peripheries had something like a colonial relationship to the core.

At mid-century, the pattern of agricultural specialization was not very different from what it had been at the turn of the twentieth century. The Midwest still had its Corn Belt, where corn was grown primarily to feed to animals that were sold for meat. There was still a Cotton Belt, although its center of gravity had continued to move westward, as many worn-out cotton lands in the Atlantic states were abandoned. Great Plains and Canadian Prairies settlement had expanded greatly since 1900, and those areas specialized in wheat, a crop that did rather well under relatively dry conditions. In the United States, perhaps the most prominent change in agriculture since 1900 was the expansion of irrigated agriculture in the West, where dozens of irrigation projects had allowed a tremendous increase in the area under cultivation. Arguably, the most impressive of these was in the Central Valley of California, where specialized fruits and vegetables were grown for a national market. Within the U.S. core, the first five decades of the

Settlement Patterns before 1950

20th century saw the maturation of an intense belt of vegetable and fruit production along the northeastern seaboard, where farmers could prosper by taking advantage of their proximity to the country's largest market and the high prices offered there.

The year 1950 is a logical point to end this historical treatment, for it approximates the start of some fundamental changes in the geography of the United States and Canada. The following decades saw many changes in the nature and location of industry, population, political power, and other important aspects of life. To illustrate, 1950 was about the time of the start of a mobility revolution that, among other things, led to the true suburban revolution, the growth of the U.S. Sun Belt, and the rise to prominence of the Vancouver area in Canada. The rest of this text is devoted to the many, often dramatic, geographic changes that have taken place since that year.

Summary

In 1492, the native population of North America was far larger than has previously been thought, but this population was soon decimated by Old World diseases. American Indians lived in highly varied physical environments, which they had modified substantially, and they had a great variety of cultures.

Several European nations established colonies in what became the United States and Canada, beginning with the Spanish in 1565 in Florida. Spaniards had a more lasting impact in the Southwest, where they began to settle along the upper Rio Grande in 1598. French settlement began early in the 17th century, especially along the St. Lawrence River. The search for furs led the French far inland to the Great Lakes and the Mississippi River Valley, where they established settlements. The Dutch settled around New York, but eventually both the French and Dutch colonies were taken over by the British.

British settlement was by private companies, leading to highly varied colonies. Based on their economic and cultural similarities, the 13 colonies fell into three broad groups. The mainly agrarian Chesapeake Bay culture hearth was based on the growing of plantation crops and, eventually, the use of slaves. Southern New England was dominated by Puritans, who established near-subsistence agricultural communities inland and coastal trading cities. The Midland region, centered on southeast Pennsylvania, specialized in grains grown by small farmers.

After the Revolutionary War, the United States rapidly expanded westward, and by 1853, all the area that became the contiguous 48 states had been acquired. By 1865, three major economic regions had evolved, with the Northeast seaboard specializing in manufacturing and having the largest cities; the South emphasizing plantation crops, such as cotton and tobacco; and the Midwest concentrating on grains and animals. Between the end of the Civil War and 1931, most of the remaining agricultural land of the United States and Canada was settled, though not always successfully.

After 1880, the origins and destinations of European immigrants to the United States shifted dramatically, with an increasing flow from Southern and Eastern Europe to the manufacturing cities of the Northeast and Midwest. The ghetto concentrations of newcomers led to the call for immigration restrictions, which took effect in the 1920s. The interruption of immigration by World War I gave African-Americans the opportunity to find jobs in northern factories, and a large outflow from the South began that continued until well after 1950.

Following the Civil War, the outlines of the Manufacturing Belt began to take shape, as manufacturing spread from its Northeast seaboard concentration to the areas on and near the southern Great Lakes. By 1950, the geography of both the United States and Canada could be described in terms of a core area around this Manufacturing Belt and a periphery that supplied the core with raw materials.

Key Words

Chinese Exclusion Act	metes and bounds system
core	open-range cattle ranching
cotton gin	periphery
culture hearth	seigneurial system
Cumberland Gap	sharecropper
Erie Canal	township and range system
Gentlemen's Agreement	Treaty of Greenville
long lot	voyageur

Gaining Insights

1. What were several impacts of the coming of Europeans on American Indians following 1492?
2. What is a culture hearth? What were the three main culture hearths in the British colonies, and what were their major characteristics?
3. Where were the earliest industrial regions of the United States? Why were these areas favored for industry?
4. Why did the early Midwestern agricultural surplus usually reach its market through New Orleans? What developments led to great changes in this pattern by 1860?
5. Where was the Erie Canal? Why did it have such a great impact on national transportation patterns? What cities were especially favored by the canal?
6. What development made the growing of upland cotton very profitable? What impacts did this development have on the geography of cotton production?
7. Why was the expansion of the settled area of Canada so limited before 1865?
8. What is open-range ranching? Describe where it began, to where it expanded, and why it virtually disappeared.

9. What were the main changes in European migration to the United States between 1880 and 1924? Why did they occur?
10. Explain the major characteristics of black migration within the United States between 1865 and 1950.
11. Where was the Manufacturing Belt located in 1870? Where did it expand between 1870 and 1910, and why there?
12. What was the core-periphery relationship in 1950?

Selected References

Brown, Ralph H. *Historical Geography of the United States.* New York: Harcourt, Brace and Company, 1948.

Butzer, Karl W. "The Americas before and after 1492: An Introduction to Current Geographical Research," *Annals of the Association of American Geographers* 82 (1992): 345–68.

Conzen, Michael P. "The Progress of American Urbanization, 1860–1930." In *North America: The Historical Geography of a Changing Continent,* edited by Robert T. Mitchell and Paul A. Groves. Savage, Md.: Rowman and Littlefield, 1990.

Cronon, William. *Nature's Metropolis: Chicago and the Great West.* New York: W.W. Norton & Co., 1991.

Denevan, William M. "The Pristine Myth: The Landscape of the Americas in 1492," *Annals of the Association of American Geographers* 82 (1992): 369–85.

Doolittle, William E. "Agriculture in North America on the Eve of Contact: A Reassessment," *Annals of the Association of American Geographers* 82 (1992): 386–401.

Earle, Carville. "Regional Economic Development West of the Appalachians, 1815–1860." In *North America: The Historical Geography of a Changing Continent,* edited by Robert T. Mitchell and Paul A. Groves. Savage, Md.: Rowman and Littlefield, 1990.

———. "Pioneers of Providence: The Anglo-American Experience, 1492–1792," *Annals of the Association of American Geographers* 82 (1992): 478–99.

Groves, Paul A. "The Northeast and Regional Integration, 1800–1860." In *North America: The Historical Geography of a Changing Continent,* edited by Robert T. Mitchell and Paul A. Groves. Savage, Md.: Rowman and Littlefield, 1990.

Harris, R. Cole, and John Warkentin. *Canada before Confederation.* New York: Oxford University Press, 1974.

Hornbeck, David. "The Far West, 1840–1920." In *North America: The Historical Geography of a Changing Continent,* edited by Robert T. Mitchell and Paul A. Groves. Savage, Md.: Rowman and Littlefield, 1990.

Lemon, James T. "Colonial America in the Eighteenth Century." In *North America: The Historical Geography of a Changing Continent,* edited by Robert T. Mitchell and Paul A. Groves. Savage, Md.: Rowman and Littlefield, 1990.

Lewis, Pierce. "America between the Wars: The Engineering of a New Geography." In *North America: The Historical Geography of a Changing Continent,* edited by Robert T. Mitchell and Paul A. Groves. Savage, Md.: Rowman and Littlefield, 1990.

McCann, Larry D., ed. *Heartland and Hinterland: A Geography of Canada,* 2nd ed. Scarborough, Ontario: Prentice-Hall of Canada, 1987.

Mitchell, Robert T. "The Colonial Origins of Anglo-America." In *North America: The Historical Geography of a Changing Continent,* edited by Robert T. Mitchell and Paul A. Groves. Savage, Md.: Rowman and Littlefield, 1990.

Nostrand, Richard L. "The Spanish Borderland." In *North America: The Historical Geography of a Changing Continent,* edited by Robert T. Mitchell and Paul A. Groves. Savage, Md.: Rowman and Littlefield, 1990.

Popper, Deborah E., and Frank J. Popper. "The Fate of the Plains." In *Reopening the Western Frontier,* edited by Ed Marston. Washington, D.C.: Island Press, 1989.

Ward, David. *Cities and Immigrants: A Geography of Change in Nineteenth Century America.* New York: Oxford University Press, 1971.

———, ed. *Geographic Perspectives on America's Past: Readings on the Historical Geography of the United States.* New York: Oxford University Press, 1979.

———. "Population Growth, Migration, and Urbanization, 1860–1920." In *North America: The Historical Geography of a Changing Continent,* edited by Robert T. Mitchell and Paul A. Groves. Savage, Md.: Rowman and Littlefield, 1990.

Wishart, David J. "Settling the Great Plains, 1850–1930." In *North America: The Historical Geography of a Changing Continent,* edited by Robert T. Mitchell and Paul A. Groves. Savage, Md.: Rowman and Littlefield, 1990.

Yeates, Maurice. *Main Street: Windsor to Quebec City.* Toronto: Macmillan of Canada, 1975.

Zelinsky, Wilbur. *The Cultural Geography of the United States,* rev. ed. Englewood Cliffs, N.J.: Prentice-Hall, 1992.

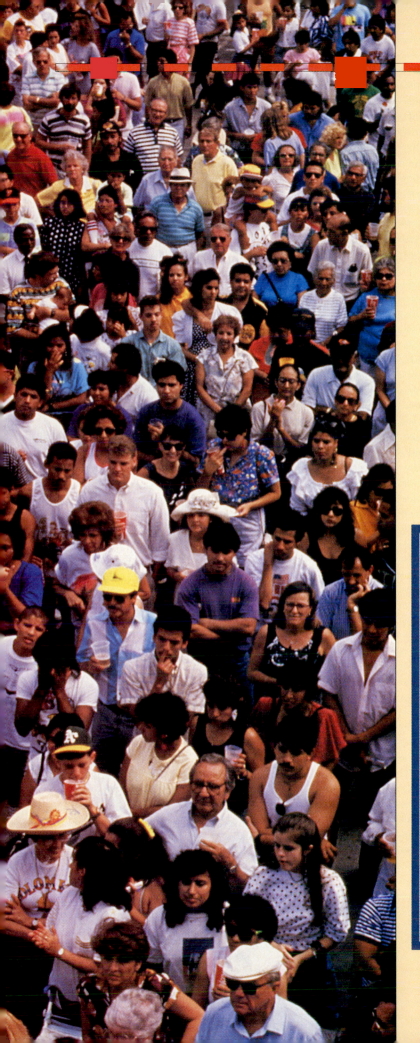

CHAPTER

4

Population Geographies of Canada and the United States Since 1950

Few events in the history of the world have had the demographic impacts of World War II. The victims of the war numbered in the millions, but that is not what we have in mind. By 1950—shortly after the end of the war—two trends were clearly in place that were precipitated by the war itself and continue to reverberate throughout the United States and Canada. One was the dramatic rise in the birth rate, termed the baby boom; the other was the decline in death rates in less developed countries, which also led to rapid increases in their populations and subsequent migrations to the United States and Canada, among other places. In this chapter we trace some of the causes and consequences of these trends to see how demographic change has permanently altered the cultural landscapes of both countries. We begin by defining population geography, and then we described the overall pattern of population growth and distribution since 1950.

Population geographers study the distribution of people across the earth's surface. Two of the questions they ask about Canada and the United States are: (1) where do the people of the two countries live, and (2) why there? Three closely related questions that they might also ask are (1) which parts of the countries have gained the largest number of people in recent decades, (2) where do the millions of immigrants coming into Canada and the U.S. tend to settle, and (the most important question) (3) what accounts for these geographic patterns?

Population Growth Since 1950

In 1950, 151 million people were living in the United States. That was enough people so that even if the population had stopped growing at that moment, the United States would still be the fifth most populated country in the world in 2000, behind China, India, Indonesia, and Brazil. The Soviet Union used to be the third most populous nation, but its breakup left its largest republic, Russia, at sixth place on the list. Of course, the population did not stop growing in 1950. As Table 4.1 shows, an additional 28 million people were added between 1950 and 1960, bringing the total in the latter year to 179 million. This number would have made the United States fourth on the list in 2000, even if population growth had ceased in 1960. But growth did continue, as more than 20 million were added in each successive decade. The numbers in Table 4.1 are not adjusted for census under enumeration (which averages about 2% in each census), nor do they represent the presence of all illegal immigrants. Some studies have indicated that only about half of illegal immigrants were counted in recent censuses.

The pattern of population growth in Canada since 1950 has paralleled that of the United States. As was true of its southern neighbor, the post–World War II decade of the 1950s was a time of extraordinary growth in Canada, as more than four million people were added to the population. Thereafter, there was a consistent slackening in the increase until 1991. The decade of the 1990s, however, again saw more than 4 million people added, a level not seen since the 1950s.

On July 1, 2000, the population of the United States stood at approximately 276 million, ranking it third highest in the world (Table 4.1). This figure was 27 million higher than the 249 million in 1990, showing a growth for the decade of just over 10%. On the same date, Canada's population was about 31 million (ranking it 35th in the world), or 15% higher than at the last census in 1991. Historically, Canada's population has been about 10% the size of its southern neighbor, but in recent decades its growth rate has exceeded that of the United States. Its July 1, 2000, population was about 11% that of its neighbor. The total population of the two countries, 307 million, accounted for about 5% of the world's people. They occupied approximately 14.5% of the earth's land area.

Much of the growth during the post-1950 decades in both countries was via natural increase. The **baby boom** is usually defined as lasting from 1946 to 1964, or about 18 years. In particular, the decade between 1950 and 1960 was a time when both the United States and Canada experienced significant increases in their birth rates. In the United States, the average annual rate of growth that decade was 1.7%, compared to 1.35% during the previous ten years. A closer look at Table 4.1 shows that during each decade between 1960 and 1990, the United States added about the same number of people. Since the country was building on an increasingly large demographic base, this meant that the average rate of growth was declining. Between 1980 and 1990, for example, the average rate of increase of the resident population was just under 1% per year. Indeed, by the early 1990s, the United States would actually have been approaching a zero rate of growth had it not been for the influx of international immigrants—a topic we return to later in the chapter. After 1990 the yearly rate of growth increased again, largely because of immigration.

Nearly the same comments apply to Canada except that, as mentioned earlier, its rate of growth has been consistently higher. In the 1950s, when the population of the United States was increasing 1.7% per year, it was going up by 2.6% in Canada. But the rate dropped more sharply over time in Canada than in the United States, and in the decade between 1980 and 1990 Canada's rate had slowed almost to the U.S. level. Thus, by 1990 Canada, like the United States, would have been nearing **zero population growth,** were it not for continued immigration. That is, the population would be in equilibrium, with births plus immigration equaling deaths plus emigration. However, in the 1990s Canada's yearly rate of population growth increased to about 1.10%, mostly because of immigration.

Regional Patterns of Growth

As we saw in Chapter 3, the early European settlements in what would become the United States and Canada were established principally in the eastern parts of the continent,

Table 4.1
Population Growth in the United States and Canada

	United States				Canada		
Year	Resident Population (in thousands)	Increase during Previous Decade (in thousands)	% Growth per Year during Preceding Decade	Year	Resident Population (in thousands)	Increase during Previous Decade (in thousands)	% Growth per Year during Preceding Decade
1950	151,236	19,161	1.35	1951	14,009	2,503	1.97
1960	179,323	27,997	1.70	1961	18,238	4,229	2.64
1970	203,302	23,979	1.26	1971	21,568	3,330	1.68
1980	226,546	23,243	1.08	1981	24,343	2,775	1.21
1990	248,710	22,164	0.93	1991	26,800	2,457	0.96
2000 (est)	276,000	27,290	1.00 (est)	2000 (est)	31,000	4,200	1.10 (est)

Sources: U.S. data from Statistical Abstract of the United States, 1998. *U.S. Bureau of the Census, 1998 and U.S. Bureau of the Census, International Data Base 1998; Canada data from W. Kalbach and W. McVey,* The Demographic Bases of Canadian Society, *2d ed., table 1.4. McGraw-Hill Ryerson Limited, Toronto, 1979; and Population Reference Bureau World Population Data Sheets for appropriate years; 1996 Census of Canada,* Population and Dwelling Counts, *1997.*

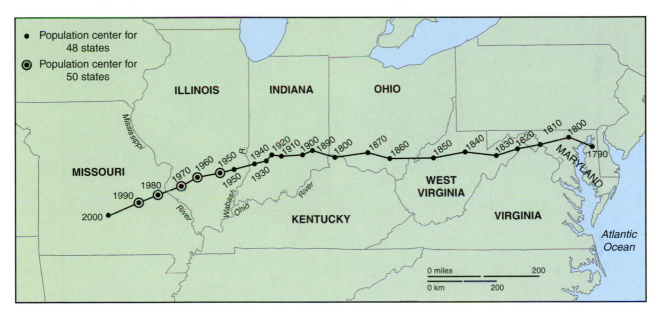

Figure 4.1
The population center of the United States has been drifting westward since the founding of the country. The two different locations for the population center in 1950 and the symbol change indicate the geographical pull on the center of population exerted by the admission of Alaska and Hawaii to statehood. Note how the post-1950 trend reflects population growth in the Sun Belt.

the areas closest to Europe. Almost from the beginning, however, people began to move westward. At first this meant the valleys west of the Atlantic seaboard; then the Plains states and provinces were settled; and well before the end of the 19th century, the Pacific Coast regions had become popular destinations.

Continuing this pattern, the populations of both countries have had a clear westerly drift since 1950. In the United States the drift has also been southerly (Figure 4.1),

but this pattern is not emulated in Canada, as the vast bulk of the population already lives close to the southern border. Figure 4.2 shows the states with the most rapid *numeric* increases in population between 1960 and 2000, while Figure 4.3 depicts the *rate* of population growth by state during that forty-year period. Both maps show a strong tendency for the western and southern states (with a few exceptions) to have the largest increases in both percentage and absolute terms. In other words, the figures reflect

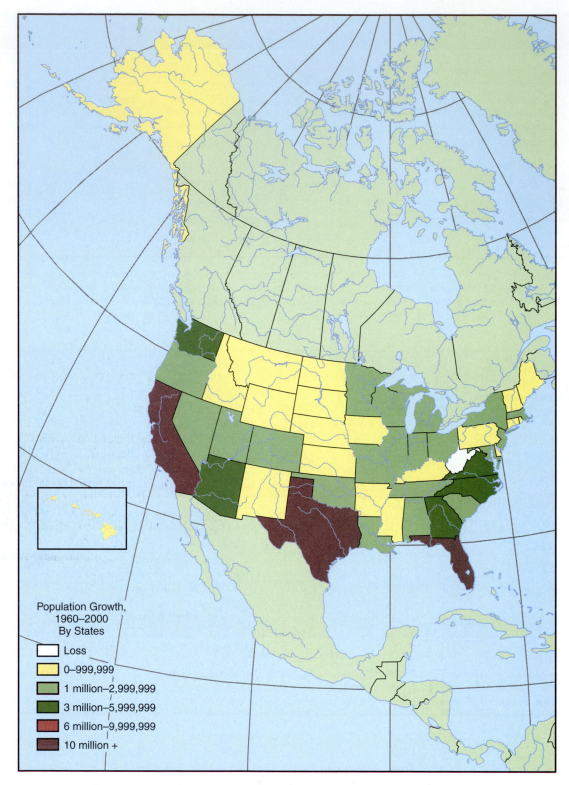

Figure 4.2

The absolute population increase has been most rapid in the Sun Belt states, 1960–2000.

Calculated from Census Bureau statistics and forecasts.

Population Growth,
1960–2000
By States

- Loss
- 0–999,999
- 1 million–2,999,999
- 3 million–5,999,999
- 6 million–9,999,999
- 10 million +

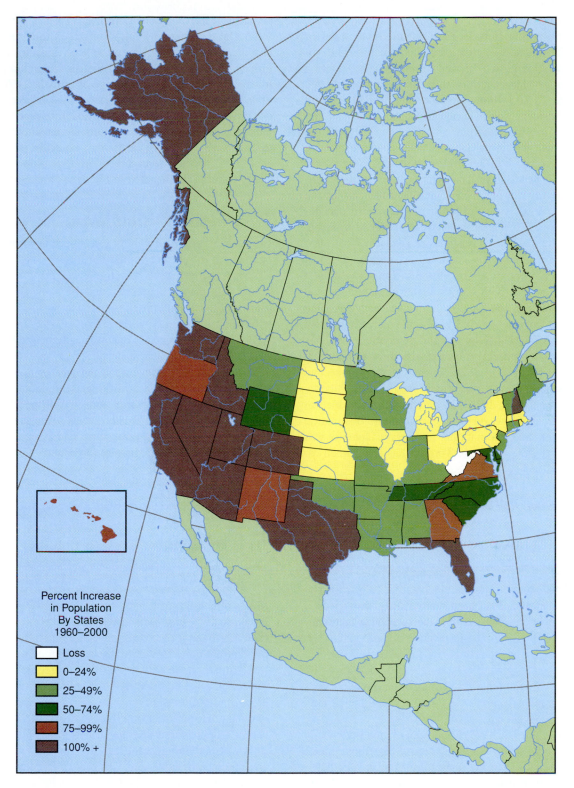

Figure 4.3
The rate of population growth has been most rapid in the Western and Sun Belt states since 1960.

Calculated from Census Bureau statistics and forecasts.

Percent Increase
in Population
By States
1960–2000

- Loss
- 0–24%
- 25–49%
- 50–74%
- 75–99%
- 100% +

the increasing concentration of population in the western states and in the Sun Belt. In fact, just the three states of California, Texas, and Florida accounted for fully 39% of national population growth between 1960 and 2000. During those 40 years each added more than 10 million people, led by California's 16.8 million. No other state added as many as 4 million residents in the same period.

North Carolina, Georgia, and Arizona followed the three states just mentioned as leaders in absolute growth, each adding from 3.2 to 3.9 million people between 1960 and 2000. The Mississippi River can be seen as the continental divide for migrants; as a general rule, east of the Mississippi River they go to Florida or Georgia (especially the Atlanta area), whereas those west of the Mississippi tend to move to California, Arizona, or Texas.

Canadians are already about as far south as they can get, so there is no equivalent of the Sun Belt. The closest they come to an area with a mild winter climate is coastal British Columbia in the Far West. In fact, between 1966 and 1996 that province grew by a higher percentage than any other (98.8%), almost doubling in population. Still, this was achieved by a fairly modest absolute population growth of 1.9 million on a rather small base. The second highest provincial growth rate was in neighboring Alberta, which grew by 84.3%, fueled in part by an abundance of natural resources. However, the largest single destination for both internal and international immigrants actually was the province of Ontario, where Toronto has acted as an es-pecially powerful magnet. In the 30 years from 1966 to 1996 Ontario grew by 54.5% to 10.8 million, and that province accounted for almost 43% of Canada's population growth. In the same 30 years, British Columbia accounted for the second highest percentage of growth, 21%. The portions of Canada that actually grew at the fastest rates were not provinces, but the two territories in the far north. Both Yukon and the Northwest Territories grew about 121%, but that increase was on such a small base that together they accounted for less than 1% of total population growth. These patterns are depicted in Figures 4.4 and 4.5.

Population Distribution

How can we generalize about the pattern of location, or geography, of the current populations of the United States and Canada? One effective way to answer this "where" question is with maps. Figures 4.6 and 4.7 show the two countries' traditional patterns of population. In this case the distributions are highly generalized, using just the three relative density categories of low, medium, and high.

In the United States, despite many decades of southward and westward movement, the area of highest density still corresponds to the traditional Manufacturing Belt of the northeast. Roughly, this zone is outlined by straight lines connecting greater Boston, Milwaukee, St. Louis, and Washington, D.C. Here are many of the most populated states, and a substantial share of the country's largest cities.

HOW TO INFLUENCE THE FUTURE: POPULATION POLICY

Projections into the future tell us what we think might happen under a given set of circumstances. One reason the future is often different than the forecast is that the act of projecting or forecasting can in and of itself lead to policy changes that alter the course of events. If the projection shows an undesirable scenario, then it may be possible, within certain limits, to institute social changes to ensure that the future looks a bit different than might otherwise have been the case.

Many countries have explicit population policies designed to influence the future course of demographic events. Neither the United States nor Canada is in that group, however. Both countries have *implicit* population policies, but nothing coherent. Implicitly, the United States and Canada both support the ability of people to maintain low fertility through the availability of contraceptives, voluntary sterilization, and abortion. The governments of both countries also encourage legal and discourage illegal immigration.

Explicitly, the closest that either country has come to a population policy was when President Richard Nixon in 1969 proposed a commission, which Congress then legislated into existence, to examine the growth of population in the United States, assess the impact of that growth on the nation's future, and recommend how the country should cope with those impending changes. The commission spent three years gathering evidence and data before presenting its findings in 1972, when it essentially proposed that the United States pursue a policy of population stabilization. Among several specific recommendations was that a permanent policy organization be established, but it was never implemented. Thus, the various recommendations, such as legalizing abortion, ratifying the Equal Rights Amendment, and limiting illegal immigration, were pursued as independent activities, outside the framework of an organized national population policy.

Canada
Population Growth,
1966–1996

- Less than 50,000
- 50,000–99,999
- 100,000–499,999
- 500,000–999,999
- 1 million +

Figure **4.4**
Canada's population growth, 1966–1996, was concentrated in Ontario, Québec and the West.

Calculated from data from the Census of Canada for the appropriate dates. (www.statcan.ca)

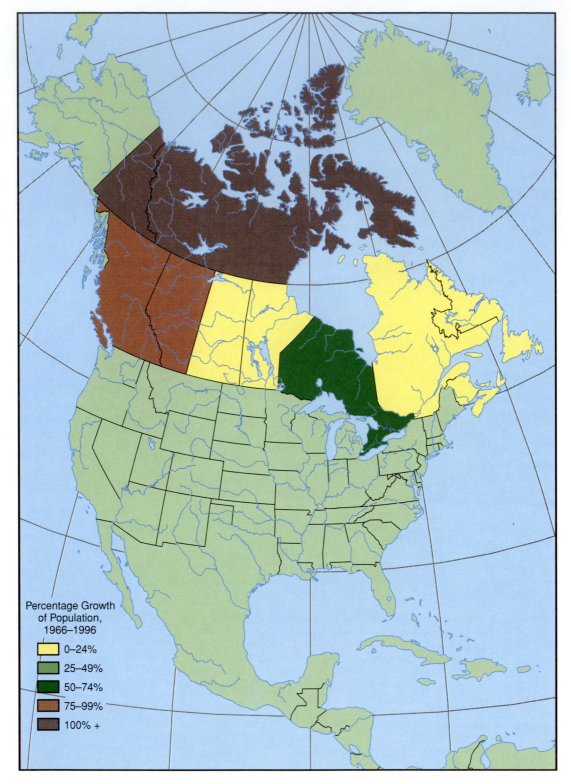

Figure **4.5**

Between 1966 and 1996, the provinces with the highest growth rates were British Columbia and Alberta. However, the two northern territories had even higher rates.

Calculated from data from the Census of Canada for the appropriate dates. (www.statcan.ca)

Percentage Growth
of Population,
1966–1996

- 0–24%
- 25–49%
- 50–74%
- 75–99%
- 100% +

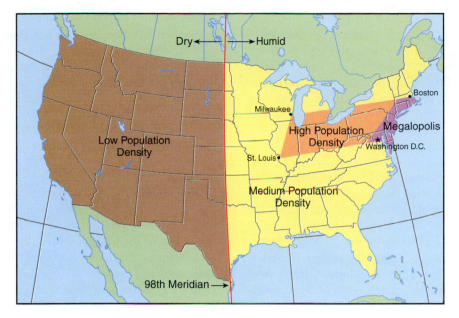

Figure 4.6
Simplified map of U.S. population density regions.

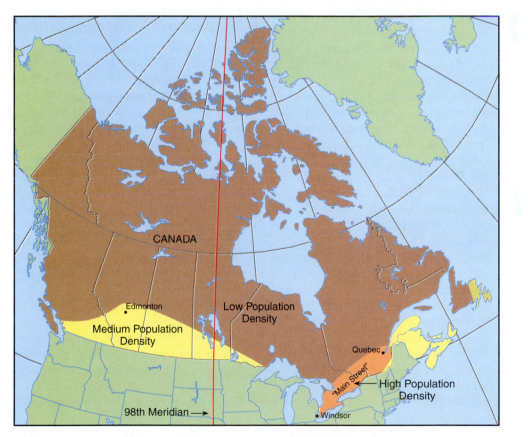

Figure 4.7
Highly generalized map of Canada's population density regions.

For example, this zone has three of the top five, and six of the top ten metropolitan areas. Still, in the second half of the twentieth century, the area's population grew slowly, increasing from about 68 million to 73 million, or a mere 7%. During the same fifty years, the country's total population grew 81%. In 2000, the region had 27% of the national population, versus the 45% who lived there in 1950. Despite its slow growth, population geographers still classify the area (with adjoining parts of Canada) as the fourth largest regional concentration of people in the world, with only the East Asia, South Asia, and Europe clusters ranking ahead of it.

Within this region, one area stands out. This is the approximately 450 miles of largely continuous urbanized area between metropolitan Boston in the north and greater Washington in the south, including New York, Philadelphia, and Baltimore. Often called **Megalopolis,** it has about 50 million people, or approximately 18% of the national total, living on only 1.5% of the land. No other area of comparable size in North America comes close to containing that share of the total population.

The two remaining population zones are divided by the 98th meridian, which separates the more humid eastern half of the United States from the drier west. This meridian approximates the 500 millimeter (20 in) rainfall line, with regions to the west receiving less than that amount annually. Most of the area east of the 98th meridian has enough rainfall to support stable and intensive agriculture without the need to irrigate, and in the past this fostered the development of a fairly dense pattern of both farms and cities. This long-standing pattern remains largely valid today, but decreasingly so, as parts of the West are experiencing impressive population growth.

Beyond the 98th meridian is the area most closely identified with the historic West. In the past, people's activities there were greatly limited by water availability, and water is still a major force in the geography, economy, demography, and politics of the West. Here are America's wide-open spaces, with low relative humidity and abundant sunshine. Most of the population is concentrated in just a handful of major metropolitan areas that take up a tiny proportion of the area, notably Denver, Salt Lake City, Las Vegas, Phoenix, Tucson, San Diego, Los Angeles, San Francisco, Sacramento, Portland, and Seattle. The construction of huge water-movement systems has made it possible to develop most of these population clusters. The coastal zone of northwestern California, Oregon, and Washington is an exception to this western dryness, but it forms only a small part of the total area.

Similar population zones are found in Canada (Figure 4.7). The highest concentration of people, by far, occurs in the corridor from Québec City to Montréal in Québec, then continuing through Ontario to Toronto, Hamilton and Windsor, the latter across the Detroit River from Detroit. This is Canada's "Main Street," its portion of the traditional Manufacturing Belt. It contains about 60% of Canada's population, a figure that has not changed much over the past few decades. It has the best farmlands in eastern Canada, and historically in North America it was such areas that contained most of the largest cities. To the north Main Street stops abruptly, with vast areas of low population density in the poor farmland and difficult climate of the Canadian Shield. West of the Shield, in the southern parts of the rather dry Prairie Provinces of Manitoba, Saskatchewan, and Alberta, with their fine soils, and in southern British Columbia, the population density is moderate. A large proportion of the people in these provinces lives in the four metropolitan areas of Winnipeg, Edmonton, Calgary, and Vancouver.

East of the Canadian Manufacturing Belt is another area of moderate population density. This zone includes the more habitable portions of New Brunswick, Nova Scotia, Prince Edward Island, and Newfoundland. In the rest of Canada conditions are too harsh to support many people. Overall, about 72% of Canadians live in the warmer southern part of the country within 242 kilometers (150 mi) of the United States border.

Urban and Rural Populations

Another way to answer the question, where do Americans and Canadians tend to live, is to say "in cities." Most people in both countries are classified as "urban," and this has been the case for many decades. In the United States, about 76% of the population in 1995 was urban, while 24% was rural. In Canada, the comparable figures were 77% and 23%. The urban percentage has been increasing for a long time, essentially continuously since the first censuses were taken in each country. The initial United States census was carried out in 1790, when only about 5% of the population lived in cities.

Why do most people in Canada and the United States live in cities? One obvious response would be to say that is where most jobs are, but that answer begs the question. A more effective answer is to give much of the credit to **economies of agglomeration.** Literally, the term refers to the advantages (economies) that come when related business and other activities cluster (agglomerate) in one place. It can be defined as the savings, in time and/or money, which accrue when mutually interdependent activities locate near each other. Thus, businesses and people can usually operate more efficiently in a large urban center than in a small town or rural area because many of the goods and services they need to be successful are available locally. A manufacturer who locates in a metropolis is likely to find close at hand such needs as an ample supply of skilled workers, banks willing to provide him/her with capital, suppliers of parts, people who can fix the photocopier quickly, fire protection, tax consultants who specialize in that particular industry, and many trucking firms competing for business. The chances of having all, or even many, such needs easily available in a rural area are small to nil. Note that the larger the urban area, the wider the variety of goods and services likely to be found, and thus the greater the economies of agglomeration. Great cities also contain large numbers of consumers (a market), and proximity to market is the single most important determinant of location for most businesses.

Compare that to a small town of say 1,000 people, located well away from any large city. Here it may be impossible to recruit an adequate number of skilled workers, to say nothing of possible difficulties in obtaining duplicating services, temporary (overload) employees, overnight delivery systems, or a whole range of other requirements. Further, the town provides only a miniscule market. In sum, economic activities tend to locate in the larger cities because, usually, they can operate most effectively there. It is no wonder that people in the technologically advanced countries are concentrated in cities, and particularly in the biggest cities. All these economic advantages are in addition to the well-known social advantages of large cities over the small, such as much greater choices in entertainment, a greater range of social institutions, and (usually) better schools.

There are also diseconomies that come with larger size. That is, over time the biggest cities tend to build up not only advantages but also considerable disadvantages, such as high land costs and rents, traffic congestion, pollution, and sometimes high crime rates. Despite these problems, to the present time it is apparent that the advantages of the larger places outweigh the disadvantages. This can be inferred from the continuing concentration of the populations of Canada and the United States into cities, particularly the few largest metropolises. To illustrate, in 1990, for the first time, more than half the U.S. population was located in the 30 metropolises with more than a million people. Put another way, by that year only 30 dots on the map accounted for over half the country's population.

Defining Urban and Rural

When discussing the terms "urban" and "rural" (or any other term), it is critical to have a clear definition in mind. The United States Census Bureau classifies every concentration of 2,500 people or more as urban. This number is a surprise to many students, as most come from large metropolitan areas and would hardly consider a place with 2,500 people as really urban.

Why does the Census Bureau use such a small figure? The answer is found in history. When the first census was taken in 1790, a place with 2,500 people was unusually large and logically could be called urban. For consistency, this number has been used ever since. In Canada, with far fewer people in a larger national territory, the census office requires only 1,000 people to make the same distinction.

Because the figure 2,500 is not very satisfactory today, the Census Bureau has come up with some additional concepts. In particular, it has defined **Metropolitan Statistical Areas** (MSAs), which consist of a core large city (in simplified form, a place of at least 50,000 people) plus the surrounding county or counties (called parishes in Louisiana). The only exceptions are in Alaska, which has no counties, and New England, where MSAs consist of towns and cities, not counties. In the rest of the country, the entire county, sometimes with extensive rural areas, is always included in the MSA to reflect the fact that the "real" (functioning) city extends well beyond the political boundaries of the central city; that is, it should encompass the suburbs and commuters from more outlying areas. An MSA can include additional contiguous counties if they have a high degree of economic and social integration with the core county. One good measure of such close ties would be if the majority of the working population of those adjacent counties commutes to the central city-county to work. MSAs, then, always have a large core city and consist of one or more adjoining counties. Despite some imperfections, this concept is widely accepted. In this book, when metropolitan populations are given, they usually refer to the MSA. In 2000, there were approximately 275 MSAs in the United States, and they contained about 80% of the country's population.*

Many MSAs are adjacent to each other, and when they form continuously built-up areas they constitute some of the country's largest metropolises. The New York metropolitan area, for example, consists of many MSAs that stretch across four states (New York, New Jersey, Pennsylvania, and Connecticut), but the whole complex is focused on New York City. The Census Bureau recognizes these supermetropolises as Consolidated Metropolitan Statistical Areas (CMSAs), which can be thought of simply as the country's largest metropolises. In rank order, the six largest CMSAs are New York (almost 20 million people), Los Angeles (15.5 million), Chicago (8.5 million), Washington-Baltimore (7.2 million), San Francisco (6.6 million), and Philadelphia (6 million).

If 76% of the U.S. population is urban, then rural people must account for the other 24%. Contrary to popular impressions, few rural dwellers are farmers. In fact, in 2000, only about 1.7% of the population was classified as rural-farm. This is a remarkably small figure, especially considering how well fed the country is and how much of the total farm output is exported. It is also quite a contrast to the 1790 census, when about 95% of the people fell into the "rural-farm" category.

The rest of the rural population, about 22% of the United States total, is rural-nonfarm. How do these many

---

*Because some counties are large, a certain percentage of the MSA population consists of rural people (farmers and other non-urban dwellers). This fact accounts for the difference between the rural (24%) and nonmetropolitan (20%) national populations. Primary Metropolitan Statistical Areas (PMSAs) are those counties that are parts of MSAs that have less than 50% of their workers employed in a different county.

Figure 4.8
West Virginia has a large proportion of its population classified as "rural nonfarm."

people, more than one person in five, make a living? Some classified as rural-nonfarm are retired, including many who looked for an inexpensive place to live. Others work in small-scale manufacturing or in service jobs (such as grocery clerks, motel employees, and food service workers) in communities with fewer than 2,500 people. Still others work in mining or lumbering, or are on welfare. All these categories, however, would not come close to accounting for over one-fifth of the country's population. Most of the others are people who live beyond the built-up urban areas who commute long distances to cities to work. Many Americans prefer to live in a rural environment and are willing to pay the price for this choice in long drives to work.

Rural-nonfarm populations (Figure 4.8) make up an especially large proportion of people in certain states, such as West Virginia and Idaho. In West Virginia, it is importantly a matter of people with strong emotional ties to their traditional rural communities, where many families have lived for generations, who are reluctant to move. These West Virginians will not hesitate to travel great distances to work so that they can live on their family farms or in other rural communities. Those who live this way may be classified "rural" according to where they live, but in practical terms they are part of a widespread urban workforce.

Changing Location of the Population

How has the location of the United States population been changing in recent decades? We have previously used the phrase **Sun Belt,** but without defining it. This term is constantly used informally by people in everyday situations, as well as by the media, but it is seldom defined. There is no universally accepted definition; in fact, the term seems to

have almost as many meanings as there are users. In this chapter, Sun Belt refers to the area south of a line that starts, on the east, at the Virginia–North Carolina border and then extends westward along the northern boundaries of Tennessee, Arkansas, Oklahoma, New Mexico, and Arizona (Figure 4.9). From the northwest tip of Arizona the boundary goes in a straight line to just north of the San Francisco Bay Area. Roughly then, it refers to the southern half of the United States. Hawaii is also a part of the Sun Belt. Since about 1970, this area has experienced remarkable population growth.

Why have so many people moved to the Sun Belt? Certainly a powerful reason is that it is viewed by most Americans (and some Canadians) as an area rich in **amenities.** Put simply, amenities are those things about an area that are attractive to people, those that contribute to the "good life." These are usually, but not solely, seen as the physical assets of an area, such as abundant sunshine, mild winter weather, and attractive beaches, mountains, or lakes. One common denominator, for many people, would be an area's recreational possibilities. In the public mind today, such amenities are especially associated with the Sun Belt, and states like Georgia, Florida, Texas, and Arizona have grown rapidly as a result. Still, in the final analysis, amenities are very much a personal thing. To someone who loves top quality plays, and who goes to them whenever possible, New York City, with its Broadway theater district, is an amenity location. The person who loves forested mountains, cool weather, and clean air, may find Vermont, the Pacific Northwest, or coastal British Columbia especially attractive.

The importance of the Sun Belt can be gauged by the large number of people who have been moving there. Between 1990 and 2000, the region's population grew from about 105 million to 119 million; in that decade it accounted for 54% of the country's population growth.

The southern half of the country also contains most of the booming MSAs. Of the 50 fastest growing MSAs between 1990 and 1996 (Figure 4.10), 37 (74%) were in the Sun Belt. The most impressive individual MSA, by far, was Las Vegas, as its population grew 41% in just six years to a total of 1.2 million. Most of the top 50 were small to medium-sized places, such as Austin and Laredo, Texas; Naples, Florida; and Las Cruces, New Mexico, as it took a much smaller absolute increase for such MSAs to show impressive growth rates. Dallas–Fort Worth was the largest of the top 50 MSAs, with a 1996 population of 4.6 million. The 50 leaders also included the major Sun Belt metropolises of Atlanta (3.5 million) and Houston (4.3 million). In contrast, not a single fast-growth MSA was in the old northeastern Manufacturing Belt.

Still, not all of the fastest-growth MSAs were in the Sun Belt. Thirteen of the top 50 were outside this zone. Most rapid-growth cities outside the Sun Belt are in the West; among the larger such places are Salt Lake City;

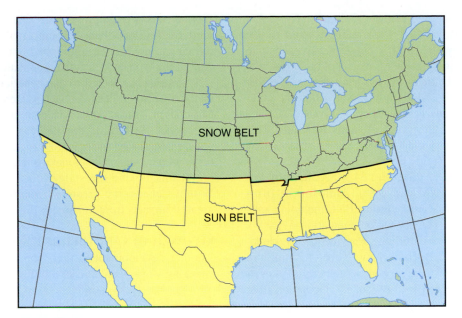

Figure 4.9

The line separating the Sun Belt from the rest of the country, as used in this text, largely usually follows existing state borders.

● and blue type = The 50 fastest growth MSAs, 1990–1996
● and red type = The 34 MSAs that lost population, 1990–1996

Figure 4.10

The 50 fastest-growing MSAs, 1990–1996, and the 34 MSAs that lost population in the same period.

Data from Census Bureau at www.census.gov/population/estimates/metro-city/ma96.08.txt

Utah; Portland, Oregon; and Denver, Colorado. The fastest growing non-Sun Belt MSA was medium-sized Boise, Idaho, which grew by almost 26% between 1990 and 1996 to 373,000. Other rapidly growing metropolitan areas with rather cold winters included Provo, Utah; Reno, Nevada; and Medford, Oregon. These findings show that the inter-mountain west, particularly the states of Idaho, Utah, and Nevada, have become the goals of many migrants. This area was long avoided by most Americans, except as a way to get to and from the Pacific Coast.

At the other end of the scale, 34 MSAs actually lost population between 1990 and 1996. Almost four-fifths were in the traditional Manufacturing Belt, including such larger metropolises as Toledo, Ohio (population 611,000), Hartford, Connecticut (1.1 million), Providence, Rhode Island (1.1 million), and Buffalo, New York (1.2 million). The largest MSA in this category was Pittsburgh, Pennsylvania (2.4 million), which lost .6% of its population from 1990 to 1996.

Developments That Facilitated Sun Belt Growth

In recent decades, there have been a number of changes in the U.S. and Canadian economies that help to explain the growth of amenity locations. Put another way, certain developments within these economies have given an increasing number of people considerable freedom when choosing where to live. With this newfound flexibility, many decided to move to amenity locations in the Sun Belt. In Canada, the major corresponding impact has been the movement of people to the west, particularly to British Columbia. As an attractive urban area, Toronto has also experienced rapid growth. Three developments that have favored the growth of the Sun Belt are discussed below.

Retirement Incomes

One relevant factor in Sun Belt growth is the increasing number of people on regular retirement incomes. In the United States, social security legislation was first passed in the 1930s, and the first disbursement took place in 1940. Not until the 1950s, however, did a substantial number of people begin to collect benefits. A key geographic fact about social security, and most other retirement plans, is that recipients can have their monthly checks sent anywhere they want. Within the limits of their incomes, this gives retirees tremendous freedom of location, and many have chosen to move to the Sun Belt. This has led to other developments that reinforced the flow of people, including the construction of retirement communities (with minimum age limits, commonly 55) and of specialized medical services for the elderly. There is also a "friends and relatives effect," as those who move first urge others to follow in their footsteps. The movement of retirees to the southern half of the country was first noted in the 1950s, when

Florida started its impressive growth that has continued to the present. Since then the number of people with regular retirement incomes has multiplied many fold, as the "graying" of the U.S. and Canadian populations attests.

Florida has been the main beneficiary of this trend, but many other Sun Belt states, such as Georgia, South Carolina, Mississippi, Texas, California, and Arizona have shared the experience. In Florida this process has progressed to the point where, in 2000, about 20% of the population was over 65 years old, versus a national average of 14%. Not surprisingly, that is also the state with the highest death rate. Within Canada, the greater Vancouver area has seen the greatest impacts from this trend (Figure 4.11).

Growth of Quaternary Activities

Another development to favor the Sun Belt has been the changing ways people make a living in modern economies.

(a)

(b)

Figure 4.11
Senior citizens with steady retirement incomes can have considerable geographic flexibility. They have contributed to the growth of certain favored areas, such as (a) Vancouver in Canada, with its mild winters. Florida has also attracted large numbers of retirees since the 1950s. This group of seniors is exercising in (b) Miami Beach.

For many decades the fastest growing part of the national economies has been the **tertiary,** or the service, sector. Approximately 80% of the jobs in both countries these days are in the services, ranging from sales employees to elementary school teachers to bank tellers and tax consultants. This is in stark contrast to conditions around 1800, when 95% of the populations depended on **primary** occupations. The latter are defined as jobs based on the direct exploitation of earth resources, such as farming, fishing, cutting timber, and mining. Another category is the **secondary,** which includes all types of manufacturing. Secondary activities account for less than 20% of jobs, a figure significantly smaller than just a few decades ago.

More recently, a fourth level of economic activities, the **quaternary,** has been identified. These are defined as primarily involved in the gathering, manipulation, and dissemination of information. Quaternary activities certainly are also services, and could be considered a part of the tertiary sector. However, in the past few decades the information sector has grown so rapidly that there is general agreement it should be treated separately. An increasing number of businesses are information-based; that is, their main input needs are information, and their product, or output, is also information (such as recommendations, or findings). One good example is a stockbroker. Typically, she or he spends a great deal of time gathering information about stocks and bonds (among other things), summarizing, analyzing, and digesting that information (manipulation phase), and then reaching conclusions about which ones to recommend to the customer to buy, hold, or sell (dissemination phase). Other examples of quaternary activities include educational and environmental consultants, and corporate headquarters. The headquarters of a global firm can be seen as the place where information about its (and competitors') widespread operations is gathered and analyzed. The "output" of a headquarters is orders, or decisions. After receiving the information, management typically sends out instructions about such items as where to increase or decrease production, where to establish new warehouses, and where to change pricing or marketing strategies.

As should be obvious from the above, for quaternary activities excellent access to information is vital to success. In recent decades, the technology for storing, manipulating, and moving information has improved tremendously, especially with the widespread availability of computers and the Internet. Much of the world has been linked into a vast web that can move massive amounts of information cheaply and almost instantaneously. In one sense, anyone who has access to a telephone, to a cable system, or to a satellite network, can use this system, and there are few populated areas in the United States and Canada today that cannot provide such access. It is true that for massive users of information, the larger cities with their teleports and superior transmission technologies are still preferred, but such cities are widely scattered across the two countries, including in the Sun Belt. For people with less exacting access requirements, such as many consultants, their needs can be met in most locations.

These developments have had important geographic impacts. Since most places where people are likely to want to live have ready access to information, they are all potential locations for quaternary activities. In other words, information-based activities, especially those with modest information requirements, should be able to locate successfully just about anywhere. This is in direct contrast to earlier times, when only the largest cities had good (fast and cheap) access to information (word-of-mouth, telephone, the best newspapers). In that era, most quaternary activities, such as corporate headquarters, almost had to locate in the largest cities. Today it is possible for a major global company like MCI WorldCom (Figure 4.12) to be headquartered in Jackson, Mississippi, a place that as little as thirty years ago would have been considered far too remote from the "action" to be viable. Now there are a number of successful stockbrokers and mutual funds, with a national customer base, located in such formerly relatively isolated states like Vermont, Colorado, Arizona, and Montana.

For quaternary activities, the impact of this new geographic freedom is predictable. People with a choice often will move to amenity locations, and in the United States they frequently identify such places with the Sun Belt. The rise of the quaternary activities, then, has contributed significantly to the growth of the southern half of the United States, and of select places in Canada.

A word of caution is appropriate. This trend has only *allowed* people greater geographic freedom, it has hardly *caused* them to move to the Sun Belt. In fact, most businesses are located where the founder happens to be living,

Figure 4.12
The headquarters of MCI/WorldCom in Jackson, Mississippi.

and many quaternary activities can be successful just about wherever that happens to be. The relevant point is that if the owner ever decides to move from that original location, certainly she or he is free to select an amenity location. Usually that means a place with mild winter weather, ample sunshine, and perhaps a beach or lake, but it can also be a ski resort, or a major cold-winter city like Chicago, Toronto, Montreal, Minneapolis, Boston, or New York with rich cultural resources, or a growing non-Sun Belt smaller city in the intermountain West (Figure 4.13). When the founder of perhaps the most successful new company of recent decades, Bill Gates of Microsoft, decided on a permanent location for his firm, he moved it from the Southwest to his boyhood home of Seattle.

It is also easy to overstate the locational freedom of quaternary activities. Environmental consultants, for example, may locate successfully in small towns with many amenities, but there can be a price. Occasionally most will have to travel to meet with clients, or clients may come to them. Most long distance business travel these days is by air, and for such trips the largest cities still have great advantages over small places in terms of frequency of service and fares. It is worth repeating, too, that the best access to information-moving technology is still concentrated in the largest cities. Other possible disadvantages of the smaller towns could include lack of cultural opportunities, few or no high-grade restaurants to entertain clients, relatively high food prices, and mediocre schools.

High-Tech Manufacturing

Another development that has allowed more people to live in the Sun Belt involves the secondary sector, or manufacturing. Although the importance of this sector has declined markedly in the past few decades, one portion has grown impressively. This is what is loosely called **high-tech manufacturing,** firms which turn out products that have high values in relation to their weight (i.e., a high value per pound). Typically, the products are rather small and light, and therefore easy to transport. Good examples of such manufacturing include the making of computer chips,

Figure 4.13
Boise, Idaho is a good example of a rapidly growing, non-Sun Belt city in the intermountain West.

highly specialized medical instruments, and parts for spacecraft. To give one example, the American space shuttle includes over a million parts, most of which would meet the definition of high-tech; thousands of firms are involved in producing the craft. Their typical products weigh just a few pounds or less, yet they can be worth hundreds of thousands of dollars or even more. For such commodities, transportation costs are only a tiny fraction of the final selling price, and they play either a minor role in the decision of where to locate, or no role at all. This is in great contrast to manufacturers of bulky or low value-per-weight items like refrigerators, steel construction beams, or automobiles, where transport costs constitute a much larger percentage of the sales price. Companies that manufacture these latter items have to make careful calculations about the costs of getting raw materials and parts to their plants and of reaching the consumer, and to locate accordingly. Often they locate centrally to their suppliers and customers, in order to minimize transportation charges, which in the past often meant in the Manufacturing Belt. But this does not apply to high-tech firms, which have considerably greater geographic freedom.

The implications of these facts are by now familiar. Where do high-tech companies tend to locate? Because of their considerable freedom, many have chosen to be near amenities, especially in the Sun Belt. To such firms, a coast-to-coast transportation charge of $3,000 for a product worth, say $3 million, is only a small consideration, as it adds only 0.1% to the final price. To a refrigerator manufacturer, whose average product may sell for $1,000, even a $300 transport cost looms much larger. The latter must be sensitive to transportation charges when choosing a location, and some places are definitely much better than others. The producer of the $3 million item, on the other hand, can be quite choosy about where she/he will locate. One logical choice might be in a medium-sized city on a beach, with warm winters, strong cultural resources (entertainment, museums, churches, a university), and a reputation for good schools for the children of the employees. In such places workers are likely to be satisfied with their environment, resulting in low employee turnover. More importantly, perhaps, the company president will be where he or she likes to live!

Overall, these three changes have enabled an increasing number of people to live in amenity locations. Of course, many high-tech companies, quaternary activities, and retirees are not located in the Sun Belt and are quite content. The question is, if they ever decide to move, for whatever reason (congestion costs, high rents, high labor turnover), where would they choose to go? Clearly, in recent decades many have chosen the southern half of the United States and parts of the Canadian west. Thousands of such individual decisions have contributed to major changes in the population geographies of the two countries. This trend started in the 1950s, and accelerated considerably after 1970.

Two Additional Considerations

In the United States, two other considerations have contributed to this population shift. One was the development of relatively cheap air conditioning, which is usually dated to the 1960s. Without this change, the larger-scale movement of people to the Sun Belt simply would not have happened. To take only the most obvious case, it is impossible to imagine almost three million people as permanent residents of Phoenix in the year 2000 without affordable air conditioning. In its absence, undoubtedly Phoenix would be quite small today, except possibly seasonally during the mild winter months. Similarly, how many elderly retirees would put up with a Florida summer without air conditioning, or would even survive such hot and humid conditions? Before widespread and relatively cheap air conditioning, many retirees went to Florida just for the winter, returning to their cooler homes in the north each spring.

Another critical factor, but one that only applies to the Deep South, was the end of legal segregation (segregation sanctioned by law), also starting in the 1960s. Initially this change led to great social turmoil, and must have discouraged many people from moving there. But eventually there was considerable progress, and today legal segregation is a thing of the distant past. As long as the South had this social problem, many people and companies did not consider moving there. As the problem receded, the region was opened to the impressive developments that have since taken place. The end of legal segregation, then, was critical to opening the floodgates to the massive movement to the South of manufacturing, quaternary activities, and retirees.

Geography of Fertility, Mortality, and Life Expectancy

Fertility

In almost every society, minority groups—no matter how *minority* is defined—tend to have higher-than-average fertility (the number of children that women bear). Three readily definable sets of minority groups in the United States and Canada are: (a) racial/ethnic (such as blacks, Hispanics, Asians, and First Nations); (b) national origin/linguistic groups (especially French speakers in Canada); and (c) religious groups (such as Mormons in the United States and Hutterites in Canada).

Figures 4.14 and 4.15 show areal differences in birth rates in the United States and Canada. Those states in which the percentage of Hispanics is high—the southwestern states—have a higher-than-average birth rate, although the highest statewide birth rate by a good margin is found in Utah (which has a high proportion of Mormons, who favor large families). Other states that are well above the national average in birth rates include Alaska (which has a high proportion of American Indians, Inuit and Aleuts), Idaho (with a substantial Mormon population), and those in

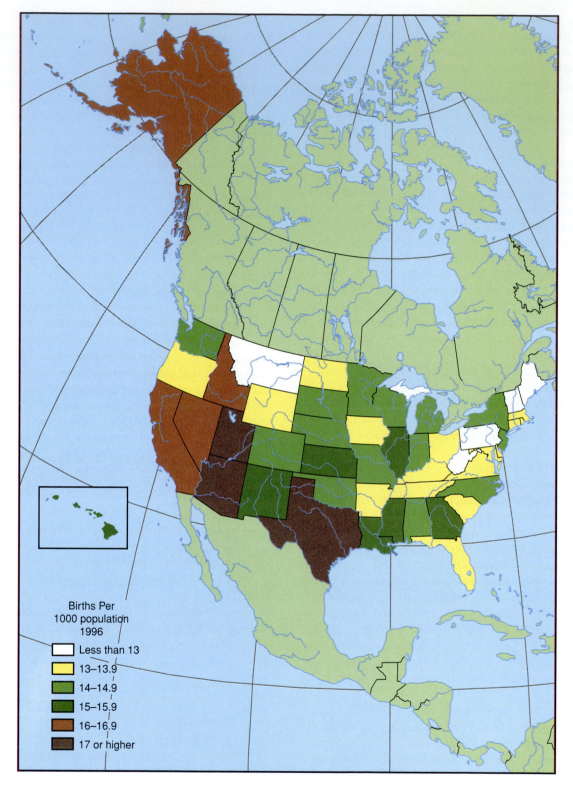

Figure 4.14

The birth rate, or the number of children born per 1000 population, for the states in 1996.

Monthly Vital Statistics Report, Vol. 96, No. 1 (5) 2, September, 1997.

Births Per
1000 population
1996

- Less than 13
- 13–13.9
- 14–14.9
- 15–15.9
- 16–16.9
- 17 or higher

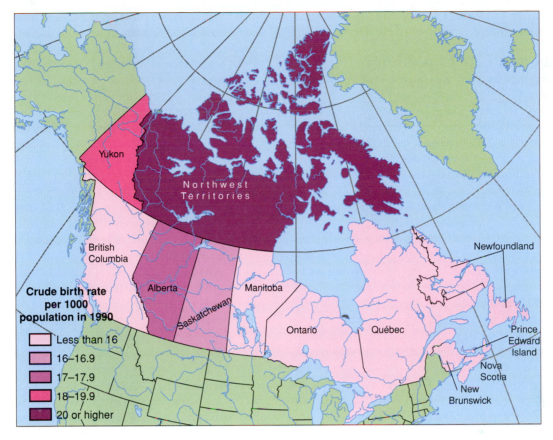

Crude birth rate per 1000 population in 1990

- Less than 16
- 16–16.9
- 17–17.9
- 18–19.9
- 20 or higher

Figure 4.15
Birth rates for Canadian provinces and territories, 1990. The northern territories, with high percentages of minority groups, had the highest birth rates.

Source: Data from *Canada Yearbook 1992*, Statistics Canada.

which there is a higher-than-average percentage of blacks, such as Illinois (with a high fraction in Chicago), Georgia, Louisiana, and Mississippi. Conversely, the lowest rates are found in Maine, Vermont, and West Virginia, all states with a high percentage of non-Hispanic whites.

The historically most significant fertility differential in Canada was between the French speakers, especially concentrated in Québec, and English speakers in the other provinces. In the 17th and 18th centuries, the high fertility of women in New France was legendary, and French speakers in Canada maintained higher-than-average levels of fertility until the 1960s. When the baby boom peaked in Canada in 1959, Québec's fertility was still slightly higher than the rest of the country, but during the next decade the birth rate declined so rapidly that by 1970 it had the lowest birth rate of any of the provinces. The indigenous populations of the northern regions maintain high levels of fertility, but elsewhere in Canada, the differences in fertility between groups have diminished greatly.

In virtually every country, farm women have more children on average than do women who live elsewhere. Data from the 1992 Current Population Survey of the United States confirmed that this pattern persists. For example, among American women aged 35–44 (essentially at the end of their childbearing years), the average woman living on a farm had given birth to 2.3 children, compared to 1.9 among nonfarm women. Furthermore, 17% of nonfarm women at that age were childless, compared to 14% of farm women. These differences have little impact on the national averages, however, because farm women represent less than 2% of all women of childbearing age.

Given the very small farm population in the United States, a more useful distinction is between those living in metropolitan areas and those living in nonmetropolitan areas. In 1992, among women aged 35–44, those living in nonmetropolitan areas had an average of 2.1 children, compared with 1.9 among those in metropolitan areas. Nonmetropolitan women accounted for one in five women of childbearing age, and only 12% of them were childless, compared with 19% of the metropolitan women who were still childless at ages 35–44. Be aware, however, that within metropolitan areas, the central city areas have higher fertility levels than do the suburbs. This is largely due to the fact that minority-group women (especially nonwhites) are more likely to live in central cities and tend to have slightly more children on average, as mentioned previously.

Mortality and Life Expectancy

We are among the first people in the history of the earth to be able to take a long life pretty much for granted. As recently as the year 1900, **life expectancy** at birth in the United States and Canada was only 47 years, a level comparable to Somalia and Mozambique today. There was little difference in death rates for males and females in 1900, but there was a big difference by race. In the United States, life expectancy for white females, for example, was 48.7 years, whereas for nonwhite women, it was only 33.5 years (about the same as historically premodern societies).

The first half of the 20th century witnessed a tremendous amount of control over mortality through a combination of controlling the environment (public health measures) and learning to cure diseases (medical measures). Thus, by 1950, the average life expectancy in both the United States and Canada was 68 years (about the same as Russia today, but lower than Mexico). The improvement was more rapid for females than males, and nonwhites in the United States were also able to close part of their gap with whites during the first half of the century. In 1950, life expectancy at birth for white women was 72.2 years, and had risen to 62.9 for nonwhite females.

Mortality conditions have continued to improve since 1950. By 1997, nonwhite females in the United States had added 11.8 years to life expectancy, while nonwhite males also gained ground with an additional 8.1 years (Table 4.2). White women added 7.7 years during this time, to reach a life expectancy at birth of 79.9 years—among the highest in the world, although not as high as the 81 years for Cana-

dian females. White males in the United States fared a little better than white females in terms of overall improvement, gaining 7.8 years of life expectancy between 1950 and 1997. Canadian males gained 9 years during that period.

An important but nearly invisible ingredient in both the lower fertility of American and Canadian women and their greater freedom to participate in the labor force has been the decline in the risk of death among infants and children. Although the bulk of this decline took place before 1950, it has continued since then. For example, in 1915, the **infant mortality rate** (deaths under one year of age per 1000 live births) was 99.9 in the United States—about the same as in India today. By 1950, it had dropped to 29.2—about the level of Argentina in the 1990s. By 1997, it had fallen to 7.2—very low by world standards, although not quite in the "top 20" countries with the lowest levels of infant mortality. Until fairly recently, Canada's infant mortality was higher than in the United States. In 1950, for example, it was 38.5, but it had dropped to 5.3 by 1997, placing Canada in the "top ten" countries with the lowest levels of infant mortality.

Locationally, the differences in infant mortality in the United States are almost entirely traceable to the geographic distribution of African-Americans. Infant mortality rates have been persistently higher for blacks than for any other group, including recent immigrants from Asia and Mexico. The high rates are due largely to the above-average number of low-birth weight babies among blacks, but we do not know for sure why black women are so prone to bearing low-birth weight babies. A number of theories have been advanced, including high poverty levels and poor access to health care, compounded by youthful motherhood and drug use during pregnancy. None of these, however, convincingly explains the problem. Most of these same difficulties

Table 4.2
Changes in Life Expectancy after 1950 in the United States and Canada, by Race and Gender

Race and Gender	1950*	1997	Years of Life Added between 1950 and 1997
United States			
All females	71.1	79.4	8.3
All males	65.6	73.6	8.0
White females	72.2	79.9	7.7
White males	66.5	74.3	7.8
Nonwhite females	62.9	74.7	11.8
Nonwhite males	59.1	67.2	8.1
Canada	**1951**	**1992**	**Years of Life Added between 1951 and 1992**
All females	71	81	10
All males	66	75	9

Sources: U. S. data from "Annual Summary of Births, Marriages, Divorces, and Deaths: United States, 1997" in Monthly Vital Statistics Report, *1997. National Center for Health Statistics; Canada data from* Canada Yearbook, *1992, table 4.1. Statistics Canada at www.statcan.ca.*

have also been confronted by Asian and Hispanic immigrants without generating the same high levels of poor reproductive outcome.

Since African-Americans are disproportionately located in the South and in the larger cities outside the South, those areas have the highest infant mortality rates (Figure 4.16). Regionally, rates are highest in the South (with Mississippi recording the highest state rate in 1998), followed by the Midwest (where Chicago and Detroit have substantially above-average rates). The western states generally have low rates. The lowest infant mortality rates by state, however, are found in the Northeast—especially in Maine and Massachusetts. Significantly, if we include Washington, D.C., among the major states and divisions, then we find that its infant mortality rate is almost twice the national average. This is due partly to its high fraction of residents who are black, but that is not the entire explanation. Even the spatial variability of infant mortality rates among blacks is higher in Washington, D.C., than elsewhere in the nation, and black babies, like others, are less likely to die during the first year of life if they live in the Northeast than if they live in the South.

Babies tend to die of communicable disease; younger adults tend to die of socially produced causes (such as suicide, homicide, and AIDS); and older people usually die of degenerative diseases. Indeed, since 1950, an ever-increasing fraction of all deaths has been occurring at the older ages, as most causes of death are brought increasingly under control at the younger ages.

By the time people move into their 50s, the chances of dying start to increase at an accelerating rate. At ages above 60, by far the largest number of people of either gender die of cardiovascular diseases. It is probable that many of the deaths from cardiovascular causes represent biological degeneration, although social factors, especially stress and smoking, are frequently implicated.

Running a distant second as a cause of death among the elderly is cancer, or malignant neoplasms. They account for 13% of all death of females and 16% of males aged 60 and older. Among the elderly in the United States, the cancers that produce the highest proportion of deaths are those of the digestive organs, especially the stomach, intestines, and pancreas. These account for one-third of all deaths from cancer. The second most frequently encountered cancers are those of the breast and genitourinary system, which account for an additional one-fourth of all cancer deaths among the elderly. Lung cancer is an important cause of death, but it tends to hit early rather than late. Lung cancer death rates peak in the late 50s and early 60s.

Death rates from diseases like cancer are higher in those areas of the country where greater fractions of the population are at the older ages. This explains why Florida's cancer death rate is higher than that of any other state. It is not something in the water supply in Florida; it is just that Florida has a higher percentage of older people than any other state. As a region, the northeastern states have the highest percentage of elderly, and so, outside of Florida, the highest death rates are there, followed by the Midwest, and then the South. The western states, with the youngest populations, also have the lowest overall death rates from cancer.

The death rate from specific causes of death can be misleading, of course, if we do not take into account the age distribution of the population. Life expectancy is the measure most often used to control for the differences in the likelihood of death at different ages, and Figure 4.17 shows the variability from state to state in life expectancy. Note that the Northern Plains states (especially Minnesota) have the highest life expectancy in the United States, whereas the southern states have the lowest.

Geography of Migration

Migration involves the movement of people from one residential location to another. When that relocation occurs across political boundaries, whether state lines or international borders, it affects the population structure of both the origin and the destination jurisdictions.

Migration within the United States and Canada

Migration within the United States and Canada has noticeably affected the geographic distribution of the population. Until about 1950, migrants within the United States tended to head west, as already noted. They also had been moving out of the southern states and into the northeastern and north-central states. This generally represented rural-to-urban migration out of the economically depressed South into the industrialized cities of the North, such as New York, Baltimore, Cleveland, Detroit, and Chicago. In the 1950s, the pattern of net out-migration from the South began to reverse itself, and the northeastern and north-central states increasingly were migration origins rather than destinations. This trend towards the South became substantially stronger after 1970. Such cities as Atlanta, Nashville, and Charlotte have grown considerably as a result of this turnaround. Still, the strong westward movement has remained the prevailing direction of migration (Figure 4.18).

In Canada, the stepwise motion westward has seen Québec fade proportionately as the population shifted in the direction of Ontario. Then Ontario was overtaken as a destination by the Prairie Provinces. Since 1950, British Columbia has emerged as the most rapidly growing province, followed by adjacent Alberta. Nonetheless, we do not want to exaggerate the westward movement in Canada. In 1950, British Columbia accounted for 8% of the country's population; that had increased to 13% by 1996, but 62% of all Canadians still lived in Ontario and Québec.

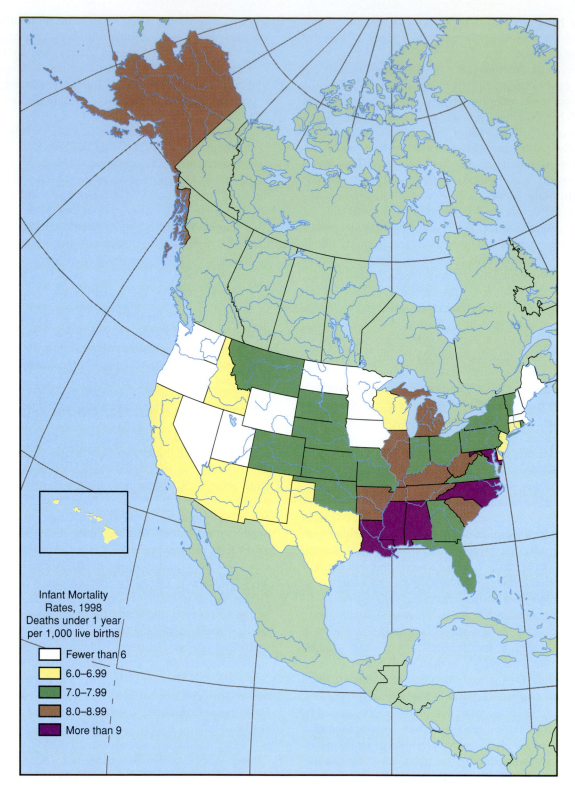

Figure 4.16

Infant mortality rates are higher in the eastern half of the country than in the West, and are highest in Mississippi, Louisiana, and Alabama.

Source: National Vital Statistics Systems. (www.cdc.gov)

Infant Mortality
Rates, 1998
Deaths under 1 year
per 1,000 live births

	Fewer than 6
	6.0–6.99
	7.0–7.99
	8.0–8.99
	More than 9

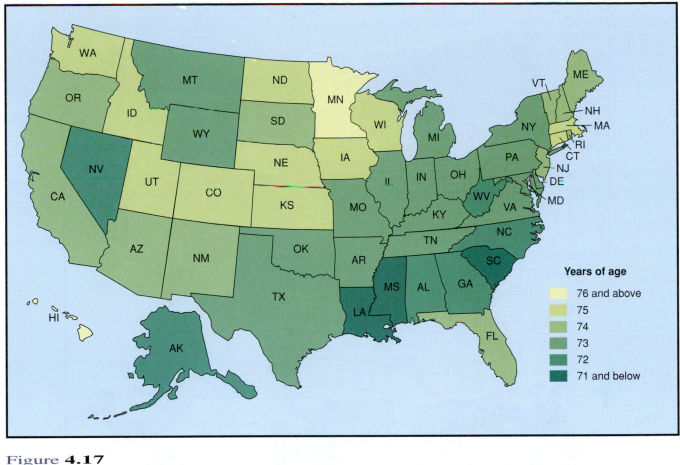

Figure 4.17
Life expectancy varies geographically in the United States.

Most Americans and Canadians are descended from immigrants, and migration has persisted among them as an important strategy for dealing with life. More than people almost anywhere else on the globe, Americans and Canadians tend to move when they perceive opportunity to exist somewhere else. Like a storm moving onshore, people blow into places that are prospering; conversely, many are not shy about bailing out when the going gets rough. For well over a century, California has been an area of opportunity, and migrants flocked there in droves. However, the recession of the early 1990s momentarily hit California somewhat harder than many other parts of the country, and around that time there was a net migration of U.S. residents out of California (Figure 4.19). People headed farther north, especially to such places as Seattle, Washington, and Boise, Idaho, or a bit east to Las Vegas and Phoenix. Soon thereafter, the state's economy recovered, and in-migration from other parts of the country resumed.

The net out-migration of U.S. residents from California did not mean, however, that the state was no longer receiving migrants. Migrants from other countries, especially Mexico and the Philippines, continued to pour into California, the most popular destination by far for people entering the United States from abroad. In fact, among immigrants entering the United States in 1996, three of the top ten metropolitan areas of intended residence were in California: Los Angeles was the most common choice, but San Diego and San Francisco were also popular. The other seven were, in order, New York, Miami, Chicago, Washington, D.C., Houston, Boston, and Newark, New Jersey.

International Migration

Immigration peaked in the first decade of the 20th century, when nearly 9 million people entered the United States and 1.5 million entered Canada. In that decade (1901–1910), immigrants accounted for more than one in ten persons in the United States and one in four Canadians. The enormity of that migration stream, coupled with fears after World War I that the countries would be overrun by Eastern and Southern European refugees, led both nations to pass restrictive immigration laws. As a result, immigration bottomed out during the Great Depression of the 1930s. At the

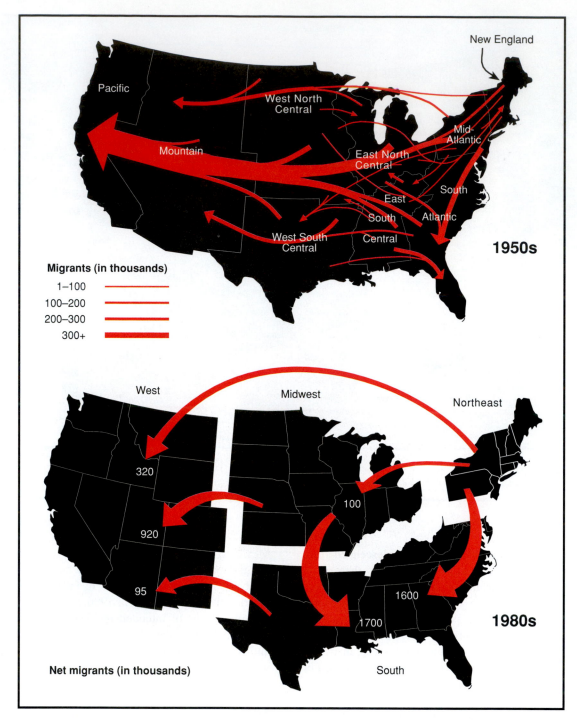

Figure 4.18

Migration patterns in the 1950s and 1980s. In the 1950s, the U.S. population flow was west and south. In the 1980s, the population was heading south.

end of World War II both countries, but especially the United States, had a rather small number of first-generation immigrants. However, the door opened to waves of refugees after the war, including people fleeing the Cuban Revolution (which permanently changed the cultural landscape of southern Florida).

In the mid-1960s, the ethnically discriminatory aspects of Canadian and American immigration policy ended, although certain kinds of restrictions remained. The Immigration Act of 1965 in the United States ended the nearly half-century of national origins (or preferred nationalities, as they were called in Canada) as the principal

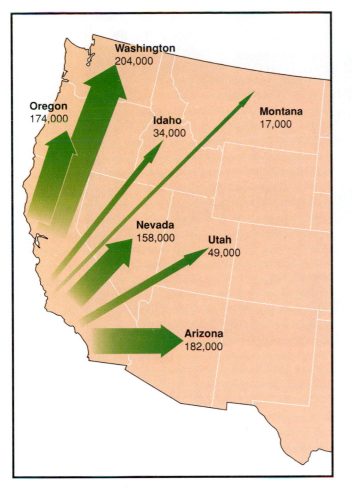

Figure 4.19

The number of Californians moving to each state, 1985–1991.
These years saw a reverse flow of population out of California into other western states. Fleeing a stagnant economy and troubled cities, migrants moved not only to large cities, such as Portland and Seattle, but also to smaller communities in, for example, central Oregon, eastern Washington, and southern Idaho.

Source: Data from Census Bureau, California Department of Finance.

determinant of who could enter the country from non-Western Hemisphere nations.

Although the criterion of national origins is gone, restrictions on the numbers of immigrants remain, including a limit on those from Western Hemisphere as well as Eastern Hemisphere countries. The 1965 act set annual limits of 120,000 persons from the Western Hemisphere and 170,000 from non-Western Hemisphere nations, with no more than 20,000 allowed from any single country. Congress retained the right to grant exemptions from those limits for any special groups, such as Vietnamese refugees. A system of preference was retained but modified to give first crack at immigration to relatives of American citizens—the so-called "family reunification" feature of immigration policy. Parents of U.S. citizens can migrate regardless of quota. Nonrelatives must obtain certification by the Labor

Department that their skills are required in the United States.

In 1976, the immigration law was amended so that parents of U.S. citizens had the highest priority for legal entry only if their child was at least 21 years of age. This was designed to eliminate the fairly frequent ploy of a pregnant woman entering illegally and bearing her child in the United States (making the child a citizen and eligible for welfare benefits), then applying for citizenship on the basis of being a parent of a U.S. citizen.

Illegal immigration has gained notice largely as a result of the quotas on the number of Western Hemisphere immigrants set by the 1965 Immigration Act, which followed closely the demise of the *bracero* program. Under this program, set up in the 1950s, Mexican farm laborers (*braceros*) were brought to the United States to work on contract in U.S. fields. By the early 1960s, the need for such low-wage, unskilled labor seemed to have slackened, and there was a good deal of public outcry about the conditions under which many of these immigrants worked and lived. But the immigration streams were now in place. People who had come from Mexico to work in the United States now knew where jobs were and knew others who knew about jobs, and the word spread back to the interior of Mexico. The United States had created a demand for Mexican labor that could not be stemmed simply by setting numerical quotas on legal immigration.

Furthermore, the new requirements for legal entry came at a time when Mexico's population was beginning to grow much more rapidly than its economy could handle, creating distinct push factors. One such push factor was that wages in Mexico were only one-sixth those in the United States. Literally, Mexicans working in the United States often could earn in one day what it took them a whole week to earn south of the border. It is reasonable to assume that as long as such a large gap exists, there will be people who will want to migrate (often illegally and temporarily) from Mexico to the United States (Figure 4.20). Further, they will take considerable risks, and be willing to put up with great privations, to be able to earn the higher rate. Thus, in the 1970s many illegal (undocumented) immigrants began to slip across the U.S–Mexico border, a phenomenon for which there is no end in sight.

As undocumented migration from Mexico has continued, and as Central and South Americans have been increasingly added to the stream, the cry is often heard that the southern border of the United States is "out of control," and it is widely believed that undocumented workers are taking jobs from U.S. citizens and draining the welfare system. Reality appears to be less dramatic. In the first place, the number of illegal immigrants in the United States is large, but less than many people have imagined, partly because they frequently come to work only temporarily and then return to their home country. The Immigration and Naturalization Service estimated that there were about 5 million illegal immigrants in 1996, about 40% of whom

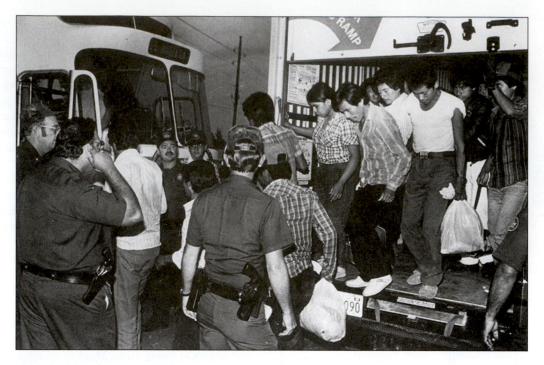

Figure 4.20
Suspected illegal aliens being unloaded from a truck and placed into custody in Chandler, Arizona.

were in California. Mexico was the source of about 2.7 million of these "illegal alien residents," or 54% of the total.

Studies of labor market impacts have found that the effects of immigrants (both legal and undocumented) on the wages and earnings of other labor force groups are either nonexistent or small. Studies of welfare use and abuse by illegal immigrants present a more mixed picture: some show a major impact, while others do not. We simply do not have enough information to know for sure what is happening.

The widespread public perception of negative consequences of undocumented migration helped impel Congress to pass the 1986 Immigration Reform and Control Act. From the standpoint of the undocumented immigrant, this was "good news/bad news" legislation. The good news was that the law offered amnesty—relief from the threat of deportation along with the prospect of legal resident status—for undocumented workers who had been living continuously in the United States since at least January 1, 1982. The bad news was that to curtail new workers from entering the country without documentation, it became unlawful for an employer knowingly to hire such a person. The teeth of the law include fines for employers beginning at $250, with much larger fines and even jail sentences for those who repeatedly hire undocumented workers. Aimed entirely at curtailing illegal immigration, this law has had no known impact on the pattern of legal immigration. Did it slow down illegal immigration? Yes, but only temporarily. A few years after passage the level of illegal immigration

appeared to be as high as it ever was, and new demands were being voiced to do something about it.

It should be apparent that the immigration policies of both the United States and Canada have focused on **gatekeeping** or keeping out people who are considered undesirable. Those generally considered most desirable are professionals, such as physicians, scientists, engineers, professors, and lawyers. The liberalization of the immigration policy in the mid-1960s encouraged such persons to enter from less-developed nations, and has, as a consequence, contributed to the "brain drain" from Asia, Latin America, and Africa. With the American gatekeeper being less restrictive than in the past, the composition of ethnic origins of immigrants has changed drastically since the mid-1960s. For example, the number of Europeans entering the United States declined by 45% between 1965 and 1986, whereas the number of Asians grew by 1200%.

From 1961 to 1970, Europeans accounted for nearly one-third of all immigrants to the United States, led by people from the United Kingdom, Italy, and Germany (in that order). Only Mexico, Canada, and Cuba sent more immigrants during that period. Twenty years later, in the 1981–1990 period, the picture was very different. Fewer than one in ten immigrants were from Europe (the United Kingdom, Poland, and the Soviet Union were the top sending countries), and immigration was dominated by people from Mexico, the Philippines, Vietnam, China, Korea, and India (in that order). With the number of immigrants in the 1990s approaching the levels of the beginnings of the century, we are seeing a new pattern of non-European ethnic diversity.

Lands of Great Ethnic Diversity

Because of this large stream of immigrants, by the end of the 20th century, the United States and Canada were racially and ethnically diverse. In both countries, however, the population was still predominantly composed of people who are identified as non-Hispanic whites, the numerically dominant group, compared to which other groups are "minorities" or "persons of color." In the year 2000, non-Hispanic whites accounted for 71% of the U.S. population, while blacks were 13%, Hispanics 11%, Asians 4%, and American Indians, Inuit and Aleuts, just under 1%. Between 1990 and 2000, the non-Hispanic white population declined as a fraction of the total from 75% to 71%, while all other groups were increasing.

In Canada, which uses the term "visible minorities," these groups constituted 11.2% of the 1996 population of 28.5 million. The largest component was the Chinese, who numbered 860,000 that year, or 3% of the national total. They were followed, in descending order, by South Asians (2%), blacks (2%), and Arabs-West Asians (1%). Many of the Chinese had come from Hong Kong, South Asians tended to be from India, and the black newcomers often arrived from Caribbean Islands. A large proportion of these visible minorities emigrated from former British colonies, many of which (like Canada) are members of the British Commonwealth.

In the United States, racial and ethnic diversity is more obvious in some regions than in others. Figure 4.21 shows the distribution of Hispanics, blacks, Asians, and American Indians separately for 1990. The patterns are quite distinct. Despite decades of out-migration, blacks were still more heavily concentrated in the southeastern states than anywhere else. Hispanics, on the other hand, are located in basically three geographic nodes—those from Mexico are especially found in the southwestern states (Figure 4.22), those from Cuba are concentrated in Florida, and those from other Caribbean countries and from Puerto Rico are in Florida and the New York–New Jersey area. Asians were especially numerous along the Pacific Coast, although there is also a concentration in the New York City area. Mentally putting those four maps together reveals that in California, Florida, and New York we find the tremendous racial and ethnic diversity that is becoming manifest in the United States, but that is less apparent in other parts of the country.

The diversity of the population can also be seen by examining data on ancestry and on the foreign-born population in the United States. The 1990 census revealed (Table 4.3) that more Americans considered themselves to be of German origin than any other ancestry—58 million people (23% of the population). In and of itself, this is evidence of diversity since it was immigrants from England who were most instrumental in creating the nation. People who consider themselves to be of English origin numbered 33 million (13% of the population), but more people indicated Irish origin (39 million, or 15% of the total). This latter figure is particularly remarkable considering that the population of Ireland today is about 4 million; roughly ten times as many people in the United States considered themselves Irish as there are residents in Ireland.

The ancestry data in Table 4.3 also show that blacks accounted for the fourth most populous ancestry group, while Mexicans were seventh. It is noteworthy that nearly 9 million people said that their ancestry was American Indian, although only 2 million indicated that their "race" was American Indian. The census question on race asks individuals to name their specific tribal affiliation if they indicate that their race is American Indian. However, many people whose ancestry is at least partly American Indian are descended from tribal members who left a reservation and married outside of the tribe, thus blurring the lineage. It is an important sign of the times that people are proud to indicate their American Indian heritage. It has not always been so.

Ancestry may be a vague and imperfect concept, but place of birth is a more clear-cut expression of a person's roots. In 1990, more people living in the United States had been born in Mexico (over 4 million) than in any other foreign country, and most of these resided in California (Table 4.3). Indeed, people born in Mexico accounted for more than one in five (22%) of all foreign-born residents of the United States. Second on the list of countries of birth was the Philippines (900,000 Filipinos, residing largely in California), followed by Canada (spread geographically), Cuba (concentrated especially in Florida), and Germany.

In Canada the 1996 census showed almost five million people who were foreign-born, or about 17% of the population. In contrast to the United States, about 47% of these immigrants were born in Europe. About one-fifth had arrived in Canada before 1960, when most newcomers were still coming from Europe. The leading country of birth for immigrants was the United Kingdom, with 656,000 people, or about 13% of the total foreign-born population. By region, the other leading areas were Eastern Asia, 12%, Southern Asia, 7%, Central and South America, 6%, and Caribbean-Bermuda, 6%. Ontario was the leading place of residence for these immigrants by far, with 55% of the total. The second leading province was British Columbia, with a distant 18%, while Québec was third, with 13%.

When just recent immigrants are considered, the picture is very different. In fiscal 1997–1998, 59% of the roughly 200,000 immigrants to Canada came from Asia, with many countries represented. That same year only 21% of the newcomers came from Europe; a mere 10% of the European total was from the United Kingdom. An additional 8% of immigrants came from Africa, while 6% were from the West Indies and 3% from South America. These

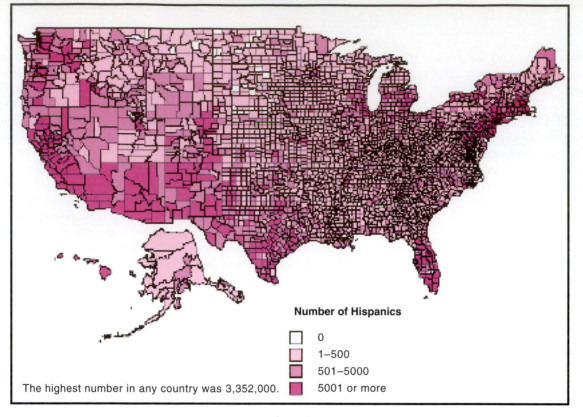

Number of Hispanics

- [ ] 0
- [ ] 1–500
- [ ] 501–5000
- [ ] 5001 or more

The highest number in any country was 3,352,000.

Figure 4.21a
Hispanic population of U.S. counties, 1990.

Source: Prepared by Decision Demographics from 1990 Census Summary Tape File 1C.

Note: Hispanics may be of any race.

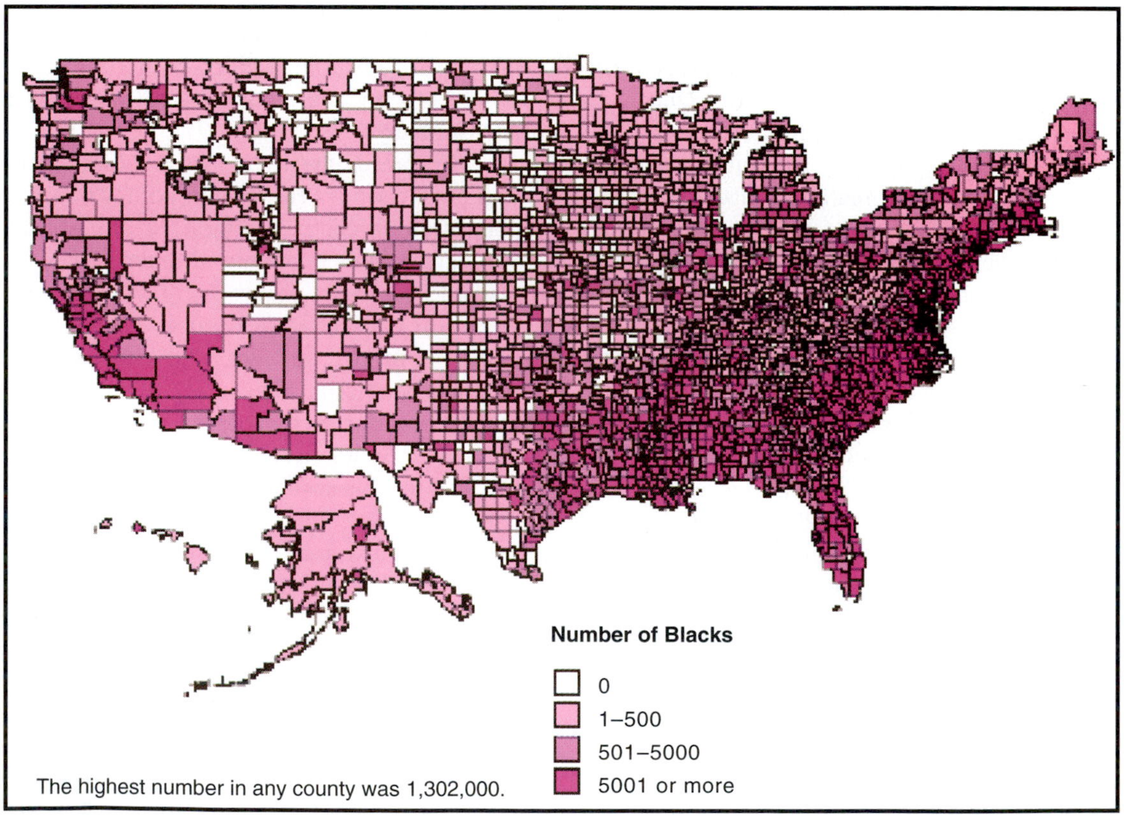

Number of Blacks

- [ ] 0
- [ ] 1–500
- [ ] 501–5000
- [ ] 5001 or more

The highest number in any county was 1,302,000.

Figure 4.21b
Black population of U.S. counties, 1990.

Source: Prepared by Decision Demographics from 1990 Census Summary Tape File 1C.

Note: Excludes Hispanics.

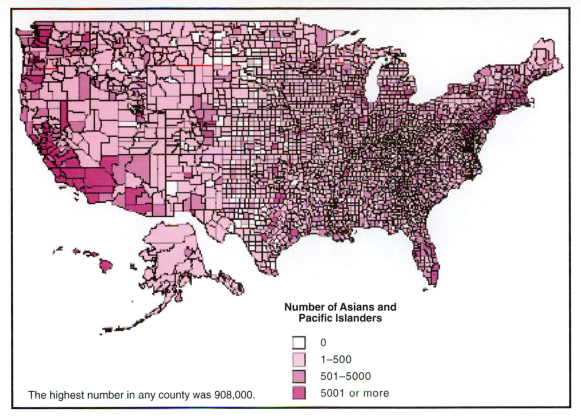

**Number of Asians and
Pacific Islanders**

- 0
- 1–500
- 501–5000
- 5001 or more

The highest number in any county was 908,000.

Figure **4.21c**
Asian and Pacific Islander population of U.S. counties, 1990.

Source: Prepared by Decision Demographics from 1990 Census Summary Tape File 1C.

Note: Excludes Hispanics.

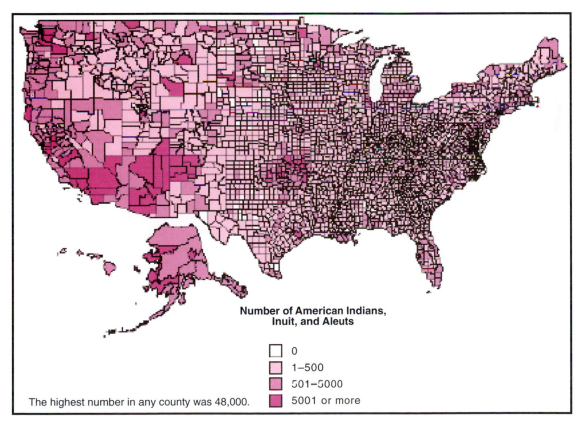

**Number of American Indians,
Inuit, and Aleuts**

- 0
- 1–500
- 501–5000
- 5001 or more

The highest number in any county was 48,000.

Figure **4.21d**
American Indian, Aleut, and Inuit population of U.S. counties, 1990.

Source: Prepared by Decision Demographics from 1990 Census Summary Tape File 1C.

Note: Excludes Hispanics.

Figure 4.22

The Hispanic population of many American cities is so large that sometimes signs appear just in Spanish. This banner in Orange County, California where a quarter of the population is Hispanic, advertises "Easy Financing, Immediate Delivery."

figures show that most immigrants to Canada are adding to the country's visible minorities.

While immigrants have been streaming into the region, the American Indian (First Nations in Canada) population has been rebounding from the catastrophic depopulation caused by European contact. It is likely that there were between 2 and 4 million indigenous people north of the Rio Grande when Columbus reached America in the late 15th century. His discovery was followed by one of the greatest demographic disasters in the history of the world, as the natives succumbed to Old-World diseases to which they had little or no immunity. (Remember that this was not a one-way street. European explorers took plenty of new diseases back home, from which Europe took a long time to recover.) This resurgence in American Indian and First Nations populations is adding to the ethnic mix of the continent.

Future Population and Ethnic Patterns

What is the likely future for Canada and the United States in terms of population and ethnicity? In both countries, the outlook is for fairly rapid population growth, a changing geography of people, and considerably greater ethnic diversity. In 1996, the United States' Bureau of the Census published a study that projected major trends for these three variables for the period 1995–2025. For the latter year, the Bureau predicted a national population of 335 million, which is about 59 million above the figure for 2000.

To simplify the study of the changing geography of population, the nation was divided into just four regions, the Northeast, Midwest, South, and West. Among these four, the West is expected to experience the fastest growth, more than twice the national average. For the South, the outlook to 2025 is growth at slightly higher than the national average. In contrast, the Midwest and Northeast should grow at less than the average rate. Together, the South and West will absorb 82% of the nation's population growth.

During 1995–2025, seven states, all in the South and West, are projected to gain more than 2 million people each. They are California, Texas, Florida, Georgia, Washington, Arizona, and North Carolina. These seven will account for 58% of the country's population growth. Just three of them, California, Texas, and Florida, are projected to account for 45% of the increase, a percentage only slightly lower than in the years immediately preceding 1995. Florida is expected to grow fast enough so that by 2015 it should replace New York as the third most populous state.

In terms of race and ethnicity, the Census Bureau projects that non-Hispanic whites will have the slowest growth rate, and that their numbers will grow by 16 million during the study years. The black population is projected to have the next lowest rate, which should lead to about 12 million more blacks in 2025 than in 1995; 64% of them are expected to live in the South. Asians are likely to experience the highest growth rate. This group should grow by 25 million (267%) by 2025, more than half of whom will be living in the West. In terms of absolute numbers, the

Table 4.3
Origins of Immigrants to the United States, 1990

Twenty Largest Ancestry Groups: 1990

Top Twenty-Five (Population Rankings of Foreign-Born Persons by Place of Birth: 1990)

1990 Rank	Ancestry	Number	Percent	Rank	Place of Birth	Population	Percent
	Total U.S.	248,709,873	100.0	Total	Foreign Born	19,767,316	100.0
1	German	57,947,374	23.3	1	Mexico (1)*	4,298,014	21.7
2	Irish	38,735,539	15.6	2	Philippines (7)	912,674	4.6
3	English	32,651,788	13.1	3	Canada (3)	744,830	3.8
4	Afro-American	23,777,098	9.6	4	Cuba (6)	736,971	3.7
5	Italian	14,664,550	5.9	5	Germany (2)	711,929	3.6
6	American	12,395,999	5.0	6	United Kingdom (5)	640,145	3.2
7	Mexican	11,586,983	4.7	7	Italy (4)	580,592	2.9
8	French	10,320,935	4.1	8	Korea (10)	568,397	2.9
9	Polish	9,366,106	3.8	9	Vietnam (12)	543,262	2.7
10	American Indian	8,708,220	3.5	10	China (11)	529,837	2.7
11	Dutch	6,227,089	2.5	11	El Salvador (28)	465,433	2.4
12	Scotch-Irish	5,617,773	2.3	12	India (16)	450,406	2.3
13	Scottish	5,393,581	2.2	13	Poland (8)	388,328	2.0
14	Swedish	4,680,863	1.9	14	Dominican Republic (19)	347,858	1.8
15	Norwegian	3,869,395	1.6	15	Jamaica (18)	334,140	1.7
16	Russian	2,952,987	1.2	16	Soviet Union (9)	333,725	1.7
17	French Canadian	2,167,127	0.9	17	Japan (13)	290,128	1.5
18	Welsh	2,033,893	0.8	18	Colombia (23)	286,124	1.4
19	Spanish	2,024,004	0.8	19	Taiwan (33)	244,102	1.2
20	Puerto Rican	1,955,323	0.8	20	Guatemala (39)	225,739	1.1
				21	Haiti (29)	225,393	1.1
				22	Iran (24)	210,941	1.1
				23	Portugal (14)	210,122	1.1
				24	Greece (15)	177,398	0.9
				25	Laos (42)	171,577	0.9

Source: From Census and You, *28(2): 1993. U.S. Bureau of the Census.*

*Note: 1980 rank in parentheses.

largest increase (although second in growth rate) by far will be among Hispanics; of the 72 million people added to the United States population in these 30 years, 44% (32 million) are expected to be Hispanics. Because of such growth, the Hispanic population is expected to pass the black population in about 2005 to become the country's largest minority group.

California, already virtually a laboratory for a population mix unprecedented at such a scale in world history, will become even more ethnically complex. Between 1995 and 2025 California should grow by 56%, adding about 17.7 million people, or nearly the 1995 population of New York state. About eight million of these new residents are expected to be immigrants, as the state will receive about one-third of all newcomers. California's Hispanic population should more than double to 21 million by 2025, at which time it will contain 36% of the national total of this ethnic group. It will also have far more Asian-Americans than any other state, with 41% of the total. While California contained 12% of the U.S. population in 1995, it is projected to have 15% thirty years later.

In 1995, New Mexico and Hawaii were the only states where non-Hispanic whites were not the majority. That year, New Mexico was 50% white, and Hawaii only

Measuring population growth requires that we have information from one or more censuses, as well as from vital statistics (births and deaths) and, if possible, data on migration. We can calculate a simple measure of the overall rate of population growth by comparing the results of two censuses. For example, the 1990 census of the United States counted 248,709,873 people, whereas the 1980 count had been 226,545,805. We can calculate the percentage change in population in the following manner:

$$\frac{248,709,873 - 226,545,805}{226,545,805} \times 100 = 9.8\%$$

We typically prefer to talk about the *rate of growth* on an annual basis, rather than on a per decade basis, however. Therefore, we would divide 9.8% by 10 (the number of years between the censuses) and get 0.98% per year as the average annual rate of growth of the U.S. population between 1980 and 1990.

Although the overall rate of growth is of considerable interest, we typically also want to know about the components of growth. How much of that increase was due to natural increase (the excess of births over deaths), and how much was due to the net effect of migration? Since we were expressing the annual growth as a rate, we can do the same with births, deaths, and migration in the following fashion.

The **crude birth rate** (sometimes called simply the *birth rate*) is the annual number of live births per 1000 population. In the United States, in 1991, for example, we find the following:

$$\frac{4,111,000 \text{ births}}{252,688,000 \text{ midyear pop.}} \times 1000 = 16.3 \text{ per 1000 population}$$

The **crude death rate** (sometimes called simply the *death rate*) is calculated in an analogous way.

$$\frac{2,165,000 \text{ deaths}}{252,688,000 \text{ midyear pop.}} \times 1000 = 8.6 \text{ per 1000 population}$$

Now we can find the **rate of natural increase** of the population by subtracting the crude death rate from the crude birth rate (*natural* means that increases and decreases due to migration are not included). Thus, in the United States in 1991, the rate of natural increase was $16.3 - 8.6 = 7.7$ per 1000.

30%. By 2000 they had been joined by California, where the percentage non-Hispanic white was about 48%, down from 53% in 1995.

At the start of 2000, nearly one U.S. resident in ten was foreign-born, or about the same percentage as in 1850. Historically, the percentage of foreign-born residents peaked in 1880, when it was 14.8 percent, and reached bottom in 1970, at just 4.7%. During the 1990s foreign-born residents increased nearly four times as fast as native-born. A census in 1998 revealed that the country contained 25.2 million foreign-born, or 9.3 percent of the population, versus 7.9% in 1990. That amounted to an astonishing increase of about 5.4 million, or 28%, in just eight years. The absolute growth was largest for foreign-born Hispanics, whose numbers grew by 34% during the same period, increasing from 8 million to 10.7 million. Among Asians the increase was from 4.6 million to 6.4 million, an even higher increase of 39%. In fact, by 1998 foreign-born Asians outnumbered native-born people of Asian descent.

Religion

The majority (nearly 60%) of Americans are at least nominally Protestant Christians. So pervasive is this that many people take that fact for granted and when asked about their religion they will respond with the name of their particular branch of Protestant Christianity, such as Lutheran or Baptist. Catholic Christians are the second most numerous, comprising about 25% of the U.S. population, Jews constitute about 2%, and Muslims make up approximately another 2%. The remainder of the population is divided among Mormons (who, though Christians, do not consider themselves Protestants), Hindus, Buddhists, and others, as well as the nearly 10% of the population that professes no religious affiliation.

Figure 4.23 shows the approximate geographic distribution of the U.S. population by religious preference. Note first that various Protestant groups dominate most of the country, and that perhaps half the United States has a considerable mixture of Protestants. The largest single denomination is the Methodists, who are particularly numerous in a sizable belt near the center of the country. Also noteworthy is the considerable Lutheran concentration in the north central section, centered in Minnesota. This area saw an especially large influx of immigrants from Scandinavia, and Lutheranism is the dominant form of Protestantism in all Scandinavian countries. The southeastern states, on the other hand, are dominated by the Southern Baptist and Methodist branches of Protestantism.

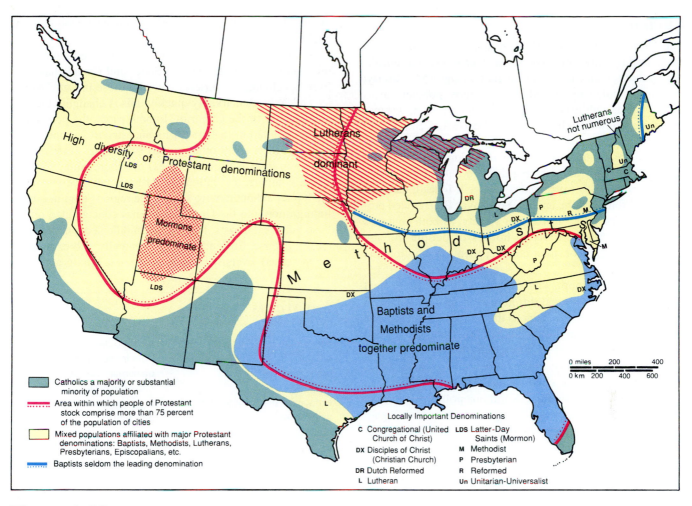

Locally Important Denominations

C Congregational (United Church of Christ)
DX Disciples of Christ (Christian Church)
DR Dutch Reformed
L Lutheran
LDS Latter-Day Saints (Mormon)
M Methodist
P Presbyterian
R Reformed
Un Unitarian-Universalist

Catholics a majority or substantial minority of population

Area within which people of Protestant stock comprise more than 75 percent of the population of cities

Mixed populations affiliated with major Protestant denominations: Baptists, Methodists, Lutherans, Presbyterians, Episcopalians, etc.

Baptists seldom the leading denomination

Figure 4.23
Religious affiliation in the conterminous United States.

Source: Redrawn from "Christian Denominations in the Conterminous United States" in *Historical Atlas of the Religions of the World,* edited by Isma'il R. al-Faruqi and David E. Sopher, Macmillan, New York, 1974.

The Catholic population has three noteworthy concentrations. One is in the Southwest, near the Mexican border, where the large Hispanic population is usually Catholic. Southernmost Florida, with its large Hispanic population, also has a cluster of Catholics. The northeastern concentration, on the other hand, is associated especially with the migration of large numbers of European Catholics to the area, particularly from Ireland, Italy, Poland, and Germany, although Puerto Ricans also add to that number. The Mormon concentration is focused on Utah, but extends as well into adjoining states, especially southeastern Idaho. Large numbers of Mormons are also found in the major cities of California.

A good deal of demographic attention has focused on the comparison of birth rates between Catholics and Protestants. During the past century, Catholics have tended to have more children than have Protestants. Recent evidence from both the United States and Canada suggests that these differences have virtually disappeared in the aggregate. Re-

ligion, race, and ethnicity are intimately bound up with one another, however, and the data do suggest that the highest fertility levels in the United States right now are found among Hispanic Catholics, whereas non-Hispanic whites with no religious preference have the lowest fertility. The data show that among whites, Protestants now actually have slightly higher fertility levels than do Catholics. As mentioned earlier, in Canada, fertility in Québec (predominantly Catholic) is lower than in any other province.

Religious beliefs can also lead to higher than average fertility. The Mormon religion encourages large families, and these "Latter Day Saints" have a much higher than average birth rate. Mormons dominate the social, political, and demographic fabric of Utah, explaining the facts that (as noted earlier) Utah's birth rate is the highest among any of the states, and that the city of Provo, which has especially many young adults as it is the home of Brigham Young University, has the highest birth rate of any U.S. city.

Summary

The United States is the third most populous country on earth, trailing only China and India. This is a rather remarkable achievement, considering that in the 15th century, China and India were already the two most populous areas of the world, with tens of millions of people each, whereas the population of what became the United States and Canada was only a few million indigenous people. The two countries are a source of constant attraction to international migrants, but they are growing through natural increase— the excess of births over deaths—as well. Indeed, the baby boom that followed the end of World War II generated profound changes in the structure of society, as both countries had to adjust sequentially to hordes of people moving through grade school, then college, then into the job and housing markets.

The baby boom was followed by the baby bust (1965–1977), but regional differences in fertility persist. It is highest in those areas, such as the Southwest, that have a fairly high proportion of new immigrants from Mexico and Asia, although regions dominated by Mormons actually have the highest fertility. Farm women in the United States continue to have more children than nonfarm women, but there are few farm women. Nonmetropolitan women (a broader category than those living on farms) have higher fertility than women living in metropolitan areas, but within metropolises, fertility is lower in the suburbs than in the central cities.

Migration patterns within both the United States and Canada generally have been westward over time. Patterns in the U.S. are complicated by the rural-to-urban migration out of the South before the 1950s, followed by a migration into the South out of the Northeast in more recent decades.

In the United States, the years after 1950 saw a substantial shift of the population towards amenity locations, especially the Sun Belt. This trend has been facilitated, in part, by three developments. An increasing number of people are retiring with respectable incomes, and such people have a great choice in where to live. Many of them have chosen amenity locations in the Sun Belt, with Florida by far the leading choice. In Canada, British Columbia is seen by many as an amenity province, with similar consequences. The rise of quaternary activities has also favored the Sun Belt. Many such firms can locate just about wherever they can plug into the Web, and some of them have taken advantage of this freedom to move south. Finally, high-tech manufacturing has contributed to Sun Belt growth.

Because of economies of agglomeration, most people in both the United States and Canada live in cities. About three-quarters of the population north of Mexico is classified as urban; within that category, the majority live in metropolitan areas of more than one million people. Despite distinct diseconomies that come with larger size, the populations of both countries are increasingly concentrating into the few largest cities, those with the greatest economies of agglomeration.

In contrast to earlier times, in recent years most immigration into both the United States and Canada has been from developing countries, especially those in Latin America and Asia. Remarkably quickly, this is changing the ethnic diversity of the countries. In both cases, the percentage of the population that traces its ancestry to Europe is decreasing, even as all other groups (including American Indians, or First Nations peoples) are increasing. If these trends continue, the outlook is for great ethnic diversity in the future, although more so in some regions than in others. Immigration into the United States has concentrated on the East and West Coasts, leaving the middle sections less ethnically diverse. Still, all regions eventually will experience the increased diversity brought by the new immigrants.

Key Words

amenities	megalopolis
baby boom	metropolitan statistical area
bracero	primary activities
crude birth rate	quaternary activities
crude death rate	rate of natural increase
economies of agglomeration	secondary activities
gatekeeping	Sun Belt
high-tech manufacturing	tertiary activities
infant mortality rate	zero population growth
life expectancy	

Gaining Insights

1. Generalize about the population geographies of the United States and Canada in the year 2000.
2. Where is the Manufacturing Belt? Megalopolis? Main Street? How important are these areas in terms of population today versus in 1950? Why?
3. Explain the term "economies of agglomeration." Apply the concept to your local area.
4. Define MSA. What are the strengths and weaknesses of this concept?
5. What are amenities? What are the amenities of your area? How competitive is your area in terms of amenities for attracting businesses and people? Explain.
6. How has the growth of high-tech industry reinforced the rapid growth of the Sun Belt?
7. What are quaternary activities? Why have they contributed to population growth in the Sun Belt?
8. Which parts of the United States have the highest fertility rates? Why?
9. Where is infant mortality highest in the U.S.? Why there?
10. Why has illegal immigration become such a large problem for the United States? Why are illegal immigrants so difficult to count?

11. How have the ethnic and racial compositions of the United States and Canada changed since 1950? Where in the United States is racial and ethnic diversity most evident? Why there? Where do you think it would be most evident in Canada? Why?

12. Generalize about the geography of religion in the U.S. today.

Selected References

Ahlberg, D. and C. De Vita. "New Realities of the American Family," *Population Bulletin* 47, no. 2 (1992).

Bean, Frank, W. Parker Frisbie, Edward Telles, and B. Lindsay Lowell. "The Economic Impact of Undocumented Workers in the Southwest of the United States." Chapter 12 in *Demographic Dynamics of the U.S.–Mexico Border,* edited by John R. Weeks and Roberto Ham-Chande. El Paso: Texas Western Press, 1992.

Beaujot, R. "Canada's Population: Growth and Dualism," *Population Bulletin* 33, no. 2 (1978).

Bianchi, S., and E. McArthur. "Family Disruption and Economic Hardship: The Short-Run Picture for Children." *Current Population Reports,* Series P-70, no. 23 (1991).

Bouvier, L. and C. De Vita. "The Baby Boom—Entering Midlife." *Population Bulletin* 46, no. 3 (1991).

Campbell, Paul R. *Population Projections for States by Age, Sex, Race, and Hispanic Origin: 1995 to 2025.* U.S. Bureau of the Census, Population Division, PPL-47, 1996.

Dawson, D. "Family Structure and Children's Health." *Vital and Health Statistics,* series 10, data from the National Health Survey, no. 178 (1991).

London, K. "Cohabitation, Marriage, Marital Dissolution, and Remarriage: United States, 1988." Advance data from *Vital and Health Statistics* of the National Center for Health Statistics, no. 194 (1991).

National Center for Health Statistics. *Vital Statistics of the United States, 1988.* Vol. 2—Mortality, part A. Hyattsville, MD: National Center for Health Statistics, 1991.

Population Reference Bureau. This office has a wide range of information on the Web (www.prb.org) on such topics as health, mortality, migration, marriage and race/ethnicity.

Snipp, C. Matthew. *American Indians: The First of This Land.* New York: Russell Sage Foundation, 1989.

U.S. Census Bureau. *Statistical Abstract of the United States, 1998.* Washington: U.S. Census Bureau, 1998. On the Web at www.census.gov/statab/www/.

U.S. Department of Health and Human Services, National Center for Health Statistics, *Monthly Vital Statistics Report.* The Web address (www.cdc/nchs/) has much information on such topics as births, deaths, marriage, divorce, and life expectancy.

Warren, R. and J. Passel. "A Count of the Uncountable: Estimates of Undocumented Aliens Counted in the 1980 United States Census," *Demographcy* 24, no. 3 (1987).

Weeks, J. *Population: An Introduction to Concepts and Issues.* 7th ed. Belmont, CA: Wadsworth, 1998.

Americans and Canadians have an intense interest in politics, as evidenced by the amount of time devoted to the topic on news programs and in the space devoted to it in newspapers and news magazines. In both countries, political science is a popular university subject, and it is required under some curricula. Political scientists study a wide range of topics, such as the process of forming new political parties, the courts' changing interpretation of "one person, one vote," and the voting behavior of minority groups. Other topics they may find of interest include the changing fortunes of political parties, the international relations of countries, and the development of the post–Cold War "world order."

Traditionally, political geographers focused on geography-based power relations among countries, a subfield called *geopolitics*. They studied such subjects as the resource bases of various historic European power blocs or the emergence of Japan as a world power in the first part of the 20th century. In this chapter, we pay little attention to geopolitical relations between Canada and the United States. Rather, we emphasize political-geographic developments within these countries, especially the United States. Additional materials on Canadian political geography are found in Chapter 14, "The Canadian Difference."

How is political geography defined? We will use the following broad, yet simple, definition: political geography examines the interaction between location and political activity. Under this definition, we could include such topics as which geographic areas supported the 1996 Ross Perot presidential candidacy, the changing geography of support for the Liberal Party in Canada, the way different states have been subdivided into districts for elections to the House of Representatives, or differences among states in voter turnouts for presidential elections (Figure 5.1). In all cases, the emphasis is on addressing the "where" and the "why there" questions.

In this chapter, we will take a geographic look at American and Canadian (1) governance and (2) politics. That is, we investigate how space in the two countries is organized for governance, and some geographic patterns of political behavior. In the process, we assert that *where* people grow up, work, and live, can profoundly influence their views on issues. In both the United States and Canada, then, place often plays a large role in election outcomes and in the development of public policy.

States and Nations

Before proceeding, we need to define two terms. This is important because political geographers (and political scientists) use the words "nation" and "State" quite differently from the way they are used in everyday language.

A **State** is a political unit that occupies a precisely defined, permanently populated territory and that has full control over its internal and foreign affairs (i.e., it is independent). As recommended by the United Nations, the term is capitalized in order to distinguish it from "state," which is a lower-order political unit (such as the 50 states of the United States). The State is the dominant form of political organization in the world today, and almost all of the earth's land area is included in such units. The only major exception is Antarctica; it has neither a permanent population nor an established government, so is not a State. Nevertheless, parts of Antarctica are claimed by several countries. Colonies, which were once common but only a few of which remain today, are also not States. As used in this chapter, "State" is synonymous with "country."

Political geographers use **nation** to refer to a reasonably large group of people with a common culture that occupies a particular territory, bound together by a strong sense of unity arising from shared beliefs and customs. Thus, nations share one or more important cultural traits

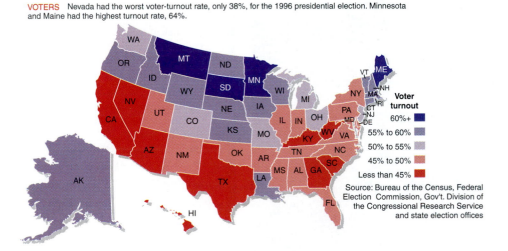

VOTERS   Nevada had the worst voter-turnout rate, only 38%, for the 1996 presidential election. Minnesota and Maine had the highest turnout rate, 64%.

Voter turnout

- 60%+
- 55% to 60%
- 50% to 55%
- 45% to 50%
- Less than 45%

Source: Bureau of the Census, Federal Election Commission, Gov't. Division of the Congressional Research Service and state election offices

Figure 5.1
State-by-state variations in voter turnout for the 1996 presidential election. Minnesota and Maine had the highest turnout rate, 64%, and Nevada the lowest, 38%.

Source: Time Magazine *Jan. 1, 2000.*

such as religion, language, history, values, and political institutions. They have an attachment to a particular territory, and their identity is intimately associated with that territory. A multicultural State consists of several nations; India is a good example of such a country. Some nations, such as the Kurds of the Middle East, do not have their own State, but are scattered across several countries. The Palestinians are a nation, but only now are they again beginning to have a State of their own.

The **nation-State** is the ideal form to which most nations and States have traditionally aspired. Here, the territory of the State is occupied by only one distinct nation or people, so there are no important minority groups. Japan, Denmark, and Poland are often used as good examples of nation-States. This arrangement minimizes conflict and makes for strong States, for tensions between different nations within the same country are common. On the other hand, there are strong States even with sizable minorities; a key feature of such States seems to be that there are safe ways for minorities to express themselves, perhaps because of the presence of decentralized political organization that allows their culture and traditions to be maintained.

Canada, with its English- and French-speaking populations, is often called two nations within one State. As we defined these terms, this is true, as the two groups have distinct cultures. One might also ask whether the original inhabitants of Canada should be identified as separate nations. Canada's use of the term "First Nations" to refer to indigenous groups would argue in favor of this. Not all political geographers would agree, however, as that depends on how one defines the "reasonably large group of people" that is required of a nation.

Centrifugal and Centripetal Forces

Some States are stronger than others. That is, some countries have the loyalty of almost all of their people, while others are fragmented and weak. What makes for strong and weak States?

Introducing two terms, centripetal and centrifugal forces, can help to answer this question. In political geography, **centripetal forces** are those that bind a country together, the unifying factors (such as language and religion) that give people a shared, positive vision of what their country is all about. Normally, they include a strong sense of shared history and values. Importantly, they almost always include a shared language. Centripetal forces can be reinforced by external aggression, which is often highly effective in unifying a population.

Historically, the United States has had strong centripetal forces, such as shared beliefs in the ideals of democracy and economic freedom, which have united the country, especially at critical times. These positive forces helped the State to survive such severe tests as early doubts about the country's viability, a devastating Civil War, and the disruptions caused by enormous immigration, yet eventually to emerge as a world power. They include such iconography and political pageantry as the flag, the recitation of the Pledge of Allegiance, celebrating the Fourth of July, and inspiring stories and myths about George Washington and other national heroes. Also important are a sense of shared history and struggle, such as World Wars I and II. Centripetal forces are fostered by an education system that socializes children into a national ideology, and mass media, which inculcate the population into believing in the essential rightness of the government in power.

Centrifugal forces, of which the most powerful tend to be the presence of more than one language or religion, work in the opposite direction. That is, they tear a country apart. Another powerful force of this type is perceived political and economic inequality, such as when the economically disadvantaged feel that they do not have access to decision makers, that their aspirations are ignored, or that they are being exploited by the ruling class. Centrifugal forces can also be strong when the central government is identified in people's minds with a different ethnic or cultural group, and when economic gains are expected to flow from separation. Such negative forces are often strongest in the outlying regions (peripheries) of the State, which are frequently areas of concentrations of minority populations.

Some political geographers argue that, in recent decades, centrifugal forces have been gaining in the United States. One factor that weakened the country was the Vietnam War, with very strong feelings on the parts of both "doves" and "hawks"; for several years the State was deeply split, and the damage has not yet been fully repaired. The continuing controversy over abortion (largely a non-issue in Canada) has also deeply divided the population; it seems impossible to find a compromise between such diametrically opposed, and strongly held, points of view. Since 1970, the influx of large numbers of Spanish-speakers into various regions has introduced another potentially significant centrifugal force. Until recently, the virtual universality of English was a strong unifying factor. Many of the newcomers seem intent on keeping Spanish as their primary language. If sustained, this trend could lead to a split that could destabilize the country.

Historically, nation-building forces in Canada have been weaker than in the United States. As several authors have pointed out, perhaps the strongest centripetal force in Canada has been a negative one; Canadians do not want to be Americans. As one well-known Canadian put it, in Canada such negative feelings about the United States comes with mother's milk. This attitude probably originated at the time of the American Revolution, when many loyalists moved to Canada to remain under the British Crown. The presence of such a confident, brash, and highly populated nation next door certainly contributed to discomfort, resentment, and a sense of alienation. Canadians fear both cultural and economic domination by their much larger neighbor. One way they have attempted to counter this is to limit the amount of American programming that can appear on Canadian television.

Such attitudes are reinforced by the evident fact that Canadians are, generally, more socially conservative than Americans. A hint of this difference can be seen by comparing the generally polite and self-controlled crowds at Toronto Blue Jays baseball games with the boisterous (or worse) crowds at places like Yankee Stadium (New York) or Wrigley Field (Chicago). Although Canada's negative attitude has decreased somewhat in recent times, its persistence makes it difficult to take seriously the occasional discussions in the media about a possible merger between the two States.

Language is Canada's strongest centrifugal force, for about one-fourth of the population considers French, not English, its mother tongue. Many French Canadians see themselves as a nation separate, and strongly advocate the establishment of an independent French republic in North America, based on the Province of Québec. Certainly they do not identify with the British Crown, which has acted (at least in the past) as a centripetal force among substantial numbers of Canada's English-speakers. So far the separatist movement has not succeeded, in part because other French Canadians feel that their identity is safe within the country's federal framework, and that secession would bring economic and political disadvantages.

Canada's population distribution also works against a strong, positive sense of "Canadianess." In particular, it largely consists of four or five separate areas of settlement, all of which have easier access to nearby U.S. regions than to other parts of Canada. Not surprisingly, these widely scattered population clusters often see their self-interests quite differently. The Western Provinces, for example, with their emphases on agriculture and natural resources, may well find it difficult to identify with the people of distant Ontario, to say nothing of French-speakers in Québec. All these people, in turn, may see little in common with the distant Atlantic Provinces of New Brunswick, Prince Edward Island, Nova Scotia, and Newfoundland.

Unitary and Federal States

The world's countries can be divided into two broad categories, the **unitary** and the **federal States.** The word unitary derives from Latin *unitas,* meaning unity, which implies the oneness of the State and a high degree of internal homogeneity and cohesiveness. Unitary States are organized around a single political center, the national capital; the whole country is under the direct control of the central government.

All States are divided for administration into units, so that a local government can deal with local matters. In unitary States, the subnational governments (like provinces and counties) owe their very existence and authority to the national government. The latter creates these administrative units and determines their number, boundaries, authority, and operations. In many unitary States, even the first-order civil divisions (the largest general-purpose administrative

or government subdivisions within a State, such as states and prefectures) have no legislative or judicial functions and little decision-making power. They operate as administrative districts, designed to make the workings of the central government easier and more effective. The national government generally finances them, largely or completely, and appoints their chief officials. A unitary State can be a monarchy, democracy, or dictatorship in form of government.

France is an excellent example of a unitary State. Here the first-order civil divisions, the provinces, have governors that are appointed in Paris. They have very limited powers, other than to carry out the laws mandated by the national government. In such States, services like health, education, and police are carried out uniformly throughout the country.

Unitary States tend to be small and/or culturally homogeneous. Most Arabic, Latin American, and African countries have such centralized arrangements. Other examples include the United Kingdom, Sweden, Japan, and New Zealand. The largest unitary State today is China. There are far more unitary than federal governments in the world. In 2000 there were more than 170 unitary States, but fewer than two dozen with federal structures.

In federal States, the responsibilities of government are divided formally between the central authorities in national capitals and lower levels of government. The word "federal" derives from the Latin word for league, and implies alliance, contract, and the coexistence of the State's diverse regions and peoples. Under this system, then, there are a variety of power centers. In a weak federation (such as Switzerland) the power of the constituent states (Cantons) is large and that of the central government is small. In a country like Germany, with a strong federal government, the power tends more toward the center at Berlin. Both Canada and the United States have federal structures, as do, for example, Mexico, India, and Brazil.

Under a federal framework, the central government represents the first-order divisions within the State where they have common interests, such as in defense, foreign affairs, and communications. At the same time, this system allows the various first-order units to have their own identities, laws, policies, and customs in certain fields. In Canada and the United States, these first-order entities are the provinces and the states, and they have considerable powers (Figure 5.2). The power to tax is probably the single most important measure of "sovereignty" at this level. The powers of states and provinces have been exercised in many different ways; thus there are considerable variations among them in such things as divorce laws, minimum driving ages, educational systems, environmental regulations, certification requirements for teachers, and motor vehicle codes. In Canada, Québec has taken advantage of this latitude to foster a separate French-language-based society, which has (among other distinctive features) a legal code based on the Napoleonic model rather than on English

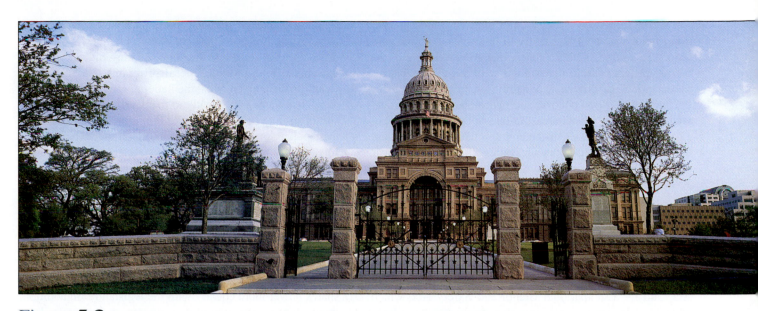

Figure 5.2
In the United States and Canada, the provinces and states have considerable political power. In Texas, much of that power is projected from the state capitol building at Austin.

common law. Theoretically, large and very large States, as well as heterogeneous countries, are especially well-suited to a federal structure.

Today, there seems to be a tendency for unitary States to move towards a greater decentralization of power. Ironically, at the same time many federal Sates, like the United States, are going in the opposite direction, towards a greater centralization of authority. In fact, some federal States are so highly centralized they function almost like unitary States.

The U.S. and Canada as Federal States

Under their respective constitutions, the responsibilities of governance in the United States and Canada are divided between the central authorities in Washington and Ottawa, and the provincial and state governments. The latter, in turn, delegate powers to lower levels of government, such as to municipalities and counties. Thus, the powers of cities and counties depend entirely on the role assigned to them by the various states. In a sense, then, states and provinces behave somewhat like unitary States.

When the United States was formed in the late 18th century, considerable differences among the former colonies made a unitary State impossible and undesirable. At the time, federation was seen as the solution, as it could preserve unity while allowing considerable diversity.

The United States Constitution, the basic framework for American democracy, granted a number of executive, legislative, and judicial powers to the national government. For example, it assigned foreign affairs, the coining of money, and the national defense, to Washington. The Tenth Amendment to the Constitution says that those powers not specifically given to the federal government are reserved "to the states, respectively, or to the people." Among the most important of the latter are education, police powers, and health care. Within their boundaries, states are free to do as they please, provided they do not violate the federal Constitution, federal laws, or treaties. Conflicts between national and subnational governments are resolved by the Supreme Court.

The situation in Canada is the reverse. Here the Constitution assigned specific powers to the provinces, and all the powers not so assigned are reserved for the central government. This suggests that the government in Ottawa is much more powerful that its counterpart in Washington, but this is not the case. Why Canada's central government is not stronger is a complicated issue, and a full answer is beyond the scope of this chapter. However, it is evident that the provinces gained power as the central government found this the only way to reconcile the conflicts among them. Further, the British North America Act of 1867, which created Canada, gave the provinces the power over "property and civil rights," a broad term that seems capable of almost endless enlargement. There have been numerous federal-provincial conflicts over mineral rights, fishing rights, land, and other "properties."

Originally, the American government in Washington, D.C., was weak, but the powers exercised from there have increased greatly over time. This change began in earnest during the 1930s, when the Roosevelt Administration started to move towards "big government," called the New Deal, as a way to get the country out of a severe depression.

The Tennessee Valley Authority (TVA), an organization founded in the 1930s to try to improve economic conditions in certain southern states, is a good example of the increased federal role that started with the Roosevelt administration. In this case, Washington invested in water transportation, flood control, and hydro-electric power, and later in many additional areas, in one of the most economically depressed regions of the country. The states were not necessarily enthused about federal encroachment on their soil, but economic conditions improved rapidly and the TVA remains active to this day.

After 1960, Washington also began to enact many laws regulating state and local activities. That is, the federal government began to assert increasing authority over environmental issues, education, health and civil rights, and other spheres formerly under state control. This legislation included the Coastal Zone Management Act, the Clean Air Act, and the Occupational Health and Safety Act, which addressed emerging problems that could not have been foreseen by the writers of the Constitution.

Simultaneously, state and local governments began to depend increasingly on federal revenues, such as those for highways. These funds often came with considerable federal strings attached, increasing the power of the national government. In one case, when Washington wanted a nationally uniform minimum drinking age of 21, it made the granting of federal highway funds to states contingent on compliance. Despite some grumbling, all states with lower drinking ages soon fell into line, as they did not want to lose their funds. Many federal programs use "matching funds" to accomplish what the national government wants.

The states vary greatly in many areas, such as in their administrative structures. In addition, certain states exercise their police, or regulatory, powers much more than others. In the past, for example, a few states closely regulated trucking and air transportation within their borders, but others practiced little or no regulation.

Such state-to-state differences are nicely illustrated in the amount of power given to governors. Some allow the governor to appoint important state officials, while in other states these same officers are elected or appointed by boards. Certain states give the governor the dominant power in setting their budgets, while in others he or she shares power more equally with the legislature. In addition, there are states that permit their governors to exercise line-item vetoes, a provision which greatly increases the executive's power over the state budget. This variability in the roles of governors is illustrated in Figure 5.3, which shows the geography of their powers.

As implied by the above, state constitutions differ considerably. They all tend to be lengthy, however, because they spell out in detail how government powers are to be limited.

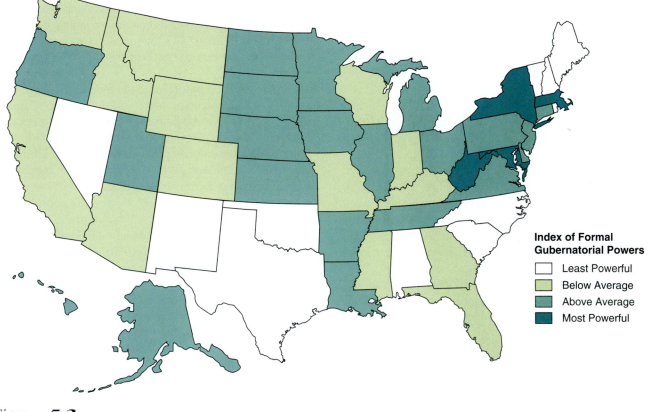

Index of Formal Gubernatorial Powers

- Least Powerful
- Below Average
- Above Average
- Most Powerful

Figure 5.3
The geography of the distribution of formal gubernatorial powers.

Local Governments

There are three levels of government in the United States and Canada. Below the national and state levels, in the United States, there are more than 80,000 units of local government. All such governments share a number of features. Each has jurisdiction over a specific area, and most have the authority to levy taxes in order to provide services.

Under the American federal system, local governments have rights bestowed on them by state governments. Legally they are creations of the state, and any change in their authority requires state approval. There are four major categories of local government; these are counties, townships, municipalities, and special districts. As creations of the various states, these lower-order civil divisions vary considerably in terms of such variables as taxing authority, other powers, and structure.

Among the United States' best-known general-purpose governments are the more than 3,000 counties (called parishes in Louisiana). They are the basic units of local government below the state level, and most of the United States (exceptions include the District of Columbia and Alaska) is covered by them.

The functions of counties are assigned to them by the states. These functions can vary considerably. Traditionally, the more rural counties have been concerned primarily with such basic services as law enforcement, the local administration of justice, building and maintaining highways, supervising county property, assessing and collecting taxes, recording vital statistics and legal documents, directing welfare, and conducting elections. Today, many densely populated suburban counties have become virtually urban governments in both functions and organization, providing a wide array of additional services like libraries, sewage disposal, water supply, flood control, and fire protection.

The states vary greatly in the number of counties they have within their borders. The range is from three in Delaware to 254 in Texas. In area they vary from 62 square kilometers (24 square mi) to 52,000 square kilometers (20,000 square mi), the latter in San Bernardino County, California. That county is large enough to encompass the four smallest states plus the District of Columbia. Alaska has boroughs, whose functions resemble those of counties in other states. Connecticut and Rhode Island have deprived their counties of all functions, and they survive only as statistical and judicial units.

In New England the original unit of settlement was the Town, and today it performs many of the functions that counties do in other parts of the country. The Town system spread south to the Middle Atlantic states and then west to the Midwest and even to parts of the state of Washington, so today there are nearly 17,000 such units nationwide. However, outside of New England, New York, and Wisconsin, they are called townships.

Many townships were established in conjunction with the rectangular land survey system under the Land Ordinance of 1785 (see Chapter 3). They form squares six miles on a side, and so consist of 36 square miles. Today such townships exist in 21 states. There is great variation among the states in the powers and activities of townships. In some they are atrophying and may eventually disappear.

Special purpose districts, which are created to serve specific public functions, are the most numerous form of local authority. Nowhere are special purpose districts so varied and prolific as in the United States. Historically, such districts were created in rural areas to administer drainage, irrigation, flood control, fire protection, soil conservation, mosquito abatement, and the like. Since World War II they have grown phenomenally within urban areas, and newer ones are largely associated with such urban needs as airports, libraries, law enforcement, housing, hospitals, courts, air and water pollution control, and utilities. Commissions or boards usually operate them. Currently, there are about 30,000 such districts. They are unevenly distributed, with California and Illinois having the largest number, and Alaska and Hawaii the fewest (Figure 5.4).

The most numerous special districts are the approximately 14,000 school districts. Several decades ago there were many thousands of additional such units, but because of the advantages of consolidation, they have been merged into so-called unified school districts. Despite cost reductions from consolidation, education still accounts for nearly half the expenditures of local government and 60 percent of all state aid to local government units. The biggest push towards consolidation has been the great increase in the cost of building and running schools; most abolished districts were unable to finance the needed expansions and improvements. Typically state governments give their school districts considerable administrative, fiscal, and curricular freedom. Most are administered by unsalaried, elected boards.

Why don't other minor civil divisions perform the functions of special districts? Sometimes they lack the resources or do not cover the needed area, their expansion may be restricted by law, their record of quality of service may be poor, or special interests may want new districts established to get the best results. Such districts can hire experts and are generally nonpolitical, although many are run by elected officials. Most have the independent power to tax, borrow, or levy service charges. In some states, such as California, counties serve as tax collectors for special districts.

The smallest general-purpose administrative units in the United States are the cities, or municipalities. There are more than 19,000 such units, ranging in population from fewer than 100 people to 7 million in New York City.

Legally, incorporated cities are "municipal corporations" that are established through a state charter that grants them powers of local self-government. Thus, they have the power to enact ordinances or laws that are in effect only within their borders. Such laws are usually uncontroversial,

Political Geography

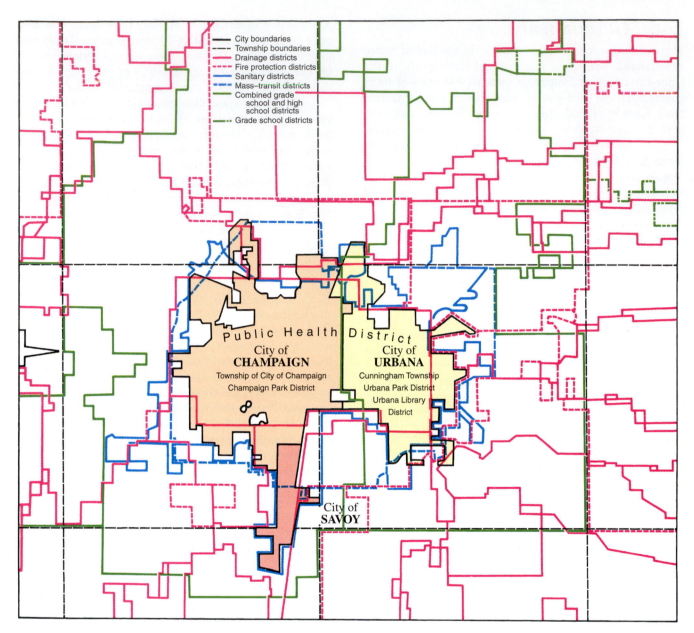

Figure 5.4
Political fragmentation in Champaign County, Illinois. The map shows a few of the independent administrative agencies with separate jurisdictions, responsibilities, and taxing powers in a portion of a single Illinois county. Among the other such agencies forming the fragmented political landscape are Champaign County itself, a forest preserve district, a public health district, a mental health district, the county housing authority, and a community college district.

but in recent years some cities have tried to enact ordinances in nontraditional areas, such as trying to regulate ATM charges. These attempts have been challenged in the courts and are unlikely to be upheld.

The traditional way to provide urban services to people in an urbanizing area around a city has been **annexation,** the incorporation of additional areas into a city. Through annexation, many cities' boundaries have changed dramatically. This is only true, however, in those states that grant their cities considerable powers of annexation. Such

actions tend to be highly politicized, especially when cities want to annex high-income areas. Some states like Texas and Oklahoma grant broad powers of annexation, but in others, especially in the Northeast and the Great Lakes states, the power to annex is very limited. Frequently, suburban areas incorporate as cities specifically to avoid annexation to the central city.

There are three types of municipal government in the United States. Under the mayor-council system, used by most large cities (over 80% of those with half a million or

TOO MANY GOVERNMENTS

If you are a property owner in Wheeling Township in the city of Arlington Heights in Cook County, Illinois, here's who divvies up your taxes: the city, the county, the township, an elementary school district, a high school district, a junior college district, a fire protection district, a park district, a sanitary district, a forest preserve district, a library district, a tuberculosis sanitarium district, and a mosquito abatement district.

Lest you attribute this to population density or the character of Cook County politics, it's not that much different elsewhere in Illinois—home to more governmental units than any other state in the United States. According to late-1980s figures from the U.S. Bureau of the Census, there were 6626 local government units in Illinois. Second-place Pennsylvania had 4956 governments—1670 fewer than Illinois—and the average for all states was 1663.

Along with its 102 counties, Illinois has nearly 1300 municipalities, more than 1400 townships, and over 1000 school districts; but the biggest factor in the governmental unit total are single-function special districts. These were up from 2600 in 1982 to nearly 2800 in 1987, with no end to their increase in sight. Special districts range from Chicago's Metropolitan Sanitary District to the Caseyville Township Street Lighting District. Most of these governments have property-taxing power. Some also impose sales or utility taxes.

This proliferation is in part a historical by-product of good intentions. The framers of the state's 1870 constitution, wanting to prevent overtaxation, limited the borrowing and taxing power of local governments to 5% of the assessed value of properties in their jurisdictions. When this limit was reached and the need for government services continued to grow with population, voters and officials circumvented the constitutional proscription by creating new taxing bodies—special districts. Illinois' special districts grew because they could be fitted to service users without regard to city or county boundaries.

Critics say all these governments result in duplication of effort, inefficiencies, higher costs, and higher taxes. Supporters of special districts, townships, and small school districts argue that such units fulfill the ideal of a government close and responsive to its constituents.

Adapted with permission from J.M. Winski and J.S. Hill, "Illinois: A Case of Too Much Government," *Illinois Business,* Winter 1985, pp. 8–12. Copyright 1985 by Crain Communications, Inc.

more people), the mayor is elected citywide, while the council members typically are elected by wards or council districts. In many cases, other officials, such as the city attorney, assessor, and auditor, are also elected. The mayor, in what is often called a "strong mayor" system, is responsible for appointing and supervising the heads of departments (such as trash collection and street repairs), which provide services to the city. This type of city government is found especially in the Northeast and the Midwest.

The commission and council-manager systems were the result of a reform movement that was active early in the 20th century. Under the latter, voters elect a city council which, in turn, selects a mayor from its membership and appoints a city manager. Sometimes the mayor is elected separately, but usually these are weak mayors whose functions (other than as a part of the city council) are largely ceremonial. The manager runs the day-to-day affairs of the city, while the council passes ordinances and makes other political decisions. About half of all U.S. cities have adopted the council-manager plan. Typically, medium-sized cities prefer this approach.

Under the city commission plan, a board of commissioners is elected, and they are directly responsible for the administration of city services. Only about three percent of U.S. cities have adopted this form of government.

Are there too many units of local government in the United States? Many social scientists think so, especially in the largest metropolitan areas, where there may be hundreds of such units. Here, they argue, the resulting fragmentation of government impedes the efficient providing of services and the making of sound regional policies.

Those who feel this way sometimes advocate metropolitan government. That is, in some areas a two-tier approach to government is advocated, where a metropolitan government performs some area-wide functions, while pre-existing cities perform more local functions. Thus, a metropolitan government handles some needs, such as public transportation and pollution control. The latter, presumably, can provide these services more efficiently than if they were spread over many governments. Other functions, however, such as control over land uses, are retained by the individual cities within the metropolitan area. The retention of such land use controls by local governments can, however, make it difficult to find sites for certain unpopular, but necessary, region-wide needs, such as jails, dumps, and sewer plants. Some critics of metropolitan government point out that the present fragmented system has several advantages, including better citizen access to decision-makers, and greater citizen influence on the decisions of local government.

Metropolitan government has not found much support in the United States, but it has been more successful in Canada. The prototype was the Municipality of Metropolitan Toronto, which was created in 1954. Metro Toronto provides some services for the whole region, such as water, sewage, police, public transit, and major roads. It also funds and provides day care, ambulance services, business licensing and the like. The idea has been copied by several Canadian cities, including Edmonton and Winnipeg.

One of the few adopters of this system in the United States is Miami-Dade County in Florida. The 26 municipalities in that region still exist and retain jurisdiction over certain local matters. More common has been a somewhat related concept, the merger of county and city governments. This practice has been adopted, among other places, in Jacksonville (Florida), San Francisco (California), and Oklahoma City.

Philosophies of Virtual and Particular Representation

Democracy is not practiced uniformly around the world, but comes in a variety of forms. For example, in the United States we find both direct and representative democracy. Under **direct democracy,** all eligible voters can personally cast their ballots on issues of public interest. The well-known New England Town Meeting is the archetype of American direct democracy, for here decisions were made by all of the Town's voters. Another form of direct democracy, especially popular in California, is the ballot initiative, where citizens or interest groups can place proposed laws directly on the ballot for consideration by the voters.

In **representative democracy,** which is the norm these days, there is a two-step process in decision making. Here voters first elect their representatives, such as members of city councils or senators, and the latter then meet to enact legislation.

There are two major traditions, or philosophies, of representative government. Under the philosophy of **virtual representation,** each member elected to a governing body, be it national or local, sees himself or herself as representing the interests of the whole body politic, not just of the particular area where the legislator lives. The English tradition of virtual representation was in force at the time of the American Revolution, and seems to have contributed to

the unrest. While the colonists were chafing under unpopular laws and cried of "taxation without representation," the view in the United Kingdom was that members of parliament were elected to represent all interests, including those of the colonists. As is well known, the colonists were not satisfied with this view and, eventually, they revolted.

Perhaps because of their dissatisfaction with virtual representations (which ended in the UK in 1832), the framers of the United States Constitution chose a different path. Under the philosophy of **particular representation,** sometimes called areal representation, which is generally practiced in the U.S. and Canada, a person is elected as the representative of a specific geographic area, such as a city ward, a state or province, or a congressional district. With this approach, each legislator is seen as accountable primarily to the residents of a territorial constituency. In Canada, however, party loyalties tend to be stronger than in the United States; at times, they may take precedence over the politician's loyalties to his or her constituency.

Virtual representation is not completely gone in the United States and Canada. Some municipalities today elect their council members city-wide, rather than on the basis of particular geographic districts.

The Geographic Basis of Representation

Early in American history, territoriality emerged as the explicit basis of representation and as a basic principle of American democracy. Well before the American Revolution, every colony had legislative districts for its assembly, and each legislator was seen as accountable primarily to the residents of that district. This geographic link between representative and constituent remains a fundamental component of American and Canadian democracy. Many political geographers believe that the inherently geographic qualities of this system are among the most important reasons for its success. In particular, it has made it easy for individuals and interest groups to approach and to command the attention of decision makers.

In both countries, it is presumed that the person elected will represent the interests of his or her constituency, no matter how varied that constituency may be. The representative from Oklahoma or Manitoba, therefore, is expected to represent the interests of her/his state or province, and to vote accordingly. At other times, of course, such as in national emergencies, the representative is expected to vote in the national interest, however he or she may interpret it.

One reason that geography became the basis for representative government was the prevailing attitude, strongly felt even in colonial times, that people should have easy access to government. Early Americans thought it a matter of right, and not just of personal comfort, to have a centrally located local government. In the 1600s and 1700s, when poor transportation was the rule, great efforts were made to make government as accessible as possible. The prevailing notion was that a man should be able to get to his county seat, and back home, on a horse or in a stage, in a single day. This meant that counties had to be rather small, certainly by later standards. Thus, ease of access to governmental officials and policymaking bodies was an important feature of American democracy virtually from the beginning, and it was seen as an important corollary of particular representation.

Following the Revolution, the question of the location of the new national capital became a major public issue. Centrality to the population was an important, and perhaps the main, guiding principle in making that choice. Another consideration, however, was the north-south split between slave and free states; the new capital should be well located to serve both these sections.

There was considerable sectional debate about exactly where the capital should be. Although they did not have the data to support it at that time, the politicians' sense of centrality was correct. In the later 1700s, the center of the U.S. population was in Maryland, somewhat west of Baltimore and close to the site eventually chosen for Washington. Maryland was also the northernmost of the slave states. George Washington was a strong supporter of the site that bears his name today, along the Potomac River in Maryland, but there was also considerable backing for a location further north along the Susquehanna River in southern Pennsylvania. Thus, both major choices were reasonably close to the center of population, as well as to the Mason-Dixon Line. As a part of compromise legislation, the present site was chosen on a close vote (Figure 5.5).

Geographic centrality has also been important in the location of state capitals. In fact, many capitals were moved from their original locations to more accessible sites, sometimes several times. Iowa presents a good example, as the capital was first located at Burlington (on the state's eastern edge), but then it was moved somewhat closer to the center at Iowa City, and finally to highly central Des Moines (Figure 5.6).

Selection of Representatives

As mentioned earlier, in the American and Canadian political process space is usually divided into territories, which form the basis for electing representatives. That is, voting districts are established, and people are elected to represent those geographic areas.

In democracies, the method of selecting representatives by electoral districts can vary considerably. In the United States, Canada, and many other English-speaking countries, single-member representation is used. Under this system, the candidate with the highest number of votes in a district, whether a majority or a plurality, wins the seat. This is also called the winner-take-all approach. When it is combined with a two-party system, a simple majority of votes usually is enough to be declared the winner.

Single member district representation implies a direct geographical link between representatives and their constituents. This relationship between voters and representatives, in turn, has contributed to close ties between elected officials and the people. Americans often talk about "their" representatives, state legislators, or city council members, for they see a direct link between the representatives and themselves.

Such a system usually operates where there are strong, stable, two-party systems. In the United States, for example, Democrats and Republicans have been dominant for about 150 years. On the negative side, under this system it is very difficult for third parties to gain any measure of success. Thus, single member district representation promotes political stability, but often at the expense of full representation of minority interests within the country.

Other democracies, especially in non-English-speaking States, practice proportional representation. Here, a voting district elects more than one legislator, and seats are allotted to political parties in proportion to the votes they garner. In this system, people cast their votes for parties and party platforms rather than for individual candidates.

Under one possible scenario, each voting district elects 10 representatives to a legislature. If the leading party receives 60% of the votes, it will send six members from that district to the capital. If each of four other parties receives ten percent of the vote, they would send one member each from the same district. Under such a system, smaller parties have a better chance, but the strong territorial link between constituents and legislators may be missing. Proportional representation encourages the formation of many political parties, for even relatively minor parties can gain some power. Under the winner-take-all system, as noted earlier, it is extremely difficult for a third party to win an election. Instead, a wide variety of viewpoints usually are encompassed within the two leading parties.

The system of territorially based single-member representation, as practiced in the United States and Canada, is subject to abuse. Its explicitly geographic nature implies that where the district boundaries are drawn can have a critical impact on election results. Put another way, voting district boundaries can be manipulated to favor one political party at the expense of another. Such manipulations can be a part of the strategy of the party in power to retain control.

The deliberate distortion of boundaries for political purposes has been around almost as long as particular representation. Early American politicians soon learned that voting district boundaries could be manipulated to favor their party in future elections. Such contortions of political space often led to strangely shaped voting districts.

Gerrymandering

Gerrymandering is defined as the deliberate manipulation of political district boundaries to achieve a particular electoral outcome. It is named after an early governor of

Figure 5.5

The site for the national capital, on the Potomac River in Maryland, was chosen on a close vote. The capitol building and Washington monument are in the background towards the left. The Potomac River is in the foreground, with the Lincoln Memorial on its banks to the right.

Figure 5.6

Iowa's capital was moved twice, each time closer to the center of the state. This is the capital building today in Des Moines.

Massachusetts, Elbridge Gerry, who blatantly practiced it in the 1800s (Figure 5.7).

When the responsibility for drawing electoral boundaries is in the hands of elected representatives, it can be used as a weapon in the fight to stay in power. As a political geographer once noted, thereby the geography of politics becomes the politics of geography. Claims that the districting process has been biased against the interests of particular parties, political factions, or ethnic groups have been made almost from the time districting first began.

To explain the strategy in gerrymandering, it is useful to note that, in all elections, the votes cast can be classified into one of three categories. **Effective votes** are those needed to win a seat, and in a two-candidate race half the total number of votes cast plus one are effective. Those cast above what is needed to win are called **excess votes,** while **wasted votes** are cast for losing candidates. To win as many seats as possible, each party tries to ensure that as many of its votes as possible are effective, rather than excess or wasted.

In gerrymandering, the objective of the party in power is to draw district borders so as to force the opposition to

THE GERRY-MANDER. (Boston, 1811.)

Figure 5.7
The original gerrymander. The term *gerrymander* originated in 1811 from the shape of an electoral district formed in Massachusetts while Elbridge Gerry was governor. When an artist added certain animal features, the district resembled a salamander and quickly came to be called a gerrymander.

cast as many excess and wasted votes as possible. Two techniques commonly are used to do this, and which is employed depends on the geographic distribution of voters. **Opponent-concentration gerrymandering,** also known as excess-vote gerrymandering, forces opposing parties to cast many excess votes. This is done by gerrymandering opposition voters into as few districts as possible. In this approach, the opposition party wins its (relatively few) seats easily by large majorities, but it loses more seats elsewhere. **Opponent-dispersion gerrymandering,** or wasted-vote gerrymandering, disperses the opposition's vote between districts. That is, borders are drawn so as to force the opposition party to lose many close elections. In practice, usually both types of gerrymandering are applied simultaneously. The outcome, in either case, is that majorities within the electorate are magnified in the legislature. Put another way, the party in power ends up with a larger percentage of elected seats than its proportion of all voters.

The practice infringes on the notion of one person, one vote, and is inherently undemocratic. In effect, gerrymandering makes one person's vote worth more than that of another (Figure 5.8).

When parties are relatively equal in strength, the one whose members are dispersed across space enjoys advantages over another whose adherents are clustered. Thus, gerrymandering is easiest to apply to areas where there are large concentrations of supporters of one party, such as in large cities, relative to opponents whose members are spread more evenly, perhaps in the rural parts of the state. In particular, the clustering of support for one party facilitates the strategy of opponent-concentration. In the United States, the concentration of many Democrats in inner-city areas has afforded Republicans advantages where they have been able to gerrymander most opposition voters into a few districts. On their part, Democrats have used the concept to carve up suburban areas to their advantage. It is worth

Political Geography

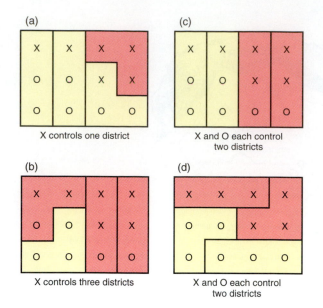

(a)

X controls one district

(b)

X controls three districts

(c)

X and O each control
two districts

(d)

X and O each control
two districts

Figure 5.8
Alternative districting strategies. Xs and Os might represent
Republicans and Democrats, urban and rural voters, blacks and
whites, or any other distinctive groups.

noting that the spatial concentration of voting blocs increases the chances for racial minorities or ethnic groups to get at least a few representatives elected.

The Courts and the Reapportionment Revolution

The United States Constitution mandates that a census of population be taken every ten years to determine representation to the House of Representatives. That is, every decade the 435 seats to the House must be reapportioned among the states according to the new population figures. In this process, some states will gain representation, while others will lose or maintain their numbers. After the 1990 census, for example, fast-growing California gained seven seats to the House of Representatives, while slower-growing New York lost three seats. However, the Constitution says nothing about how this reapportionment is to be done. Almost from the start of the republic, therefore, there were controversies over the methods used to apportion representatives among the states. Eventually, a method that has worked reasonably well was adopted.

As the number of seats allocated to individual states changed, it became necessary to redraw district boundaries within them. The Constitution also says nothing about how such districts, or other constituencies (such as state senate districts) are to be delimited, or how often that should be done. The states, therefore, had wide latitude in choosing methods for redrawing voting district boundaries, and in deciding when to update them.

During the first six decades of the twentieth century many states put off redistricting as long as possible, and eventually some had delayed action for 50 years or more. For example, in 1960 Tennessee had not redistricted since 1901, Idaho since 1911, Louisiana since 1912, and six other states since 1931. Even though the number of representatives for some states may have remained constant, the population distributions within these states often changed drastically. In particular, in this period the populations of most of the larger cities boomed, while rural populations were either stagnant or declining. Because of the failure to redistrict, the declining rural areas retained their political power, while urban areas gained people but not political representation. Among the serious consequences of this practice were that rural districts were heavily overrepresented relative to urban districts and that the workloads of legislators varied greatly.

The result was **malapportionment,** an imbalance in the number of voters in the various constituencies. In Alabama, districts for the lower house varied in population from 236,000 to 635,000, in Connecticut from 319,000 to 690,000, and in Texas from 216,000 to 952,000. Thereby one voter in a rural area carried two or three times the political clout of his or her counterpart in the city. Malapportionment was maintained by political power, and it was a powerful factor in perpetuating incumbents in office.

All this began to change in 1962 when the Supreme Court, in a case involving Tennessee's General Assembly, declared malapportionment unconstitutional. Specifically, the Court ruled that a state's failure to redraw malapportioned districts violated the equal protection clause of the Fourteenth Amendment to the Constitution. In this landmark decision, which was soon extended to Congressional districts, the court ruled that, in the future, redistricting had to be guided by the principle of "one person, one vote." Two years later the Supreme Court ruled that state senates also had to be reapportioned on the basis of population, and not on the basis of county or some other unit. The result was that many voting districts had to be redrawn to get as close as possible to equal representation. Although they did not say it this way, in essence the courts ruled that political fairness dictated that no party could be forced to cast a large number of excess or wasted votes compared to its rival party.

These rulings led to the reapportionment revolution, with far-reaching consequences. Since then, state and federal courts have ruled in thousands of additional cases that challenged voting districts drawn for elections to the House of Representatives, to state legislatures, and for local offices. A large number of redistrictings had to be performed, often leading to a substantial redistribution of political power from rural areas to the cities and their suburbs.

The courts went beyond malapportionment in setting rules for redistricting. In an attempt to decrease gerrymandering, they ruled that formal redistricting criteria should include contiguity, compactness, and respect for the integrity

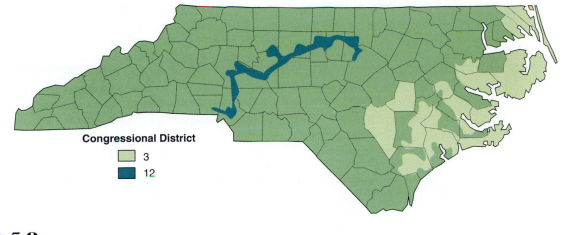

Congressional District

- 3
- 12

Figure 5.9

After 1990, the state of North Carolina created minority-majority districts to try to concentrate the African-American population.

of existing subnational political units. The contiguity criterion implies that a single, continuous line can be drawn around the district; that is, there are no split districts or outliers. Compactness struck at the heart of gerrymandering by outlawing grotesque shapes for districts. The reference to the integrity of existing subnational political units means that the district borders should, so far as is possible, follow existing political boundaries, such as those of counties and cities. This is because such local governments are well-known units with identifiable goals and needs, and drawing districts without splitting these government units tends to foster that sense of community that is needed for effective territorial representation. What the courts want to avoid, then, is excessive splitting of counties and other well-known boundaries when drawing voting districts for state and national offices.

Despite the malapportionment revolution, it soon became clear that even strict adherence to these redistricting criteria did not automatically eliminate the potentials for gerrymandering. While malapportionment is a thing of the past, the manipulation of territory for political purposes is still widely practiced. Thus, the reapportionment revolution did not eliminate all controversies about districting.

When it became clear that the prescribed formal criteria could not eliminate gerrymandering, and that not all criteria could be satisfied simultaneously, political geographers began to advocate the most explicitly geographic criterion of all. Their recommendation, which has found considerable judicial favor, was to establish district borders based on the principle of **community of interest.** The latter is a territorially defined group of people with common economic, social, political, or cultural interests.

In this view, districts should be drawn based on collective interests that come from the identity of citizens with real places. Electoral districts should not be just arbitrary aggregations of people for conducting elections, but they

should be meaningful units with legitimate common interests. Cities and surrounding rural areas are good examples of communities of interest, for usually there is a strong interaction between them based on mutual dependence. That is, rural areas depend on the nearby cities for markets and for buying certain goods and services, while the rural areas help to support the economies of the cities.

Increasingly, the courts have recognized community of interest as a legitimate basis for redistricting. The idea that a legislator should represent a community of interest fits well with the philosophy of particular representation underlying U.S. representative democracy.

In recent years, there has been considerable controversy about whether it is legal to draw district borders on the basis of race or ethnic identity, with the intent of forming "minority-majority" districts. In other words, do race or ethnicity constitute a legitimate community of interest? This question arose in North Carolina when, following the 1990 census, that state redrew voting districts with the explicit intent of concentrating African-American voting strength. This was defended by the state as a way to promote minority representation, for North Carolina had not sent an African-American to Congress in many decades. The two resulting districts were far from compact, but the state defended them as promoting minority representation. Still, on a map they resembled the most blatant of the early forms of gerrymandering. One of these districts included many eastern counties with a high concentration of rural blacks, while the other constituted a thin strip along a major highway that contained a high percentage of North Carolina's urban black population (Figure 5.9).

This process was challenged in the courts. Eventually, the question facing the Supreme Court was whether it was legitimate to draw voting districts strictly on the basis of race, ignoring other widely recognized criteria such as compactness and the use of existing political boundaries. In

The irregularly shaped Congressional voting districts shown here represent a deliberate attempt to balance voting rights and race. They have been called extreme examples of racial gerrymandering, however, and at least in part ruled unconstitutional by the Supreme Court.

All of the districts shown contain a majority of black voters. They were created by state legislatures after the 1990 census to make minority representation in Congress more closely resemble minority presence in the state's total population. Specifically, they were intended to comply with the federal Voting Rights Act of 1965, which provides that members of racial minorities shall not have "less opportunity than other members of the electorate . . . to elect representatives of their choice." In the 1980s, the Justice Department prodded states to create districts designed to give black voters representation in rough proportion to their numbers in the population.

In North Carolina, for example, although 24% of the population of that state is black, past districting had divided black voters among a number of districts, with the result that blacks had not elected a single Congressional representative in the 20th century. In 1991, the Justice Department ordered North Carolina to redistrict so that at least two districts would contain black majorities. Because of the way the black population is distributed, the only way to form black-majority districts was to string together cities, towns,

and rural areas. The two newly created districts had slight (53%) black majorities.

The redistricting in North Carolina and other states had immediate effects. Black membership in the House of Representatives increased from 26 in 1990 to 39 in 1992; blacks constituted nearly 9% of the House as against 12% of blacks in the total population. Within a year, those electoral gains were threatened as lawsuits challenging the redistricting were filed in a number of states. The chief contention of the plaintiffs was that the irregular shapes of the districts were a product of racial gerrymandering and amounted to reverse discrimination against whites.

In June 1993, a sharply divided Supreme Court ruled in *Shaw v. Reno* that North Carolina's 12th Congressional District might violate the constitutional rights of white voters and ordered a district court to review the case. The 5–4 ruling gave evidence that the country had not yet reached agreement on how to comply with the Voting Rights Act. It raised a central question: should a state maximize the rights of racial minorities or not take racial status into consideration? A divided court provided answers in 1995, 1996, and 1997 rulings that rejected Congressional redistricting maps for Georgia, Texas, North Carolina, and Virginia on the grounds that "race cannot be the predominant factor" in drawing election district boundaries, nor can good-faith efforts to comply with the Voting Rights Act insulate redistricting plans from constitutional attack.

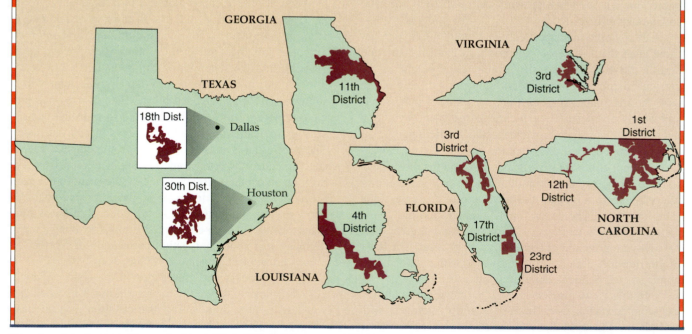

Taken from *Introduction to Geography,* 7th ed, (Getis, Getis & Fellmann, pp. 340–41)

1993 the Court declared these districting plans void, calling their boundaries irrational, and saying that their sole purpose was to segregate voters according to race. A year later, however, a federal court upheld the state's claim that the two unusually shaped districts could be justified on other grounds, and made specific reference to community of interest. One was called a rural community of interest, and the other was referred to as an urban one. Despite this ruling, it appears that the drawing of minority-majority districts is on the wane.

The Geography of Elections

As previously noted, in the United States, political races essentially involve a competition between the two major parties. While the Democrats and Republicans are the main parties today, this has not always been the case. Also, occasionally a third party has been formed to try to challenge the two-party system, but with little success (Figure 5.10). In addition, there are many minor parties, with no real chance of electing anyone to a position of power.

We have already mentioned that the two leading American parties can each accommodate a wide variety of viewpoints. After all, we use such labels as "liberal" and "conservative" to refer to both Republicans and Democrats. The parties can also greatly change their positions on issues over time. It has sometimes been pointed out that today, Democrats and Republicans hold roughly the opposite positions on important issues from those they held just a few decades ago. At one time, for example, Republicans tended to be against free trade while Democrats favored it. Today, the Republican Party is especially identified with this philosophy, while many Democrats oppose it.

One way that the two major parties differ today, which is of interest to geographers, is that their strengths are concentrated in different parts of the nation. Thus, there is a geography of partisanship, or of party support. Scholars have noted that adjacent states often vote in similar fashion on major issues or for candidates for president. When such adjacent states with similar voting records are combined, we can define regional political alignments. Such geographic alignments have been a fact of U.S. political life since early in the history of the republic. This tendency for groups of adjacent states to vote similarly on political issues is sometimes called **sectionalism** (Figure 5.11).

One part of political geography studies the voting patterns of elections, or of a series of elections. Since the 1960s, electoral geography has become a recognized subfield of political geography, with particular emphases on studying contested elections for state legislatures, Congress, and the presidency. In the following discussion, we emphasize presidential elections.

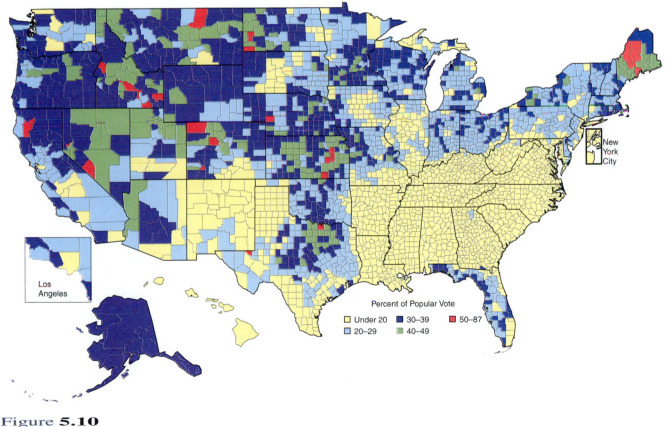

Percent of Popular Vote

□ Under 20   ■ 30–39   ■ 50–87
□ 20–29   ■ 40–49

Figure **5.10**
The popular vote, by county, for independent presidential candidate Ross Perot, 1992.

Political Geography

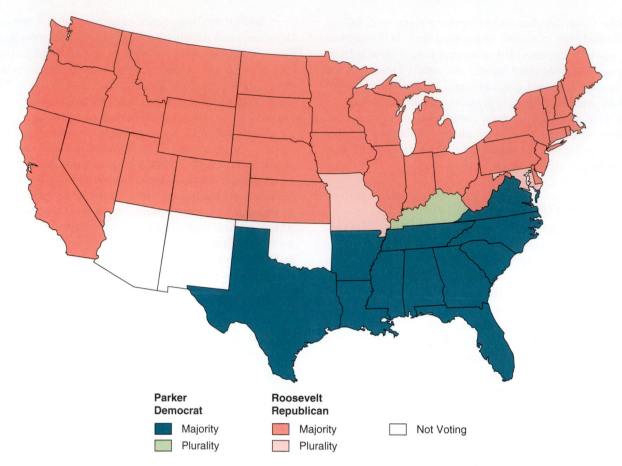

Parker
Democrat

■ Majority

■ Plurality

Roosevelt
Republican

■ Majority

■ Plurality

□ Not Voting

Figure **5.11**

The 1904 election for president is a good example of sectionalism, with the West and Northeast voting solidly Republican, and the South voting Democratic.

The Geography of Presidential Elections

The nineteenth century witnessed the birth of today's two major political parties. It also saw the development of basic geographic, or regional, cleavages in voting. That is, three distinct sections emerged, each with its own particular voting pattern. These electoral regions were the Northeast, the South, and the West, and they are still characteristic of contemporary U.S. politics, although decreasingly so.

At any given time, then, the voting patterns of each of these sections was reasonably predictable. Thus, the outcomes of presidential elections can be studied in terms of the varying successes of the two major parties in the Northeast, South, and West.

Such distinct regional patterns of voting behavior first developed in the 1850s. By 1880, for example, the South was strongly in the hands of the Democrats, largely because Republicans had been discredited through their identification with Northern interests during the recent Civil War. Simultaneously, the Republicans dominated most of the Northeast and the West, while intense competi-

tion between the parties was typical in the rapidly industrializing central states. The outcomes of presidential elections in the late 19th century can be seen, therefore, as depending on the candidates' abilities to appeal to voters in the volatile central states while maintaining strength in the sections normally identified with their party.

When such a regional voting pattern remains in place for many years, it is called the **sectional normal vote.** This term refers, then, to the expected voting behavior of the country's three regions. Sectional normal votes are not static, but tend to change slowly over time.

Analyzing the Sectional Normal Vote

There are vast amounts of data available on historic presidential elections. This information can be used to analyze the geography of elections. The results of successive elections, for example, can be used to identify underlying trends in voting patterns.

Such analyses clearly identified the sectional normal vote that emerged before 1900 and then reemerged shortly after the turn of the century. During this period, the South voted as a bloc for the Democratic presidential candidate, while Republicans dominated the Northeast, and the West was politically somewhat volatile. The same unpredictability was found in some industrial states in the Northeast, including New York, New Jersey, Ohio, Indiana, and Illinois. These states and the West, then, were the major battlegrounds in presidential elections. This pattern continued until after World War II, when the interior and the Far West became consistently Republican.

After 1960 a new sectional normal vote emerged. By then the once solidly Democratic South was becoming increasingly Republican in presidential politics. The election of 1948 was pivotal in moving the region in this new direction, as the old Solid South began to break up with the Democratic Party's espousal of civil rights. By the 1960s the Democratic Party's increasing identification with opposition to the Vietnam War also cost it many Southern voters, who supported an anti-Communist intervention-oriented foreign policy. In that decade, which saw the nation polarized by both civil rights and Vietnam, the South's presidential politics shifted strongly towards the Republicans. At the same time, Republicans dominated the West, while Democrats gained strength in the Northeast. In 1964, for example, Goldwater, a conservative, won most of the Solid South for the Republicans. By that year the sectional normal vote saw Republican strength in the Deep South while Democrats were strongest in the Northeast, weak in the West, and weakest in the core South. Republicans also were strong in the suburbs, which saw great population growth after World War II. By 1990 nearly half the country's voters were in the suburbs.

Since 1976, when Jimmy Carter was elected, U.S. presidential races have been characterized by another pattern, the so-called "conservative normal vote." By that year, the West had emerged as the most strongly Republican section of the country, while the Northeast was strongly Democratic. The old Solid South has become the country's pivotal section in presidential elections. In other words, whichever party makes the strongest inroads in the South has a good chance to be the winner in the next presidential contest. Since 1976 Democrats have won in 1976, 1992, and 1996, and in each case their tickets were headed by moderate Southerners who gained strong support in that region. Still, such sectional patterns of political behavior are not as predictable today as they were even three decades ago. Recent voting results show a more complex voting pattern than what would be expected from the sectional normal vote.

The Geography of Presidential Elections, 1992–1996

Clinton's election victories in the presidential races of 1992 and 1996 can be studied in terms of this conservative normal vote. He won in 1992 by sweeping the recession-impacted Northeast, even gaining several traditional Republican bastions like New Hampshire and Maine. He also won all the Middle Atlantic and Great Lakes states, except for the Republican vice-presidential candidate Dan Quayle's home state of Indiana. Clinton helped himself by cutting into Republican strength in the suburbs, the Sunbelt, and the West. Clinton's strong showing in the suburbs was especially important. In the process he carried such suburban-dominated states as Maryland, New Jersey, and Connecticut, and many Western and Southern states that had been reliably Republican in the 1980s.

Overall, Clinton held onto Democratic areas while cutting into normally Republican areas in the South and West, and in the suburbs. He did this, in part, by successfully exploiting the economic issue at a time of severe recession. His opponent, George Bush, did best among a minority who felt that current economic trends were favorable for them.

The 1992 elections put to rest any idea of a solid Republican South, and showed a continuing willingness on the part of many working class white voters in that region to support Democratic candidates. Others who voted strongly for Clinton were African-Americans and urban residents.

The same geographic patterns largely reappeared in 1996, but now Clinton's chances were buoyed by being the incumbent, and he was the beneficiary of a strong economy. Historically, the electorate has been loath to "change horses" when the economy was doing well, as it was in 1996.

The Political Geography of the Future

What are some possible changes in the political geography of the United States and Canada during the first decades of the new millennium? This is a difficult question, as almost all those who have made longer-run predictions in the past have found out. In truth, despite all the impressive progress in science over the past few decades, we are almost nowhere in our ability to predict the future.

One possibility, suggested by a few geographers from time to time, is to discard existing U.S. state borders and to delimit new, more rational, ones. Doing this can be justified in several ways, including the obvious fact that many existing state borders are virtually arbitrary, simply drawn as straight lines on a map. As such, they divide areas that have things in common as much as they unite them. Thus, many states in reality contain "other states" within their borders, as shown by strong differences within them in political culture, economics, demography, levels of urbanization, and physical geography. This makes governance more difficult, often because of a lack of community of interest.

These differences have led to strong sectional conflicts within some states. In a few, a major municipality, such as New York City, is pitted against the smaller cities and the rural areas of the rest of the state. Similar situations exist in Michigan, Illinois, Maryland, and Georgia, where there are recurring conflicts between metropolitan and more rural regions. In still other cases, there are strong, and sometimes even bitter, rivalries between two major cities, such as in California (Los Angeles and San Francisco), Oklahoma (Tulsa vs. Oklahoma City), and Pennsylvania (Philadelphia vs. Pittsburgh).

Some states cut across distinct physical regions, contributing to differences in community of interest. These include states with both wet and dry regions, such as Washington, Texas, and California, and highland-lowland divisions such as found in the southern states of Tennessee, North Carolina, Alabama, and Arkansas (Figure 5.12). Still other states have cultural divides, such as in Louisiana with its Protestant north and Catholic south.

At the extreme, there are recurrent movements for parts of states to secede, either to form new states or to be attached to another. These include occasional secessionist tendencies in New York City, the Massachusetts islands of Martha's Vineyard and Nantucket, and western Kansas and Nebraska. One well-known example is the periodic threat

(a)

(b)

Figure 5.12

North Carolina is one of several states that has a marked physical contrast between lowlands and highlands, with some political implications. These views show (a) a coastal area with marshes and (b) the Blue Ridge Mountains in the west.

of northern California to secede from the south in order to form a new state. In this case, the proposed division of California is based on differences in economy, culture, and physical factors. There have also been occasional threats by the people of Michigan's Upper Peninsula to secede and to form the new state of Superior.

Such threats give a hint that the existing states might not be the best units for solving pressing problems. There may also be too many states, as they were formed in an era of much poorer transportation that encouraged smaller political units. Is it possible that today's borders are badly suited to tomorrow's needs, and that much better state units could be established?

These questions form the background for several proposals in recent decades that the United States be reorganized into a new set of relatively homogeneous states. A few geographers have suggested the need for such reorganization, perhaps based on the idea of community of interest. The proposed new states would be organized around many of the country's leading metropolises, which would form the foci for communities of interest.

Figure 5.13 illustrates one possible scenario for such a future United States. It represents what one political geographer calls "macrolevel administrative units," drawn on the basis of strong regional identities rather than existing state boundaries. One of the strengths of this scheme is that it recognizes that many important spheres of influence areas today extend across more than one existing boundary. The proposed capitals for each administrative unit would be responsible for addressing the concerns of its residents and the problems of its area.

Another possibility is to redraw legislative districts based on the idea of one person, one vote. Today, the United States Senate is the only major legislative body in the country whose members are not elected from districts with substantially equal populations. Under the present system, where each state is entitled to two senators regardless of population, this measure of fairness is wildly disregarded. At least one political geographer has suggested that new senatorial districts be drawn to eliminate this unfairness. In drawing such borders, existing state boundaries were often ignored. Figure 5.14 represents just one of literally millions of ways senatorial districts could be drawn on the basis of equal representation.

The roadblocks to establishing such new states and senatorial districts are enormous, and change cannot be expected any time soon. Just the legal complications of reordering states would be daunting. Further, it must be recalled that equal representation in the Senate for all states was chosen deliberately as a way to protect the interests of the smaller, less populated states. Yet such proposals for reorganization cannot be dismissed out of hand, especially as they are logical, adhere well to current patterns of behavior (communities of interest), and/or are fairer to voters.

For Canada, the basic question is whether this State will survive as one, or whether it will become two or more

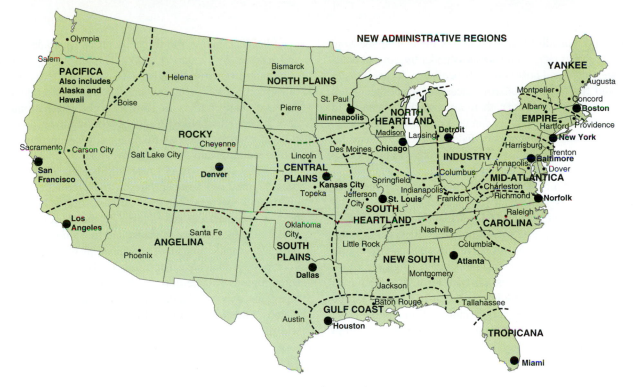

NEW ADMINISTRATIVE REGIONS

Figure 5.13
Proposed future "administrative units" for the United States, based on communities of interest.

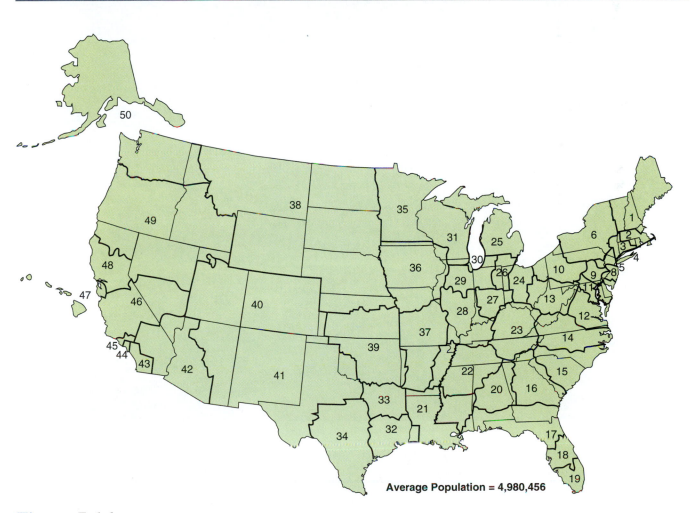

Average Population = 4,980,456

Figure 5.14
Proposal for new U.S. Senate districts, based on equal representation by population using 1990 census data.

countries. Periodically, a large portion of the French-speaking population shows signs of restlessness under the current political arrangement and pushes for independence. If that population would find a *cause celebre,* secession would become a distinct possibility. However, there are also disadvantages to going it alone, and uncertainty can be a powerful force for retaining the status quo. Canada's federal structure has so far held the fort, for it has allowed Québec considerable latitude to retain and to develop its distinctiveness.

The last time there was a referendum on independence, in the early 1990s, it lost by a close vote. There is no assurance that it will lose again in the future. One sobering way to look at the question is to note that the vote for independence can lose many times, yet just one positive vote probably would lead to a permanent change (Figure 5.15).

Another political question with strong geographical dimensions involves **NAFTA,** the North American Free Trade Agreement, whose participants are Canada, the U.S., and Mexico. The main aim of this compact, which began to be implemented on January 1, 1994, is to achieve virtually open borders among the three countries for trade in goods and services by January 1, 2004. This means the step-wise elimination of virtually all tariffs on goods (e.g., automobiles, computers, furniture, and agricultural products), so that eventually trade will flow as if the international borders did not exist. NAFTA also is working towards the free flow of services, such as trucking, advertising, consulting, and accounting, and of foreign investments.

Theoretically, NAFTA should benefit all three countries, as they will concentrate on the production of those goods and services that they do best; that is, where they have a "comparative advantage." In the process, it should lead to a great increase in trade, and to economic growth in those geographic areas that specialize in what each country does best or that have some other advantages. Of course, some regions will be hurt, as their "comparative disadvantages" are exposed, leading to closed factories and the loss of jobs. One well-publicized example in the United States was the clothing industry, which was subject to lower cost (especially lower labor cost) producers in Mexico. In fact,

Figure 5.15
Québec City is the capital of the Province of Québec. If Québec votes for independence, it could become the capital of a separate, French-speaking country.

between 1994 and 1998 Mexico's apparel exports to the U.S. increased greatly, and in the latter year it passed China to become the leading source of imported apparel.

Concerns over its negative impacts were mirrored in the geography of the vote on NAFTA in the House of Representatives in 1993 (Figure 5.16). In general, the Northeast (sometimes called the Rust Belt) felt vulnerable to new competition from Mexico and opposed the legislation. This region had seen declining competitiveness, especially in the 1970s and 1980s, with the loss of hundreds of thousands of jobs to the Sun Belt states and to foreign competitors. Democrats, who have become increasingly identified with protectionism and opposition to free trade, usually voted against NAFTA, particularly those from the Northeast. On the other hand Republicans, who tend to espouse free trade, generally supported the bill. This was especially true of Southern and Western Republicans, regions where most people thought that NAFTA offered new opportunities for sales to Mexico. Although these generalizations are valid, the map of the vote shows that the geography of the decision on NAFTA was quite complicated.

As predicted, trade among the three member countries increased greatly after implementation. During the first five years trilateral trade grew by 71%, or by an impressive 11% per year. Between 1994 and 1998 Canada's trade with the United States increased by 47%, and with Mexico by 91%. Mexico benefited especially during this period, as trade with its NAFTA partners increased by 137%. At the same time, U.S. exports to Mexico increased by 110%, and Mexican exports to the U.S. rose by 140%. The latter increase was enough to elevate Mexico to the United States' second largest trading partner (after Canada), replacing Japan. One government source claimed that U.S. exports to its NAFTA partners supported more than 2 million jobs, although it was not clear how many of those jobs had been created since 1994. On the other hand, in 1999 one labor source claimed that NAFTA had cost the U.S. 378,000 jobs. The definitive study to answer this question of job gains versus job losses has yet to be made.

No matter what the effect on jobs, NAFTA will have a measurable impact on the geography of the future. In particular, the border zone between the United States and Mexico should see considerable growth in the next few decades. In fact, this should be one of the United States' leading growth regions, and perhaps the leader. This trend, which had started before NAFTA, should have important positive benefits on some Sun Belt states, particularly California, Arizona, and Texas.

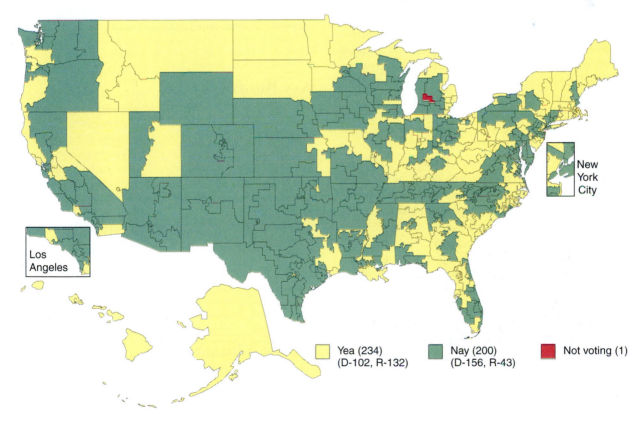

Yea (234) (D-102, R-132) Nay (200) (D-156, R-43) Not voting (1)

Figure 5.16

The North American Free Trade Agreement Implementation Vote produced a complicated geographic pattern, with Southern and Western Republicans generally supporting it, and Northeast Democrats opposed.

Summary

Political geographers study a wide variety of topics. Most of the world's land area is covered by States, only a few of which qualify as nation-States. Some States are strong while others are weak, depending on the degree to which centripetal and centrifugal forces are active within them. Historically, the United States had strong centripetal forces, but in more recent times centrifugal forces seem to have gained ground. There has always been a weak sense of "Canadianness." In that country, the presence of a large French-speaking minority, geographically concentrated in Québec, threatens to split the country through secession.

The world's countries are organized either as unitary or federal States. Both Canada and the United States have federal structures, with a division of powers between the national and the state/provincial levels. Federal States have unity at the national level, yet they allow for considerable variations in laws and customs at the more local level. In the United States, the powers of Washington seem to be increasing, while in Canada the provinces have been gaining power.

There are more than 80,000 local governments in the United States, including over 3000 counties. With powers assigned to them by the states, counties perform a wide array of functions, the number of such functions varying with the population density. In New England and a number of other states, Towns and townships are important forms of local government. There are also many special purpose districts, the most numerous of which are school districts. Cities, or "municipal corporations," are the smallest general purpose administrative units. Metropolitan government operates in several places in Canada, but the concept generally has been resisted in the U.S. in favor of more local control.

Direct democracy is relatively rare today, for it has been largely replaced by representative democracy. Under the philosophy of particular representation, which is dominant in the U.S. and Canada, the shape of electoral districts can play a crucial role in the outcomes of elections. Thus gerrymandering has been widely practiced.

Because of the failure of states to update electoral districts for many decades, malapportionment was widespread by the 1960s. Since then, the courts have found consistently for "one-person, one-vote," and thousands of redistrictings were mandated. These have eliminated the most blatant forms of malapportionment and gerrymandering, but the latter continues at a reduced scale. Increasingly the courts have favored redistricting plans based on "community of interest," a somewhat vague concept that is subject to a number of interpretations.

One subfield of political geography is concerned with election results, including presidential elections. Electoral geographers have identified the sectional normal vote, a pattern of regional voting that can persist for many decades. Presidential elections can be studied in terms of these sectional normal votes; that is, the outcomes of elections can be explained by the candidates' abilities to hold sections of the country that normally vote for their parties, and to make inroads into areas not usually associated with that party.

Since 1976 the conservative normal vote has seen Republican strength in the West, Democratic strength in the Northeast, and a volatile South. While the concept of sectional normal vote remains somewhat valid, recent elections suggest that a more complex geographic voting pattern is emerging.

One possibility for the political geography of the future is the rejection of the current organization of the United States into fifty states, perhaps by the designation of larger units based on the idea of community of interest. Changing electoral districts for senators to conform to one-person, one-vote have also been suggested. Any such changes will be difficult to implement, but the payoffs for such radical moves could be considerable. In Canada, the main political-geographic question is whether that country can remain one unit, or whether it will split into two or more States. So far the forces of unity have prevailed, but there is no guarantee the same situation will hold in the future. Finally, NAFTA has a great potential to influence the future geographies of Canada and, especially, the United States. The most likely locus for positive impacts is a wide band adjacent to the U.S.-Mexico border, especially in California, Arizona, and Texas.

Key Words

annexation	opponent-concentration
centrifugal forces	gerrymandering
centripetal forces	opponent-dispersion
community of interest	gerrymandering
direct democracy	particular representation
effective votes	representative democracy
excess votes	sectional normal vote
federal States	sectionalism
gerrymandering	special purpose districts
malapportionment	State
NAFTA	unitary States
nation	virtual representation
nation-State	wasted votes

Gaining Insights

1. What are the differences between the terms State, state, and nation?
2. What are nation-States? What are the advantages and disadvantages of nation-States?
3. What is the difference between centripetal and centrifugal forces? What are some centripetal and centrifugal forces current in the United States? In Canada?

4. Describe the United States as a federal State. Discuss the functions of the first two levels of government.

5. Has the power of the U.S. national government increased or decreased in recent decades? Explain.

6. Discuss the following quote: "The widespread availability of home computers, in conjunction with television, may make direct democracy more practical again in the future."

7. How common are virtual and particular representation in the U.S. and Canada today? Explain.

8. Discuss the advantages and disadvantages of single member and proportional representation.

9. What is gerrymandering? Why was it once so widely practiced?

10. What was the background to the reapportionment revolution that started in the 1960s?

11. What is community of interest? Give some examples of community of interest from the area or region where you live.

12. What is the sectional normal vote? What are its traditional regions? How can the concept be used to analyze the outcomes of presidential elections?

13. Discuss the pros and cons of the options discussed in the section, the "Political Geography of the Future."

Selected References

Glassner, Martin I. *Political Geography*. 2nd edition. New York: John Wiley & Sons, Inc., 1996.

———. and De Blij, Harm J. *Systematic Political Geography*. New York: John Wiley & Sons, 1980.

Martis, Kenneth C. and Elmes, Gregory A. *The Historical Atlas of State Power in Congress, 1790–1990*. Washington, D.C.: Congressional Quarterly, 1993.

Merrett, Christopher D. "Research and Teaching in Political Geography: National Standards and the Resurgence of Geography's 'Wayward Child.'" *Journal of Geography,* Vol. 96, No. 2, March–April, 1997, 50–54.

———. "Nation-States in Continental Markets: The Political Geography of Free Trade." *Journal of Geography,* Vol. 96, No. 2, March–April 1997, 105–112.

Shelley, Fred M., Archer, J. Clark, Davidson, Fiona M., and Brunn, Stanley D. *Political Geography of the United States*. New York: The Guilford Press, 1996.

Short, John R. *An Introduction to Political Geography*. 2nd ed. London: Routledge, 1993.

Webster, Gerald R. "Geography and the Decennial Task of Redistricting." *Journal of Geography,* Vol. 96, No. 2, March–April, 1997, 61–68.

CHAPTER

6

*Agriculture, Gathering,
and Extractive Industries*

"The fertility of the soil in the Union, taken as a whole, is remarkably great. With the exception of a few rough mountains, nearly all of which are storehouses of the metals and coal, there is scarcely an acre throughout this wide domain that cannot be cultivated in grain or in grasses for cattle or sheep. . . . Everywhere the value of the soil is incalculable, and the people can secure its fertility and productiveness for evermore by their diligence and care."

(From Jacob Harris Patton, *The Natural Resources of the United States.* American Book Company, New York, 1879)

Although no comparable claim of universal soil productivity was made by Canadian observers, by the last quarter of the 19th century in both Canada and the United States, farming had spread westward discontinuously to the Pacific. In both countries, national policy was to give away public land for farming and ranching and to subsidize canal, railroad, and road construction to make those lands accessible and attractive for farm and town development (Figure 6.1).

Farmers, along with miners and foresters, benefited from policies of easy acquisition of public lands and the encouraged exploitation of the natural wealth of the continent. Indeed, the soil that Jacob Patton praised was by no means the first North American resource to attract attention. Before the first gardens were tended by English colonists in Virginia or by Spanish conquistadores in Florida and the Southwest, and long before the first recorded Canadian farmer—Louis Herbert—planted grain, vegetables, and Normandy apples on land near Québec City in 1617, North America was viewed as a land of natural bounty, with riches of forests, fish, and furs and treasures of precious metals to be quickly, if not easily, garnered.

By the 1530s, French fishermen were busy on the rich fishing grounds and banks near Newfoundland; in 1534, Jacques Cartier was trading with natives for furs in eastern New Brunswick; and English voyagers on a 1584 reconnaissance to America ordered by Sir Walter Raleigh exchanged trade goods for furs on Roanoke Island. And always there was the lure of gold "more plentiful than copper is with us," the English believed. It enticed the first venturers to Virginia in 1607, and all early colonial charters were granted on the assumption that gold and silver would be found, with clauses reserving a fixed amount for the sovereign.

Resources and Primary Activities

By the middle 1800s, the agricultural wealth and promise of North America were widely known at home and abroad, as were the riches of its waters and forests. For two centuries and more, as Chapter 3 summarizes, the "inex-

Figure **6.1**

The head of the line outside the Guthrie, Oklahoma, land office on the day it opened for business in May 1889. Faced with the refusal of most squatters to pay even the $1.25 per acre fee the U.S. government initially required for public land purchase, the Homestead Act of 1862 gave anyone title to 160 acres when the applicant lived at least part-time on the claim and cultivated a small portion of it. Conveniently located land offices registered claims and transferred ownership. After Confederation (1867), Canada was equally generous, giving away vast areas of the Prairie Provinces and advertising its free land policy widely in the United States and Europe. During the Civil War, the U.S. government also began giving away land to railroad companies—6 square miles of land for every mile of track. Altogether, 155 million acres were donated for this purpose.

haustible" storehouse of North America drew immigrants to its shores and its residents farther into and across its interior. The attractions of fishing, farming, and forestry were supplemented by that of mineral resources—metals and fuels—when industrialization began in earnest in the 1840s and 1850s. Indeed, each of the **primary activities** (enterprises that harvest or extract something from the earth) contributed to the economic development that transformed Canada and the United States from subsistence societies into two of the richest and most advanced countries of the world. The primary activities are discussed in this chapter; the secondary, tertiary, and quaternary activities are the subject of Chapter 7.

Natural Resources

The first European visitors to North America were interested in immediate gain, not long-term settlement and investment. Although the terminology would have been unfamiliar, they sought easy exploitation of the continent's **natural resources,** naturally occurring materials that their culture and avarice perceived necessary or desirable. Resources are substances understood by the observer to have value, and that is a cultural, not purely a physical, circumstance. With the passage of time, European newcomers had different resource perceptions than did the indigenous inhabitants. As noted in Chapter 1, American Indians may have viewed the resource base of Pennsylvania, Virginia, or Kentucky as composed of forests offering shelter, fuel, and food (game animals). European settlers viewed the forests as the unwanted covering of the resource that *they* perceived of greatest value: soil for farming. Still later, of course, industrialists appraised the underlying coal deposits, ignored or unrecognized as a resource by earlier occupants, as the item of value to their economy.

At the outset, the two cultures were not far apart. American Indians and Europeans both sought sustenance or profit from the fish of salt and fresh water, from the forests that so densely covered eastern North America, and from the minerals that they knew and valued (salt and native copper, for example), and—more for the Europeans than the American Indians—the soil supporting gardens and farms. The eventual difference was that Europeans used those materials as the basis of specialized industries: fishing, forestry, mining and quarrying, and agriculture. Together, these are the recognized *primary activities* that form the foundation of modern industrial-commercial economies.

Fishing and forestry are **gathering industries** based on harvesting the natural bounty of renewable resources. Mining and quarrying are **extractive industries** based on removing nonrenewable metallic and nonmetallic minerals, including mineral fuels, from the earth's crust. Both types, along with the **agriculture** dependent on the soils extolled by Jacob Patton, make up the initial raw-material production phase of the industrial (*secondary*) activities discussed in Chapter 7.

All primary activities depend on exploiting resources of varying value and permanence. The **renewable resources** that support the gathering industries and agriculture are materials that can be used or consumed, then replenished relatively quickly by natural or human-assisted processes. That renewal is not assured or even possible, of course, if the resource is used to destruction or beyond its ability to support the pressures placed on it. Soil, for example, can be eroded to bare rock and never replaced; forests can be totally cut over, never to regenerate with the original species or luxuriance; and fish stocks or animals can be exploited to the point of extinction. Soil erosion, forest overcutting, and fish stock exhaustion are, in fact, current North American concerns troubling not only environmental conservationists, but also economists, citizens, and government agencies.

When exploitation to exhaustion occurs, the renewable resource becomes a nonrenewable one. **Nonrenewable resources** exist in finite amounts and either are not replaced by natural processes—at least not within a time frame of interest to their exploiters—or are replaced at a rate slower than the rate of use. The groundwater of the Ogalalla aquifer (see Figure 12.4), for example, is so exploited it is now considered by many a nonrenewable resource. More conventionally, the nonrenewable resources are metallic and nonmetallic minerals and **fossil fuels** (coal, petroleum, and natural gas). Although the endowment in these nonrenewable resources was rich and varied, their intensive exploitation in the United States has depleted reserves and forced the country to depend on foreign sources for many of the minerals, including petroleum, that it once had in abundance. While Canada still has plentiful nonrenewable minerals and fuels, concerns are growing in both countries about depletion of stocks and the impact of their exploitation on their environment.

Primary Industries

The more rapid settlement, population growth, and industrialization of the United States encouraged earlier intensive exploitation of its natural wealth than was the case in Canada. Ocean resources were immediately recognized in both countries, and furs and hides were marketable commodities from the beginning. In the utilization of other resources, however, and the development of primary industries based on them, divergence was soon evident. The United States took the lead in agriculture, and farmers found forests an obstacle to be cleared from the land. Forests were also swiftly cut as the domestic market for lumber expanded with the urbanization and industrialization of the U.S. economy (Figure 6.2). That same industrialization led the United States more quickly than Canada to intensive extraction of its mineral riches.

Natural riches were the foundation of national wealth and economic development in the United States and Canada. But with the growth and diversification of their economies, raw materials declined in relative importance in each country's gross domestic product, labor force involvement, and other measures of economic significance. Near the middle of the 19th century, primary activities including agriculture accounted for about half the gross domestic product of both countries. In the mid-1990s the share contributed by forestry, fishing, and mining together dropped to less than 6% for Canada and 3.2% for the United States; farming alone yielded only 1.6% of U.S. gross domestic product and about 2.5% of the Canadian.

Agriculture

Although early permanent settlers quickly set about transforming the wilderness to cultivated landscapes, that transformation did not make all modern Americans or Canadians farmers or convert their countries into unending

Agriculture, Gathering, and Extractive Industries

croplands. Nor are their economies dominated by agriculture. Although farming was the major way of life in the last century in both countries, in the late 1990s, less than 2% of Americans and less than 4% of Canadians were farm dwellers, and only 2.5% of the U.S. labor force and less than 4% of the Canadian was involved in agriculture.

Patton's 19th-century enthusiasm for the wealth of American soils overlooked the reality of physical limitations on farming. At present, less than 60% of the conterminous (excluding Alaska and Hawaii) United States is put to any sort of agricultural use, and no more than 16% is actually in crops in the average year. Less than 4% of Canada's land area is in crops and only 7% in any agricultural use; climate, geology, and topography limit that area to the southern third of the country, generally within 320 kilometers (200 mi) of the U.S. border. Moreover, despite agriculture's former overwhelming dominance in the economic structure of the United States and Canada, in this modern era of urban industrial and service economies, it contributes far less than 5% of the gross domestic product of either country, though both are net exporters—and important ones—of farm commodities. In fact, agricultural exports amount to between 9% and 12% of total U.S. exports by value and 8% to 10% of Canadian.

Figure 6.2
The forests of these Virginia hills are second-growth replacements for the more luxuriant native stands of different species encountered by first settlers over two centuries ago. Some of the original cover was treated as a removable obstacle hiding the soil resource wanted by farmers; later, more was destroyed by large and small coal mining operations seeking still a different resource base. With soil loss and coal depletion, both farmers and miners have let the land revert to less than its original forested wealth.

In the beginning, of course, exports were minimal and the farm economy was largely **subsistence** (producing food solely or primarily for family consumption) in nature and local in orientation. The rich soils later extolled by Jacob Patton lay isolated and remote in the interior of the continent, lacking efficient, cost-effective connections with domestic or export markets. Only with the advent of improved transportation and farm equipment innovation in the 19th century could agriculture's full commercial promise be realized.

Nineteenth-Century Agriculture

After 1800, land—rich, virgin soil seemingly unlimited in extent—was the great magnet for one of the largest migrations in history to and across North America. Between 1790 and 1850, the United States grew from 13 to 31 states. By mid-century, although the Great Plains were still largely unpopulated, statehood had crossed the Mississippi from Iowa on the north southward to Texas, and even distant California was part of the Union.

The effective economic development of the interior, however, depended on more than settlers farming the land. It demanded efficient, low-cost transportation to connect the new farming districts and their towns with national and world markets. And it required innovations in farming techniques and equipment to manage a grassland environment so vastly different from the forestland and soils of the East. Both were forthcoming during the course of the 19th century.

Transportation Innovations

The isolation of the interior was unacceptable to its settlers. Through their growing population and voting strength in Congress, the new states focused national atten-tion on the need for transportation routes to connect them with the outside world. The need was partially satisfied as new private and public roads, including the federal National Road completed to Illinois in 1852 (Figure 6.3), supplemented the rivers and streams that earlier carried rafts of agricultural goods and farm supplies downstream. Steamboats early in the century made the rivers two-way routes, and the many canals dug in the East and eastern Midwest by 1820 gave access to new farming districts. It was, however, the railroads after 1850 that turned great areas of waste and grazing lands into valuable commercial agricultural districts by providing an outlet to the markets of the East. The United States was spanned by its first transcontinental railroad in 1869, Canada was crossed in 1885, and the interiors of both countries became newly attractive and economically viable.

Early Mechanization

Fed by the railroads, the continental interior began to fill following the Civil War (1861–1865) in the United States and Confederation (1867) in Canada. But grassland settlement became feasible only with such necessary innovations as barbed wire fencing to mark and protect fields and the windmill and dry-farming techniques to compensate in part for semiarid conditions and lack of surface water. In addition, a revolution in horse-drawn farm machinery made possible the large-scale commercial farming that the growing eastern urban and expanding foreign markets required. The steel **moldboard plow** (a moldboard is the curved face of the plow that lifts and turns the soil) to turn the stubborn prairie grasses, the mechanical reaper and threshing machines to harvest wheat, and seed drills to speed planting

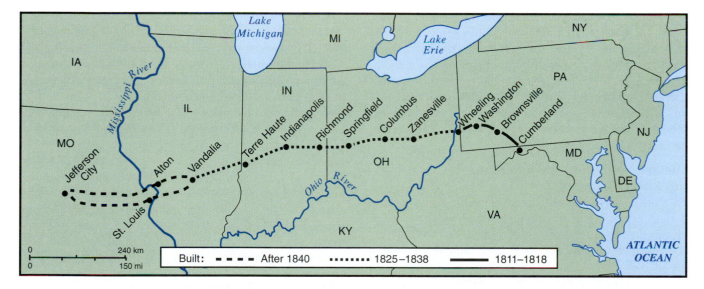

Figure 6.3

The Great National Road. Most road building in the late 1700s was private, for profit. In 1806, Congress voted to fund a public road from Cumberland, Maryland, to Wheeling, on the Virginia side of the Ohio River. That section completed in 1818, the National Road was continued to Columbus, Ohio, and eventually to Vandalia, Illinois, by 1852. For nearly 20 years, the National (or Cumberland) Road was a main land route to the interior, cutting travel time from Baltimore to Wheeling from eight to three days. Property values along it rose, towns and villages were built along it, and for merchants and farmers, it offered a route for shipping merchandise westward and farm products—particularly, livestock trailed along it—eastward at greatly reduced costs.

Agriculture, Gathering, and Extractive Industries

From the beginning, American agriculture was shaped by the apparently endless supply of unoccupied land available for those newly arriving and those, restless or dissatisfied, who wanted to move on. Land was cheap and considered valueless unless settled and cultivated as rapidly as possible. The object to its occupants was to clear it, use it, and move on when it no longer could produce in the easy bounty wanted.

Methods of colonial agriculture were so crude that knowledgeable European observers in the 18th century criticized the colonials for their "land butchery" as they cultivated plots year after year until the fertility was exhausted or the soil lost. No attempt at fertilizing was made and in newly settled forested areas, no provision for fallowing. Rather, new land was cleared to replace the old, and the process of soil exhaustion was repeated. The colonists were not ignorant or maliciously wasteful; they simply practiced a form of land management sensible in

an area and time when land was cheap and labor scarce and costly. Clearing dense forest from additional land, laboriously eradicating stumps, to follow the European system of two- or three-field rotation farming would have been a waste of energy needed for more pressing tasks.

George Washington described the treatment of land in Virginia as an alternation of corn (maize), wheat, and untended fallow ". . . without any dressing, till the land is exhausted," when it was abandoned without "any method taken to restore it." The waste and discard of land practiced by tobacco farmers and, later, their cotton-growing successors in the Southeast was characteristic of farmers in the northern states as well and carried into the interior in the 19th century. A Missouri observer in 1849 commented, "Farming is here conducted on the regular skinning system; most of the farmers . . . in this country scratch over a great deal of ground but cultivate none."

---

were among the many innovations that lifted farming from subsistence to commercial status.

Mechanization touched nearly all farm products. Hay (increasingly needed for the growing number of farm horses and mules), wheat, and corn were most affected; by 1890, about 80% of all wheat grown in the United States was cut by machine, and mechanical harvesting of corn increasingly replaced hand labor. The cotton-seed planter, cotton-stalk cutter, and improved **cotton gin** (to separate seeds, hulls, and foreign matter from the fiber) advanced the efficient production of that crop. By the end of the century, even dairying was affected, with improved cream separators, churns, and other apparatus that effectively removed cheese and butter making from farm to factory and milk and cream production from every family farm to specialized dairy operations.

Commercialization, Specialization, and Rural Change

Railroads and mechanization encouraged regional specialization in farming as environmental and production-cost advantages began to allow distant source areas effectively to compete with or replace local farm suppliers in eastern city markets. The wheat of the prairies and western range livestock provided the cheap and plentiful foodstuffs demanded by growing eastern populations and European markets. Even the distant West Coast developed competitive specializations. A wheat bonanza stimulated by East Coast demand dominated California agriculture during the 1870s and 1880s, soon to be replaced by irrigation farming and specialty crop production. The refrigerated railroad car was

patented in 1867, and by 1870, such cars were transporting fruits and vegetables from the Pacific Coast to compete with local produce in eastern markets. Refrigerated summer shipments of pork and beef soon made midwestern specialized livestock producers effective competitors of eastern general farmers.

Until 1870, agriculture was the occupation of the majority of working North Americans, but the need for rural workers increasingly declined as farm laborers were displaced by horsepower and machinery. With the loss of rural job opportunities, a farm-to-city flight of labor was inevitable. At the same time, farmers grew more efficient. In 1840, when farm work occupied nearly 70% of the U.S. working population, the average rural worker supplied food and fiber for 3.7 Americans (including the farm worker) and 0.2 persons abroad. In 1900, with only 37% of workers in agriculture, each farmer or farm worker fed 5.2 Americans and 1.7 others. Mechanization and draft animals explain the increase (Figure 6.4). Between the middle and the end of the 19th century, the human labor required to produce a bushel of corn declined from 4.5 hours to 41 minutes, and for a bushel of wheat, from slightly more than 3 hours to 10 minutes.

Twentieth-Century Shocks and Challenges

At the start of the 20th century, American and Canadian agriculture seemed mature, its continuing prosperity assured. More than a third of employed workers were engaged in farming (see "Farm Work and Farm Wages").

Figure 6.4
Horsepower and machines transformed U.S. and Canadian agriculture. Colonial and early 19th-century farmers relied nearly exclusively on hand tools and human energy. Scarce labor and cheap land fueled interest in mechanical and animal power to work the lands of the interior. The decades after 1830 were crowded with machine inventions and improvements—powered by ever-increasing numbers of work animals—that forever altered farming practices and productivity. By the end of the century, steam and later gasoline engines began a near-total replacement of work animals in American farming. Shown is a multiple-horse hitch to a large reaper.

Domestic markets, fueled by urbanization and industrialization, continued to grow. Food exports in 1900 were higher than at any previous time, and output appeared balanced with domestic and foreign demand. Instead of a predictable future, however, farming was entering a century of unpredictable change—from assured demand to disappearing markets and from farm prosperity to depression and back to prosperity.

Boom, Depression, and Preparation for Change

The century began well. From 1900 to the end of World War I, commodity prices increased, and farming seemed a stable enterprise with gradually rising land values and increasing efficiencies through mechanization and specialization. Expanding domestic and foreign demand culminated in an explosive growth of output during the war (1914–1918) and beyond to about 1920. Then, worldwide protectionist policies, increasing international competition, and domestic overproduction created a devastating agricultural depression that lasted until a renewed World War II boom of the 1940s.

Depression did not mean stagnation. The 1920s and 1930s saw changes on the farm and in national policy and programs that prepared agriculture for its later wartime expansion and early postwar role of major world food supplier. On the farm, the biggest changes involved the replacement of horses and the addition of equipment made possible by mechanical power. The changes initiated by 1940 prepared the way for an explosion of progress after the war.

FARM WORK AND FARM WAGES

In 1820, slightly more than 2 million persons of both sexes were engaged in agriculture in the United States; in 1840, the number was 3.7 million. By 1880, there were 7.7 million, and by 1900, 10.3 million. The rise in number accompanied a drop in proportion of U.S. workers in farming: from 83% in 1820 to 44% in 1880 and 35% at the turn of the century. No matter that the United States is a land of immigrants, foreign-born farm workers were not prominent by 1900, when only 8.5% of all white agricultural laborers were immigrants. Women, too, were in the minority by the beginning of the century; they comprised 11% of all employed farm workers in 1900. "Employed" is the key word, for no statistics record the incalculable labor contribution of women to the family farm of that or any other date.

By contemporary standards and money value, no laborer was richly rewarded in 1900, and farm laborers were no exception. At the beginning of the 20th century,

farm hired hands had annual earnings of some $170 per year in 1900 dollars, or about $14 a month with room and board. Manufacturing workers in 1900 received almost $490 per year, about $41 per month, but of course provided their own food and lodging. As an observer of the time noted, ". . . the farm laborer . . . really gets, in comparison with his situation as it would be if he lived in the city, . . . very likely a larger amount than he would be likely to earn in any occupation open to him in the city." The advantage, of course, was the room and board or, if these were not provided, the customary rent-free house and the chance to grow needed foodstuffs or purchase them at farm wholesale prices. Yet farm labor was not considered underpaid or the farm worker poor. When a 1909 survey asked whether it was reasonably possible for farm laborers, considering their expected income levels, to save enough to buy a farm that would support a family, respondents felt that more than 70% could do so.

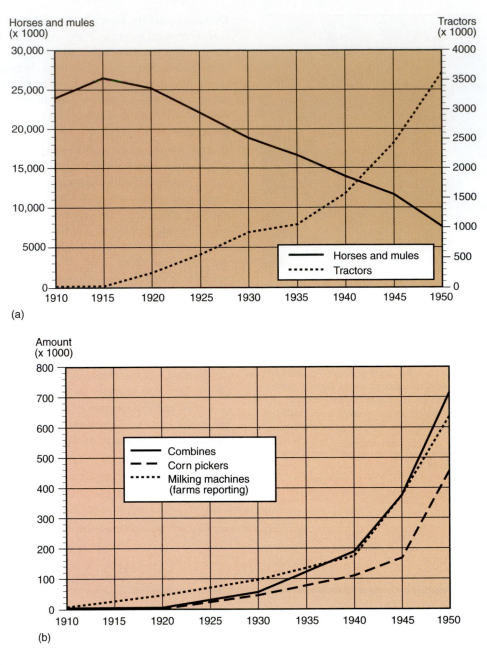

Horses and mules (x 1000)   **Tractors (x 1000)**

(a)

Amount (x 1000)

(b)

Figure **6.5**

Work animals and tractors on farms, 1910–1950 (a); farm machinery increases, 1910–1950 (b); farm output and labor input, 1910–1950 (c). About 1940, modernization and productivity began to accelerate, building on the previous 30 years and pointing to impressive postwar improvements in all phases of agricultural activity.

(a), (b) *Source:* From *Historical Statistics of the United States.* (c) *Source:* Arthur R. Raper, *Rural Trends,* U.S. Department of Agriculture, 1952.

Already in 1910, 1000 gasoline tractors were in use on U.S. farms. Their superiority to animal power was so immediately apparent that a quarter-million were working on farms by 1920. Between 1915—the peak year for farm horses and mules—and 1940, nearly 13 million work animals were replaced, and land earlier devoted to producing more than 200 million bushels of grain and 7.5 million tons of hay was freed for other crops or livestock. The automobile and truck improved farmer mobility and saved time and money moving crops and livestock to market. Efficient but expensive field and farmstead equipment reduced farm labor needs; there were almost 2.5 million fewer farm workers in 1940 than in 1920. **Hybrid seeds** (the product of parent plants of two different varieties imparting consistent superior characteristics to the seed) and improved varieties of food and industrial crops, animal breeding advances, and new chemical fertilizers, pesticides, and herbicides all greatly increased production potential while adding further to farm capitalization and operating expenses (Figure 6.5).

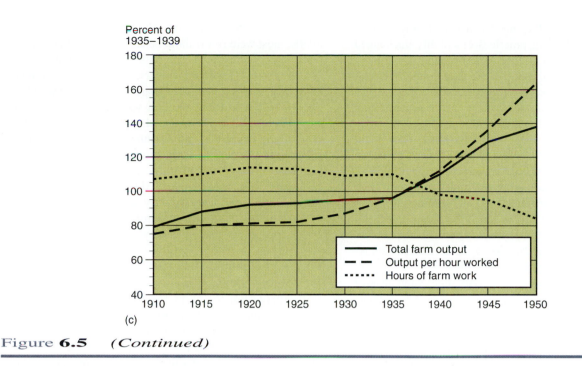

Percent of
1935–1939

Legend:
— Total farm output
- - Output per hour worked
···· Hours of farm work

(c)

Figure 6.5 *(Continued)*

Overproduction and declining prices after 1920 led to government programs designed to ease farm problems and failures. Acreage allotments, commodity loans, crop insurance, marketing agreements and quotas, food stamp programs, soil and water conservation programs and payments, and other similar governmental actions offered rural relief and assistance. Public subsidies for rural electrification began in 1935; within five years, 25% of all U.S. farms were electrified, improving the quality of rural life and enlarging the range of motorized equipment available. These and other innovations, programs, and modernizations culminated in a structure of American agriculture and conditions of farm life far removed from those of 1900. These changes, moreover, were indispensable in shaping the new agricultural patterns and practices of the last half of the 20th century.

The American Farm and Farm Life: 1935–1940

There has never been a "typical" American farm. Variations in location, environmental conditions, crop and market specialization, accessibility, farm size and capitalization, and a myriad of other circumstances have always given American farms regional, local, and individual characteristics. Earlier countrywide conditions of rural isolation and depression were ending by the late 1930s. World War II (1939–1945) and its aftermath fueled a truly remarkable transformation of farm life and farm economy that essentially erased meaningful contrasts between the rural and urban way of life.

Those contrasts were stark before the war. Farmers in the late 1930s—even more than their urban counterparts—were victims of prolonged national hard times. Agricultural

depression had shattered their financial security and kept from them the modern improvements and conveniences beginning to be available to the public at large. The average farm family had less than 60% of the total annual income available to city families, and of that amount, a full 40% was noncash receipts in the form of farm-furnished food, housing, fuel, and other products. The farm family produced about two-thirds of its own food, and subsistence was very much a part of the usual farm economy. Although farm-produced food, fuel, and shelter did give Depression-era rural families a degree of basic security denied their city kin, their low money incomes prevented them from shopping for the goods and services readily available to urban dwellers of comparable annual earnings.

Among the goods that they lacked were the many conveniences—even "necessities"—enjoyed by city folk. Although (with the exception of white and black sharecroppers of the Southeast) the farm families likely lived in houses that had at least the approved standard of one room per person, only one out of ten of the homes had an indoor toilet of any type, chemical or flush. Indeed, only 16% of farm families had any water at all in their houses, and most of that was delivered by hand pump. Less than 10% had any piped water to kitchen or bath, and only 8% had both hot and cold water. It was unlikely the farmhouse had electricity; only 25% did in 1940, and almost all of those had had it less than five years. Kerosene lamps were the rule, wood-fired cook stoves the norm, and only one house in eight had any form of central heating. In much of America, then, the light, warmth, sanitary facilities, and home conveniences almost universal in cities were rare or nonexistent in the countryside.

What the farm family needed it usually supplied by its own efforts, following practices essentially unchanged from the 19th century. While the farm husband would plow the garden plot, it was the farm wife's and children's task to plant, weed, and harvest its produce and to store and preserve all that was needed for the winter. Women also tended the poultry and by tradition sold surplus eggs for "pin money"—often to a local farm town store. They might also milk the one or two family cows and always made the butter—some of which was also sold. More than 95% of farm women regularly canned fruits and vegetables. They also were responsible for rendering lard, making sausage—of meat they ground and animal intestines they cleaned and stuffed—and canning meat, female adjuncts to the autumn male tasks of slaughtering, butchering, and salting or smoking the meat of one or more hogs (Figure 6.6).

Although subsistence was a continuing component of family farms, most were primarily commercial operations. Regional specialties were frequent—grain, dairy products, cotton, and the like—but many farms were "general" in the sense of selling a variety of crops and livestock. In 1939, those sales averaged less than $2000. Commercial family farms in 1940 operated about three-fourths of all cropland in the United States, and during the Depression years of the 1930s, most were run without benefit of the earlier-common hired man, who typically ate with the family and was considered nearly one of its members. Such laborers were paid on average $28 a month with board and another $8 without.

Whether working alone or with hired help, the average Midwestern farmer operated no more than 73 hectares (180 acres), a size that had remained relatively constant for many years. In the general farming areas of the East and Midwest, the single family that lived on and worked the farm probably also owned it, though perhaps bearing a heavy mortgage. For the country as a whole, however, the odds were that the occupant family was a tenant—a renter or sharecropper of land owned by others.

Forces of Change: 1940–1980

Whatever the Depression-era isolation and apparent material deprivation the farm family suffered, its circumstances soon radically improved. Farming after 1940 began a dramatic transformation that would, within a single generation, alter nearly every previous generalization about agriculture and rural life. That transformation stemmed from rapid increases in farm incomes, continuing development and improvement of farm and household machinery, and the application of accumulating scientific knowledge about seeds, fertilizers, pesticides, soil, and nutrition. Part cause and part effect of the transformation were great gains in productivity and increasing commercialization and specialization of farm output. The one element that did not change was the continuing dominance of the family farm. Most farms continued as family businesses with most labor supplied by the farm family. In the United States, 88% of farms were family-operated in 1980; in Canada, 91% were.

Improving market and income prospects of farmers and better farm production and processing equipment encouraged conversion of traditional general farming to commercial specialization. For the United States and Canada, their agriculture became the world's most productive, their consumers spent less on food than those in any other industrialized countries, and exports of agricultural products became their largest single source of foreign exchange earnings by 1980. And all was accomplished by a diminishing number of farms and farm families.

Profitability and Productivity The rapid increase in farm income during World War II reflected the sudden expansion of wartime food and industrial crop demands of the United States, Canada, and their allies, demands that continued during postwar European reconstruction and the Korean conflict (1950 to 1953). As a result, farmlands, crops, and livestock increased in value, the net worth of farm families grew, and mortgages were liquidated. One reflection of the new prosperity was the sharp decline in number of farm units operated by tenants: from 2.4 million in 1940 to 1.4 million in 1950 and to 280,000 in 1980 in the United

Figure 6.6
Caring for chickens and gathering eggs for the table and off-farm sales were just a small part of the innumerable indoor and outdoor tasks that made women indispensable partners in the operation of the 1930s-era family farm.

States. Tenancy rates were reduced to 12% in 1980 (from 39% in 1940). In Canada, only 6% of farms were tenant-operated in 1981. The average farm family that in 1940 had less than 60% of the money income of city families had in 1980 some 88%. That figure was even higher when farm laborers and small and part-time farmers were excluded. Such exclusion is proper, for in 1980, large farms—$40,000 or more in sales—accounted for virtually all farm-generated net income; farms with sales of less than $10,000 (49% of all farm units, but mostly part-time operations and rural residences) actually had negative farm-generated income.

In large measure, increases in farm profitability reflected the tremendous gains in productivity between 1940 and 1980. Total output rose by almost 150%, while total inputs increased less than 10%. The key was technology. Mechanization, hybrid and improved seed varieties, chemical fertilizers, pesticides, herbicides, and irrigation all increased productivity of farm labor. Between 1940 and 1960, productivity per farm labor-hour on average increased some 6% a year, a far higher rate than in industry. Between 1960 and 1980, labor requirements dropped by more than half on roughly constant amounts of land.

Mechanization and Science Improvements in labor efficiency and farm productivity owed much to the increased number, variety, and versatility of farm machines. In 1940, only one farmer in four owned a tractor. By 1960, tractors outnumbered farms by more than a million, and by 1974, they were so common that no further census count was made. Tractors were just the starting point. They led the way for new, bigger, and better machines that not only reduced labor requirements but greatly increased the acreage that a farmer could handle and that had to be operated to justify the cost of the machinery. And the more efficient, specialized, and costly the machinery, the greater was the increase in farm specialization.

Machinery improvements and specializations during and after World War II built on a base extending back, as we have seen, to post–Civil War introductions of reapers, threshers, seed drills, and improved plows that made grass-land agriculture possible. The **combine** (a machine that harvests, threshes, and cleans grain while moving over the field), for example, began to dominate grain harvesting during the war and later was used on soybeans as their importance increased. By 1960, more efficient (and expensive) self-propelled combines accounted for 20% of all combines. The cotton picker that was introduced in 1942 became common in the 1950s and 1960s, cutting cotton production labor-hours from 150 to about 25 per acre. Rice harvesters, machines to dig and load potatoes, pickers for canning tomatoes, and machines to strip tobacco leaves, shake nuts from trees, and harvest peanuts—indeed, specialized machines to speed every phase of production of nearly every crop—were the rule by 1980.

Equipment improvement also benefited irrigation farming as better pumps (plus the low cost of energy) and improved sprinkler systems permitted expansion of ground-water irrigation in Texas, Oklahoma, Kansas, Nebraska, and elsewhere on the Plains, augmenting the traditional gravity irrigation of the western states. Each year showed an increase in area of irrigated land, from some 8.5 million hectares (21 million acres) in 1940 to more than 20 million (50 million acres) by 1980. By the early 1950s, sprinkler irrigation became common, affecting 810,000 hectares (2 million acres) in 1955 and increasing to more than 6 million (15 million acres) by 1980. Eventually, most sprinkler systems were of the center pivot design, giving rise to the distinctive circular green field patterns so visible from above against the semiarid background of the Plains (Figure 6.7).

Productivity also increased dramatically because of aspects of the Green Revolution, as farmers adopted high-yielding hybrid seeds and improved crop varieties, liberally added fertilizers to maximize their genetic potential, and used a wide range of chemical pesticides and herbicides to shelter the crop from insects and weeds. Wheat-strain improvement and selection had a long history, dating to the 1860s and 1870s with continued development to the present (see "Canada's Wheat Fields"). Hybrid seed corn was first introduced in 1926 by Henry Wallace, later secretary of agriculture and vice president of the United States. The double-crossed strains he developed brought a quick increase from an average of 40 bushels to 100 to 120 bushels per acre; hybrid varieties gradually were widely adopted in the Corn Belt (Figure 6.8). More quickly, grain sorghum production was converted to hybrids in a span of four years in the late 1950s. By 1980, all commercial crops had benefited from genetic manipulation to increase yields, uniformity, harvestability, shipping or canning qualities, and the like.

Similar advances were made in livestock genetics, beginning in the 1930s with swine selection for larger litters, faster weight gains, and—later—reduced fat ratios. Dramatic increases in milk production were achieved through artificial insemination and breed improvement—from less than 4500 pounds per cow in 1940 to more than 12,000 pounds in 1980. In 1950, broiler chickens were being commercially produced with a feed efficiency of about 3 pounds of grain ration per 1 pound of broiler live weight. By 1970, both chicken genetics and feed research reduced the ratio to 1.8 pounds, and by 1980, it dropped still further. For livestock of all kinds, manufactured feeds supplemented with proteins, vitamins, and often antibiotics became available in the late 1940s.

Insecticides and herbicides contributed to the remarkable increase in crop yields after 1945. After DDT became available in 1943, a variety of pest-specific poisons for all commercial crops was introduced, as were true herbicides, products of World War II research in chemical warfare. The agricultural chemical industry claims that without its products, food production would decline up to 50% and overall food prices would increase by 36%. Although that claim may be disputed, the positive effect of fertilization on yields

Agriculture, Gathering, and Extractive Industries

Figure 6.7
Center-pivot irrigation creates a distinctive circular pattern of fields and crop production as shown here in western Nebraska. More than 12.5% of Nebraska's farmland is irrigated, much of it by center-pivot systems.

is well documented (Figure 6.9). In their successful search for improved yields, U.S. farmers registered a four-fold increase in commercial fertilizer consumption between 1940 and 1970.

Specialization and Commercialization Multiple lines of evidence point to the reduction of "general" farming after World War II. Increases in farm size, the growing share of very large farms in total agricultural production and income, introductions of specialized and expensive farm machinery, and the application of science to all phases of agriculture are among the indicators of the trend toward product concentration and commercialization between 1940 and 1980.

Farm size is a telling indicator of the increasing concentration of production. By 1980, the largest 1% of United States farms—with sales of $500,000 and more—accounted for two-thirds of net farm income. The largest 6% of Canadian farms in 1981 (sales of $160,000 and more) claimed 38% of the gross value of farm production; the smallest 50% (sales of less than $22,000) took in only 7%.

Concentration of land among large landowners was also apparent: the largest 5% of owners held nearly 25% of the farmland in the Corn Belt and Great Lakes states in 1978, 49% in the Southeast, and 71% in the Pacific Coast region.

Size and specialization went together as increasingly sophisticated machinery required ever-greater capital investment that could be justified only by ever-larger operating units with ever-growing gross sales and net profits. Smaller and middle-size farm operations were less able or unable to compete; farms had to expand by buying or renting additional land. The percentage of operators who rented at least part of their land tripled (to 30%) between 1940 and 1980 in the United States. In the Prairie Provinces, part owned/part rented farms represented 42% of all cereal farms and 57% of the total grain producing area in 1981. Because of the rapid rise in farmland prices during the 1970s, both land indebtedness of farm owners and absentee ownership by nonfarm investors greatly increased.

Specialization and commercialization were accompanied by increasing ties between producers and purchasers

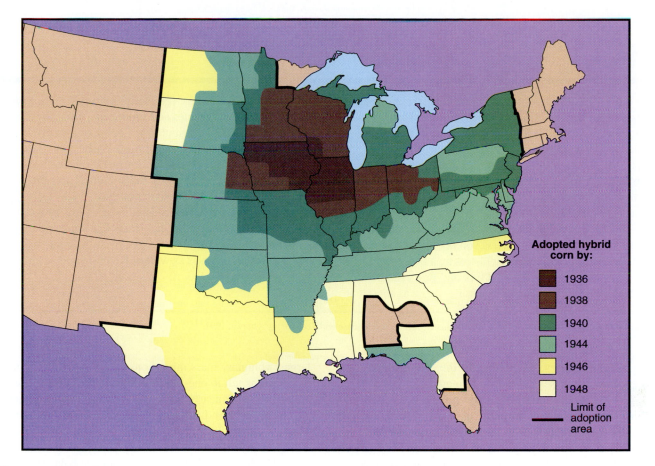

Figure 6.8

The adoption of hybrid corn proceeded rapidly following its effective introduction in the Corn Belt in the early 1930s. The map patterns show areas in which 10% or more of the corn acreage was planted to hybrid seed in the specified years.

Source: Redrawn from data in Zvi Griliches, "Hybrid Corn and the Economics of Innovation" in *Science* 132(3422):277 (July 29, 1960), Figure 3.

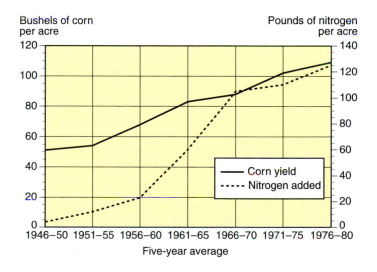

Figure 6.9

Fertilizer use and corn yields in Illinois. Average corn yields doubled when 150 pounds of nitrogen an acre a year were added, though part of the gain was also attributable to improved hybrids and adequate phosphorus and potassium. By 1980, fertilizer was being used on 96% of corn acres in the United States.

Source: Redrawn from *Our American Land: 1987 Yearbook of Agriculture,* p. 310.

In the 1840s in Peterborough, Ontario, a farmer named David Fife chose a strain of wheat from a variety of samples. He called it Red Fife, and it became the main variety of wheat grown in Canada in the 19th century.

The wheat grew well, even in Canada's short growing season, but it was a gamble on the Prairies where the weather was drier and the growing season even shorter.

In 1907, Charles Saunders, chief wheat researcher at the Experimental Farms and son of William Saunders, the farm's director, began field-testing a new type of wheat. It was a hybrid of Red Fife and Red Calcutta, called Marquis. The summer it was tested was short, but even in harsh conditions, the Marquis ripened eight days earlier than other varieties and produced a good harvest.

In 1911, it was distributed to Canadian farmers. That year the country produced 75 million bushels of wheat, three times as much as in 1901. Farmers' incomes jumped by $100 million a year and by 1918, Canada was the world's second largest wheat exporter.

Marquis was a farmer's dream and nearly 90% of Prairie farmers relied on it; but in 1916, a wheat rust epidemic revealed the wheat's weakness. Marquis did not resist the plant diseases as well as Red Fife and scientists began formulating other rust-resistant strains of wheat. . . .

Source: From *Canada Yearbook 1992,* p. 353. Minister of Industry, Science and Technology, Ottawa, 1991.

of farm commodities. Beginning in the 1950s, specialist farmers and businesses developed strategies for standardizing agricultural products to reduce purchasers' handling and processing costs and for averting farmer risk in production and sales. Processors began to demand uniformity of product quality and timing of delivery. Canners of tomatoes, sweet corn, and other vegetables demanded delivery of raw products of uniform size, color, and ingredient content and in volume and on dates that accorded with cannery and labor schedules.

And farmers wanted the support of a guaranteed market at an ensured price to minimize the uncertainties of their specialization and stabilize the return on their investment. The solution was contractual arrangements or **vertical integrations** uniting contracted farmer with purchaser-processor. Vertical integration signifies corporate control and ownership of all phases of production, including the provision of seed, livestock, and feed, with the farmer receiving a predetermined unit price for delivery of finished commodities. Broiler chickens of specified age and weight, cattle fed to an exact weight and finish, wheat with a minimum protein content, popping corn with prescribed characteristics, potatoes of the kind and quality demanded by particular fast-food chains, and similar product specification became part of production contracts between farmer and buyer-processor. The percentage of total farm output produced under contractual arrangements or by vertical integration (where production, processing, and sales were all coordinated within one firm) rose from 19% in 1960 to 30% in 1980.

Emptying of Rural Areas

Massive population losses and relocations accompanied the post-World War II agricultural changes in North America. Within the United States, four-fifths of the farm population abandoned farming between 1940 and 1980, one of the largest voluntary mass migrations ever recorded. The number of farms fell to one-third of their mid-1930s peak as average farm size nearly tripled. Canada experienced nearly the same kind and degree of change: farm populations dropped from 27% to 4% of the total, and farm numbers fell by nearly 60%.

The farm population loss was not age-balanced or uniformly distributed areally. Young people led the exodus. Both the mechanization that decreased farm labor requirements and the expanding farm size and capitalization requirements reduced the opportunities for younger persons to acquire sufficient land or capital to enter commercial farming.

Regionally, the South suffered the greatest decline as northern industrial jobs continued to offer alternative livelihoods to its small farm sharecroppers and tenants displaced as cotton farming became mechanized (Table 6.1). Many also lost their land as leases were terminated or sharecropper farms bought so that former cotton land could be used for cattle or sold to timber companies. Black farm operators were the most affected. From 1940 to 1980 nationally, the number of white farm operators dropped by 60% and black operators by 92%. Farm abandonment—and reversion to woodlot and forest—was common in the Northeast as well, a response to off-farm job opportunities and low returns from marginal farmlands.

In all regions, farm consolidation and displacement of farm families reduced the population and purchasing-power support necessary for the continued prosperity of small farm towns and their merchants. In all agricultural regions, farm town stores and theaters closed, doctors and other professionals and service workers left, and shops and offices were shuttered. Not only had most customers and clients gone, but any still remaining now traveled in their cars and pickups over improved roads and expressways to more distant communities offering a greater range of services and shops.

Table 6.1
Regional Change in Farm Population

Region	Percent Change
United States (1940–1980)	−80
Northeast	−82
Midwest	−71
South	−86
West	−70
Canada (1941–1981)	−67
Atlantic	−87
Central	−70
Prairie	−59
Pacific	−42

Sources: U.S. Bureau of the Census *and* Statistics Canada.

1. Dairying and market gardening
2. Specialty farming
3. Cash grain and livestock
4. Mixed farming
5. Extensive grain farming or stock raising

Figure 6.10

von Thünen's model. Recognizing that as distance from the market increases, the value of land decreases, von Thünen developed a descriptive model of intensity of land use that holds up reasonably well in practice. The most intensively produced agricultural crops are found on the land close to the market; the less intensively produced commodities are located at more distant points. The numbered zones of the diagram represent modern equivalents of the theoretical land use sequence von Thünen suggested over 150 years ago. As the metropolitan area at the center increases in size, the agricultural specialty areas are displaced outward, but the relative position of each is retained.

And their home freezers and refrigerators made the frequent trip to the farm town grocery unnecessary. Prosperity was ensured in a few farm towns by their having regional consolidated schools, farm implement and automotive dealerships, and the like.

The New Agricultural Landscape

Agriculture in the United States and Canada at the start of the 21st century bears the imprint of its earlier stages of development and diversification and incorporates the multitude of scientific and marketing changes appearing since mid-century. It also reflects both enduring natural environmental features and changing market locations and demands. Because of the diversity of those features and forces, agriculture is by no means a uniform activity. Even though, as we shall see, a rational set of generalized agricultural regions may be recognized, the farming landscape differs widely in product mix, farm size, financial and management arrangements, and a host of other variables.

A Model of Agricultural Location

The evaluation of market opportunities, transportation costs, and environmental circumstances has always affected farm production decisions. Early in the 19th century, Johann Heinrich **von Thünen** (1783–1850), a Prussian landowner, observed that lands of apparently identical physical properties were used for different agricultural purposes. Around each major urban market center, he noted, developed a set of concentric land use rings of different farm products (Figure 6.10). The ring closest to the market specialized in perishable commodities that were both expensive to ship and in high demand. The high prices they could command in the urban market made their production an appropriate use of high-valued land near the city. Rings of farmlands farther away from the city were used for less perishable commodities with lower transport costs, reduced

demand, and lower market prices. General farming and grain farming replaced the market gardening of the inner ring. At the outer margins of profitable agriculture, farthest from the market, were found livestock grazing and similar extensive land uses.

To explain why this should be so, von Thünen suggested that the uses to which parcels were put were a function of the differing "rent" values placed upon seemingly identical lands. Those differences, he claimed, reflected the cost of overcoming the distance separating a given farm from a central market town. The greater the distance, the higher was the operating cost to the farmer, since transport charges had to be added to other expenses. When a commodity's production costs plus its transport costs just equaled its value at the market, a farmer was at the economic margin of its cultivation. A simple exchange relationship ensued: the greater the transportation cost, the lower the rent that could be paid for land if the crop produced was to remain competitive in the market.

In the simplest form of von Thünen's model, transport costs are the only variable, and the relationship between land rent, distance from market, and crop selection is decisive. Crops that have both the highest market price and the highest transport costs will be grown on high-priced land nearest to the market. Less perishable crops with lower production and transport costs will be grown on land of lower rental value at greater distances away. The

Agriculture, Gathering, and Extractive Industries

Figure 6.11

A schematic view of the von Thünen zones in the sector south of Chicago. There, farmland quality decreases southward as the boundary of recent glaciation is passed and hill lands are encountered in southern Illinois. On the margins of the city near the market, dairying competes for space with livestock feeding and suburbanization. Southward into flat, fertile central Illinois, cash grains dominate. In southern Illinois, livestock rearing and fattening, general farming, and some orchard crops are the rule.

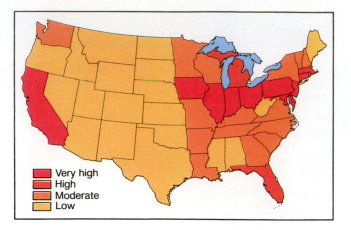

Figure 6.12

Relative value per acre of farmland and buildings. In a generalized way, per acre valuations support von Thünen's model. The major metropolitan markets of the Northeast, the Midwest, and California are in part reflected by high rural property valuations, and fruit and vegetable production along the Gulf Coast increases land values there. National and international markets for agricultural goods, soil productivity, climate, and terrain characteristics are also reflected in the map patterns.

Source: From *Statistical Abstract of the United States.*

von Thünen model may be modified by introducing ideas of differential transport costs or variations in topography or soil fertility, but with or without such modifications, his analysis helps explain the changing crop patterns and farm sizes evident on the landscape at increasing distance from major cities (Figure 6.11). Farmland close to markets takes on high value, is used intensively for high-value crops, and is subdivided into relatively small units. Land far from markets, commanding lower price, is used extensively and in larger units. The presumed effect of these relationships for the United States is seen in Figure 6.12.

Farming Regions

With the passage of time, growth in market size, improvement of transportation access, and modernization of equipment, specialized farming replaced general and subsistence cropping in nearly all parts of the United States and Canada, and some of the expectations of von Thünen's model became evident on the landscape. When a majority of farmers adopted the same specializations in response to favorable environmental circumstances, market conditions, and costs of land and transportation, recognizable agricultural regions or "belts" gradually emerged. Indeed, such terms as Corn Belt, Cotton Belt, Dairy Belt, and others became popularly recognized regional identifications for large sections of the two countries.

Such belts or "types-of-farming regions" were identified and defined by the U.S. Department of Agriculture in a series of studies from the early to the middle 20th century. In the 1950—and final—version, the department divided the United States into 9 major agricultural regions, 61 subregions, and 165 generalized type-of-farming areas.

Distinctive and tightly bordered "types-of-farming areas" are less prominent today than they were a half-century earlier, and farm specialization is too intricate even for the large number of type areas earlier recognized. Indeed, such "areas" and "belts" are by no means permanent descriptions of rural land use. They have markedly changed—in some instances disappeared—in the past and are still changing today. Yet, farmers continue to respond comparably to similar natural and economic considerations, and broad belts or districts of regional specialization are still recognizable. These are shown on the map of generalized agricultural regions (Figure 6.13) and are discussed on the following pages.

Truck Crops, Fruit, and Mixed Farming   Fruits and **truck crops** (vegetables grown for market) should be classic examples of von Thünen's principles of farm responses to crop value, perishability, and transportation costs. Indeed, the summer appearance of roadside fruit and vegetable stands and farmers' markets in towns throughout the United States and Canada are enduring evidence of the attraction and competitive marketability of local produce for specialized market gardens or diversified farms. Yet supermarket availability of off-season Chilean fruit or season-extending Mexican and Central American vegetables tells us that with modern facilities for refrigeration and speedy transportation—and with grower selection of crop varieties that can be picked early

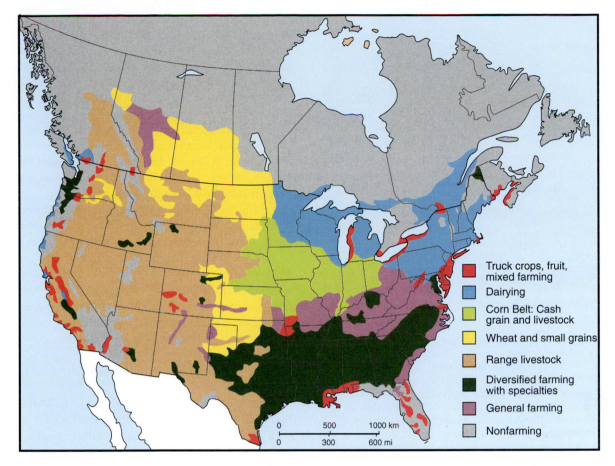

Figure 6.13
Generalized agricultural regions.

and kept long—close spatial association of truck-crop production and local market no longer is determining.

The dispersion of fruit and truck-crop farming in the United States today, as Figure 6.13 suggests, is less tied to the distribution of urban centers than it is to regional conditions of climate and soil and local circumstances of slope, drainage, and the like. Even in those districts where the value of fruit and vegetable crops—and their occupied areas—is sufficient to warrant map recognition, fruit or truck farms are frequently intermingled with farms of other specializations or types, often dairy or poultry farms. The map generalization of fruit and truck cropping suggests common characteristics and requirements that are not, in fact, universal.

Deciduous fruits (apples, pears, peaches, and so on), for example, are associated with either protected valley or mountain slope locations, as in Washington, California's Sacramento and San Joaquin Valleys, the Ozark Plateau, and the Cumberland and Shenandoah Valleys from Pennsylvania into Virginia. In these locales, slopes provide the air drainage that helps avert late spring or early fall frosts. Similar protection is afforded by major water bodies, and important deciduous fruit and truck cropping districts are found along the eastern shore of Lake Michigan and the southern

shores of Lake Erie and Lake Ontario. In the main, comparable protected growing districts are limited or absent in Canada. As Figure 6.13 indicates, they are restricted to small areas in British Columbia, southern Ontario and Québec, and the New Brunswick–Nova Scotia region.

Climate and—if necessary—irrigation water availability are the chief concerns for growing citrus fruits and vegetables for the winter and early market. Southern California and southwestern Arizona, the lower Rio Grande, and central Florida meet the requirements for citrus (Figure 6.14); southern Florida has concentrated on winter vegetables, and specialized truck and mixed farming are found along the Gulf Coast.

In all areas of specialized fruit and truck cropping, associated warehousing, packing, canning and freezing, and—in the case of citrus fruits—juice canning and concentrating are well-developed and important parts of the local industrial and seasonal labor pattern.

Dairying Northeastern United States, southeastern Canada, and the Great Lakes states were long known as the Dairy Belt of North American agriculture. The name remains, and dairying still is widespread within the zone identified with it

Agriculture, Gathering, and Extractive Industries

Figure 6.14
Oranges are Florida's leading money crop, followed by greenhouse products and tomatoes. The state produces some 70% of the country's orange crop and 80% of its grapefruit. Its citrus groves, like those here in the lake district of central Florida, and its winter vegetable fields can be severely damaged by killing frosts. In recent years, citrus acreage has declined and shifted southward, in part reflecting severe winter frost damage of the 1980s and 1990s.

on Figure 6.13, but the regional activity the name describes has undergone great recent changes. Among these are a great decrease in the number of farms with a few dairy cows for sideline or household milk production, a decline in the number of commercial dairy farms with a large increase in the herd size on those that remain, and significant expansions in milk production in West Coast, Southwest, and Florida locations to satisfy growing regional market demand.

These shifts emphasize the fact that milk production, once almost universal on American and Canadian farms, has become a commercial farming specialization. Between 1960 and the 1990s, the number of dairy cows in the United States dropped by more than one-half, but the production of milk increased by nearly 20%, and milk output per cow expanded by more than 130%. The Canadian experience is much the same: the number of dairy cows decreased by 42% between 1971 and 1991, and the number of farms with dairy cattle declined by 73%. Even in Québec, which contains most of Canada's Dairy Belt, farms with dairy cattle declined by two-thirds over the 20 years. At the same time, dairy specialization has increased on the reduced number of farms; more than half of the province's commercial farms are now classed as dairy farms.

To reduce milk production support payments, the U.S. government has purchased and destroyed entire dairy herds, many of them from smaller operations in the traditional Dairy Belt. Yet dairy products remain by value of off-farm sales the leading farm income earner—or among the top four commodities sold—in all the Northeastern and Middle Atlantic states and lead farm sales in Michigan, Wisconsin, and Minnesota. These three Great Lakes states alone account for nearly 30% of U.S. milk production; the Northeast accounts for another 20%. For these reasons, the recognition of a distinct multistate northeastern and north central dairy region still remains valid. More localized dairy specialization is encountered in Washington, Oregon, and California, where dairy products are also major farm income earners. Some fluid milk, of course, is produced in every part of the country for local markets.

The traditional Dairy Belt states have physical conditions that are generally well-suited for their recognized specialty and not, for example, for cereal crops. They lack the length of growing season and the rich prairie soils of the Midwest Corn Belt or the soils and level terrain of the northern Wheat Belt states. Their cool climate with ample, well-distributed rainfall, however, is favorable to hay and pasture development. In addition, the Dairy Belt has major nearby urban markets for fluid milk, outlets also important in West Coast dairying. Those parts of dairying areas not well located for fluid milk sales produce milk for cheese and condensed-milk processing.

Increasingly, the traditional Dairy Belt has become a diversified farming belt. Dairy operations are combined with or interspersed with poultry, fruit, or truck-crop farming in southern New England, the Hudson Valley area of New York, and the Pacific Northwest. Elsewhere, feed grains, wheat, and general farming are allied activities. In all sections, the future of dairying is in some doubt. The dietary trend toward low-fat milk and margarine threatens traditional markets and products; the success of 5000 to 10,000 cow operations in the South and Southwest and the evident trend everywhere toward fewer and larger commercial dairying operations suggest that the small-herd, self-contained, family-operated dairy farm may be more part of the past than of the future.

The Corn Belt: Cash Grain and Livestock The agricultural heart of North America is the Corn Belt, arguably the richest farming area of comparable size in the world. Stretching across the interior of the United States from central Ohio to central Nebraska and into parts of adjacent states, the Corn Belt covers only about 10% of the area of the country but accounts for some two-thirds of U.S. sales of corn and soybeans, crops that represent the major source of farm income (Figure 6.15).

Livestock—hogs and cattle—were formerly key parts of the farm economy in nearly all parts of the Corn Belt, fed from the crops produced on the farms as a form of on-site grain conversion. By 2000, however, a transformation to a near-exclusive **cash grain** (off-farm sales of grain alone) economy had spread throughout much of the region, reflecting universal adoption of high-yielding hybrid corn (Figure 6.8) and improved bean varieties. In addition to

Figure 6.15
Open storage of 1 million bushels of Iowa corn. The productivity of Corn Belt farms in some years exceeds both the capacity of storage facilities to hold the bounty and of the market to quickly absorb the surplus.

continuing labor needs, livestock requires expensive additional investment in equipment and facilities for breeding hogs and feeding cattle. It has remained an important component of the farm economy primarily on the margins of the region.

The importance of the corn and soybean combination in the Corn Belt economy is a reflection of productivity and profitability, not of inability to grow other farm products. In fact, with the exception of subtropical crops, every major American farm product can be—and is—grown within the Corn Belt; Ohio and Indiana, for example, rank second and third among the states in tomato sales. But the region's specialties provide maximum return for labor and capital input. The key is the natural endowment of the Corn Belt, wisely but intensively used. The land is generally level, permitting easy use of efficient, specialized machinery. The soils are deep and fertile, rich in organic matter and benefiting from the superb structure imparted by millennia of grassland

cover. Sufficient rainfall well-distributed throughout a lengthy growing season is coupled with the hot days and warm nights ideal for corn. Taking advantage of this endowment, farmers have put more than 75% of all farmland into crops, and to maintain the fertility and productivity of that cropland, they apply about one-quarter of all commercial fertilizer and agricultural chemicals used on U.S. farms.

Chemical fertilizers and the decision of many Corn Belt farmers to concentrate solely on cash sales of grain have reduced the need for manures and leguminous hay crops. The older three-year crop rotation cycle of corn, small grains, and hay has been replaced by a cash grain, two-year corn-soybean rotation. Since soybeans are planted later and harvested earlier than corn, the two crops lend themselves to management by a single operator well-supplied with very expensive equipment. Capitalization needs plus relatively low unit prices for crops in most recent years have induced farmers to enlarge their operations,

Agriculture, Gathering, and Extractive Industries

usually by renting additional land near or on the margins of their owned properties. More than half of the farmland in the Corn Belt is now included in "part-owner" operations. While absentee land ownership has increased, family farm operations are still the Corn Belt norm, even though farm size in many instances has increased to 405 or more operated hectares (1000 or more acres).

Concentration on a two-year rotation of row crops with elimination of grasses and leguminous hay and no return of animal manures has altered the landscape and taken its toll on Corn Belt soils. The elimination of livestock has also erased the need for field fences and the large, all-purpose barns of the horse-power era. The slatted corn cribs to hold and dry ear corn disappeared with the advent of field shelling of corn by combine. Although "no-till" farming has begun to grow in popularity, most land is still plowed, cultivated, and planted at high density in closely spaced rows. Soil loss, even in this deep soil region, has become a concern. Without careful observation of soil conservation practices, Iowa land continuously in corn shows soil losses of between 33 and 42 metric tons per hectare (15 and 19 tons per acre). Eleven metric tons per hectare (5 tons per acre) is generally considered the soil-loss tolerance level. Without careful management, North America's richest farmland slowly washes and blows away.

Wheat and Small Grains   Rainfall decreases in amount and reliability westward from the Corn Belt. Small grains, particularly wheat, become the cash grain replacement for corn and soybeans, not because they find their optimum environment in the semiarid Plains but because no other crops grow as well there or yield comparable income. The Wheat Belt, extending discontinuously from the Prairie Provinces southward into Oklahoma and west Texas, in general has a broad level to undulating surface well-suited to the large-scale mechanized operations of extensive wheat farming. Favorable, too, are the deep, dark grassland soils, precipitation between 35 and 75 centimeters (15 and 30 in) a year, a growing season ranging from 90 days in the north to 200 days in the Texas panhandle, and dry and sunny summers suited for ripening and harvesting wheat. Where the terrain is too irregular for machine operation, the land is left as range for beef cattle.

Hard **winter wheat**—planted in the fall and harvested in June or July—occupies the southern part of the Wheat Belt centering on Kansas. Grain and forage sorghums, members of the grass family that look much like corn, are the other major crops, planted in spring and harvested in fall. The northern part of the Wheat Belt is the **spring wheat** region, with wheat and associated grains (oats and barley, particularly) planted in spring for fall harvesting. In area, this is the larger portion of the Wheat Belt and the dominant farming region of the Prairie Provinces, which contains 80% of all farmland in Canada and 98% of Canadian wheat acreage and centers on Saskatchewan. A third major wheat region, producing both spring and winter

wheat varieties, lies in the Columbia River Basin of the Pacific Northwest.

Farms throughout the Wheat Belt are large; nearly half the farms in Saskatchewan, for example, are more than 405 hectares (1000 acres). The average farm in Kansas is more than 405 hectares (1000 acres), and in North Dakota more than 525 hectares (1300 acres). Machinery needs are great, but farm operator labor inputs may be relatively small considering the size of farm units. The planting period, which lasts about a month, represents about half of the year's work. Wheat harvesting is concentrated in at most a two-week period and frequently is handled by contractors operating giant, self-propelled combines that begin the winter wheat harvest in late May in Texas and move northward at a rate of about 16 kilometers (10 mi) a day, working continuously until they complete the fall harvest of spring wheat in the Dakotas and Canada (Figure 6.16). Unless other grains or livestock are part of the farm operation, not much is required between planting and harvesting, and the isolated farmsteads are frequently unoccupied. Off-farm residence and shorter work-period stays are evidence of the "suitcase" or "sidewalk" farmer[1] increasingly characteristic of the Wheat Belt.

Range Livestock   Range land in the western United States covers about 285 million hectares (700 million acres)—some one-third of the land area of the conterminous states. Range occupies a much smaller total and proportionate area of Canada, again in the western districts (Figure 6.13). Just as the Wheat Belt replaced corn and soybeans on the dry margin of the Middle West, range livestock replace extensive wheat and small grain farming in that part of the West where soils, elevation, rugged topography, and climate make grain cultivation no longer feasible. Range livestock—primarily cattle—in a sense are the product of last resort, a land use that is profitable when growing crops is not possible. The exception is cropland farming under irrigation systems along some river valleys.

In the Southwest from Texas to southwestern Nevada, winters are mild enough to permit year-round grazing, though supplemental feeding may be necessary in dry years and some late winter periods. Feeder cattle and lambs are generally shipped to outside areas for fattening, though some finishing is done locally, usually in irrigated districts. To the north over the rest of the Range Livestock region, grazing is seasonal under widely varying conditions of temperature, rainfall, and topography (Figure 6.17). In

---

[1]"Suitcase farmers" own or rent land in several locations, moving with their equipment between holdings—and changing their short-term residence—during planting and harvesting seasons. "Sidewalk farmers" reside in urban communities, storing their equipment on the rural land they cultivate. These "absentee" operators are nearly exclusively in grain farming, where a daily management presence throughout the crop year is not necessary.

Figure **6.16**

Modern contract harvesters follow the ripening wheat northward through the plains of the United States and Canada. During the 1880s, the harvester-thresher, or combine, was first introduced in the wheat fields of California and soon found its way to the Wheat Belt of the Plains. The huge machine was pulled by 20 to 40 horses and could complete all operations from reaping through final bagging of the wheat, harvesting from 25 to 45 acres a day. The combine drawn by steam tractor appeared in the 1890s and the self-propelled combine after World War II. The great capital investment in those machines—and the brief period they are needed on any one farm—make attractive the use of contractors operating their own equipment.

Figure **6.17**
Cattle grazing on the range in Utah.

Agriculture, Gathering, and Extractive Industries

the Intermontane district, for example, migratory grazing is the rule, with winter, spring, and fall grazing on plateaus and lower-lying grassland and brush areas and summer grazing in the mountains. Along the eastern portion of the region, nonmigratory grazing is the norm. A few areas in Kansas, Oklahoma, and Texas can support year-long grazing, but farther north into North Dakota and South Dakota and westward through Wyoming and Montana, two to four months of winter feeding is the rule. In the Canadian portion of the range country, grazing is confined to the middle and lower elevations, and winter feeding is always necessary.

The essentials in the range economy are land for grazing, water, and winter feed. Because the marginal lands of the West have low carrying capacity and much is available only seasonally for grazing, large expanses are required for reasonable ranching efficiency and security. Even a small family-operated ranch may have 4860 or more hectares (12,000 or more acres), and many are far larger. Only a little more than half of the land used for pasture and grazing in the western range area of the United States is privately owned; more than 40% is public land, and 6%—but still more than 16 million hectares (40 million acres)—is reserved for American Indians.

Federal land, most of it administered by the U.S. Forest Service or the Bureau of Land Management, may be used by ranchers under lease or permit. Water in semiarid range country is a continuing concern, for range cattle require 20 to 35 liters (5 to 9 U.S. gallons) of water a day. To supply it, ranchers depend on dammed streams or well water, and water holes or tanks are spaced across the ranch. Winter feed may be cut from ungrazed ranch land or, if the rancher is fortunate, from hay grown on irrigated land.

Although the casual tourist passing through the Range Livestock region may see barrenness and desolation, in reality the journey is through an actively used and productive major agricultural region of North America. The widely spaced, relatively small ranch headquarters are not often seen from the main highways. The infrequent small towns offer few amenities to the traveler, but house social and business services for the sparse range country population and, in their corrals and loading chutes, show their close functional relationship to the region's economy.

Diversified Farming with Specialties By and large, the diversified farming with specialties category in Figure 6.13 contains the areas of the southeastern and south central United States that, until 1950, the Department of Agriculture designated as the Cotton Belt. Cotton is still part of the crop complex of that former belt, but has been displaced in both location of production and relative regional importance in the past half-century. When cotton was truly "king" in the middle of the 19th century, the South produced some 80% of the world's cotton for export. The crop suffered setbacks from the Civil War and the loss of slave labor, the spread of the boll weevil across the region in the early 1900s, soil exhaustion and erosion,

and changes in clothing fashion that reduced the domestic market for the fiber. Cheaper, more efficiently produced irrigated cotton from Texas, Arizona, and California began to replace that from the Old South, though a modest revival came during the 1980s, particularly in the lower Mississippi Valley (Figure 6.18).

The replacements for cotton in the old Cotton Belt are of two types: areal and income. In area terms, much former cotton land has been converted to woodland or to pasture, and much as well has simply remained idle. Pulp and paper companies have purchased or leased large tracts throughout the Southeast for pulpwood production, and the volume of standing sawtimber—hardwood and softwood—has increased by more than 200% since the early 1950s. Idle and pasture land also are important replacement uses. In this region of year-round grazing, the number of landholders having off-farm full-time jobs and untended cattle on pasture has rapidly increased.

Since 1945, soybeans replaced cotton as the leading crop of the region. However, for every state but two in this southern Diversified Farming area, cattle and broiler chickens rather than crops are the leading farm income producers. Beef cattle are roughly uniformly distributed over the region, while broiler production shows a spotty concentration in nearly every state but takes first place in farm receipts only in Georgia, Alabama, Mississippi, and Arkansas. Peanuts are important in southwestern Alabama, throughout southern Georgia, and in eastern Virginia and North Carolina. Rice production is concentrated along the Gulf coastal prairies of southeastern Texas and southwestern Louisiana and along the Mississippi River in Arkansas and Mississippi. Tobacco, usually associated with only two or three southern states in the popular view, is actually rather widely grown. It is cultivated, for example, in Pennsylvania, Wisconsin, and southern Ontario in the North and northern Florida in the South, but its major occurrence is in northern Kentucky, south central Virginia, and eastern North and South Carolina. Its relative importance nationally, however, is declining. Between 1980 and 1994, tobacco production dropped by about 18%.

Outliers of the Diversified Farming district are found as far away as Maine and New Brunswick, where potatoes, livestock, and dairying are emphasized, and in the Puget Sound–WilloMette Valley area of the Pacific Northwest, where grains, cherries and berries, hops, dairying, and market vegetables are among the variety of farm interests on the relatively small amount of available agricultural land.

General Farming The largest contiguous area of General Farming is found south and east of the Corn Belt where climate and soils permit production of many different kinds of crops and livestock, but favor none in particular. Broken, hilly topography limits the extent of any single activity and keeps much of the land in pasture. Small fields of corn, oats, hay, soybeans, truck crops, and orchards are interspersed

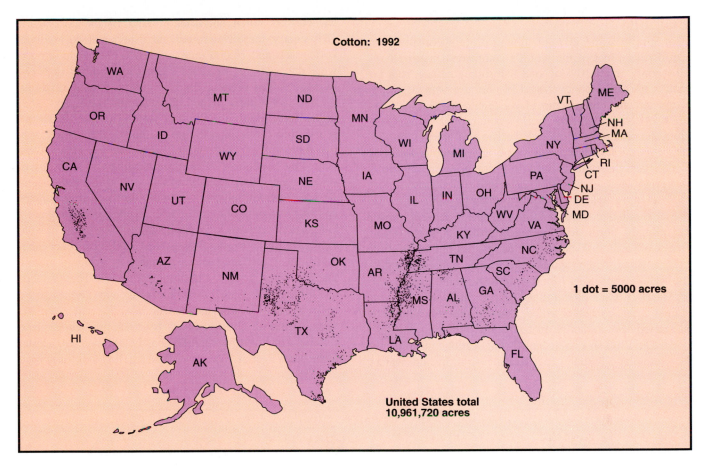

Cotton: 1992

1 dot = 5000 acres

United States total
10,961,720 acres

Figure 6.18

Cotton acreage in 1992 has shifted westward and become much more localized than was the pattern when the Cotton Belt spread across all of the southeastern states.

Source: From *1992 Census of Agriculture,* Vol. 2, Subject Series, Part 1, *Agricultural Atlas of the United States,* page 164. U.S. Department of Commerce, Bureau of the Census, 1995.

with a good deal of idle land and woodlot and, near larger cities, nonfarm rural residences.

An observer of the agricultural scene has commented: " 'General farming' is a euphemism for 'nonfarming' as far as most of the Appalachian and Ozark uplands are concerned." This observation also serves as a useful reminder that a map pattern sketched or a farming category assigned by no means implies either a total agricultural use or a uniformity of farming types and activities within the described district.

Gathering and Extractive Industries

The gathering and extractive industries—fishing, forestry, mining and quarrying—depend on the availability of natural resources whose occurrence and wealth are in part reflections of global physical geographic patterns and in part the result of fortunate geological accident. In both respects, Canada and the United States were bountifully endowed, originally possessing a remarkably high proportion of the world's natural wealth. At least for the United States, much of that endowment has been dissipated. For both countries,

however, it made possible their ascent as the world's most advanced industrialized and affluent national economies.

Both the United States and Canada are now considered postindustrial in their economic structures, and, as we have seen, the share of their gross domestic products contributed by gathering and extractive activities has plummeted over the last century. Both countries, however, are major contributors to world trade in primary materials; exports of fish, fuels, minerals, and metals constitute 16% of Canada's and 6% of the United States' merchandise exports. They are as well major consumers of the primary materials on which advanced economies depend. Between them, Canada and the United States, with some 5% of world population, consume 27% of the world's fossil fuels, 20% of its logs, and major shares of its metals—more than one-quarter of refined aluminum, a quarter of lead, 17% of tin, and 15% of crude steel, for example. Many of the commodities on which particularly the U.S. economy depends must be supplied by other countries—importantly, by Canada. The two countries, then, figure prominently in world trade in non-agricultural as well as agricultural primary materials.

Agriculture, Gathering, and Extractive Industries

Fishing

Fish supply some 20% of human protein consumption worldwide. Most of the world's commercial marine fisheries (only about 10% of total fish supply comes from inland waters) are concentrated in northern waters, where warm and cold currents mix above continental shelves (Figure 6.19). About half the world's commercial catch comes from the waters of the North Pacific and North Atlantic, major portions of which are within the economic zones of Canada and the United States. Both countries are important in world trade in fish and fish products.

Fishing sparked an early economic interest in the Western Atlantic. Firm evidence indicates pre-Columbian venturing by Portuguese and French fishermen to the Grand Banks off eastern Canada, and documents tell of English, French, Spanish, Portuguese, and Basque fishermen working waters from Long Island to Newfoundland early in the 16th century. Cod, plentiful and easy to catch, was the most sought species; dried or salted, it would keep for months to be carried back to the European market.

The English established fishing stations in protected Newfoundland harbors, and the catch from inshore waters was cleaned, salted, and dried on land. By the close of the 16th century, these Newfoundland fisheries alone involved 200 vessels and 10,000 men and boys. Their fishing ports were the basis for England's territorial claims to Newfoundland. The French, England's chief fishing rivals by the late 16th century, fished on the Grand Banks and other productive, relatively shallow waters off the coast and salted their catch aboard ship. Although the French, too, had shore-based headquarters, their ownership claims were denied by subsequent treaties between the two countries.

Farther south, from Massachusetts southward, fishing was an important adjunct to farming in colonial economies and provided not only subsistence supplies for the colonists but important exports to England and the West Indies as well. Boston began to export fish in 1633, and soon every other port in New England had its fishing fleet bringing in cod, mackerel, bass, herring, and other marine fish. These American fishermen expanded their hunting grounds northeastward to the coast of Nova Scotia, the Gulf of St. Lawrence, and the Grand Banks off Newfoundland. By 1731, the New England fisheries employed between 5000 and 6000 men. Gloucester alone had a fleet of 70 vessels. Whaling, too, took early root.

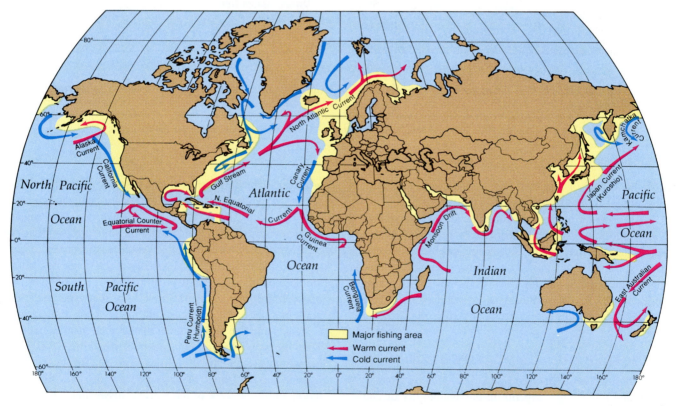

Figure 6.19

Major commercial marine fisheries of the world. More than 98% of the world marine catch is taken within 320 kilometers (200 mi) of shore. North America's long coastline gives it abundant and rich fishing potential, the target of commercial fishing fleets from Asian and European countries. U.S. coastal waters alone account for nearly one-fifth of the world's annual fish and shellfish harvests. During the two-month salmon fishing season in the North Pacific, massive fleets from Japan, South Korea, Taiwan, Poland, and Russia converge to capture huge quantities using *driftnets*—massively destructive nets that entrap wanted and unwanted species. Control and use of coastal fisheries is in constant dispute. In the late 1970s, both Canada and the United States extended their territorial claims seaward to include the rich Georges Bank fishery east of Cape Cod in the Gulf of Maine. The dispute went to the World Court for settlement.

As early as 1614, Captain John Smith visited the New England coast "to take whales," and by the close of the 17th century, Plymouth, Salem, Nantucket, and eastern Long Island ports had a profitable whaling trade. By the American Revolution, nearly 400 vessels were sailing out of American ports seeking whales as far as the coast of Brazil and the Arctic Ocean. In 1770, dried fish alone made up 14% of all British North American colonial exports by value; whale products—oil and fins—added another 3%.

Fish and fish products did not retain their relative export importance over the long haul. In the late 1990s, they accounted for less than 1% of United States and about 1% of Canadian exports by value. The two countries are, however, major sources of fish products in international trade as the first (United States) and second (Canada) largest exporters of fish products in value terms. More than half of Canada's exports go to the United States.

Despite their long coastlines and thousands of rivers and lakes, the United States and Canada—like the world at large—face difficulties maintaining the volume or species of fish catch traditional to their market expectations (Figure 6.20). Although the total North American marine catch, fin fish and shellfish combined, nearly doubled between 1970 and 1989, catches began declining substantially after 1993, victim to destructive overfishing of many of the most wanted and valuable species (see "Why Overfishing Occurs").

Fisheries of North America face serious losses, attributable in part to the destructive methods used by domestic fleets (Figure 6.21) and by "distant water" countries, such as Japan, Spain, Russia, and Taiwan, whose long-range fleets land much of the world's catch. Their factory ships damage breeding grounds and deplete stock within or just outside territorial waters even of countries such as Canada and the United States that control rights to fish within 200 nautical miles of their shores. In 1992, for example, Newfoundland's small-boat cod fishery effectively ceased to exist, costing 20,000 jobs. The U.S. Fisheries Service reports that 65 commercial species are overfished in American waters, caught faster than they can replace themselves. Fishing moratoriums have been selectively applied; although some success at stock regeneration is reported, full recovery of depleted stocks is doubtful. In 1993, in an attempt to save the Atlantic salmon in North America, the United States began a two-year buyout of the salmon-fishing industry of Greenland to reduce the allowed catch of Greenland fishermen and restore breeding stock.

Forestry

Forests originally covered slightly more than half of the United States and Canada. Growing densely to the water's edge all along the Atlantic Coast, they were both

---

WHY OVERFISHING OCCURS

Overfishing is partly the result of the accepted view that the world's oceans are common property, a resource open to anyone's use with no one responsible for its maintenance, protection, or improvement. The result of this "open seas" principle is one expression of the so-called **tragedy of the commons.** That phrase summarizes the economic reality that when a resource is available to all, each user, in the absence of collective controls, thinks he or she is best served by exploiting the resource to the maximum, even though this means its eventual depletion.

Steady increases in total ocean catches from the 1950s through the 1980s led to inflated projections of the eventual total annual catch and to the feeling that the resources of the oceans were inexhaustible. Quite the opposite proved true as both overfishing and the pollution of coastal waters seriously reduced the availability in some waters of traditional food species. Overfishing contributed to the collapse of the California sardine fishery in the 1950s, for example, and in the North Pacific off Alaska, landings of pollack decreased during the late 1980s from the same cause. By the early 1980s, serious depletion was evident in the stocks of Atlantic cod and herring in the North Atlantic and salmon and Alaska king crab in the Northwest Pacific. Some 30% of the species or fish stocks caught by U.S. fishermen have declined since the late 1970s, and it is now estimated that 64% of U.S. fish species are fully or overexploited, with 18% overexploited. Overfishing, pollution of coastal waters, and loss of breeding-area wetlands have, as well, seriously depleted most commercial shellfish species. In Chesapeake Bay, one of the world's most productive estuaries, commercial harvests of oysters and crabs (as well as of such commercially important fish as striped bass) have fallen sharply. Between 1965 and 1985, oyster catches declined 50%, and the oyster population now stands at 1% of its 1870 level.

Increasingly, coastal countries have been claiming a 200-nautical mile (370 km) exclusive economic zone within which they can regulate or prohibit foreign fishing fleets. The United States and Canada made such claims in 1976 and 1977. Since the bulk of the world's distant water fishing is off the North American coasts, such claims—though often defied or circumvented by foreign fleets—should help preserve remaining stocks and provide opportunity, under domestic fishing quotas, for their replenishment.

Agriculture, Gathering, and Extractive Industries

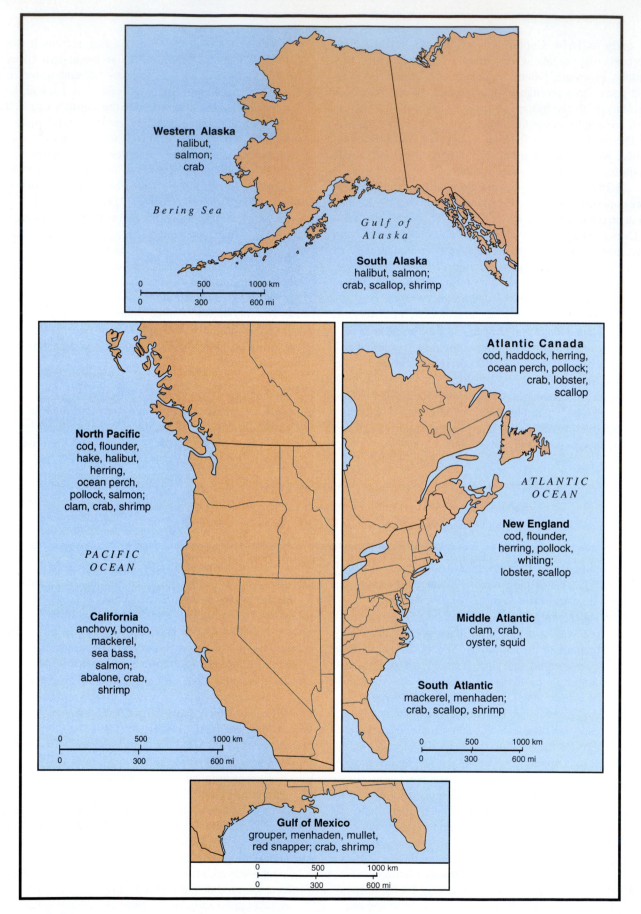

Figure 6.20

Marine fishing regions of the United States and Canada. Species names indicate traditional leading catches by volume and value. Domestic and foreign overfishing by the middle 1990s had altered—and in some cases eliminated—the familiar species sought and caught. Listings are alphabetical by leading fish and shellfish varieties.

Figure 6.21
Shrimp trawlers off the coast of Galveston, Texas. Fleets of boats like these are based in every port, large and small, along the Gulf and southeast coasts of the United States. The second most valuable (salmon is first) U.S. landing, the shrimp catch is endangered by its harvest characteristics. Trawlers like these churn up spawning beds and feeding banks, adversely affecting shrimp reproduction and food supply. The trawler nets are nonselective; for every ton of shrimp kept, trawlers catch and discard several tons of trash fish.

an immediately visible resource to be exploited and an obstacle to the agricultural settlement of the land. Logging was probably the first commercial resource exploitation by Europeans on this continent. Leif Eriksson, we learn from old Norse accounts, brought back a cargo of timber from a land across the North Atlantic around A.D. 1000. Much later, sawmills were in operation in Virginia and Maine before 1640, and suitable yellow and white pines of New England were marked and reserved for export for the use of the British navy. Logs and lumber were regularly exported in specially designed vessels before and after 1790.

As an obstacle to farming, forests were cleared as weeds in the American colonies, which also had growing populations with insatiable demands for wood for construction, fencing, fuel, railroad ties, shipbuilding, and a myriad of other needs. Although voices were raised in favor of conservation and forest reservation as early as the beginning of the 19th century, commercial lumbering spread without pause through the eastern forests to the Great Lakes states, then southward into the flatland pine stands and westward across the Rocky Mountains to the great coniferous forests of the Pacific slope. In the years since

first settlement, the United States has lost about 45% of its original forested area, and Canada's original forest has decreased more than 10%.

The original and remaining forests are not of a single kind or commercial quality (Figure 6.22). Alaskan forests and those of Canada—covering some 44% of that country—are largely old-growth coniferous stands. Old-growth forests also dominate the timberlands of the U.S. Rocky Mountain and Pacific Coast regions. Most of the 75% of forestlands of the conterminous United States that lie east of the Great Plains are secondary-growth forests, ranging from mixed and hardwood stands in the north and along the eastern uplands to southern pine stands along the coastal plain. Sixty percent of Canada's inventoried timberlands are considered commercial—that is, available for growing and harvesting forest crops; two-thirds of U.S. forestland is considered commercial, but only some 40% of all U.S. forests contribute to the annual commercial harvest.

Ownership patterns differ between the two countries. In Canada, 91% of commercial timberland is in public—federal or provincial—ownership; only 9% is privately owned. In the United States, more than 70% of forestland is in private hands, mainly in small farm woodlots that supply

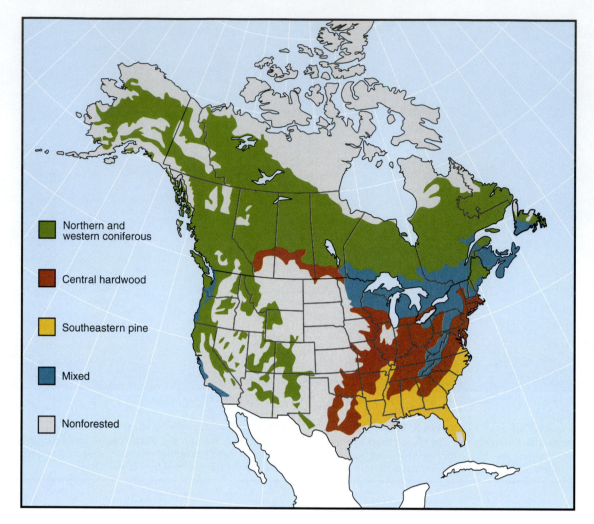

Figure 6.22
Original U.S. and Canadian forests.

Legend:
- Northern and western coniferous
- Central hardwood
- Southeastern pine
- Mixed
- Nonforested

nearly half the country's total annual timber cut, although more than 27 million hectares (67 million acres) are owned by timber industries (Figure 6.23). Some 28% is publicly owned, much of it in 171 national forests managed by the U.S. Forest Service.

The forest industry in Canada is concentrated in British Columbia, Québec, and Ontario. Together, these provinces contribute between 85% and 90% of total earnings; British Columbia alone accounts for more than one-half. The West Coast emphasis is continued in the United States: more than half of its timber harvest comes from the Rocky Mountain and Pacific states, nearly all as softwood. More than 35% comes from the southern states from Texas to Virginia, of which nearly 80% also is softwood. The remainder of the U.S. commercial harvest, primarily in the form of hardwoods, comes from the northern states from New England westward through the Middle West.

For both countries, lumber products are big business. The United States is a voracious consumer of hardwoods and softwoods and such forest industry products as millwork, veneers, plywood, pulp, and paper. Despite its own very large volume of production, the country regularly must import between 15% and 20% of its domestic consumption. Canada, on the other hand, consumes only about 20% of its own lumber production and is the world's leading exporter of forestry products, including about 30% of the world's wood pulp and almost 60% of its newsprint. Forestry products of all types account for upwards of 15% of total Canadian export earnings. Between 80% and 85% of Canadian exported lumber is purchased by the United States.

Mining

Through geological good fortune, North America was endowed with a wide variety and great wealth of metallic and

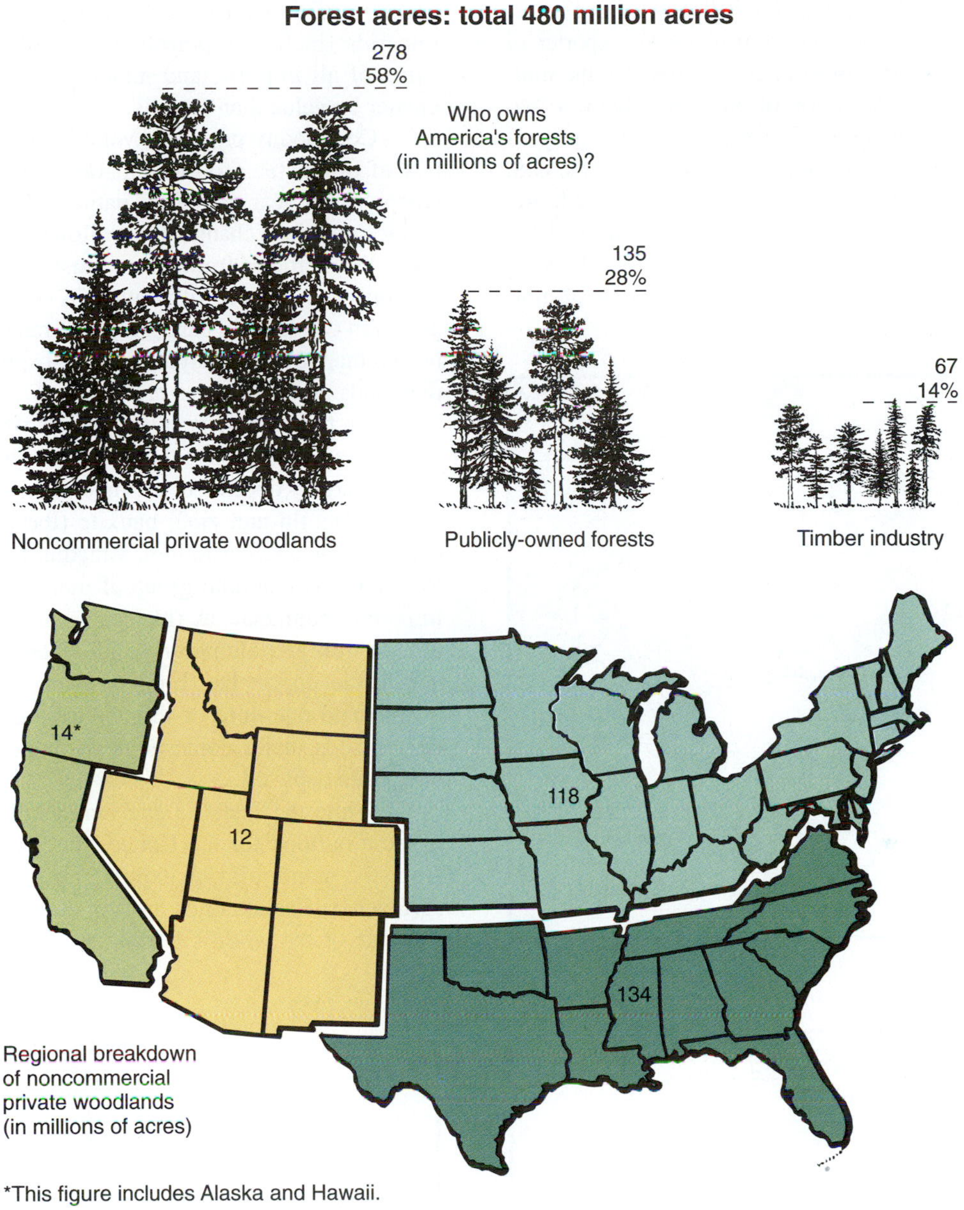

Forest acres: total 480 million acres

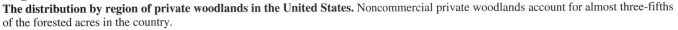

278
58%

Who owns
America's forests
(in millions of acres)?

135
28%

67
14%

Noncommercial private woodlands

Publicly-owned forests

Timber industry

14*

118

12

134

Regional breakdown
of noncommercial
private woodlands
(in millions of acres)

*This figure includes Alaska and Hawaii.

Figure 6.23

The distribution by region of private woodlands in the United States. Noncommercial private woodlands account for almost three-fifths of the forested acres in the country.

nonmetallic minerals, including the fossil fuels. Figures 6.24 and 6.25 indicate visually the variety and distribution of important fuel and metallic mineral deposits in the United States and Canada.

Agriculture drew settlement into the continental interior and lured immigrants from abroad, but until after World War II, the United States owed much of its prosperity and growing wealth to its great abundance and diversity of metals, minerals, and energy resources. Iron ore, copper, lead, zinc, ferroalloys, nonmetallic minerals, anthracite and bituminous coal, and extensive deposits of petroleum and natu-

ral gas not only supplied the country with resources sufficient for its own ever-expanding needs but also made it an important minerals exporter. The discovery of these minerals drew miners to areas not yet settled by farmers and created, temporarily or permanently, thriving local and regional economies while strengthening the country as a whole.

The United States drew lavishly on its resource base, creating by the first decade of the 20th century the world's leading industrial economy. Continued growth and expansion—and the demands of two world wars—took their toll on the minerals that fueled that growth.

After World War II, the United States began to lose its resource self-sufficiency, moving from being an exporter of metals and minerals to becoming an importer. By the middle of the 1970s, the importation of nonrenewable raw materials was increasing by 1% a year. That change profoundly affected the place of mineral production in the U.S. economy and the country's balance of payments. Mining dropped from 3.3% of the gross domestic product in 1947 to 1.4% in 1996. As the U.S. merchandise trade balance changed from positive to heavily negative, imports of mineral commodities assumed greater importance. By 1997, minerals (including petroleum) equalled 9% of the total value of all imports, and mineral imports were six times greater by value than exports.

Commodity prices are volatile, and quantities and values of domestic production, imports, and exports vary sharply from year to year as national and international economic conditions change. Even allowing for those variations, some 90 of the 100 minerals most important to the U.S. economy are domestically either in short supply or are imported in whole or in part. Very few mineral commodities—boron, magnesium, molybdenum, phosphate rock, industrial diamonds are examples—are regularly available for export. Although the country is still nearly or wholly self-sufficient in a few major metals and minerals, such as copper, it is totally or largely dependent on imports for such essential commodities as tin and zinc; bauxite (the ore for aluminum); cobalt, tungsten, chromium, vanadium, and other steel alloys; and the platinum group of metals (Figure 6.26). Such important nonmetals as fluorspar, essential in the manufacture of iron and aluminum, and—increasingly—petroleum add to the import list. Imports even supply between one-quarter and one-third of the domestic consumption of iron ore, as such major domestic deposits as the Mesabi Range have been depleted.

While the metallic and nonmetallic mineral and petroleum outlook for the United States continues to worsen from the standpoint of export earnings or national self-sufficiency, Canada finds itself in quite a different situation. It is one of the world's leading producers and exporters of

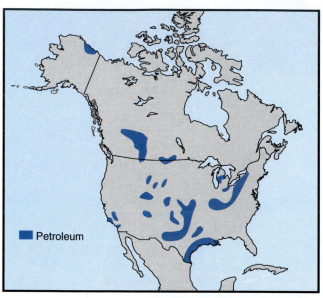

(a)

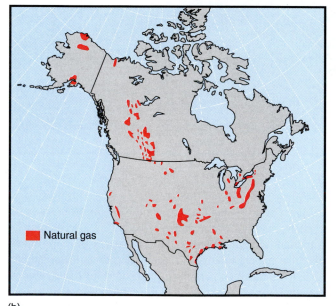

(b)

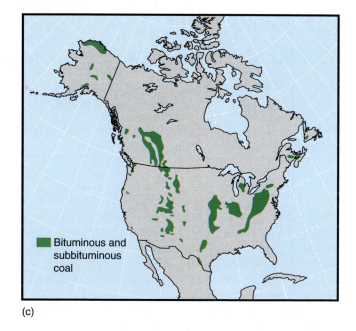

(c)

Figure 6.24

The distribution of fossil fuel resources in the United States and Canada: petroleum (a), natural gas (b), and coal (c). The locations of anthracite and lignite coal deposits, of lesser economic importance today, are not shown on (c).

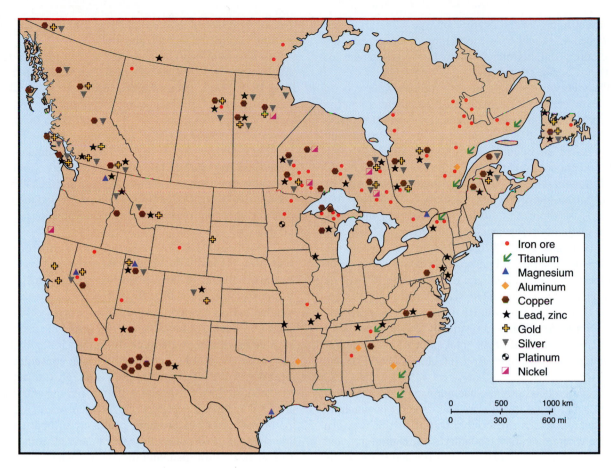

Figure 6.25
The distribution of selected economically important minerals.

Source: U.S. Geological Survey.

Mineral		Percentage imported, 1997	Major foreign sources
Bauxite and alumina	100		Australia, Guinea, Jamaica, Brazil
Columbium	100		Brazil, Canada, Germany
Graphite	100		Mexico, China, Brazil, Madagascar
Manganese	100		South Africa, France, Gabon
Mica (sheet)	100		India, Belgium, Brazil, Japan
Strontium	100		Mexico, Spain, Germany
Tantalum	76		Germany, Thailand, Australia
Tungsten	85		China, Bolivia, Germany, Peru
Cobalt	80		Zaire, Zambia, Canada, Norway
Chromium	78		South Africa, Turkey, Zimbabwe, Yugoslavia
Tin	85		Brazil, Bolivia, China, Indonesia, Malaysia
Potash	76		Canada, Israel, former USSR, Germany
Nickel	54		Canada, Norway, Australia, Dominican Republic
Barite	71		China, India, Mexico, Morocco

Figure 6.26
U.S. reliance on foreign supplies of minerals. Like other industrially developed countries, the United States is a leading consumer of the world's mineral resources. Imports account for 50% or more of national consumption of the minerals indicated.

Source: Data from Bureau of Mines, U.S. Interior Department.

both nonfuel minerals and energy sources. Minerals represent between 4.5% and 6% of total gross domestic product and (including coal, petroleum, and natural gas) account for one-fifth of Canada's exports.

Throughout Canada's history, minerals have been one of the chief spurs of economic development. As in the United States, the discovery of extensive mineral deposits drew mining settlements and industrial investment to even remote parts of the country. Now, Canada produces more than 60 mineral commodities and is the world's largest producer of uranium and zinc and the second largest source of gypsum, potash, nickel, cobalt, titanium concentrates, and asbestos. It is among the top five producers of molybdenum, the platinum-group metals, sulfur, copper, lead, cadmium, silver, and gold. With the exception of bauxite, chromium, manganese, and phosphate, Canada meets all its own domestic mineral needs and exports about 80% of its production, mostly to the United States, Japan, and Western Europe. Since mineral wealth is related to geology, Canada's mineral industry is unequally concentrated. Ontario has about one-third of the total value of nonfuel mineral production, followed by British Columbia and Québec, each with about 15%.

Canada produces more crude oil, natural gas, and coal than it uses and is the world's leading exporter of uranium. Its energy resources are concentrated in the western provinces, where most of the country's oil, natural gas, and uranium are located. Canada had an estimated 0.7% of the world's proved petroleum reserves in 1995, 2% of its natural gas, and less than 1% of its coal. Much of its production is consumed domestically, with well over half of Canada's energy requirements in 2000 met through petroleum and natural gas.

Summary

The natural resource wealth that drew settlement to and across North America formed the basis of rapidly growing farming, fishing, forestry, and mining activities. Together, these primary industries fueled the economic growth of Canada and the United States to their present rank among the world's richest and most developed countries.

Colonial subsistence farming was from the start buttressed by commercial agriculture for the domestic and export markets. Commercial farming increased dramatically during the 19th century as domestic urban markets expanded, railroads opened up the rich continental interior for settlement, and mechanization increased the efficiency and reduced the costs and labor requirements of crop and livestock production. The rural prosperity of the late 19th and early 20th centuries was broken by a post–World War I farm depression that lasted until 1940. By then, farmers had fallen far behind their urban compatriots in money income and in access to the amenities and necessities of modern life. Although farmers commonly owned automobiles and trucks, most lacked indoor plumbing, running water, central heating, electricity, and telephones.

That rural deprivation ended with a revived demand for farm products and rising prices during and after World War II. Improvements in seed, fertilizers, pesticides and herbicides, and farm machinery increased crop yields and reduced labor requirements. The capital costs of those improvements required larger farm units and the spread of contractual arrangements between the still dominant family farms and corporate farm-product purchasers. Commercial specialization increased, and old "farming belts" were in some cases intensified and in others readjusted to compensate for changing regional competitive positions and crop economies.

Fishing and forestry, early New World commercial activities, gradually declined in their share of contribution to the economies of the United States and Canada as they diminished by exploitation the resources on which they depended and as the modern industrial-service economy reduced their relative importance. The economic contribution of mineral resources, particularly in the United States, has also been lessened by intensive domestic consumption and pre–World War II export trade. Although Canada is still well endowed with, especially, timber and mineral reserves, the United States has become largely dependent on imports, particularly from its northern neighbor. In both countries, the share of gross domestic product derived from exploitation of soil, forest, fish, and mineral resources has declined from more than 50% in the middle of the 19th century to far less than 15% at the start of the 21st century.

Key Words

agriculture	nonrenewable resource
cash grain	primary activities
combine	renewable resource
cotton gin	spring wheat
extractive industry	subsistence agriculture
fossil fuel	tragedy of the commons
gathering industry	truck crops
hybrid seed	vertical integration
moldboard plow	von Thünen
natural resource	winter wheat

Gaining Insights

1. What preconditions had to be established for the effective agricultural settlement of the American and Canadian interior?
2. How did mechanization change traditional farming practices and economies? What kinds of farm machinery were particularly important in the creation of commercial agriculture in what were to become the Corn and Wheat Belts?
3. In what ways did farm life of the 1930s differ from city life of the same period? What conveniences and

appliances common in cities were rare or absent in the countryside?

4. What economic advantages accrue to volume purchasers and producers of farm commodities from production contracts and vertical integrations?

5. What are the basic assumptions of von Thünen's agricultural model? What pattern of crop production and land values does it predict? Is there evidence from the United States that von Thünen's predictions are reasonable?

6. What is the distinction between a *renewable* and a *nonrenewable resource?* Under what circumstances might the distinction between the two be blurred or obliterated? Have those circumstances occurred for any resource category?

7. What economic problems can you discern that affect the viability of any or all of the *primary industries?* What is the *tragedy of the commons?* How, if at all, is that concept related to the problems you cited?

8. In what ways have the Canadian and U.S. economies differed in the rate and intensity of their use of natural resources? In what ways are those economies now complementary from the standpoint of resource availability and trade?

Selected References

Agriculture Canada. *Farming in Canada.* Publication 1296. Ottawa: Minister of Supply and Services Canada, 1983.

Cameron, Eugene N. *At the Crossroads: The Mineral Problems of the U.S.* New York: John Wiley & Sons, 1986.

Easterbrook, W.T., and H. J. Aitken. *Canadian Economic History.* Toronto: University of Toronto Press, 1988.

Ebeling, Walter. *The Fruited Plain: The Story of American Agriculture.* Berkeley: University of California Press, 1979.

Edwards, Everett E. "American Agriculture—The First 300 Years." In *Farmers in a Changing World. The Yearbook of Agriculture, 1940,* 171–276. Washington, D.C.: Government Printing Office, 1940.

Gates, Paul W. *The Farmer's Age: Agriculture, 1815–1860.* Vol. 3, *The Economic History of the United States.* New York: Holt, Rinehart and Winston, 1960.

Goreham, Gary, Cornelia Butler Flora, and Emmett P. Fiske, eds. *Encyclopedia of Rural America: The Land and People.* Santa Barbara: ABC-Clio, 1997.

Gregor, Howard F. *Industrialization of U.S. Agriculture: An Interpretive Atlas.* Boulder, Colo.: Westview Press, 1982.

Hart, John Fraser. "Change in the Corn Belt," *Geographical Review* 76, no. 1 (January 1986): 51–72.

———. *The Land That Feeds Us.* New York: W.W. Norton, 1991.

Hudson, John C. *Making the Corn Belt: A Geographical History of Middle-Western Agriculture.* Bloomington: Indiana University Press, 1994.

McCann, Larry D., ed. *Heartland and Hinterland: A Geography of Canada.* Scarborough, Ontario: Prentice-Hall of Canada, 1982.

Organization for Economic Cooperation and Development. *OECD Economic Surveys: Canada.* Paris: Organization for Economic Cooperation and Development, 1991.

Peterson, R. Neal, and Nora L. Brooks. *The Changing Concentration of U.S. Agricultural Production During the 20th Century.* 14th Annual Report to the Congress on the Status of the Family Farm. Agriculture and Rural Economy Division, Economic Research Service, U.S. Department of Agriculture. *Agriculture Information Bulletin* no. 671. 1993.

Pillsbury, Richard and John Florin. *An Atlas of American Agriculture: The American Cornucopia.* New York: MacMillan, 1996.

Schlebecker, John T. *Whereby We Thrive: A History of American Farming, 1607–1972.* Ames: Iowa State University Press, 1975.

Smith, Everett G., Jr. "America's Richest Farms and Ranches," *Annals of the Association of American Geographers* 70, no. 4 (December 1980): 528–41.

Sommer, Judith E., and Fred K. Hines. *Diversity in U.S. Agriculture: A New Delineation by Farming Characteristics.* Agriculture and Rural Economy Division, Economic Research Service, U.S. Department of Agriculture. *Agricultural Economic Report,* no. 646, 1991.

Spector, David. *Agriculture on the Prairies, 1870–1940.* Parks Canada, *History and Archeology* 65. Ottawa: Minister of Supply and Services Canada, 1983.

Statistics Canada, *The Canada Yearbook.* Ottawa: Minister of Supply and Services Canada. Annual or biennial.

Swann, Nina, and Pat Weisgerber. *Canadian and U.S. Farm Sector Comparisons.* International Economics Division, Economic Research Service, U.S. Department of Agriculture. Washington, D.C.: Department of Agriculture, 1981.

Troughton, Michael J. *Canadian Agriculture.* Budapest: Akadémiai Kiadó, 1982.

U.S. Bureau of Mines. *Minerals Yearbook.* Washington, D.C.: Government Printing Office. Annual.

U.S. Department of Agriculture. *Our American Land. Yearbook of Agriculture, 1987.* Washington, D.C.: Government Printing Office, 1987.

———. *Agriculture and the Environment. Yearbook of Agriculture, 1991.* Washington, D.C.: Government Printing Office, 1991.

Vogeler, Ingolf. *The Myth of the Family Farm: Agribusiness Dominance of U.S. Agriculture.* Boulder, Colo.: Westview Press, 1981.

Wallace, Iain, and William Smith. "Agribusiness in North America." In *The Industrialization of the Countryside,* edited by Michael J. Healey and Brian W. Ilbery, pp. 57–74. Norwich, England: Geo Books, 1985.

CHAPTER 7

Industrial and Commercial Organization

In August 1992, a famous manufacturer in Cortland, New York, announced that it was moving its manufacturing operations to Mexico. It laid off 875 of its 1300 workers, keeping only its much smaller engineering, distribution, and customer service divisions in Cortland. The company paid manufacturing workers in Cortland $11 an hour. Fringe benefits and the cost of supervision increased this to $18 an hour. The firm estimated that labor in Mexico would cost $4 an hour. Interestingly, when the firm originally opened around 1910, it imported workers from the Ukraine and Italy to upper New York state because they were willing to work for lower wages than local people.

The movement of many American factories to low-wage areas in Third World countries began in earnest in the mid-1960s (Figure 7.1). It accelerated as a result of the *North American Free Trade Agreement* (NAFTA). As its name implies, the treaty is intended to unite more than 408 million people in the world's largest trading bloc. Its passage was opposed in the United States by many labor unions, which contended that the accord would cause a massive flight of American jobs and capital to Mexico. Some of this flight did begin soon after implementation in 1994, as major companies like General Motors built large plants in Mexico. On the other hand, the treaty has also opened important new markets to many American manufacturers. It remains to be seen how the United States will adjust to the challenge and opportunity that NAFTA presents.

The United States emerged from World War II (1939–1945) as the world's dominant producer of industrial goods. Unlike in Europe, Japan, and the Soviet Union, where many industrial areas were devastated, the industrial and commercial infrastructure of the United States was undamaged, much of it was very new—after the large expansions of industrial production during the war years—and many new products were ripe for development as spin-offs of wartime military production. The war years had also been years of experimentation in the United States in developing systems that could produce quality products in mass using first-generation industrial workers who were largely unskilled—teenagers, women, and migrants. Many migrants had come from Appalachia and the rural South to work in manufacturing in the cities and metropolitan areas of the industrial heartland of America. Production processes were simplified by breaking the work into more tasks, each of which could be learned quickly and easily.

In addition to industrial production declines in the 1940s in Europe, the Soviet Union, and Japan, industrial production was small in the large areas of the world still dominated by European colonial powers, which continued to view their colonies as providers of raw materials and as captive consumers for their industrial goods. Low levels of industrial production also characterized Central and South America, India, and China. With their dollars strong in the world economy, both the United States and Canada had positive balances of trade as the foreign demand for North American goods was much larger than their demand for foreign-made goods.

In this chapter, we first identify the location of manufacturing centers in the United States and Canada. Geographers have three different approaches to explaining these

Figure 7.1

Seeking lower labor costs, American manufacturers have increasingly established component manufacturing and assembly operations along the international border in Mexico. The finished or semifinished products are brought duty-free into the United States. *Out-sourcing* has moved a large proportion of American electronics, small appliance, and garment industries to offshore subsidiaries or contractors in Asia and Latin America.

locations. We show how the three are complementary, each asking different questions and adding to our understanding in distinctive ways. We then discuss the changes in the economies of the United States and Canada in the 55 years since 1945 and how these changes were accompanied by a shift in the nature of manufacturing and commercial activities. Finally, we describe the development of new kinds of economic activities upon which many people now rely for their employment.

Location of Manufacturing in the United States and Canada

The importance of manufacturing in the United States and Canada, both as an employer of labor and as a contributor to the national income of both countries, has been steadily declining. In 1960, 28% of the labor force was engaged in manufacturing (Figure 7.2). By the late 1990s, manufacturing employment had dropped to about 16% of a much larger labor force, and industry contributed less than 20% of the gross domestic product of the two countries. Nevertheless, the importance of the United States and Canada as major forces in world industrial production is unquestioned.

Eastern United States and Canada

Manufacturing is found particularly in the urbanized sections of the two countries, but as Figure 7.3 indicates, it is not uniformly distributed. It is especially concentrated in the northeastern part of the United States and adjacent sec-

tions of southeastern Canada, the **U.S./Canadian Manufacturing Belt.** That belt extends from the St. Lawrence Valley on the north (excluding northern New England) to the Ohio Valley and from the Atlantic Coast westward to just beyond the Mississippi River. Covering less than 5% of the land area of the United States and Canada, the Manufacturing Belt contains the continent's densest and best-developed transportation network, the largest number of its manufacturing establishments, and the preponderance of heavy industry. It is also part of the continent's agricultural heartland. These interrelated factors have all influenced the past history and present structure of industry within the belt.

As noted in Chapter 3, North American manufacturing began early in the 19th century in southern New England, where water-powered textile mills, iron plants, and other small-scale industries began to free Canada and the new United States from total dependence upon European—particularly English—sources (Figure 7.4). Although lacking the raw materials that formed the industrial base of the continental interior, the eastern portion of the belt contained early population centers, a growing canal and railroad network, a steady influx of immigrant skilled and unskilled labor, and concentrations of investment capital. Although the early manufacturing advantages of the Northeast were soon lost as settlement advanced across the Appalachians into the rich continental interior, the U.S. eastern seaboard remains an important producer of consumer goods and light-industrial and high-technology products on the basis of its market and labor skills.

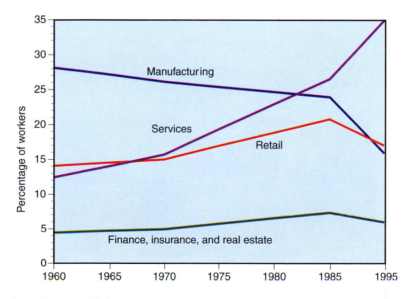

Figure 7.2

The civilian labor force and employment, United States. The percentage of workers employed in manufacturing has decreased steadily since 1960, while service-sector jobs have increased dramatically. That sector includes legal, business, health, transportation, and communication services.

Industrial and Commercial Organization

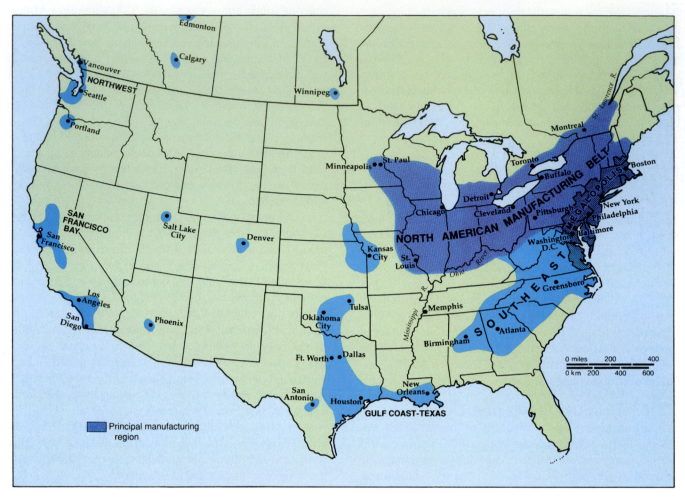

Figure 7.3
Manufacturing districts of the United States and Canada.

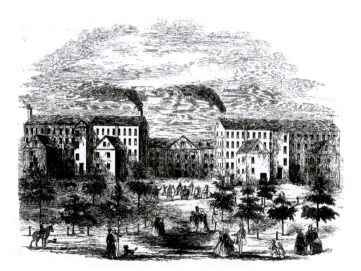

Figure 7.4
The Boott cotton mills at Lowell, Massachusetts, in 1852. The mill that Francis Cabot Lowell established in 1814 was the first in America to combine the carding, spinning, and weaving of cloth under one roof. After Lowell's death, his associates developed America's first factory town and named it Lowell. It was built in the 1820s at a water-power site on the Merrimack River.

Southern New England, where it all started, still possesses a diversified industrial base with electronic and electrical goods, instruments, apparel, and consumer goods industries in general being prominent. Its strengths lie in its technical and engineering skills, educational institutions, and proximity to the major markets of the eastern seaboard. It also is the northern end of *Megalopolis,* a city system 1000 kilometers (600 mi) long stretching south to Norfolk, Virginia: one of the longest, largest, most populous of the world's great metropolitan chains (also discussed in Chapter 4). As a major consumer concentration, it has attracted a tremendous array of market-oriented industries and thousands of industrial plants served by the largest and most varied labor force in the country. Within Megalopolis are areal concentrations and specializations.

New York City has printing, publishing, metal fabrication, electrical goods, the apparel industry, and food processing—and increasingly, financial and administrative "industries." The *Delaware Valley* district, including southeastern Pennsylvania, Philadelphia, and the Baltimore area, shares some of the same market and labor force attractions as does New York City and almost the same degree of industrial diversification. Philadelphia, however, has a richer agricultural hinterland than does New York (and therefore more food-processing plants) and far better access to the coal and steel districts farther west in Pennsylvania. Both Philadelphia and Baltimore—on the southern margin of the Manufacturing Belt—have excellent port facilities and good rail connections to the industrial-agricultural interior. Iron ore brought from Labrador and Venezuela and petroleum from Venezuela, Nigeria, North Africa, and the Persian Gulf supply steel plants and petrochemical works that show by their tidewater locations a raw material orientation at break-bulk points such as Sparrows Point east of Baltimore and the chemical plants at Wilmington, Delaware.

The heart of the U.S./Canadian Manufacturing Belt developed across the Appalachians in the interior of the continent in the valleys of the Ohio River and its tributaries and along the shores of the Great Lakes. The rivers and the Great Lakes provided the early "highways" of the interior, supplemented by canals built throughout the eastern Middle West of the United States during the 1830s and 1840s. They were followed after the 1850s by the railroads that tied together the agricultural and industrial raw materials, growing urban centers, and multiplying industrial plants of the interior with both the established eastern and the developing western reaches of the continent. The raw material endowment of the region was impressive: anthracite and bituminous coking and thermal coals, iron ore, petroleum, copper, lead and other nonferrous metals, timber resources, and agricultural land of a richness and extent unknown elsewhere in the world.

The building of the railroad system that united the manufacturing region itself gave rise to early and enduring industrial emphases—the iron and steel industry producing first iron and then steel rails and locomotive and railcar building as forerunners of metal fabricating, transportation equipment manufacturing, and the automobile industry (Figure 7.5). Agricultural settlement following railroad construction sparked interest in farm machinery and led to the development and manufacture of the reaper and the steel moldboard plow, among other implements. The region's bountiful harvests then and later fed flour mills, corn and soybean processing plants, and food and agricultural goods industries in general.

Canadian industry is highly localized. About one-half of the country's manufacturing labor force is concentrated in southern Ontario, every bit the equivalent in industrial development and mix of its U.S. counterparts in upstate New York and nearby Pennsylvania, Ohio, and eastern Michigan. Toronto forms the hub, but the industrial belt extends westward to Windsor, across from Detroit. In the Ontario section as a whole are iron and steel mills, automobile

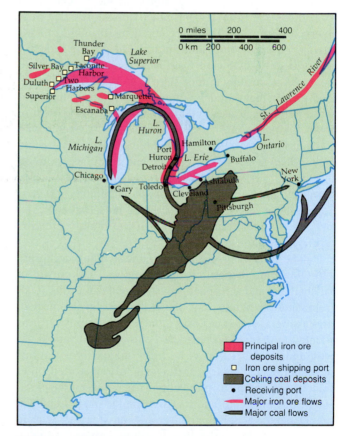

Figure 7.5

Material flows in the steel industry. When an industrial process requires the combination of several heavy or bulky ingredients, an intermediate point of assembly of materials is often a least-cost location. Earlier in the 20th century, the iron and steel industry of the eastern United States showed this kind of localization—not at the source of any single input but where coking coal, iron ore, and limestone could be brought together at the lowest price.

Industrial and Commercial Organization

manufacturing, other metal fabrication, food processing, consumer goods production—indeed, the full range of industrial activity expected in an advanced society with significant domestic and export markets and a skilled labor force. Another third of Canadian manufacturing employment is found in Québec, with Montréal as the obvious core but with important energy-intensive industries—particularly aluminum plants and paper mills—found along the St. Lawrence River and elsewhere in the province.

Other Concentrations

For much of the 19th century and as recently as the late 1960s, the Manufacturing Belt contained from one-half to two-thirds of American and Canadian industry. By 2000, its share had dropped to less than 40%. Not only was the position of manufacturing declining in the economy of both countries, but what remained was continuing a pattern of relocation reflecting national population shifts and changing materials and product orientations.

Figure 7.3 suggests the location of some of the larger of these secondary industrial zones. In the Southeast, an iron and steel industry based on local ores and coal emerged during the 1870s, centered in Birmingham, Alabama. At about the same period, the textile industry began to shift southward from New England to the Piedmont area, especially to the central Carolinas. The tobacco, food-processing, wood products, and furniture industries became important users of local resources. Big growth of diversified and consumer goods industries has come with population increase and urbanization of the southeastern states. That area's milder climate and ready access to oil and natural gas from the Gulf states helped keep energy costs low and attract footloose high-technology industry. Atlanta has emerged as the diversified industrial and commercial regional capital, but all of the rapidly growing cities of the region are beneficiaries.

The *Gulf Coast-Texas* district (extending northward into Oklahoma) benefitted from the post-1950 growth of the chemical and aerospace industries and expanding food-processing operations, and the explosive development of large metropolitan complexes attracted a full array of firms oriented to the regional market. Petroleum and natural gas provided wealth, energy, and raw materials for a vast petrochemical industry; sulfur and salt supported other branches of chemical production. Deep-water transportation through the Gulf and by barge via the intracoastal waterway and the Mississippi River system—together with an extensive pipeline and railroad system—gives the region access to the industrial core of the United States and Canada.

Farther west, *Denver* has become a major, though isolated, industrial center, while on the West Coast, three distinctive industrial subregions have emerged to national prominence. In the *Northwest,* from Vancouver to Portland, the regional and broader Asian-Pacific markets are of greater significance than the primary domestic markets of Canada and the United States, both more than 3000 kilometers (1800 mi) away. Seattle's aircraft industry has world significance, of course, but transportation costs restrict access to eastern markets for most more ordinary products. The *San Francisco Bay* district is home to Silicon Valley and the electronics, computer, and high-tech manufacturing that name implies (Figure 7.6). Food specializations (wine, for example) for a national market have their counterpart farther south in the *Los Angeles–San Diego* corridor, where fruits and vegetables are grown and packed. More important, however, is production for the rapidly growing California and western market. Automobile assembly plants, consumer goods manufacturing, aircraft and defense industries (for national and world markets), even steel and nonferrous metals plants, are increasing in number, size, and product diversity.

Concepts of Enterprise Location

In this section, we discuss three concepts that are often used to explain the location of industry in the United States and Canada. Each has its own value, but each is incomplete. Taken together, however, the three concepts help us understand how the patterns described above have come about.

Location Theory

Location theory for industry and commerce assumes rational economic behavior on the part of business people, consumers, and governments and attempts to explain the location of economic activities within largely self-sufficient national economic systems like that of the United States. In this section, we describe the key features of location theory and show how it accounts for the geographical patterns of major aspects of commercial and industrial activities in the United States in the period immediately after World War II.

Location theory aims to show how simple principles of rational decision making, when applied to given patterns of resource distributions, lead one to expect to find industrial, commercial, and agricultural activities arranged in distinct spatial patterns. Entrepreneurs pursue the goal of maximizing profits. They know the values of the factors that affect profitability and are able to calculate the amounts of each factor to use, the locations from where each should be procured, the places where the product should be marketed, and the prices that consumers will be willing to pay. The theory assumes that in commercial economies, the best measure of the correctness of economic decisions is afforded by the consumption of products by markets at prices that cover producers' costs, including cost of capital, and the equilibrium between supply and demand that market prices establish (Figure 7.7).

Location theory seeks to find least-cost production and distribution sites: locations at which the combined costs of assembling raw materials and distributing the final

Figure 7.6
The Silicon Valley in California.

Figure 7.7

Supply, demand, and market equilibrium. The regulating mechanism of the market may be visualized graphically. The *supply curve* (a) tells us that as the price of a good increases, more of that good will be made available for sale. Countering any tendency for prices to rise to infinity is the market reality that the higher the price, the smaller the demand as potential customers find other purchases or products more cost-effective. The *demand curve* (b) shows how the market will expand as prices drop and the good becomes more affordable and attractive to more customers. *Market equilibrium* (c) is marked by the point of intersection of the supply and demand curves and determines the price of the good, the total demand, and the quantity bought and sold.

product to markets are lowest. To assess the advantages of one location over another, industrialists must evaluate the most important **variable costs** (Figure 7.8). They divide their total costs into categories and note how each cost will vary from place to place. In different industries, transportation charges, labor rates, power costs, plant construction or operation expenses, the interest rate of money, or the price of raw materials may be the major variable cost. The industrialist must look at each of these and, using a process of elimination, select the maximum-profit site. The selection is particularly difficult because the market that would be supplied by any producer often varies from one potential manufacturing site to another. Consequently, a firm deciding to set up a new manufacturing plant must also make conjectures about the likely behavior of its competitors.

The concern with variable costs as a determinant in industrial-location decisions extends the **least cost theory of location** proposed by Alfred Weber (1909), who explained the optimum location of a manufacturing establishment in

Industrial and Commercial Organization

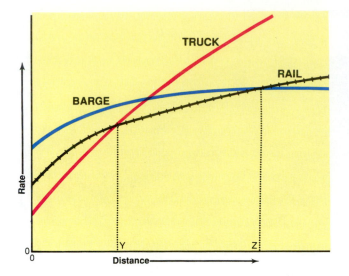

Figure 7.8

In many industries, transportation charges are a major variable cost. Different transport modes have cost advantages over differing distances. The usual generalization is that when all three modes are available for a given shipment, trucks are most efficient and economical over short hauls, railroads have the cost advantage over intermediate hauls, and water (barge) movement is cheapest over longer distances.

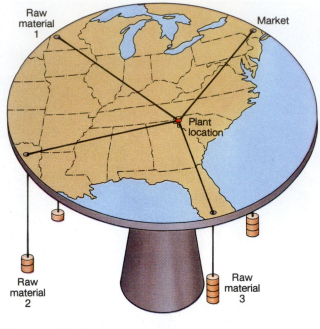

Figure 7.9

Plane table solution to a plant location problem. This mechanical model, suggested by Alfred Weber, uses weights to demonstrate the least transport cost point when there are several sources of raw materials and a single market. When a weight is allowed to represent the "pulls" of raw materials and market, an equilibrium point is found on the plane table. That point is the location at which all forces balance each other and represents the least-cost plant location.

terms of the minimization of three basic expenses: relative transport costs, labor costs, and agglomeration costs. *Agglomeration* refers to the clustering of productive activities and people for mutual advantage. Such clustering can produce agglomeration economies through shared facilities and services. Diseconomies, such as higher rents or wage levels resulting from competition for these resources, may also occur.

Weber concluded that transport costs were the major consideration determining location. That is, the optimum location would be found where the costs of transporting raw materials to the factory and finished goods to the market were lowest (Figure 7.9). He noted, however, that if variations in labor or agglomeration costs were sufficiently great, a location determined solely on the basis of transport costs might not, in fact, be the optimum one.

Assuming, however, that transportation costs determine the "balance point," optimum location will depend on distances, the respective weights of the raw material inputs, and the final weight of the finished product. It may be either *material-oriented* or *market-oriented*. As Figure 7.10 suggests, material orientation reflects a sizable weight loss during the production process; market orientation indicates a weight gain.

Many theorists have argued that the most profitable way to manufacture a product might change from one location to another. **Factor inputs** are each of the different kinds of resources needed to make a product. Raw materi-

als, labor, land, capital, and transportation costs are all important categories of factor inputs. Different locations use different relative quantities of factor inputs, each of which might have prices that vary from place to place. The economically rational producer will substitute the use of cheaper inputs for the more expensive inputs (Figure 7.11). Where labor is expensive, for example, automation of the production process might be more profitable. In this case, the use of more machinery substitutes for the use of some labor. Higher transportation costs might be incurred to locate at a site where labor costs are lower.

With substitution, a number of different points may be optimal manufacturing locations. Further, a whole series of points may exist where total revenue of an enterprise just equals its total cost of producing a given output. These points, when connected, mark the *spatial margin of profitability* and define the area within which profitable operation is possible (Figure 7.12). Location anywhere within the margin ensures some profit and tolerates both imperfect knowledge and personal (rather than economic) considerations.

Through time, the geographical pattern of factors that affect the location of an industry change, and the optimum locations then change. If there are many firms in the industry, the relocation can occur within a few years if the new

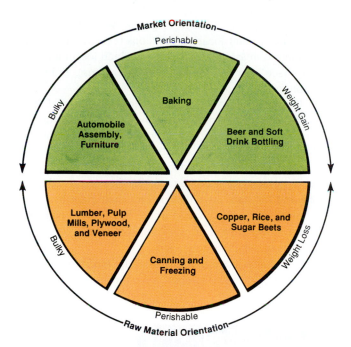

Figure 7.10

Spatial orientation tendencies. *Raw material orientation* is presumed to exist when there are limited alternative material sources, when the material is perishable, or when—in its natural state—it contains a large proportion of impurities or nonmarketable components. *Market orientation* represents the least-cost solution when manufacturing uses commonly available materials that add weight to the finished product, when the manufacturing process produces a commodity much bulkier or more expensive to ship than its separate components, or when the perishable nature of the product demands processing at individual market points.

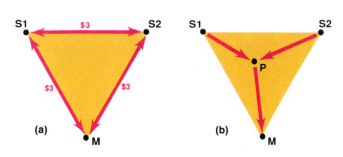

Figure 7.11

Weber's locational triangle with differing assumptions. (a) With one market, two raw material sources, and a finished product reflecting a 50% material weight loss, production could appropriately be located at S_1, S_2, or M since each length of haul is the same. In (b), the optimum production point, P, is seen to lie within the triangle, where total transport costs would be less than at corner locations. The exact location of P would depend upon the weight-loss characteristics of the two material inputs if only transport charges were involved. P would, of course, be pulled toward the material whose weight is most reduced by the manufacturing process.

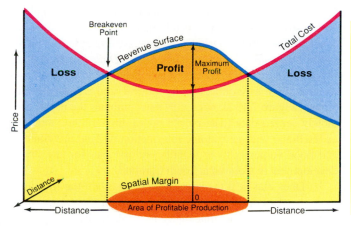

Figure 7.12

The spatial margin of profitability. In the diagram, 0 is the single optimal profit-maximizing location, but location anywhere within the area defined by the intersects of the total cost and total revenue surfaces will permit profitable operation. Some industries will have wide margins, while others will be more spatially constricted. Skilled entrepreneurs may be able to expand the margins farther than less able industrialists. It should be noted that a *satisficing* location may be selected by reasonable estimate even in the absence of the totality of information required for an *optimal* decision.

spatial pattern of factor prices is substantially different from before. If, however, the number of firms in an industry is small and costs of relocation are high, sometimes firms will agree not to relocate and attempt to set prices so that they continue to make a profit even though they are not located at the most profitable locations. Whether they succeed will depend on the cost of new entrants to the industry and their ability to devise strategies that keep new entrants from setting up production in new locations and selling at lower prices than their production costs.

As the migration of the U.S. textile industry shows, the location of industrial plants is sensitive to the costs of different factors of production (see "Location of the Cotton Textile Industry"). Costs of labor, for example, traditionally have been higher in the Northeast and North Central regions of the United States and lower in the South. Southern states have also been more active in the postwar period in developing policies to attract new industrial employment.

The ability to conduct searches for highest-profit locations has changed recently as better information about costs and availability of potential inputs at different locations is becoming available in computer data bases. Computers can search this information to identify maximum-profit locations as suggested by location theory. A new technology that can be used for such searches is known as **geographic information systems (GIS).** These are computerized systems for storing, retrieving, and analyzing geographic information. In GIS, large amounts of economic data and their locations are recorded. Sophisticated algorithms have

Industrial and Commercial Organization

been devised to simplify for the user many of the complex spatial search criteria employed in making location decisions. The U.S. Geographic Data Committee, for example, has published a *Manual of Federal Geographic Data Products* that lists thousands of computer-readable data sets describing, by location, U.S. economic, social, political, and environmental factors. Figure 7.13 shows a GIS that combines the use of remotely sensed data (data picked up via high altitude aircraft or satellites) with geographic information on land use zoning and flood vulnerability to identify optimal sites for industrial development in Columbia, South Carolina.

Does classical theory adequately explain the location patterns of contemporary North American manufacturing activities? It is clear that many multiplant firms do carefully evaluate costs and revenues at alternate locations and that the costs of transporting needed inputs to different locations and transporting the product to its markets are often very important. But surveys of people who have participated in major plant location decisions reveal a more complex pattern of reasoning than that on which location theory is based (see "Survey Method for Determining Industrial Location Factors").

The theoretical view of the process of locational decision making has been questioned by some on a number of grounds. Classical location theory assumed that labor is a homogeneous factor that can be used in the production process the same way at all locations. In the United States, however, some areas are known for having organized labor unions and for state laws that protect their right to strike. Other areas, particularly states in the South, have **right-to-work laws,** which prohibit workers from being required to join unions to obtain or hold a job.

Critics also cite unrealistic assumptions in the theory. For example, do manufacturers have the same knowledge

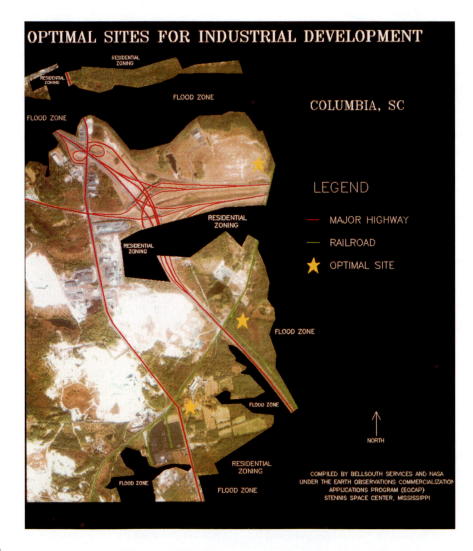

Figure 7.13

Remotely sensed data and geographic information systems (GIS) technology were combined to model potential industrial development in Columbia, South Carolina. The potential areas for development are outside the 100-year floodplain and residential zoning, adjacent to transportation routes, and contain at least 61 hectares (150 acres) of contiguous space.

One of the earliest examples of locational change in a major U.S. industry was the migration of the cotton textile industry from New England to the South between 1880 and 1930. A variety of essentially political factors prevented earlier industrial growth in the South, which in 1880 was like a developing country today. In the Piedmont area, the cotton plantation system had collapsed with the abolition of slavery in the 1860s. Many of its people were engaged in subsistence agriculture. Wages were very low. But, until the late 1870s, the South was not thought to be a safe place for northern capital. This image changed, however, following the Atlanta Cotton Exhibition of 1881, and northerners then deemed it safe to invest there. Soon, profits from the new mills became the dominant source of capital in the South for further expansion of the industry, and the area's earlier dependence on the North ceased.

The textile industry, then as now, required many relatively unskilled workers to produce the rough cotton cloth for which demand both in the United States and abroad was growing rapidly. In New England at that time, factory wages were higher than in the South, and states there were beginning to pass laws that restricted (to 60) the number of hours that women and children could work. Union strength was concentrated in Fall River and New Bedford, Massachusetts, and strikes by textile workers were far more common in New England than in the South. Between 1881 and 1894, for example, eight establishments in the South were engaged in strikes, compared with 300 in the North.

From 1881 on, the growth of the southern cotton textile industry was very great. Industrial wages were low and remained so as more and more of the surplus labor was drawn into the textile industry. In New England, the growth of other industries with their demand for labor drove manufacturing wages to rise more quickly than in the South, even though the number of workers in the industry increased more rapidly in the South. These factors, plus the more rapid adoption in the South of several technological innovations in production methods, resulted in the South outstripping the North in production and employment shortly after 1900. The New York garment industry is another example of the relocation of an industry. Although fashion apparel continues to be made there, most of the production of lower-cost apparel has moved to the Southeast to take advantage of lower labor costs there.

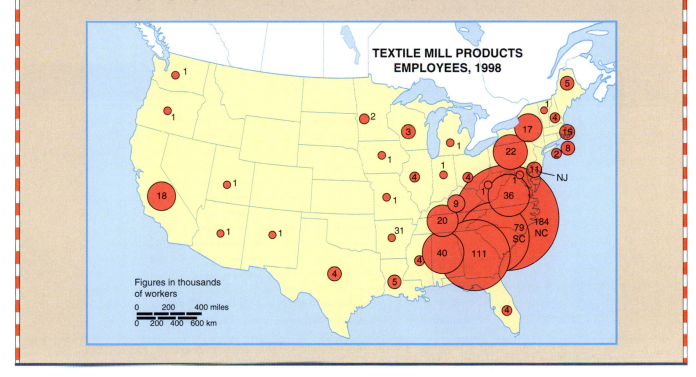

TEXTILE MILL PRODUCTS EMPLOYEES, 1998

Figures in thousands of workers

0   200   400 miles

0   200 400 600 km

Many studies have asked managers of recently opened industrial plants to identify and rank by relative importance the factors that influenced the location of their plant. A West Virginia University study made a nationwide survey of a sample of manufacturing plants that started operations between 1978 and 1988. Analysis was based on the 739 responses to a questionnaire sent to managers of 2710 randomly selected new manufacturing plants. Conclusions were that the factors that influence the locations of single-plant establishments are different from those that influence branch-plant locations. Also, plants that located in different regions were influenced by different factors. The study found that managers of many industrial plants carried out two kinds of locational search. They first searched among regions to decide which region of the country would best serve their needs, then they conducted a local search to decide where, within the region, they should locate (Table 7.1).

In their search of regions, access to markets and availability and quality of labor were the two most important factors cited, although managers of almost twice as many plants stated that the market was the single most important factor. Considerable differences occurred, however, between regions. Markets were rated first for plants that located in the East North Central region, second in New England, and fourth in the South Atlantic. Labor was the most important factor in the South Atlantic but third in the East North Central and fourth in New England. One interpretation is that management in the South Atlantic region seeks low-cost or nonunion labor as opposed to locations that are most accessible to markets.

Access to highways and vacant sites are important in the New England and East North Central regions, but availability of vacant sites is less important in the South Atlantic region. When the local search factors for locating branch plants were examined, access to schools was the most important factor in New England but unimportant in the East North Central Region and never mentioned in the South Atlantic region.

When considering the importance of different factors in the location decision, it is necessary to distinguish between the geographical scale of the firm's search—local or regional, as well as to consider that favorable factors perceived to be better in one region than another are often seen to be especially important in the selection of particular regions. Stated differently, regions with favorable factor endowments often attract plants whose managers think that these particular factors are important to them.

Table 7.1
Rankings of Factors Important in the Local Search for Locations for Manufacturing Plants

Rank	All Plants	Single Plant	Branch Plants
1	Markets	Markets	Nonunion labor
2	Nonunion labor	Highways	Markets
3	Highways	Nonunion labor	Land cost
4	Wages	Vacant site	Wages
5	Livability	Livability	Highways
6	Land cost	Wages	Property tax
7	Vacant site	Land costs	Livability
8	Property tax	Skilled workers	Water
9	Water	Previous location	Inducements
10	Skilled workers	Metropolitan area	Vacant site

From F. J. Calzonetti and R. T. Walker, "Factors Affecting Industrial Location Decisions: A Survey Approach" in Industry Location and Public Policy, edited by H. W. Herzog, Jr. and A. M. Schlottman. Copyright © 1991 University of Tennessee Press, Knoxville, Tenn. Reprinted by permission.

everywhere about product development and the technology of production? History shows how some major industries located in certain areas because of the superior knowledge and judgments of key individuals who were responsible for innovations that either changed the product in very important ways or changed the way in which the product was produced. Henry Ford, for example, is generally credited with introducing a new mode of production when he had the vision of a car for the masses, built on assembly lines, using standardized parts, by a semiskilled labor force. This was a novel method of production, not only for automobiles but for many other consumer products, too. The idea diffused out of the Detroit area and spread across the country. The product cycle approach, described below, builds on these criticisms and introduces a more dynamic element into the analysis.

Product Cycle Explanation

This approach to an explanation of site selection is based on case studies of many different industries. The **product cycle approach** describes typical changes in location patterns as a product matures and the profitability of firms in the industry declines. In this model, innovative products develop in favorable areas, often where a large number of consumers are available to critically evaluate the product and rapidly adopt features they value. This often occurs at a time of recession, when individuals, firms, and corporations are actively seeking new products to entice consumers to increase their spending. Certain areas of the country gain a reputation for being "incubator areas" where new ideas are born and become embodied in novel products. At this stage of the location-decision process, innovators are not combing the country to find least-cost locations; instead, they are busy developing the product itself. Their competitiveness in relation to potential rival companies lies in their knowledge and skill in developing the product. Particularly important in this stage in the life of the product are the product developer's ties with venture capitalists, who supply risk capital to the developer, often in return for a stake in the fledgling company. Attaining and maintaining the confidence of venture capitalists is important at this stage.

If the product line is successful, employment expands but stabilizes as the market reaches saturation and then declines as most consumers who want the product already possess it and only a few new consumers enter the market each year. Depending on the length of life of the product—approximately seven years for automobiles and 15 years for refrigerators, television sets, and many other consumer durable products—the pattern of demand never reaches the high consumer product adoption rate of when the market was new. With the maturing of the product, profit margins become squeezed as the product becomes a commodity that differs little in the minds of consumers between one producer and the other, and price becomes the distinguishing characteristic for the consumer.

With low profits, the stage is set for manufacturers to seek new locations with lower costs of production. These are usually in different areas than the original manufacturing locations. At this stage, many location searches are conducted to find the least-cost, maximum-profit location. By this time, the production process has been simplified, and the manufacturer is usually interested in finding lower-cost labor and other inputs. The result, generally, is a decentralized location pattern for the product as locations are identified with cost advantages over other locations. The product cycle thus distinguishes between product innovation at early stages of the product's development, process innovation in later stages, and, later still, further rationalization as firms strive to find the lowest-cost production sites.

Many examples exist of such shifts from early spatial concentration of production to later movement of industrial plants to lower-cost, more-dispersed locations. Radios were first manufactured in the New York region, but manufacturing later dispersed to larger plants in the Midwest and California. Canneries, first located in the San Francisco Bay area, later dispersed to the Central Valley of California. Branch plants of automobile firms decentralized outside the core area of automobile production in the Detroit area (Figure 7.14).

The garment industry was concentrated in New York in 1910, where it largely remained until the 1960s, when it began to disperse to areas within overnight trucking distance of New York. By the 1970s, much of the garment industry had relocated to North Carolina and South Carolina, where average wage rates in the industry were 45% lower than in New York, unions were less involved in the industry, and plants were closer to their raw material in the textile industry of the area. Much of the garment industry also relocated to California, where a good deal of the work is done at "piece rates" by recent immigrants from Latin America and Asia working in their homes. By 1980, only 20% of women's clothing was made in New York, down from approximately 50% in 1950.

Political Economy Explanation

The **political economy approach** explains change in industrial and commercial activities as the outcome of social and political processes that are the product of the capitalist system of production in an increasingly global economy. The social processes involve the way that labor is used in the production process and how changes in the terms of employment of labor have created the continuous development and redevelopment of new systems of industrial and commercial production through which products are manufactured and distributed at lower and lower per unit costs.

Capitalists attempt to maximize profits, while workers try to resist the demands of capitalists to work for lower wages. In this adversarial view of the relationship between capital and labor, labor demands more and more concessions from owners of firms, which increases costs of production

Figure 7.14

North American automobile and light truck plants, 1997. Motor vehicle production in North America shows far more dispersion than concentration, although the older Michigan-Ontario nucleus is still evident. Further dispersal is promised as foreign firms locate new assembly or manufacturing plants and domestic companies build replacement plants away from Detroit. NAFTA has encouraged the construction of several plants in Mexico in recent year.

Source: Data from AutoinfoBank, 1990

and decreases the firms' competitiveness in a global market in which concessions made by one firm are not automatically followed by other firms in the same industry. When Henry Ford promoted the idea of the five-day week and the idea of the weekend, he unleashed a new pattern of behavior as workers developed activities to fill their leisure time. In the 1920s, when this occurred, the automobile industry did not have to consider whether automobile manufacturers in other countries would be willing to grant comparable concessions.

Under current industrial relations and bargaining conditions, however, labor unions understand that any of their demands that increase the cost of a firm's product is likely to lead to the loss of many jobs. Employment in General Motors, for example, the largest automobile manufacturing firm in the United States and Canada, dropped from more than 500,000 in the 1970s to 229,000 in 1999 as it scrambled to reduce its costs, design more relevant products for an increasingly energy-conscious population, increase the productivity of labor, and so regain its profitability. For a short time, the third-largest U.S. automobile manufacturer, Chrysler Corporation (now Daimler-Chrysler), relied on emergency government loans to stay in business. After a few years of restructuring and reorganizing its methods of

product design, development, and manufacture, it emerged once again as a strong competitor and repaid the loans ahead of schedule. Increasingly, both owners of firms and labor understand that only by increasing the productivity of labor can plants continue in business in a global market of competing firms.

In the political economy view, the decisions of governments to sanction a given social-economic production system in a particular form of global economy control the kinds and locations of economic activities (see "Maquiladoras"). In this view, the changing locations of many activities in the United States, as well as the many decisions to move production facilities to other countries, are responses by businesses to set up different social systems of production by moving to locations where workers will accept the changes they propose. Since one manifestation of this process is that of firms moving their industrial activities from areas of well-developed manufacturing activities to areas not so well-developed, leaving behind areas of unemployment, some geographers refer to this process as the **geography of uneven development.** The political economy approach focuses on the logic behind different production systems, the tensions that inevitably arise between capitalists and workers, and the role of governments in

MAQUILADORAS

The difficulty in determining which economic-geographic theory best describes the location of an industry can be illustrated by the recent growth of **maquiladoras** (literally, mills) in Mexico. Are they an example of the political economy explanation, or do they simply represent a Weberian substitution of labor for transport costs?

Initiated by the Mexican government in 1965, the maquiladora program allowed American and other foreign-owned companies to establish plants in Mexico to produce goods for export markets. In these plants, Mexican workers assemble imported components for immediate reexport. The chief attraction for foreign companies is abundant, low-cost labor. In 1996, Mexico's minimum wage was 28 cents per hour, and the average hourly wage paid to maquiladora employees was $1.19, well below U.S. rates. In addition, the costs of land, construction, and utilities are lower in Mexico, and labor laws and environmental standards more lax. The advantage for Mexico is employment for its workers and foreign exchange for its economy.

In the approximately 35 years that the program has been in existence, more than 3800 maquiladoras employing over one million workers have been built in Mexico. About 90% of the plants are within a few kilometers of the U.S.–Mexico border, partly because their American managers prefer to live in the United States. Across the border from El Paso, Texas, Ciudad Juárez alone had 310 maquiladoras in 1995 with about 150,000 workers who assemble American-made components into television sets and other products for Zenith, Philips, and other firms. Plants on the Mexican side often have companion operations on the American side, for product finishing, warehousing, administration, and distribution.

Even given their higher transportation costs, U.S. firms with maquiladoras have been able to be more competitive in world markets by combining low-cost labor with advanced technology. For Mexico, the maquiladora industry is its second-largest source of foreign exchange earnings. Exports of maquiladora products exceed $40 billion annually, and they have been important in changing a U.S. trade surplus with Mexico into a sizeable yearly deficit.

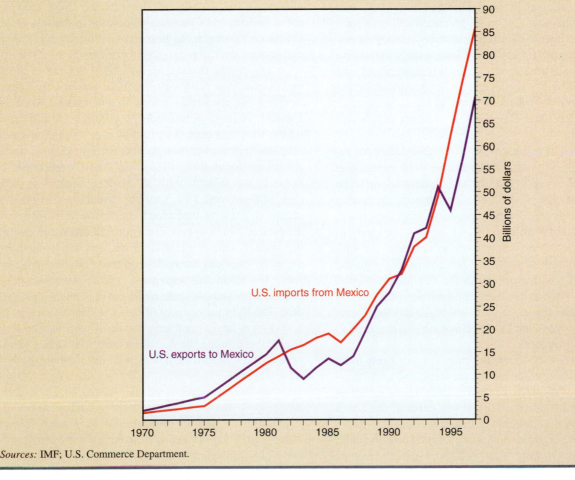

Sources: IMF; U.S. Commerce Department.

189

maintaining the underlying conditions that promote this ever-changing migration of industrial activities around North America and the world.

The political economy explanation thus focuses on the macro economic and social policies of governments and their effects on changes in the distribution of regional economic activities. It is claimed that in recent years, there has been a transition from the industrial model of mass production developed by Henry Ford to a flexible production and accumulation model. **Fordist production systems** occur in large manufacturing plants with high capital investments dedicated to making a relatively few standardized products, such as automobiles, using parts brought together in an assembly-line process. Such plants typically employ many white- and blue-collar workers (who usually belong to a labor union), with each worker responsible for a small number of increasingly fragmented and simplified tasks. These highly specialized production units were located traditionally as branch plants in a decentralized pattern in the suburbs of cities in the old industrial belts, or, more likely now, in new manufacturing places in the South and West. These branch plants were designed to reap the advantages of lower costs available to larger-scale companies.

The system of production to which North America is shifting has the key characteristic of flexibility. **Flexibly specialized production systems** are designed to efficiently produce small quantities of a product and to easily convert to producing a different product. This flexibility is reached through equipment that can be programmed and used for a variety of tasks. Instead of offering workers permanent employment, firms seek flexibility by increasingly employing temporary or part-time workers. In this new model, the hard distinction between management and workers breaks down, and people are trained to work in teams responsible for quality, safety, and efficiency in product design and manufacture.

Reductions in employment in the core industrial regions of the United States and Canada are to be expected as the Ford model is superseded by the flexible model. The developing networks of specialized producers, interacting with each other to produce new products, spur the growth of regional industrial districts. These regions are rarely in the old industrial areas, where workers' experiences with Ford-like production systems are not compatible with the new social relations that prevail in the flexible production model. In the Ford-type areas, unions are likely to be strong, wages and fringe benefits are high, and local governments are experienced in regulating employers through local planning ordinances.

The flexible production systems are more likely to be found in areas where trade unions are weak and social conditions are harmonious with their needs. These locations tend to be in the suburbs of the newer cities of the South and West; in rural communities that have climates, recreational facilities, or other amenities attractive to workers; and in the centers of large metropolises, where sweatshops employ semi- or unskilled labor. In these latter areas, occupational safety conditions are rarely observed or enforced, and employers often do not ask for proof of a worker's permit to work in the United States.

Union membership in the United States grew rapidly from the 1930s to its peak in 1953, when 33% of the nonfarm work force belonged to a union. By 1998, only 13.9% of the work force was unionized. Plant closings, employment declines in traditional, unionized industries—steel, automobiles, and coal mining, for example—and growth of new, nonunion industries, such as information processing, fueled much of this change. A second force was the deliberate migration of industry to regions of the country with no tradition of organized labor. The movement of Japanese-owned automobile plants to Tennessee, German-owned manufacturing plants to South Carolina and Alabama, and many garment manufacturers to California and South Carolina illustrates this trend.

Pivotal Changes in the Economies of the United States and Canada

The theories just discussed help shed light on the behavior of firms in commercial economies. Such behavior, however, occurs in a real world where the simplifying assumptions of the theories may not be met and the rule of life is change. In this section, we describe the major changes in the American and Canadian economies in the last 55 years before examining some important industrial and commercial sectors and the geographical changes in their activities.

During the first decade following the end of the Second World War (1945–1955), there was a large, pent-up demand for consumer goods in the United States and Canada. A large population of young men and women whose lives had been disrupted by war returned to civilian life, married, started families almost immediately, and, for the white population at least, defined the good life as a home in the suburbs with all modern amenities, such as cars, refrigerators, TVs, washing machines, and other mechanical gadgets that freed the housewife from many drudgeries. Often openly discriminated against in the housing market, minority industrial workers established themselves in inner city areas close to traditional manufacturing sites. The result was segregation of the American population on a geographic scale far larger than anything seen before. The fight for equality of the sexes on the factory floor had not yet begun, and many women who had been welcomed there during the war when labor was scarce were expected to return to their homes and resume their traditional roles (Figure 7.15).

In 1950, following the start of the cold war and the dropping of the Iron Curtain across Europe, the Korean War stimulated military production in the United States and many supporting commercial activities from Japan, now a loyal ally conveniently located on the doorstep of Korea

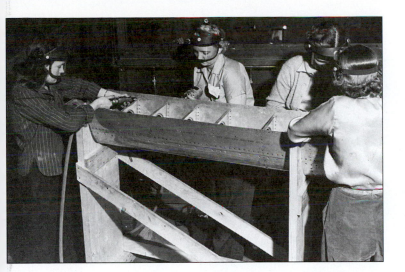

Figure 7.15
Before World War II, jobs outside the home had been considered inappropriate for wives and mothers. The manpower crisis during the war opened up work opportunities for women, more than 6 million of whom joined the labor force. "Rosie the Riveter" became a heroine, symbolizing women's dedication to the war effort.

and available as a launching site for military action there. The stage was set for a period of unprecedented prosperity as industrial and commercial activities grew in a nation in which key industrial decision makers thought first about the demands of the local market and secondly about where, within the country, would be the best location to organize the production and distribution of goods to meet the need.

This was a period, too, when many branch plants of prominent U.S. industrial firms were constructed in Canada. To avoid import tariffs, Sun and Shell Oil companies constructed refineries in Sarnia, Ontario, at the southern end of Lake Huron. Dow Chemical Company and Du Pont also built branch plants there. The "Big Three" U.S. automobile companies (Ford, Chrysler, and General Motors) all established branch plants in the Toronto–southern Ontario area. Most branch plants were located in Ontario because of its proximity to the American portion of the Manufacturing Belt. Other U.S. branch plants were located in West Europe as it recovered from the war and began, once again, to build up its manufacturing base. The founding of the European Common Market also led to increased American investment in Europe because firms locating within the Common Market avoided tariffs.

The simplified production processes developed during the war continued to be used, and the new field of "time and motion studies" made a science of assigning tasks efficiently to individuals who repetitively completed them as the production line moved forward. The decade following the war was also a period of labor unrest. Large-scale producers attempted to keep wages low, but strikes became common against giant firms in the oil, coal, steel, railroad,

and automobile industries. Many of the strikes resulted in gains for labor.

By the 1950s, the movement of people and goods within the United States and Canada was undergoing a revolutionary change. Until this time, people and materials had moved long distances by rail or water. Now, the automobile and the truck became important, and with the advent of commercial jet aircraft by the late 1950s, businessmen (and they were male almost always then) began to travel around the two countries by air. Rail and water continued to be important but were used more selectively for shipping bulky, often low-value, commodities long distances.

While trade was strong between America and Europe during this period, American-Pacific trade was not, except during the Korean War (1950–1953). American business did find important new markets for its products and, more important, foreign sources of raw materials that increasingly were becoming scarcer and more expensive in the United States. To protect these sources, the military might of the United States was used in 1953, for example, to organize coups that overthrew the governments of Iran and Guatemala. The short economic boom that began in 1954–1955 was fueled by increases in consumer demand for autos, refrigerators, dishwashers, and television sets. This was a period of concentration of American industry as firms merged to form giant companies that manufactured, in some cases, a large group of unrelated products. Many of these firms increased their investments abroad, where they saw higher returns on their capital.

The role of the U.S. government in the 1950s was to resist any suggestions that it should intervene in domestic industrial and commercial affairs. The Republican Eisenhower administration of the 1950s believed strongly in free enterprise and in leaving the economy to its own devices. Unlike most other industrial nations, the United States had no recognizable industrial policy to guide the evolution of its economy. Western European countries, for example, all of which at that time depended on imported oil, taxed oil products heavily to reduce their dependence and its impact on their balance of trade. By 1957, America's economic strength showed signs of weakness. Exports of coal, automobiles, iron, and other metals declined sharply. The importation of small cars from Europe and, later, Japan began. Imports increased and exports decreased. Soon, there were excess dollars abroad. Throughout this period, the government's social welfare, labor, immigration, trade, and monetary policies all favored the development of a global production and distribution system.

The change to a Democratic administration in 1960 brought a new policy of the government assuming responsibility for maintaining full employment and stimulating manufacturing and commercial growth. The means to achieve this were largely macro policies of increased government spending to stimulate the economy and reduce unemployment, which, at 7% in 1961, was considered too high. This period saw an increasing reliance on foreign

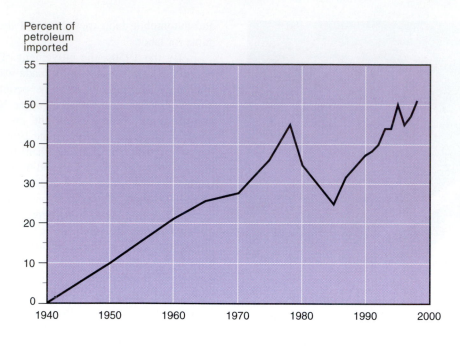

Percent of petroleum imported

Figure 7.16

Imports of foreign oil reached 45% of the total oil consumed in the United States in the late 1970s and contributed to the instability of oil markets and industrial economies. High prices, conservation, and the use of more energy-efficient vehicles reduced demand in the early 1980s, but American dependence on foreign sources of oil has increased since 1985.

sources of many essential raw materials, especially oil (Figure 7.16). There was no restraint on the movement of business capital to other countries, particularly those developing countries where the investments were regarded as politically safe—that is, safe from government expropriation and interference. The Vietnam War, although it had a destabilizing effect on American society, stimulated industrial production. Many of the defense industries prospered during this period.

In the 1970s and 1980s, it became clear that some traditional American industries were no longer competitive internationally. The products they manufactured could be bought on the world market more cheaply, and in many cases, their quality was inferior to the best made elsewhere. A leading refrain in U.S. industry during the 1980s was "maintaining international competitiveness." This generally meant that the products of U.S. manufacturing plants should sell at competitive prices on the world market and plant owners should receive reasonable profits on their capital investment.

Geographic Changes in the Steel Industry

The experience of the steel industry shows how each of the three concepts described earlier—location theory, the product cycle explanation, and the political economy approach—help us understand geographical changes in output and employment in the past 50 years.

In 1945, the steel industry was at a period of high activity oriented to supplying the large demands of an economy moving into a hoped-for era of peace and prosperity. In 1950, even when the reconstruction of the European and Japanese steel industry was almost complete, U.S. steel companies produced 46.6% of the world's steel, with the four largest U.S. producers making approximately 29%. The industry was an oligopoly, with more than 75% of U.S. steel produced by eight firms.

Between 1945 and 1960, the steel industry invested heavily in new equipment, though experts now agree in retrospect that much of this investment was made in technologically obsolescent steel-making processes. Much new steel-making capacity was developing in basic oxygen-furnace process technology, but U.S. steel makers were slow to adopt this approach, especially the dominant firms. The largest of the firms, the U.S. Steel Corporation, set prices in this period, and other firms followed its lead. The industry was oriented to meeting domestic demand and showed little concern about foreign competition. During the 1950s, a strong labor union, the United Steelworkers, fought constantly for more money and benefits for its workers.

In 1959, a 116-day strike by 500,000 United Steelworkers against all U.S. steel plants first made the country a net importer of steel, and it has remained a net importer ever since. By the 1960s, imported steel was setting the prevailing market price, and U.S. steel producers experienced a pinching of profit margins that made them appeal to the government for help against "unfair foreign competition." After Congress failed in 1967 to pass a bill that

Figure 7.17
This idled steel mill in Johnstown, Pennsylvania typifies the structural changes occurring in postindustrial America. Heavy industry is frequently unable to compete in unregulated markets with lower-cost producers from developing countries or with those receiving production and export subsidies from their governments.

would have given the president authority to work out import quota agreements with exporting nations on a product-by-product basis, the steel industry mounted an intensive lobbying campaign for import quotas on foreign steel, but there were many domestic opponents as well as foreign governments that threatened to retaliate if the United States unilaterally imposed quotas. During this debate, President Lyndon Johnson, without legislative authority, initiated talks with Japanese and European governments, which consulted their steel producers, and the result was the Voluntary Restraint Agreement of 1968.

The intent of this agreement was to give the steel industry breathing room to become internationally competitive. Instead, while all other steel-making areas of the world (including Canada) continued after 1969 to increase their investments in steel making, the United States reduced its level. Several U.S. steel manufacturers did, however, make sizable investments in foreign steel making. By 1971, U.S. steel firms were in deeper trouble than three years earlier. Not having increased their investments and carrying a large capacity less than 10 years old that was based on now obsolete manufacturing technology, their future as internationally competitive producers was most uncertain. The result was bankruptcy of former industrial giants and massive consolidation of the remaining firms. In 1979, five

states accounted for just over 70% of the nation's steel-making capacity: in order, Pennsylvania, Ohio, Indiana, Illinois, and Michigan. Since then, Indiana has become dominant over Pennsylvania and Ohio, the traditional leaders of the industry. Today Pennsylvania and Indiana are the top employers, though of a much reduced labor force.

The industry had gone through a massive reorganization with the goal of increasing labor productivity through large capital investments. The solution for the U.S. steel producers was to petition for and receive an extension of the Voluntary Restraint Agreement and then to reorganize as parts of larger conglomerates. It is now clear that steel producers during this period in which they received government subsidies (through import protection) used the income generated to diversify into other product lines and to internationalize their steel-producing facilities. In 1982, for example, U.S. Steel purchased Marathon Oil for $6.3 billion.

The stage was set for massive plant closings and the devastation of many American communities (Figure 7.17). Between 1978 and 1982, steel-industry employment fell from 449,000 to 289,000. Employment in this industry declined 61.7% from 1960 to 1986, compared with a 3.7% growth in all manufacturing employment in the same period. The decline was due to falling prices and output, foreign imports, chronic overcapacity, technical obsolescence,

Industrial and Commercial Organization

and poor management. In place of industrywide collective bargaining with the United Steelworkers, which prevailed from 1959 through 1983, each firm bargained separately with the union. Some firms emphasized new models of labor-management relations with the joint goal of becoming internationally competitive. This phase is commonly referred to as the *disinvestment* of the steel industry.

By the 1990s, the major American **integrated steel mills** were located in a belt running from the Pittsburgh area, across northern Ohio to the Detroit and Chicago areas, and a few scattered locations in the South and the West (Figure 7.18). An integrated mill is a plant that starts with raw materials (ore, coal, limestone) and produces a full range of steel products in a continuous manufacturing process. A good example was the Lackawanna plant in Buffalo (now closed), which at its peak employed 20,000 workers and made every kind of steel product.

Plant closings in the early 1980s were followed by a new phenomenon in steel making: the rise of the **mini-steel mill.** Such mills have an annual capacity of less than a million tons and rely on electric ovens, scrap metal, and nonunion labor. Their location is oriented to the sources of their raw material (scrap metal and electricity), and they are sensitive to their local prices. Note on Figure 7.19 that although some mini-steel mills are found in the northeast Manufacturing Belt, most avoid this area, where their product would compete locally with steel made by other processes.

Discussions about reasons for the decline in competitiveness cite many factors. Many observers claim that the government strategy of giving the industry temporary protection from foreign competition permitted it to disinvest as it canceled its expenditures on research and development, closed down its least-efficient plants, and used its profits to invest in other activities. From the point of view of stockholders, this was rational economic behavior, made possible by a government that had been misled into believing that the firms would, in the national interest, reinvest in steel and challenge the foreign competitors. Instead, in the pursuit of higher profits, they abandoned the industry and, in many cases, took the word *steel* out of their names—U.S. Steel, for example, became USX. Other observers believe that the manufacture in the United States of any basic commodity that contains a high labor content can no longer be competitive in a global marketplace. To them, U.S. steel firms were simply one of the earliest groups to recognize this undeniable fact.

The latest phase in the transformation of the U.S. steel industry is the direct investment in it by Japan and the resulting changes in location and methods of production. Dissatisfied with the low quality, high price, and uncertain availability of specialized steel products for its new automobile manufacturing plants in the United States, Japanese companies began in the 1980s to acquire U.S. steel plants and construct new ones. These firms have developed a spatially coordinated production chain linking integrated steel producers and their automotive assembly plants. Their intention is to minimize the costs of transporting steel and steel products and, perhaps as important, to be

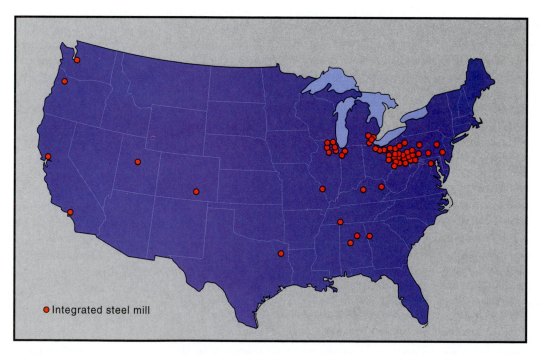

Figure 7.18
Location of major integrated U.S. steel mills, 1990s.

able to change production runs easily so that they can have **just-in-time delivery** of steel inputs to the assembly plants (Figure 7.20).

This spatial coordination accounts for a large part of the westward shift of steel manufacturing from the Pitts-burgh area to Ohio, Michigan, and Indiana. Just-in-time deliveries are made more reliably and with less risk when suppliers are located close to the factories they serve. In their projects—many of which are conducted jointly with American integrated steel mills—Japanese firms are implementing

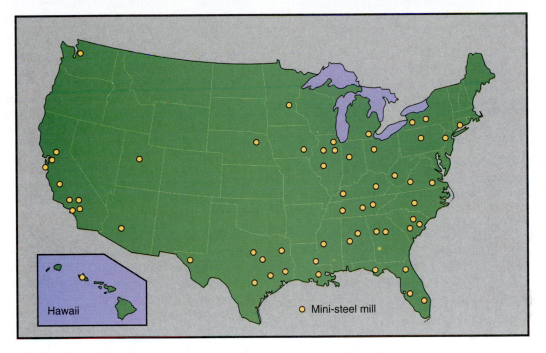

Figure 7.19
National distribution of mini-steel mills.

Source: The Institute for Iron and Steel Studies.

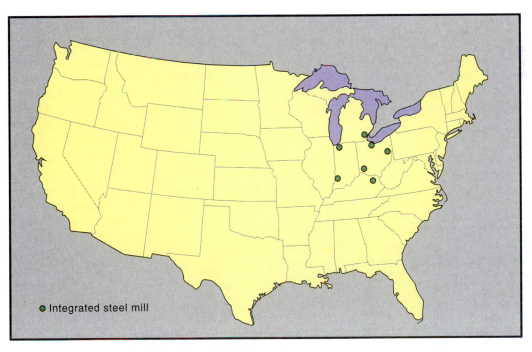

Figure 7.20
Location of Japanese-affiliated integrated steel mills in the United States.

Industrial and Commercial Organization

their own quite different method of production. Instead of the many levels of supervision, different tasks, and grades of labor found in U.S. plants, the Japanese use *team production organization,* in which responsibility is placed on the team to meet production goals. Many of the traditional social divisions between supervisors and workers of different grades are eliminated. Instead, workers are expected to work cooperatively and to suggest changes in production techniques that will lead to higher quality and greater efficiency. If these teams suggest changes that reduce the number of workers required to make a product, agreements are usually in place to protect the jobs of the workers who make the suggestion. This component of industrial change is unusual in that it is occurring in the same places and plants where Fordist methods previously endured, thus challenging the conventional wisdom that new production methods that change the social relationships in the process must be implemented in new locations where workers are not familiar with the relationships that previously prevailed.

Growth of Information-Technology Industries

While many traditional industries declined in the last two decades, new industries developed. Foremost among these are **information-technology industries,** which manufacture semiconductors, computers, communications equipment, and electronic automated machines and pursue genetic engineering. Manuel Castells has described a spatial logic that is characteristic of information-technology industries throughout the world. His basic concept is that of locations of **milieus of innovation,** places with a work culture committed to generating new knowledge, processes, and products. How do such areas come into being, and what is their structure?

Castells argued that these areas occur where the three fundamental elements of production processes converge in space and time: raw material, labor, and capital. The raw material is innovative technological information. This is found in leading universities; government-sponsored research and development (R & D) centers; R & D centers of large, technologically advanced corporations; and established small R & D centers in some existing industrial areas. The first condition for the development of information-technological industries is access to one of these sources of innovative information. The second element is access to a large pool of scientific and technical labor, which grows through time as job opportunities arise. This pool develops in places that have good academic and vocational educational institutions, high social status, and attractive urban amenities.

The third element that supports the development of a milieu of innovation is the availability of capital. Possible capital sources are the long-term R & D investments of large companies that can finance them from current profits and are willing to wait for possible payoffs, direct financing by governments or indirect financing through military contracts, and venture capitalists. Corporate R & D investments are usually made close to the firms' headquarters; military contracts for technologically advanced projects have a distinct spatial pattern favoring southern California, New York, Massachusetts, Texas, St. Louis, and Seattle. Venture capitalists are located particularly in Boston, San Francisco, and Texas. In Canada, Toronto's metropolitan area is the premier location for venture capital enterprises.

Information-Technology Specialty Areas

Silicon Valley, in the San Francisco Bay area, is the best known technology specialty area in the United States. Stanford University was prominent as early as the 1950s in developing close ties with emerging high-technology areas. Two other local universities developed excellent engineering programs, and early sources of funding were military contracts and government research grants. Venture capital supported many firms, such as Apple Computer, that later dominated their field. Many firms financed their expansion with profits or by going public and issuing shares. Because so many firms in the area were leaders in their field, Silicon Valley attracted some of the most qualified engineers and scientists in the country. The interactions between these firms, as well as between many of them and R & D centers across the world, led to the development of a distinctive local culture. This was a social as much as an economic process. The area of greater Boston known as Route 128 developed similarly, with entrepreneurial talent linked in many cases to the Massachusetts Institute of Technology and other higher educational institutions in the area.

Castells distinguishes a number of types of technologically advanced areas in the United States; Silicon Valley and Route 128 represent the first type. The second, for which he provides the examples of Los Angeles and Phoenix, is particularly linked to military research, development, and production facilities. A third type is exemplified by IBM in the New York area and AT&T in the New Jersey area, where the early advanced technological work took place within corporations. Here, the distinctive culture is quite different from the decentralized, wide-open networks of Silicon Valley. The fourth type is exemplified by Dallas, Texas, whose industry began as a relationship with corporations but soon spun off and developed a dynamic of its own. Type five, exemplified by the Minneapolis–St. Paul area, represents an early partnership between government contractors, local capital, and a group of engineers from the Navy who pursued their ideas in advanced computing and succeeded.

Production Process

The advanced information-technology industry, as noted above, developed where the spatial logic led to innovative milieus. The actual productive capacity, however, has its own spatial logic. Whereas areas of product development and research are few and efforts are concentrated in small areas, productive capacity is scattered in suburban locations across the United States—although more particularly in western states—as well as across the world. Each product is made up of parts that are manufactured at many locations. In this respect, the industry is quite different from traditional industries where most parts are manufactured either in the same location or in locations quite close.

In the information industry, parts are highly specialized, and each has distinct production needs (Figure 7.21). Because parts are valuable in relation to their size and weight, they can be moved long distances without transportation costs becoming a large part of the final cost. Therefore, design, planning, and production facilities can

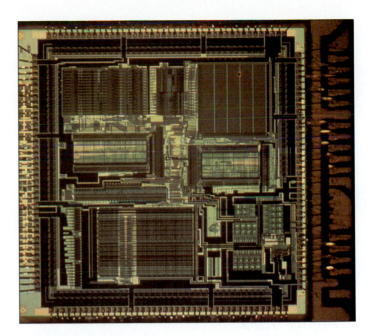

Figure 7.21
The "brains" of a computer are microprocessors, tiny slivers of silicon. A fourth-generation memory chip, the 486, has 1.2 million of the transistors that make up a computer's logic circuits. Memory chips are immensely valuable in relation to their weight and size.

be at widely scattered locations, depending on the relative advantages of each place.

With so much capital spent upfront in product development and testing, successful products need to be sold worldwide as much as possible. Strategies of market penetration are dominant, and this has persuaded many companies to scatter their productive facilities among many countries in an effort to escape tariffs and have their products adopted widely throughout the world.

Distribution Systems

A striking feature of change in the U.S. industrial and commercial system in the past 50 years is the increase in the volume and complexity of distribution systems. Henry Ford thought that **vertically integrated production** systems would ultimately prevail. He saw factories as places where raw materials, such as iron ore and coal, came in and finished products, such as automobiles, came out. Instead, the trend throughout this period has been the development in one industry after another of **horizontally integrated production** systems in which specialized products manufactured at many places are brought together through complex logistics systems and assembled in the final stage of production.

Ford formulated his ideas about vertical integration production processes in an era in which most goods moved on water or rail. He could not see the radical changes that

(a)

(b)

Figure 7.22

United Parcel Service (UPS) uses geographic information systems (GIS) to support such operations as planning routes, managing, and coordinating distribution centers. UPS has begun to replace its manual overlay system of paper maps and acetate sheets (a) with GIS technology (b).

would follow with the development of the interstate highway system, complementary state highway systems, and a highly competitive trucking industry. As this change occurred in the 1960s, industrial sites in the cities lost their principal asset—connection to the rail and water networks of the country. Although still accessible by highways, these sites lost their unique advantages, and spacious, cheaper, more accessible suburban sites became more attractive to business. Although low-value commodities are still carried in bulk in barges and railcars, most high-value products are shipped by truck, often long distances, for trucks offer highly reliable service and therefore are well-suited to just-in-time delivery.

Transportation logistics have become not only more complex but also more efficient with the advent of technological revolutions. These innovations have improved the efficiency of the geographical organization of freight movement. Mobile (cellular) telephone service, now available at most locations on the 71,000-kilometer (44,000-mi) U.S. interstate road system, permits dispatchers to communicate with drivers, monitor their movements, and change their destinations. Adoption of global positioning systems technology, which is able to find and transmit the location of a truck to a dispatcher from satellite signals, has spread rapidly in this industry since it permits central dispatching and monitoring to control directly the movement of products. Computerized transportation management systems based on geographical information systems use national, standardized spatial data sets to compute least-cost routes under a variety of conditions and pass results to dispatchers

and drivers in the field (Figure 7.22). Such systems know the contents of trucks and their capabilities as well as their locations and can combine this information with that of changing requests for service.

The increased control and flexibility offered by these new telecommunication-based technologies led to the flexible specialized production systems that grew rapidly in the 1990s. When American industries shifted away from basic manufacturing, which tended to be located in areas accessible to rail and water transportation systems suitable for transporting bulky inputs and outputs, suburban sites became increasingly suitable for the manufacture of more complex products dependent upon a large variety of high-value inputs assembled from many locations scattered throughout the country and the world (Figure 7.23). Soon, these suburban locations coalesced around main interstate routes to form corridors of development surrounded by fast-growing and often new cities.

Summary

The past 55 years have brought vast changes to the industrial and commercial geography of the United States and Canada. Key components of this change can be traced to the transition from a basically domestic economy relying little on the outside world for inputs in the production process or major markets for its products to an international economy with major dependencies on other countries for inputs and markets. The combined American and Canadian economy, as one of the three dominant economies in the

Figure 7.23

On a small scale, the *planned industrial park* furnishes its tenants with external agglomeration economies similar to those offered by large urban concentrations to industry in general. An industrial park provides a subdivided tract of land developed according to a comprehensive plan for the use of (frequently) otherwise unconnected firms. Since the park developers, whether private companies or public agencies, supply the basic infrastructure of streets, water, sewage, power, transport facilities, and perhaps private police and fire protection, park tenants are spared the additional cost of providing these services themselves. In some instances, factory buildings are available for rent, still further reducing firm development outlays. Counterparts of industrial parks for manufacturers are the office parks, research parks, science parks, and the like for high-tech firms and enterprises in tertiary and quaternary services.

world, has invested heavily in Latin America and a few countries elsewhere. Much of its trade is oriented to these countries, as well as to the European Community and Japan. In the last two decades, many traditional industries simply could not compete in the global marketplace and declined. The "Rust Belt" was a term frequently employed to capture the trail of destitution in the American Midwest and Mid-Atlantic regions that accompanied this trend.

Meanwhile, new industries appeared and grew most strongly in areas away from the traditional Manufacturing Belt. They were part of the new information-based economy and prospered in selected areas of technological leadership. The production sites for these products were scattered in suburbs and rural communities throughout the country, but especially in the West. Information technology

industries less frequently use Fordist methods of manufacture, instead favoring more flexible production processes using many related suppliers often organized in spatially integrated production systems.

Today, American and Canadian industries no longer are dominant in the world. Many are no longer competitive in the global marketplace. Others have been able to find a niche either by making specialized products where closeness to markets is essential or by assembling components manufactured in low labor-cost areas of the world in highly automated plants in the United States. There is now a new international division of labor and, with it, a new regional pattern of economic activities. The geographical patterns of commercial and industrial activities in the United States and Canada are now inextricably linked to the global economy.

Key Words

factor inputs

flexibly specialized production
   systems

Fordist production systems

geographic information
   systems (GIS)

geography of uneven
   development

horizontally integrated
   production

information-technology
   industry

integrated steel mill

just-in-time delivery

least cost theory of location

maquiladora

milieus of innovation

mini-steel mill

political economy approach

product cycle approach

right-to-work laws

U.S./Canadian Manufacturing
   Belt

variable costs

vertically integrated production

Gaining Insights

1. What is meant by the "spatial margin of profitability"? Explain how a company searching for a new location might find this spatial margin.
2. The product cycle is characterized by distinct spatial patterns of production. Name a product that has passed through the cycle, and describe the spatial patterns of its production at different times in the cycle.
3. The search for profit leads to geographically uneven development. Describe an example from the United States or Canada, and explain the link between the search for profit and uneven development.
4. Discuss the forces leading to the deindustrialization of many traditional manufacturing areas of the United States and Canada.
5. Define flexible production systems. How do they differ from Ford-type production systems?
6. What are the defining characteristics of "milieus of innovation"?
7. Describe the different kinds of information-technology specialty areas in the United States. How does their geographical pattern differ from that of manufacturing sites for their products?
8. Compare the transportation implications of vertically and horizontally integrated production systems.

Selected References

Berry, B. J. L., E. C. Conkling, and D. M. Ray. *Economic Geography*. Englewood Cliffs, N.J.: Prentice-Hall, 1987.

Bluestone, B., and B. Harrison. *The Deindustrialization of America*. New York: Basic Books, 1982.

Calzonetti, F. J., and R. T. Walker. "Factors Affecting Industrial Location Decisions: A Survey Approach." In *Industry Location and Public Policy*, edited by H. W. Herzog, Jr., and A. M. Schlottman. Knoxville: University of Tennessee Press, 1991.

Castells, M., ed. *High Technology, Space and Society*. Beverly Hills, Calif.: Sage, 1985.

———. "The New Industrial Space: Information-Technology Manufacturing and Spatial Structure in the United States." In *America's New Market Geography*, edited by G. Sternlieb and J. W. Hughes. New Brunswick: Rutgers-The State University of New Jersey, 1988.

Chapman, K., and D. Walker. *Industrial Location*, 2nd ed. New York: Basic Blackwell, 1990.

Clark, G. L. "Corporate Restructuring in the Steel Industry: Adjustment Strategies and Local Labor Relations." In *America's New Market Geography*, edited by G. Sternlieb and J. W. Hughes. New Brunswick: Rutgers-The State University of New Jersey, 1988.

———. "Restructuring the U.S. Economy: the NLRB, the Saturn Project, and Economic Justice," *Economic Geography* 62 (1986): 289–306.

Florida, R., and M. Kenney. "Restructuring in Place: Japanese Investment, Production Organization, and the Geography of Steel," *Economic Geography* 68 (1992): 146–73.

Galenson, A. *The Migration of the Cotton Textile Industry from New England to the South: 1880–1930*. New York: Garland Publishing, 1985.

Graham, J., K. Gibson, R. Horvath, and D. M. Shakow. "Restructuring in U.S. Manufacturing: The Decline of Monopoly Capitalism," *Annals of the Association of American Geographers* 78 (1988): 473–90.

Kuby, M., and N. Reid. "Technological Change and the Concentration of the U.S. General Cargo Port System: 1970–88," *Economic Geography* 68 (1992): 272–89.

MacLachlan, I. "Plant Closure and Market Dynamics: Competitive Strategy and Rationalization," *Economic Geography* 68 (1992): 128–45.

Markusen, A. R., et. al. *The Rise of the Gunbelt: The Military Remapping of Industrial America*. New York: Oxford University Press, 1991.

Scheuerman, W. *The Steel Industry: The Economics and Politics of a Declining Industry*. New York: Praeger, 1986.

Sternlieb, G., and J. W. Hughes. *America's New Market Geography*. New Brunswick: Rutgers-The State University of New Jersey, 1988.

Storper, Michael and, Allen J. Scott. *Pathways to Industrialization and Regional Development*. New York: Routledge, 1992.

Watts, H. D. *Industrial Geography*. New York: Longman Scientific & Technical/Wiley, 1987.

Wolch, J., and M. Dear. *The Power of Geography*. Boston: Unwin Hyman, 1989.

CHAPTER

8

Modern Transportation and Communication Systems

As has often been noted, today there is no such thing as a wholly Canadian or American-built automobile. All "domestic" cars are produced with components from other countries; depending on the model, such foreign components can make up a sizable proportion of the car. Thus, the engine may have been made in Mexico, the stereo in Taiwan, the tires in Japan, and the upholstery in France. The car is considered a domestic product if it was assembled in the country and if a major part of the final product was Canadian- or American-built.

The auto companies' reliance on a virtually worldwide list of suppliers is only possible because of low transportation costs (Figure 8.1). With low freight charges, production of a particular component can be concentrated in the country where it is made most efficiently, and the sum of production costs plus transportation charges to the United States or Canada is less than what it would cost to manufacture that part domestically. Note, however, that if transportation costs were to increase significantly, perhaps because of energy shortages, the current pattern of worldwide suppliers would no longer be feasible. Under this scenario, even though foreign production costs may still be lower, once transportation charges are added, it may once again be cheaper to buy many Canadian- or American-made components. The result may be higher employment in the North American auto industry, but the cost of producing a car domestically would increase, and this would soon be reflected in higher car prices. This example shows that changes in transportation can have far-ranging conse-

quences. Much of this chapter is devoted to exploring these consequences.

Transportation and Areal Specialization

Transportation is such an integral part of our lives that people seldom take any time to ponder what it is and how it may influence them. In the simplest and broadest sense, transportation can be said to consist of any kind of movement. By this definition, it would include not only such familiar modes of travel as the automobile and airplane, but also such vital functions as the circulation of blood within our bodies (transporting oxygen) and the movement of sewage to treatment plants. Viewed this way, transportation is basic to our existence, for life could not continue without it.

From the geographer's point of view, transportation's main role is to connect different places (such as a country's regions) to one another, making it possible for these areas to function as a part of a larger whole. Such a view leads to a basic notion, namely, that transportation allows **areal specialization** to occur. Because of transportation, different parts of the earth's surface can be devoted to specialized uses. Transportation ties these specialized areas together, enabling them to supply each other with those goods and services that they do not produce themselves. Examples of such areal specialization include retail and residential areas within cities and major world farming regions (for example, wheat belts) or manufacturing regions (such as Europe's manufacturing belt).

When transportation is rudimentary (slow and expensive), areal specialization can take place only locally, for transporting things any distance is simply too expensive or time-consuming. Under such conditions, people's needs and wants must be satisfied from what can be produced nearby; that is, they need to be *locally self-sufficient*. Such limited choices were universal in the distant past, when walking and carrying by hand were the only forms of transportation, but they hold in few places today.

When transportation improves, it is as if time and space were compressed, bringing people and places closer together. Another way to say this is that improved transportation reduces the **friction of distance,** that is, the impact that distance has on the intensity of interaction. It will then cost people less to go a given distance, giving them more choices of where to meet their needs and wants. With better transportation, also, those places that can produce a product most efficiently can be devoted solely to that item; in other words, greater areal specialization can take place, leading to a larger overall output and enabling everyone to enjoy a higher standard of living. On the other hand, at least in the short run, increased areal specialization can also lead to great difficulties, such as massive unemployment, in

Figure 8.1
A General Motors automobile assembly plant in Fairfax, Kansas. These cars are being assembled from components manufactured in various parts of the world, a process made economical by low transportation costs.

areas that once produced certain goods but that are no longer as competitive. In the 19th century, for example, New England farmers had to cease grain production when cheap transportation meant they could no longer compete with more efficient farmers working the fertile soils of the Midwest.

Within the past century, there have been great improvements in transportation virtually worldwide, and few places today are locally self-sufficient. In fact, most of the world now is very much a part of the global economy. This is nicely illustrated in the United States, where products from almost every country are consumed and where certain American specialties, such as some foods and aircraft, are important exports.

The concept of areal specialization can be illustrated by the overall geographic patterns of the economies of the United States and Canada today. Some parts of these countries specialize in one or more agricultural products, such as oranges and winter vegetables (Florida) and wheat (Great Plains and Prairie Provinces), while others may specialize in particular manufactured goods, such as computers (Silicon Valley), paper (Québec), and aircraft-spacecraft (southern California), and still others provide services, such as recreation (the national parks and Las Vegas) or insurance (Hartford and Des Moines). No region comes even close to being self-sufficient, and so there is a tremendous demand for transportation.

American Culture and Mobility

One of the strong characteristics of American culture is the great value we place on *mobility*. This characteristic can be gleaned, among other ways, from the large percentage of income (about 18%) that Americans devote to transportation. It can also be inferred from the tremendous importance that teenagers attach to getting their first car; to them, it means freedom, choice, the opening of whole new horizons. Nobody can pinpoint the origin of this cultural trait, but perhaps it can be traced back to the earlier pioneer phase of the country, when a large percentage of the population seemed constantly to be on the move, for the grass supposedly was always greener on the other side. Most of these generalizations also seem to apply to Canadian culture.

No matter what the trait's origins, Americans today move far more often than people in almost any other country. On average, about 18% of Americans change their residence in any given year. Many of these moves are local, within metropolitan areas, but Americans are also known for their frequent long-distance moves, especially when economic opportunities seem to beckon. In many ways, this is a positive characteristic, for Americans' willingness to move means that they quickly adjust to regional differences in job opportunities. On the other hand, this attitude also contributes to a certain lack of attachment to places, and

when people feel attached to an area, they tend to protect the quality of life there.

Transportation in the United States and Canada, 1950–2000

In 1950, the United States and Canada had well-developed transportation systems, probably the most advanced in the world. Most of the United States was served by good to excellent transportation, including extensive networks of paved roads, railroads, oil pipelines, and waterways, and there was a good start toward an extensive national airline network; the same was true for the more densely settled portions of Canada near the U.S. border. These are still the most important forms of transportation today, although their relative roles have changed somewhat.

In 1950 in the United States, the automobile already dominated the long-distance (*intercity*) movement of passengers, accounting for about 86% of national *passenger-kilometers* (when a passenger travels one kilometer, a passenger-kilometer is generated). In second place were the railroads, with about 6% of the total, followed by buses (5%) and airlines (less than 3%).

At mid-century, the railroad was the leading U.S. intercity mode for freight, accounting for 58% of domestic *ton-kilometers* (when one ton is hauled a kilometer, a ton-kilometer is generated). Trucks were in distant second place, with 16%, followed by internal water transportation (15%) and oil pipelines (12%).

The changes in transportation technology since 1950 have been unremarkable. All the technology that is widely used for transportation in the United States and Canada today was already common in 1950. To illustrate, at mid-century, the internal combustion engine was universal in highway transportation, and it was rapidly becoming dominant on the railroads and waterways. While 1950 was still the era of the piston-powered airliner, jets were widely used by the military, and civilian use was known to be just around the corner. Technological change since then has been evolutionary, not revolutionary, consisting mostly of the refinement of existing technologies.

In 1996, Americans spent almost $800 billion for passenger transportation, which was about 10.4% of the gross domestic product (GDP). As Table 8.1 shows, the private car thoroughly dominated these expenditures, accounting for 83% of the total, or $673 billion. Airlines were in distant second place, absorbing about 7% of the national passenger transportation bill. Only 5% of the total ($42 billion) went for all urban public transit (buses, streetcars, rapid transit, and taxis). All other forms of transportation, including long-distance trains and buses, accounted for a mere 2% of the bill. These percentages have stayed remarkably stable in the past few decades; the largest change since 1970 (comparable 1950 data are not available) has been the decrease in the private auto's share from 85% to 83%.

Table 8.1 Total U.S. National Outlays for Passenger Transportation (in billions of dollars and percentages)	1970		1996	
Total	$114.9	100%	$796.8	100%
Private autos	97.0	85	661.6	83
Private air	2.6	2	11.1	1
Local bus and taxi	5.4	5	38.3	5
Intercity air	6.6	6	59.3	7
Intercity rail	0.3	-	2.2	-
Intercity bus	0.8	1	1.5	-
International travel	2.2	2	19.0	2

Source: From Statistical Abstract of the United States, *1999. U.S. Bureau of the Census.*

Table 8.2 Total U.S. National Outlays for Freight Transportation (in billions of dollars and percentages)	1970		1996	
Total	$83.8	100%	$467.2	100%
Highway	62.5	75	$368.5	79
Intercity truck	33.6	40	$235.4	50
Local truck	28.8	34	$133.1	29
Rail	11.9	14	$35.1	8
Water	5.1	6	$25.0	5
Oil pipelines	1.4	2	$8.6	2
Air carriers	1.2	1	$20.1	4

Source: From Statistical Abstract of the United States, *1999. U.S. Bureau of the Census.*

The total U.S. domestic freight transportation bill in 1996 was much smaller than the passenger bill, amounting to about $467 billion, or 6.1% of the GDP. Truckers received by far the biggest share of the freight dollar—79% (Table 8.2). In second place, but with only 8% of the total, were the railroads, followed (in order) by water transportation, air freight, and oil pipelines. When we compare 1996 percentage figures with 1970 data, we see that air freight made the biggest relative gain, rising from 1% to 4%, and railroads experienced the biggest loss (14% to 8%).

When measured by the volume of intercity freight handled, the relative importance of the various forms of transportation in the United States is quite different. In 1996, the railroad was the leading freight mode for ton-kilometers handled, accounting for 38% of the total. This was followed by the truck, which carried about 27%, oil pipelines (20%), rivers and canals (13%), the Great Lakes (2%), and finally air carriers (0.4%). Ton-kilometer statistics are biased toward forms of transportation (for example, railroads, pipelines, and water) that carry heavy items, such as coal, crude oil, and chemicals over long distances, and against those (such as trucks) that handle mostly light, high-value goods over short distances. In fact, about 43% of all freight tons carried on railroads today is coal, and 9% is farm products. The largest changes from 1950 are the decrease in the railroad share from 58% to 38% and the rise in the truckers' share from 16% to 27%.

Data on intercity passenger transportation (measured in passenger-kilometers) show that the private car is highly dominant, handling about 81% of the traffic (Figure 8.2). Air transportation is in second place, accounting for 17% of the passenger-kilometers. All other forms of passenger transportation, including buses and railroads, only account for about 2%. Interestingly, in 1950, the private car was even more dominant, accounting for 86% of intercity

passenger-kilometers. This change is correlated with the rapid rise of airlines (from less than 3% in 1950 to 17%), implying that since mid-century, air transportation has replaced many long-distance car trips.

Highway Transportation

The data discussed above show that the highway mode dominates much of U.S. transportation today, as it does in Canada. The beginnings of this dominance can be traced to developments in the first decades of the 20th century, when trucks and cars began to exert some influence on geography. At that time, they were rapidly being improved technologically, and local, state, and federal governments were putting enormous resources into paving roads. At the end of the 1930s a U.S. national network of paved roads built with federal aid was nearing completion on which it was possible to drive from coast to coast (see "Route 40" on p. 207). By 1950, this highway system was basically complete. It included such well-known highways as Route 66 (from Chicago to Los Angeles), Route 1 (the Lincoln Highway) on the East Coast, and Route 101 along the West Coast. These national highways were hardly ideal, however, for they were little more than a combination of many local roads that had been strung together and went through the middle of a large number of villages, towns, and cities, which caused lengthy delays.

The inexpensive auto of the early decades of the 20th century, as represented by the Model T Ford, revolutionized many people's mobility and had great social and economic impacts as early as the 1910s and especially by the 1920s. On the eve of the Great Depression of the 1930s, the United States could already be described as a motorized country. In the process, the auto industry became a major part of the U.S. economy, and by the 1930s, it was the largest consumer of raw materials and oil.

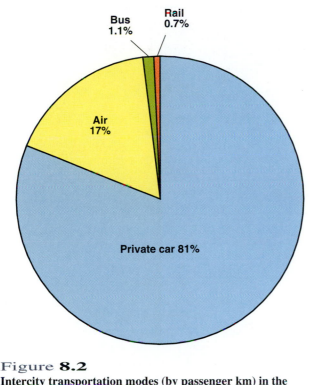

Bus
1.1%

Rail
0.7%

Air
17%

Private car 81%

Figure 8.2
Intercity transportation modes (by passenger km) in the United States, 1996.

Meanwhile, the truck was rapidly replacing horse-drawn wagons in the cities, and by the 1920s, long-distance trucking began to develop. In the late 1930s, trucks were handling more than 10% of the U.S. intercity ton-kilometers, a figure that did not change greatly during the war-dominated 1940s.

The dominant roles played by the truck and automobile today are illustrated by some impressive figures. Between 1950 and 1995, the number of cars in use in the United States increased from about 40 million to about 127 million; that is, the number of cars rose by more than 200% during a time when the population increased by about 75%. In 1950, the United States had about 3.8 persons per car, while in 1995, that figure had decreased to 2 per car. What had been a very car-oriented society in 1950 was even more so 45 years later. During the same period, the number of trucks increased even more dramatically, from about 8 million to 72 million.

Within cities (*intracity*), the bus is usually the leading form of public transportation, and buses tend to carry far more passengers than the more publicized rail systems in most metropolitan areas. Buses are well-suited to American and Canadian cities, for the road network is already in place and bus routes therefore can easily be adjusted to meet changing needs. Being relatively small, the bus is also appropriate for serving low-density residential areas, which are so common in the United States and Canada. In most cities today, buses are the only feasible form of mass trans-

portation, given the low, and often decreasing, demand for public transportation.

Interstate Highway System

Since 1950, the total number of kilometers of public roads in the United States has remained rather static, increasing by only 0.02%, to 5.4 million kilometers (3.4 million mi). These figures mask a very important change, however, for during these years, the system of *interstate highways* (the I-system, Figure 8.3) was built. The need for such a network of limited-access freeways (called expressways in some areas) had been identified in the 1930s and its development authorized in the 1940s, but actual construction did not begin until the late 1950s. To finance the system, a federal gasoline tax was passed, and the proceeds from this tax and some other highway-user payments were placed into a Highway Trust Fund. The costs of construction were then split on a 90% federal, 10% state basis.

By 1970, 51,520 kilometers (32,000 mi) of the I-system, out of an authorized total of about 66,000 kilometers (41,000 mi), had been completed. In 1980, 66,000 kilometers (41,000 mi) were finished, but by that time, the planned network had been expanded, particularly through the addition of **circumferential freeways** (also called beltways or bypass routes) around the major cities. In 1996, the system was virtually finished, with more than 72,500 kilometers (45,000 mi) completed. However, traffic has increased to such an extent since construction began that many of the "completed" sections badly need additional capacity.

Even though the interstate system accounts for only a little more than 1% of the national highway mileage, it is far more important than this figure implies, carrying about 20% of vehicle-kilometers. Further, this huge load is handled with, proportionately, only one-third as many accidents as occur on regular roads. Without much exaggeration, the I-system can be considered the circulatory system of the nation's economy, and it has vastly improved the quality of cross-country travel. Among other impacts, the system has greatly broadened people's recreational opportunities, contributing to a virtual explosion of the recreational vehicle industry.

Within most large urban areas, I-system freeways are the backbone of the transportation system, and it would be difficult to imagine what traffic conditions would be like today if they had never been built. Although not all urban freeways are a part of the I-system, the more important ones usually are. Opponents of urban freeways like to point out that, despite their huge costs, they have done nothing to decrease road traffic congestion. Realistically it should be added that *no* transportation investment, including mass transit, in recent decades has accomplished this, except possibly in the very short run. This is because transit systems tend to take the overflow from the highways, and congestion on the roads remains just as bad.

Modern Transportation and Communication Systems

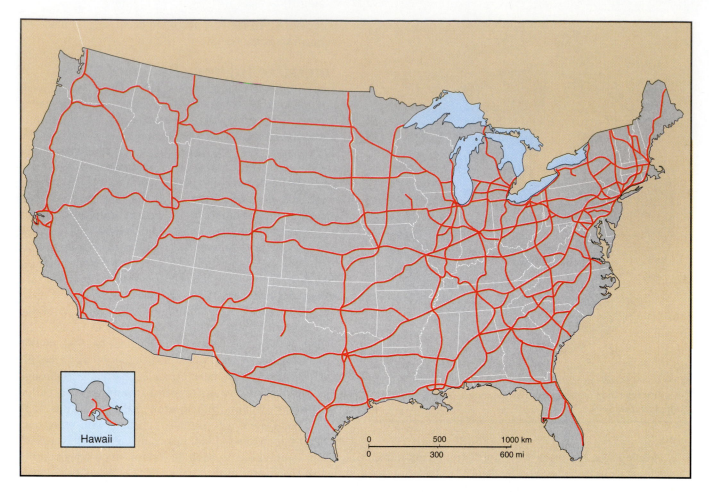

Figure 8.3

The U.S. interstate highway system. The network is relatively evenly distributed, certainly much more so than the population, reflecting its role as a true national transportation system. Because of space limitations, many urban circumferential freeways that are a part of the system are not shown.

Source: U.S. Department of Transportation.

The geography of post-1950 urban America has been profoundly affected by the location of these highways. Their construction, along with expanding auto ownership, greatly facilitated the massive *urban sprawl* that has characterized much of the growth of all large American metropolitan areas in this period. Urban sprawl was hardly caused by these highways; rather, it reflects a deep-seated American preference for living in single-family homes on large lots, as well as the high average incomes (and federal mortgage policies) that enabled a large percentage of the population to buy such housing. On the other hand, the freeway system certainly facilitated sprawl by greatly reducing the friction of distance and thereby giving individuals much more choice in where to live and work.

Freeways have also greatly influenced the geographic pattern of businesses. Retailing soon followed the population into the suburbs, sometimes abandoning center-city stores in the process. Most prominently, this change led to the construction of hundreds of major planned regional shopping centers, which first appeared in the late 1950s and soon were located almost exclusively on or near the freeway network. This was followed shortly by the dispersal of all kinds of other businesses into the suburbs, where they are often located in planned units, such as manufacturing-warehousing districts and office parks. Today, when siting factories, warehouses, office complexes, shopping malls, and many other types of retail and service establishments, decision makers often will not even consider a location that is not close to, and in the case of retailing, visible from, one of these major arteries.

Highways in Canada

Predictably, Canada's road mileage is highly concentrated in the relatively densely settled areas in the south. North of such cities as Montréal, Winnipeg, Edmonton, and Vancouver, the highway network quickly thins, and large areas have only a few (usually unpaved) roads or none at all. One important northern route is the Alaska Highway, which was

In 1950, one of the most famous of the U.S. highways wast Route 40, a road 4800 kilometers (3000 mi) long that cut across the midsection of the United States, passing through eight of the country's 30 largest cities. With end points in Atlantic City, New Jersey, and San Francisco, the road was a transect that revealed both the urban and the physiographic diversity of the country.

Route 40 began on the flat Atlantic Coastal Plain, crossed the Delaware River, then entered the eastern region of higher plains called the Piedmont. It crossed the Ridge-and-Valley section of Pennsylvania and the Appalachian Plateau. After Columbus, Ohio, the predominant view along Route 40 was of mile after mile of corn fields; the traveler was in the Midwest. West of Kansas City, as the route entered the more arid Great Plains, corn was replaced by grasslands and fields of wheat. Perhaps the most scenic portion of Route 40 came after it left Denver and crossed the Colorado Rockies, the red cliffs of the Uinta Basin, and the Wasatch Range in

Utah. Leaving the mountains, the highway traversed the Great Basin. West of Reno, Nevada, it crossed the Donner Pass through the Sierra Nevada, entered the Central Valley of California, then climbed the Coast Ranges before finally terminating in downtown San Francisco.

Route 40 was a transect across time as well as across space. Some of its pieces began as colonial footpaths or horse trails. From Cumberland to Wheeling, West Virginia, the route was part of the old National Road (built between 1811 and 1840). In Utah, Nevada, and California, the highway followed the route of the California Trail used a century earlier by emigrants in covered wagons.

Today, parts of Route 40—such as between Denver and Salt Lake City—remain a separate, numbered highway. Elsewhere, pieces of it have been absorbed into Interstate Highways 70 and 80, and in still other places, Route 40 has become merely a frontage road or local highway.

built with American aid during World War II from Dawson Creek, British Columbia, to Fairbanks, Alaska, when a Japanese invasion of mainland Alaska seemed possible (Figure 8.4). This former gravel road is now paved, but given the distances and the relative paucity of services, it remains somewhat of a challenge to driver and vehicle alike. The first paved coast-to-coast road, the Trans-Canada Highway, was completed only in 1962; largely two-lane, it extends (including two major ferry links) from St. John's, Newfoundland, to Victoria, British Columbia.

Canada does not have an equivalent of the I-system, but an important freeway route runs all along the Québec City-Montréal-Toronto-Windsor population-commercial core area, and other freeways are widely scattered across

the more populated parts of the country. Most of the largest urban centers also have freeways, although the systems are not usually as extensive or as likely to extend into the central business district as in comparably sized cities in the United States.

Truck Transportation, Location Strategy, and Traffic Patterns

As early as the 1920s, the truck practically monopolized freight transportation within cities, and no challenger has appeared since then. Early trucks greatly decreased the cost of intracity freight transportation, giving new freedom of location to manufacturers, retailers, and others. In the days

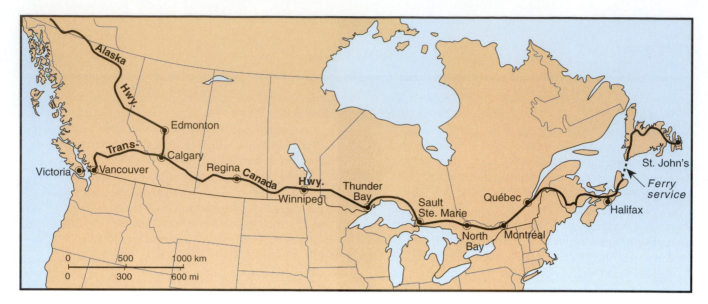

Figure 8.4
The Alaska and Trans-Canada Highways.

of horse-drawn wagons, the intracity movement of freight was so expensive that the majority of retail stores had to cluster near the downtown railroad freight depots where most of their goods arrived from the manufacturers and distributors. By the 1920s, retailers were able to take full advantage of the flexibility and low cost of being supplied by trucks to follow the more affluent population as it moved toward the edge of the city. This trend greatly accelerated after 1950, especially with the start of urban freeway construction. During this time, trucks were becoming better designed and bigger.

The development of more efficient trucks and the growth of effective suburban road networks (especially freeways) have also made it possible for many manufacturers and wholesalers, as well as warehouses, to move from old downtown locations to more spacious sites closer to the metropolitan edge. Increasingly since 1950, manufacturers have relocated to the suburbs, where land is cheaper, the work force is nearer, and truck access is easier. Many such moves were also influenced by the economic advantages of carrying out production in spacious single-story buildings, versus the multiple-story factories and warehouses that were the norm in the older locations. Many businesses that primarily serve the metropolitan market have found that distribution is most efficient from sites on circumferential freeways and expressways, and these locations have become the preferred ones.

Such developments have radically altered the traffic pattern of the entire metropolitan area. In 1950, a large percentage of jobs were located in or close to the CBD, and the overall traffic pattern was one of movement into the city in the morning and outward at night (Figure 8.5). Today, this simple pattern has disappeared, as homes and

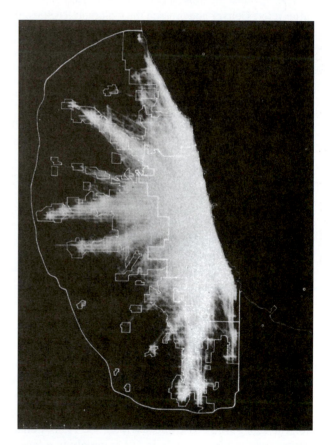

Figure 8.5
Simple commuting patterns of the 1950s. This map shows the 5 million auto trips made each day in the Chicago metropolitan area in the 1950s; at this time, the pattern was dominantly one of trips into the city in the morning and outward at night. Today, commuting patterns are much more complex, and most trips are between suburbs.

Figure 8.6

Double-stacked containers on rail cars. This type of arrangement has lowered the cost of moving containers by rail; double-stack trains are common in land bridge movements.

job opportunities are scattered all over the metropolitan area. With more than half the U.S. urban population living in the suburbs and with most job opportunities there as well, most metropolitan trips to work today take place between suburbs. Such a dispersed pattern is hardly favorable for public transit lines, especially mass transit rail systems, which do best when there is a large, concentrated traffic flow to and from one or a few major destinations.

Railroads

While the number of kilometers of highways was increasing between 1950 and 1996, the reverse was taking place on U.S. and Canadian railroads. U.S. rail route-kilometers decreased from 393,000 (244,000 mi) in 1950 to 288,000 (179,000 mi) in 1980 and 217,000 (136,000 mi) in 1996, an overall decline of 46%. Most of this change took the form of the abandonment of unprofitable branch lines that had been built in the horse-and-wagon 19th century, but which had little traffic in the truck-dominated second half of the 20th century. Even as these abandonments were taking place, railroad freight traffic increased from 628 billion to 1426 billion ton-miles. Therefore, average traffic on the remaining lines increased greatly between 1950 and 1996.

Railroad freight traffic today is heavily dominated by a few *bulk commodities,* such as minerals and grains, that tend to move in huge flows. However, **intermodal traffic** (traffic that uses more than one form of transportation between origin and destination), which handles largely manufactured goods, has also grown. This type of traffic consists of highway trailers and oceangoing *containers* that are placed on railroad cars (the latter often double-stacked) for long-distance movement (Figure 8.6). The most prominent form of intermodal traffic is the so-called **land bridge services,** in which whole trainloads of containerized imports from Asia are moved inland from West Coast ports, many for delivery to East Coast cities.

Like the population, the U.S. railroad network is relatively dense east of the 98th meridian and sparse in most areas west of it (Figure 8.7). The major rail routes tend to focus on certain metropolitan areas, especially the *gateways* where traffic is exchanged between railroad companies. The leading gateways include Chicago (the nation's number one rail center), Kansas City, St. Louis, and New Orleans, where the so-called eastern and western railroads meet and interchange traffic. There also used to be separate northern and southern railroads, which exchanged traffic at such gateways as Cincinnati, Louisville, and Washington, D.C., but mergers in recent decades have made this old regional distinction passé.

Within metropolitan areas, railroads tend to converge on the central business district (CBD), where the major passenger and freight depots were located. Along these rail lines are corridors of older factories, many of which are abandoned or no longer use rail services. The old central freight depots are gone, and many of the passenger stations have also been torn down (such as several that used to be located just south of Chicago's CBD, the Loop), while others (such as in St. Louis) have been converted into modern

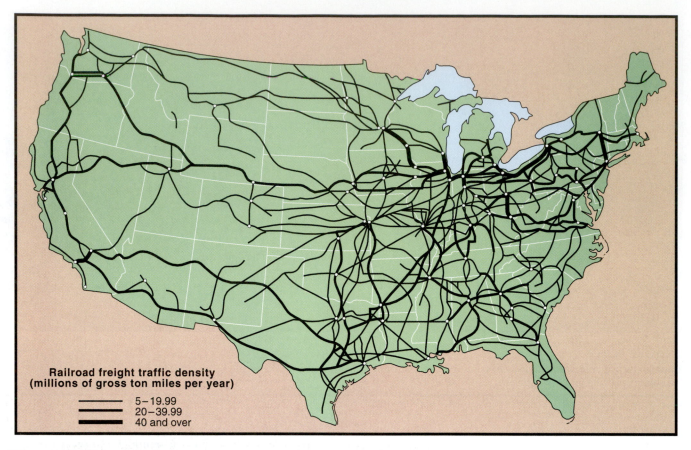

Figure 8.7

U.S. railroad freight traffic density, 1970s. Note the heavy concentration of traffic in the western part of the Manufacturing Belt. This is the country's heavy industry zone, and such industries generate much freight traffic (such as coal and new cars) that are well-suited to rail transport. Note also the heavy traffic in the zone between Chicago, Omaha, and Kansas City (the latter is the country's second largest rail center) and the relatively thin network west of the 98th meridian.

The map legend reads:

Railroad freight traffic density (millions of gross ton miles per year)
- 5–19.99
- 20–39.99
- 40 and over

shopping-hotel complexes. In Washington, D.C., the old Union Station has been magnificently restored and converted into a successful shopping-restaurant complex, even as it continues to serve a large number of passenger trains (Figure 8.8). Today, the major *rail yards,* which are extensive facilities where freight trains are organized, tend to be located near the outskirts of metropolitan areas.

The volume of U.S. intercity rail passenger traffic decreased greatly between 1950 and 1996, falling from 35 to 14 billion passenger-miles. In good part, this decrease occurred because passenger trains were heavy money losers; the general rule was that the more passengers there were, the more money was lost. Therefore, most American railroads tried to get rid of their passenger trains as quickly as possible, and by the late 1960s, they had been so successful that the end of all long-distance passenger services was in sight. At this point, the federal government stepped in, under pressure from lobbying groups, and in 1970, Congress created the federally supported National Railroad Passenger Corporation specifically to save the long-distance passen-

ger train. This company, which operates as **Amtrak** (Amtrak owns only a few hundred miles of track; mostly, it uses the tracks of the private freight-hauling railroads), has established a national, but thin, network of intercity trains that nevertheless manages to serve most major cities. On many routes, however, there is only one train daily in each direction, which means that intermediate cities often are served at inconvenient hours. In some states, such as California, Illinois, and New York, these services are supplemented by more locally oriented Amtrak trains that are state-subsidized.

The only really dense passenger services in the United States today are provided on the Amtrak-owned route between Boston, New York, Philadelphia, Baltimore, and Washington, D.C.—the heavily populated *Northeast Corridor.* For many years, the portion of this corridor between Washington and New Haven, Connecticut, was the only major electrified (versus diesel-powered) rail line in the United States. This allowed speeds up to 200 kilometers per hour (125 mph) on certain sections between New York

Washington's Union Station. This classic railroad station was magnificently restored during the 1980s and includes many chic shops as well as a huge food court.

and Washington, the most intensely used part of the route. In the late 1990s Amtrak invested heavily in the Northeast Corridor between New Haven and Boston, particularly by installing electric wires and by easing curves, to allow much higher speeds. Faster Boston–New York passenger service, using new electric equipment that can operate at up to 240 kilometers per hour (150 mph), began in January, 2000. Initially, Boston–New York scheduled times declined from a rather slow 5 hours to 4, but further reductions were planned after Amtrak gained experience with the equipment. Other high-frequency services operate between Albany and New York City and from Los Angeles to San Diego.

The same problem of passenger losses was experienced in Canada, and here also the federal government stepped in to form **Via Rail** to operate long-distance passenger services. Losses have been high, however, and the network has been reduced considerably in recent years. The most frequent services are provided between Montréal and Toronto. Ontario and British Columbia subsidize some additional long-distance passenger services northward into the less-populated reaches of those provinces.

In certain U.S. metropolitan areas, such as around New York, Boston, Philadelphia, and Chicago, extensive railroad commuter services exist. Such services were never threatened with abandonment, for they continued to carry a large number of passengers. They were all losing money, however, and could no longer be subsidized by the private railroads. Therefore, all such commuter operations have been taken over by local governments, so that now the public is providing the subsidies. Because of increasing highway congestion, new rail commuter routes have been opened in several large cities in recent years, including Washington, D.C., Miami, and Los Angeles. Because these routes almost always radiate from the downtowns, they can serve only a fraction of the trips generated in the metropolitan areas.

Rail commuter services in Canada are also government-subsidized; those provided by Go Transit in the Toronto area have acquired an especially positive reputation and have been emulated, to some extent, by several U.S. operators.

Since about 1900, subways have played an important role in urban transport in some of our largest cities. They are most prominent in New York City, where surface traffic probably would be virtually impossible without them, but they are also found in other cities, including Chicago, Montréal, and Boston. Early in the 20th century, streetcars were found in all major cities and some quite small ones, but most had been abandoned by the 1960s and replaced by buses. San Francisco, Philadelphia, and Toronto were among the few cities that did not abandon all their streetcar routes.

Since the 1970s, new **heavy-rail** and **light-rail** (streetcar) systems have been built in several metropolitan areas. These two differ primarily in that a light-rail system

Figure 8.9
San Francisco's Bay Area Rapid Transit system. This heavy-rail system links San Francisco, Oakland, and the suburbs through high-speed trains that travel over a network of above- and below-ground tracks.

usually operates on streets (especially in CBDs) with fairly light equipment and therefore is relatively inexpensive, while the faster heavy-rail operates on a separate right-of-way (and typically in subways within CBDs) with larger equipment. In the United States, the renaissance of such urban rail systems began with the opening of the Bay Area Rapid Transit network around San Francisco in 1972 (heavy rail) and the San Diego Trolley (light rail) in 1981 (Figure 8.9). Among the heavy-rail systems opened since then are those in Washington, D.C. (a user's delight, with beautiful stations and no graffiti, in a city where driving is difficult); Atlanta; and Los Angeles, while new light-rail lines opened in such cities as Sacramento, San Jose, and Portland, Oregon. Like commuter railroads, they focus on downtowns. While none of these systems actually decreased highway traffic congestion, a benefit that proponents often claimed for them, it is reasonable to say that road traffic would be worse without them.

Airlines

The airlines have increased dramatically in importance since 1950, both as freight and passenger carriers. The United States and Canada are well-suited to air transportation, for they are huge and their major cities tend to be widely scattered across the continent. The recent trend of population movements to the U.S. Sun Belt and western Canada, far from the traditional population core areas, has reinforced the advantages of air travel. Between them, the two countries generate more than 40% of the world's scheduled air travel.

In 1950, U.S. airlines carried only 17 million domestic passengers, a figure that increased more than 34-fold in the ensuing 46 years, to 581 million in 1996. Passenger growth during these years was sparked by such factors as decreases in the real costs (that is, adjusted for inflation) of airline tickets, rising average incomes, great increases in the speed and size of airliners, the greater passenger comfort of jet versus piston-powered aircraft, as well as the population trends mentioned above. In the United States, passenger growth was also stimulated by airline **deregulation** in 1978, when most federal economic controls over the industry began to be phased out. This led to greater competition among airlines, resulting in lower fares.

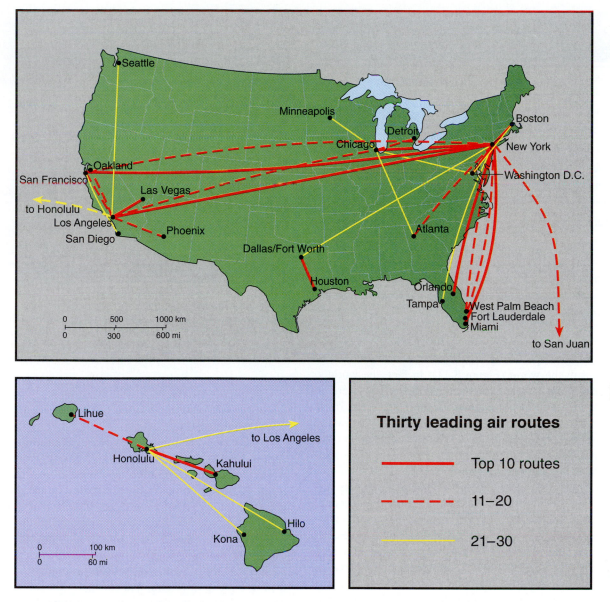

Figure 8.10

The 30 leading U.S. air traffic routes, ranked by numbers of passengers carried. Note the concentration of the top ten routes at New York City and the important air centralities of Chicago and Los Angeles. The New York area has three major commercial airports: John F. Kennedy, La Guardia, and Newark.

Source: Data courtesy of the Air Transportation Association of America.

Because of its speed, air transportation has shrunk the effective distances among countries, thereby, affecting all kinds of management, supply, and service industries. For example, it has made it much easier for large firms to exercise centralized control over their far-flung operations, whether domestic or international. Air freight has helped many companies substantially cut *inventory costs* (the cost of the money tied up in inventories); instead of keeping large inventories on hand, firms use air freight to bring in new supplies quickly when needed. With faster transporta-

tion, salespeople can cover a far larger territory than was once the case, thus increasing efficiency.

Although more than 800 airports in the United States have scheduled air service, passenger traffic is highly concentrated in those 25 largest metropolitan areas that generate at least 1% of national passenger traffic, the so-called *large hubs* (Figure 8.10). Over 70% of U.S. passenger traffic goes to or from one of these large hubs.

Table 8.3 lists the ten leading airports in the United States in 1998 for passengers boarded. That year Atlanta

Table 8.3	
Ten Leading U.S. Airports for Passengers Boarded, 1998	
	Passengers (millions)
1. Atlanta	73.5
2. Chicago O'Hare	72.5
3. Los Angeles	61.2
4. Dallas–Ft. Worth	60.5
5. San Francisco	40.1
6. Denver	36.8
7. Miami	33.9
8. Newark	32.5
9. Phoenix	31.8
10. Detroit	31.5

Source: ACI traffic data, 1998.

Figure **8.11**

In recent years Hartsfield International Airport, Atlanta, has emerged as the world's busiest for passenger traffic. It replaced O'Hare airport in Chicago, which held that title for decades.

was the largest airport for passengers not just in the United States, but in the world (Figure 8.11). In fact, the world's top three airports and six of the ten leaders that year were in the United States. Most of Atlanta's importance stems from its role as the main traffic hub for the rapidly growing Southeast. Its hub role in that region is so dominant, in fact, that Southerners say that if one of their own dies and is sent to hell, he or she can only get there via Atlanta!

For many decades, Chicago's O'Hare Field was the world's busiest airport, but in 1998 it slipped to second place. O'Hare's continuing importance stems from its role as a major traffic hub for American and United Airlines, its location near the center of the U.S. population, and the fact that it serves the country's third largest metropolitan area. It is noteworthy that two of the world's ten leading airports are in one state, California, and that traffic for the New York metropolitan area is split between three major airports (Newark, LaGuardia, and John F. Kennedy). Because of the resulting diffusion of traffic, the busiest New York airport (Newark) only ranks number 13 in the world.

Airline Deregulation and Routes

The implementation of the federal Airline Deregulation Act (ADA) of 1978 led to great changes in the geographic pattern of airline routes. The ADA transferred decision making on such basics as airline routes and fares from federal regulators to airline management. In the future, competition, not federal regulators, would determine what routes would be flown and what fares would be charged.

The airlines took advantage of the new law to change their route networks radically, replacing a complex (and sometimes seemingly random) pattern of routes that had been awarded by the federal Civil Aeronautics Board (CAB). Specifically, all major airlines realigned their networks into a *hub-and-spoke* pattern centered on certain major airports and established connecting complexes of flights (Figure 8.12).

The Denver airport is a good example of a hub. At certain times of day, perhaps 20–25 flights from various western cities converge there, all scheduled to arrive within about 45 minutes of each other, and stay on the ground for about an hour. After passengers have had a chance to change aircraft, the airplanes continue on to destinations east of Denver. One of these aircraft may have left Seattle early in the morning, with a routing to Miami via Denver. At Denver, most Seattle passengers disembark, as few of them are going to Miami, and change to the aircraft that is going to their destination. At the same time, Miami-bound passengers who originated at places other than Seattle, such as San Jose, also change airplanes. With this hub system, just ten aircraft can serve 100 (10×10) possible origins and destinations (not including Denver, which adds to the total), while 25 aircraft can serve 625 city-pairs with convenient one-stop service. The hub system helps keep average passenger loads high, greatly increasing airline efficiency. Several times during the day, this procedure is reversed, with aircraft from eastern points converging on Denver.

Such a pattern of operations is most efficiently served by airports located near the center of the country, such as those at Dallas-Fort Worth, Chicago, St. Louis, and Minneapolis-St. Paul, and at least one major airline chose to establish a hub at each of these places. Most hubs are oriented to east-west traffic flows (the nation's predominant direction), but a few, such as San Francisco and Washington, D.C., were established primarily for north-south traffic.

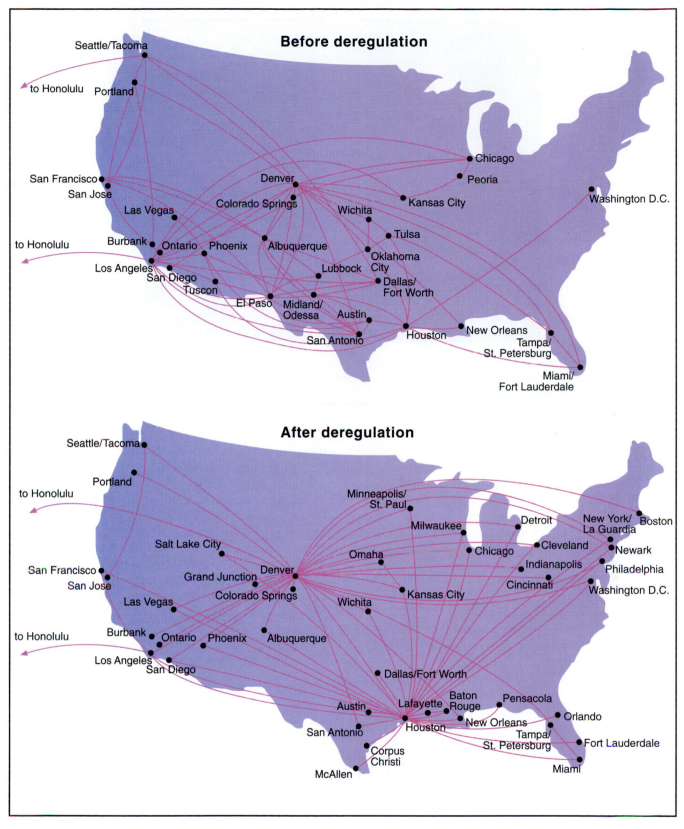

Figure 8.12

Continental Airlines route network before and after deregulation. These maps show how deregulation, which removed federal controls over domestic airline routes (as well as over mergers and acquisitions), made it possible to radically restructure airline route networks virtually overnight. Before deregulation, usually only one route at a time could be added—and only after lengthy hearings where rival airlines could, and did, object to the new service and often prevailed.

Modern Transportation and Communication Systems

Figure 8.13
Bush aircraft loading for a flight to a village. Many Alaskan villages are served by bush aircraft, where a 30-minute flight can cover the same distance as a whole day of difficult travel on the ground.

The quick rearrangement of airline networks into hub-and-spoke patterns, with its attendant economies, is one of the more successful results of airline deregulation. Before deregulation, routes were awarded in a slow and laborious process by the CAB, and such hub-and-spoke networks could take years to establish, assuming that the CAB allowed them at all. This example illustrates how government regulation, or the lack of it, can radically influence the geography of transportation routes.

Regional Patterns
Airports are key elements of the national air transport system and major foci for economic activities in large metropolitan areas. Many businesses locate near airports, such as hotels, car rental agencies, air freight forwarders, and even some manufacturers. Hotels near major hubs, such as Chicago's O'Hare, often have facilities for people who fly in from all over the country to attend a one-day business meeting and return home that night. The location of new airports is an important and controversial local question, for they are an extreme example of the *NIMBY* (not in my backyard) syndrome. Largely as a result, until the opening of the new Denver International Airport in 1996, not a single new major airport had been opened in the United States since Dallas–Fort Worth International Airport began operating in 1974.

Per capita use of airplanes in the United States is highest in the two outlying states of Hawaii and Alaska. The island fragmentation of Hawaii gives air a great advantage over surface transportation; this factor, plus a large influx of tourists, gives Hawaii the highest number of passenger boardings per capita. Alaska is in second place, but it can be argued that in some ways the role of the airplane is even greater there than in Hawaii. In Alaska, airplanes are literally the lifeline for dozens of villages, and scheduled services (often government-subsidized) are maintained to places that are far smaller than could possibly justify them in the "lower 48" (Figure 8.13). Here, small aircraft (usually single-engine) are used to bring in a great variety of products that would not be flown elsewhere, ranging from soft drinks to snowmobiles. Larger aircraft serve such isolated but important places as the North Slope oil fields, located in the Arctic extreme of the state. Air transportation's vital role in Alaska is illustrated by the fact that the state has almost twice as many public airports as California or Texas.

For similar reasons, the airplane also plays a vital role in the northern reaches of Canada, where vast areas can be reached only, in a practical sense, by aircraft. This role began in the 1920s and 1930s, when a variety of small aircraft were used in "bush" flights in support of mineral exploration and fur trappers, using primitive, unpaved airstrips that could be established at low costs. Much experience in northern flying was achieved during World War II, such as when aircraft were used in support of the construction of the Alaska Highway. With a ready market, the Canadian aircraft industry soon became world-renowned for its design and construction of aircraft appropriate for

Figure 8.14

A barge tow on the Mississippi River. This photograph was taken near St. Louis and shows the Illinois bank in the background. Crude and refined petroleum account for about 60% of inland waterway traffic; the rest consists of other bulk commodities, such as coal, fertilizers, farm products, sand, and gravel.

northern flying. Today, a network of scheduled regional routes connects the larger northern settlements with such centers as Montréal and Edmonton, and nonscheduled flights can be provided to numerous small airstrips. Small floatplanes are also used extensively to carry hunters and fishers to the many remote lakes that are scattered over this vast region.

Air freight has seen dramatic growth in the past few decades, especially in the overnight delivery of time-sensitive, high-value products and documents. Such service was pioneered by Federal Express in the 1970s, using Memphis, today the world's busiest cargo airport, as a hub. Here, flights from all over the country converge at night, their contents unloaded and sorted according to final destination, and then the reloaded aircraft return to their points of origin before morning, where final delivery is made by trucks. Several companies have successfully copied the Federal Express concept, and the amount of freight shipped this way has grown tremendously. Because of high costs, domestic general freight usually does not go by air, except in emergencies. International air freight has been growing

rapidly, for here, air has an especially great speed advantage over surface modes.

Waterways, Ports, and Pipelines

U.S. inland water transportation, which handles freight almost exclusively, is concentrated on two waterways, the Mississippi River system and the Great Lakes. The former system includes the Arkansas, Illinois, Missouri, Ohio, and Tennessee rivers and is connected to the Great Lakes near Chicago via a canal from a tributary of the Illinois River. From the mouth of the Mississippi River near New Orleans, channels 3 meters (9 ft) deep are maintained (by dredging where necessary) as far inland as Sioux City on the Missouri, Minneapolis on the Mississippi, the Pittsburgh area on the Ohio, near Tulsa on a tributary of the Arkansas, and Knoxville on the Tennessee. On all these rivers, most freight is moved via barges, which are shackled together in tows and pushed by tugs (Figure 8.14). Most of the traffic, as on all inland waterways, consists of bulk commodities, such as gravel, coal, wheat, iron ore, and fuels.

Modern Transportation and Communication Systems

Figure 8.15

A ship passing through the Welland Canal. This important waterway, wholly located in Canada, was first built in the 1800s as a bypass around Niagara Falls (located on the Niagara River connecting Lakes Erie and Ontario), and it has been improved many times since.

Transportation on the Great Lakes is mostly via ships, the largest of which are about 300 meters (1000 ft) long and can handle 50,000 to 60,000 tons. This is the maximum size that can use the *Soo Canal,* which was built between Lakes Superior and Huron to bypass rapids in the river that connects these lakes and is the busiest artificial waterway in the United States and Canada. The second major Great Lakes canal is the *Welland* (Figure 8.15), located wholly in Ontario, which was built between Lakes Ontario and Erie, bypassing Niagara Falls and some rapids on the river that connects these lakes. Because of winter freezing, the Great Lakes shipping season is about eight months long.

The relative importance of these two systems has changed since 1975, as U.S. Great Lakes traffic slipped from 194 million tons that year to 182 million in 1996 and as freight on the Mississippi waterways rose from 435 million tons to 702 million. While the Great Lakes accounted for 25% of the ton-miles on inland waterways in 1975, this had fallen to 17% 21 years later. Conversely, the Mississippi waterways carried 58% of 1975 tons, but 66% in 1996.

The different fortunes of the Mississippi and Great Lakes systems are not difficult to explain. There has been a great deal of federal investment in the Mississippi waterway system, resulting in such major changes as the construction of larger, modern locks and the extension of navigation on the Arkansas and Verdigris Rivers to just east of Tulsa in the 1970s. The new locks have allowed the operation of larger barge tows, lowering costs and attracting traffic. On the other hand, Great Lakes traffic has long been dominated by iron ore for the steel industry, and since 1975, many upper Great Lakes iron mines have closed, as the ores were depleted, and the American steel industry has gone into decline, leading to a long-term absolute decrease in traffic.

Even the opening in 1959 of the **St. Lawrence Seaway,** financed by both the United States and Canada, could not reverse the decline of Great Lakes traffic (Figure 8.16). The St. Lawrence Seaway project consisted of a series of navigation improvements in the Great Lakes and, especially, on the St. Lawrence River in Canada, which serves as an ocean outlet for the lakes. These changes enabled oceangoing ships with up to an 8.2-meter (27-ft) draft to

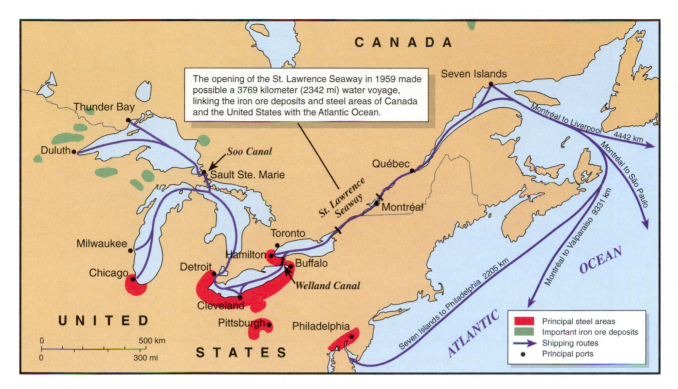

Figure 8.16
The St. Lawrence Seaway system. This waterway connects the Great Lakes with the Atlantic Ocean, making it possible for such places as Chicago, Toronto, and Duluth to serve international vessels. Despite the large investment, seaway traffic has been disappointing.

Map labels: CANADA, Seven Islands, Thunder Bay, Duluth, *Soo Canal*, Sault Ste. Marie, Québec, Montréal, Toronto, Hamilton, Buffalo, *Welland Canal*, Milwaukee, Chicago, Detroit, Cleveland, Pittsburgh, Philadelphia, UNITED STATES, *St. Lawrence Seaway*, ATLANTIC OCEAN, Montréal to Liverpool 4442 km, Montréal to São Paulo, Montréal to Valparaíso 9331 km, Seven Islands to Philadelphia 2205 km

Map textbox: The opening of the St. Lawrence Seaway in 1959 made possible a 3769 kilometer (2342 mi) water voyage, linking the iron ore deposits and steel areas of Canada and the United States with the Atlantic Ocean.

Legend:
- Principal steel areas
- Important iron ore deposits
- Shipping routes
- Principal ports

Scale: 0 — 500 km / 0 — 300 mi

enter the Great Lakes, effectively making places as far inland as Chicago and Duluth (3784 kilometers or 2350 miles from the Atlantic Ocean) in the United States and Thunder Bay in Canada into ocean ports. The seaway, offering low-cost shipping to the rest of the world, was promoted as a major stimulus for the economies of Great Lakes cities.

But this was not to be, for the seaway's timing could hardly have been much worse. Its opening coincided roughly with two major changes in ocean transportation: the introduction of huge *superships* for carrying bulk commodities and the *containerization* of general freight (Figure 8.17). The new superships, which began to dominate world trade in bulk items, had much more than an 8.2-meter (27-ft) draft and therefore could not enter the lakes. Container ships soon captured much of the world's ocean trade in high-value general commodities, which are placed in boxes (containers) that can be transferred easily between ships, trucks, and rail. The economics of container ships are such that they steam as quickly as possible between major container ports, where turnaround time is also fast. They cannot afford the time-consuming voyage into the Great Lakes. It makes much more economic sense to put high-value goods destined for export into a container at the inland factory in, say, Chicago and then to load the container onto a truck or railcar for quick movement to a U.S. East Coast port or Montréal for eventual movement to Europe. Because of these unfavorable developments, the

Figure 8.17
Loading a container ship, Port of Los Angeles. These containers are easily handled by trucks, ships, and rail, making traffic interchange between these transportation modes rapid and inexpensive. They are much too heavy for air traffic, where light aluminum containers are used.

Modern Transportation and Communication Systems

Figure 8.18
A portion of the Intracoastal Waterway.

trends of foreign tonnages on the Great Lakes have been generally downward since about 1970.

At New Orleans, the Mississippi River system connects with the *Intracoastal Waterway,* a route that carries huge tonnages for the oil refineries and other oil-related industries concentrated along the Gulf Coast (Figure 8.18). When, or if (because of environmental questions), it is ever finished, this waterway is supposed to provide a channel 3.7 meters (12 ft) deep along the Atlantic and Gulf Coasts all the way from Brownsville, in southernmost Texas, to New Jersey. Much of this route will use sheltered waters behind the offshore sandbars that parallel both coasts.

Other important waterways are along the Northeast seaboard, where canals (such as the Cape Cod and Delaware Canals) are used as shortcuts around necks of land, and the Snake and Columbia Rivers in the Northwest, which carry primarily wheat for export via Portland. In 1996, the Snake and Columbia rivers generated about 5% of U.S. domestic waterway traffic.

The foreign commerce of the United States was once directed overwhelmingly at Europe, and New York was long the largest port for foreign trade. The foreign trade rankings of ports today differ greatly according to how size is measured. When measured in tonnage, a figure that favors places that handle crude oil and oil products, the leading port in 1996 was Houston, with 79.0 million metric tons, followed by Southern Louisiana (76.0 million), Norfolk (52.3 million), and Corpus Cristi, Texas (51.4 million). All these ports except Norfolk, which is a large exporter of Appalachian coal, are dominated by oil traffic.

The 1997 port rankings for container traffic, where high-value manufactured goods dominate, are shown in Figure 8.19. For many years Europe was the country's main trade partner and New York was the national leader in imports and exports of high-value goods, but those days are long over. The current rankings reflect the more recent growth of trade with Latin America and Asia, a trend that favors such ports as Houston, New Orleans, and Miami in the Southeast and the West Coast ports. Since the ports of Los Angeles and Long Beach are adjacent to each other, they can logically be combined, making the Los Angeles area by far the largest handler of high-value goods in foreign trade. The size of the Asia trade is highlighted by the fact that Japan alone now exports more ocean-borne trade to the United States than do all of the countries of Latin America combined.

In general, the fastest growing ports are those that have adapted most quickly to the newer bulk cargo and container ships. Space and equipment to handle containers quickly are needed, as well as deep water and rapid unloading facilities for bulk (especially oil, coal, and iron ore) cargoes.

Oil and natural gas pipelines are perhaps America's best kept transportation secrets, for they are underground and out of sight, and few people are aware of the important roles they play (Figure 8.20). Oil pipelines are used primarily to ship crude oil, but oil products are also shipped this way. The overall geographic pattern of pipeline routes in the United States is that they originate in the producing areas, especially in the south-central part of the country (for

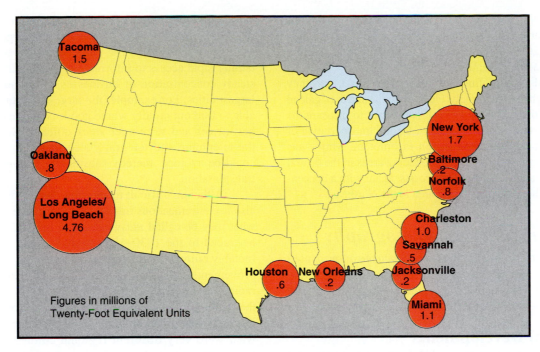

Figure 8.19
The leading U.S. ports, 1997, for container traffic.

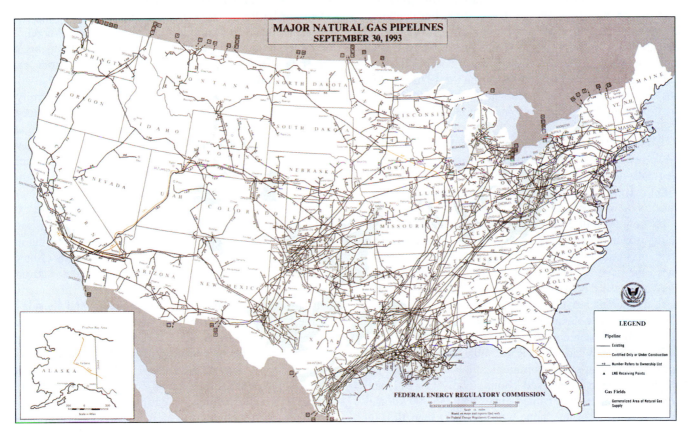

Figure 8.20
Major natural gas pipelines in the United States, 1993.

Source: U.S. Department of Energy

example, Texas and Louisiana) and extend to the major consuming areas, especially toward the Northeast. They carry about 20% of the country's ton-kilometers, silently, out of sight, and efficiently.

Natural gas pipelines are also widespread, and they carry a fuel that is extremely clean-burning and has gained vastly in importance since 1950. It is difficult to measure the relative importance of gas pipelines, for the transportation of a gas cannot be measured in ton-kilometers. Major gas pipelines extend from the larger producing areas in the central United States and Canada to both coasts, again with a concentration from the Texas-Louisiana area to the Northeast.

Communication and Geography

For the sake of simplicity, the term *communication* is used here to mean any movement of a message or an idea from place to place. Used in this way, the term illustrates the close relationship between communication and transportation.

Historical Developments

With such minor exceptions as drums, flashing mirrors, and smoke signals, until the first half of the 19th century, communication and transportation were coexistent. That is, communication between places required some physical form of transportation, and so it was slow. For centuries, letters were probably the most common form of formal long-distance communication, and in the early United States and Canada, the same horse-drawn coaches that carried passengers carried the mails. During the 1820s, it often took several days to send a letter just 100 or 200 kilometers. Speed began to improve somewhat in 1834, when railroads started to carry mail.

The first important "modern," and it would be correct to say revolutionary, communications innovation was the telegraph, which became available for public use in 1844. With the consequent rapid spread of telegraph wires, entrepreneurs, newspapers, and other businesses that had access to this new form of communication had a considerable advantage over their rivals. The first U.S. transcontinental telegraph line was finished in 1861, and it immediately replaced the far more famous, but short-lived, pony express, which took ten days to carry letters from St. Joseph, Missouri, to Sacramento, California. Ever since the start of the telegraph, the first electronic communications system, communication has been enormously faster than transportation, an advantage for which there is no end in sight.

Another important communications innovation, the telephone, began to be used commercially in 1877, and only ten years later, the United States had 235,000 kilometers (146,000 mi) of telephone wire and more than 150,000 subscribers. At first, the telephone was very expensive and usually was justified only for commercial use, but as costs dropped, it began to assume the role of virtual everyday necessity that it seems to occupy today.

In the early years, both the telegraph and the telephone were most readily available in the larger metropolitan areas. They gave these places great advantages in obtaining rapid access to information, thus contributing to the steady concentration of businesses and population in the largest cities. For example, the large-city retailer now was able to find out rapidly where goods could be bought at the best prices and to reorder stock quickly. Good access to the telephone and telegraph also meant that many trips could be avoided, thereby substituting communication for transportation.

Although difficult to measure, faster communications also had a tremendous impact on culture. The beginnings of what may be called *mass culture* in the United States can be traced to the 1830–1850 period, when transportation (notably railroads and steamboats) and communications improvements (such as faster mail and the telegram) were rapid and when powerful forces thereby were unleashed that tended to make the country more uniform culturally; that is, the nation began to move toward the convergence of what were very different regional cultures. With better transportation and communication, the first companies that did business throughout the nation (rather than just regionally) began to appear, and the country started to have nationally distributed magazines, books, and newspapers. These media forms exposed people all over the country to the same ideas, contributing to a nationwide convergence of values, attitudes, and tastes. With these developments, a small number of people, such as authors and editors, exerted their influence over a large population.

Even more powerful forces for cultural uniformity, the movies and the radio, emerged early in the 20th century. The movies (including newsreels) came first, as they began to be distributed nationally around the turn of the century. In several ways, commercial radio broadcasting, which started in 1920, was even more revolutionary than the movies. For one, it introduced simultaneity: with national radio networks, people in all parts of the country were exposed to the same idea at the same time. This medium was also able to broadcast directly into the home, magnifying its influence substantially, and, if one ignores the costs of buying the radio and electricity, and having to listen to advertising, it was free. Movies and the radio were very important in weakening some of the United States' and Canada's sharpest geographic differences, such as between rural and urban life and between North and South. They were also excellent examples of how a small group of writers, directors, producers, owners, advertisers, and actors could have a tremendous impact on national culture. Movie stars, for example, greatly influenced national, and even international, tastes in such items as clothing and hairstyles. There were, of course, movies and radio programs that reflected the highly individual views of their creators, but these did not enjoy mass audiences and therefore did not contribute greatly to the general culture.

Recent Developments

Perhaps the strongest force for producing a national culture, television, began to gain a wide audience in the United States just after World War II. There was some television broadcasting just before that war, starting in 1939, but wartime restrictions on the production of receivers delayed television's takeoff until the late 1940s. There were about 1 million TV sets in 1949, but only two years later, that total jumped to 10 million. Television expanded explosively: by 1959, there were 50 million sets nationally. Soon it was a rare home that did not have a television.

As this medium's popularity rose, viewers were offered an endless lineup of mass-oriented entertainment, commentary, and news. Virtually the whole country was exposed to the same messages and values, often simultaneously or nearly so, greatly weakening the remaining regional cultures. One such impact came with the national introduction of the teen entertainment program "American Bandstand" in the mid-1950s. This program became enormously popular, and teenagers all over the country were soon emulating the Philadelphia-area teens on the show. For example, shortly everyone seemed to be doing the dances that were popular on the show, rather than the somewhat different dances that had developed locally.

One region where the mass media have had a noticeable impact is the South, where a constant exposure to the "national norm" in pronunciation via radio and television appears to be eroding the regional accent. This is not to say that the southern accent is not still very much present (and heard on many radio and television shows), but rather that linguists are detecting a tendency for it to weaken. This is happening so slowly that it may not be noticeable to a person, even in a lifetime, but it is possible that the accent will be largely gone in 100 or 200 years; alternately, new developments could lead to a reassertion of regional distinctiveness, including the southern accent.

Today, the United States is the leading country in the entertainment field, and American movies and TV programs are seen virtually worldwide. This has led to accusations of "cultural imperialism," whereby American values and viewpoints are said to be invading other countries and eroding the local cultures. Canadians are especially sensitive to this issue, and they have taken steps to ensure that their television stations offer an acceptable amount of Canadian programming. Too much American programming was seen as a threat to those cultural features that distinguish Canadians from their much more numerous neighbors to the south. Since Canadians are (perhaps understandably, given the proximity) the original anti-Americans—this anti-Americanism coming in both mild and strong forms—such a policy was predictable.

There are limits to the influences that even a medium as powerful as television can have. It is reputed that many communist countries made television sets available very cheaply, as the governments believed that this medium was an effective way to communicate their viewpoints about the superiority of communism to the public. Further, even today some segments of society, such as the poorest ghetto areas, seem to resist adopting many aspects of the mass culture, and the general increase in ethnic identity goes against the trend toward one national culture.

Although the emphasis so far has been on communication fostering uniformity, this is not the only possibility. The same technology that has fostered uniformity can be used to foster the opposite—to encourage greater differences among people. This is because broadcasting (a message going from the few to the many) has been the norm, while *narrowcasting* has just barely begun to develop. This term refers to using communication to reach a highly select audience. For example, there could be separate Internet chat rooms and television channels for such specialized interests as stamp collecting, quilt making, hunting, and reading medieval literature. Such narrowcasting reinforces specialized interests in the audience and therefore acts to favor differentiation, not uniformity. To some degree, this has already started, but the long-term possibilities for narrowcasting have barely begun to be realized.

Today we are in the midst of a communications revolution that is unprecedented in world history (Figure 8.21). Now information, in many forms, can be moved in vast quantities virtually instantaneously to most places on the earth's surface. This revolution can bring multi-sensory experiences to large numbers of people almost regardless of location. In this way, communication improvements are continuing to make different parts of the earth more similar

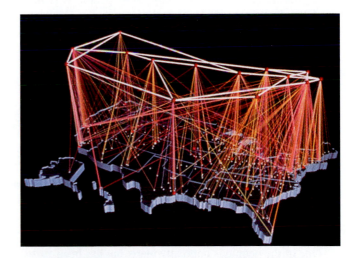

Figure 8.21
The National Science Foundation Network is part of the Internet, a maze of more than 11,000 interconnected computer networks. Millions of individuals a day use the Internet. It shrinks distance and time, enabling users to communicate very rapidly and cheaply. Computer networks give users access to vast amounts of data, distant experiments, supercomputers, and to electronic mail, meetings, and journals.

to each other, but now on a world rather than on a national scale. The potentials for narrowcasting are also greater than ever.

This revolution has been facilitated by the development of several types of hardware for moving huge amounts of information over long distances, each with its advantages and disadvantages. Traditional wires can carry little information, while coaxial cables can carry large bandwidths and are reliable, though expensive. Microwave systems also carry large bandwidths, but the signals can easily be intercepted and reception is poor under unfavorable atmospheric conditions. Satellite systems offer cheap transmission, but suffer from some continuing technical problems. Fiber-optics systems have excellent signal quality and almost unlimited capacity, and seem to be the preferred choice for the near future.

When these improved ways to move information were combined with the computer, the stage was set for the modern communications revolution. The computer started off as merely a machine to speed calculations, but it evolved first into a processor of words and graphics, and then to a new device for telecommunications.

The communications revolution, especially as carried through the Internet, is having many economic repercussions. One area that has already seen major impacts in the U.S. and Canada is retailing, with a virtual explosion of goods and services offered over the Internet. Today one can buy sweaters, books, stocks, airline tickets, mortgages, automobiles, office equipment, and just about anything else this way without ever leaving the office or the home (which themselves are often in the same location). A few years ago there was no Amazon.com; today, most people who have used the Internet, and even many who have not, know about that company. While some firms have made fortunes by using the new technology, many more traditional businesses have suffered in the process. For example, the rise of Amazon.com, along with the advent of giant national bookstore chains, has made it impossible for many small book sellers to remain in business.

Similarly, the revolution has made it easier than ever for companies to exercise control over wide areas. This, in turn, has spawned the development of the first truly global companies, which take advantage of the worldwide division of labor to put such facilities as manufacturing plants and warehouses in a wide variety of countries. North American credit card firms, as well as other financial institutions in North America, Europe, and Asia, are excellent examples of such transnational organizations. Traditionally, governments around the world exercised considerable control over financial companies and institutions, but the latter are changing so rapidly in this new electronic environment that regulators literally are having trouble keeping up with the changes.

What about the geography of communications? Contrary to widespread opinion, the advantages of the communications revolution are not distributed evenly over the earth's surface. Today, the largest metropolitan areas in economically advanced countries have excellent access to high capacity systems like microwave and fiber optics, while most underdeveloped countries, and even low population density areas in advanced countries, have much more limited capabilities, perhaps only a local telephone line. Thus, large cities continue to have significant advantages in communications services and they attract most of the economic growth linked to the communications revolution. It is in the metropolises that we find the **smart buildings** (furnished with the latest telecommunications facilities) and **teleports** (places that provide bulk access to advanced transmissions channels), which make it possible for companies to bypass local communications carriers and to obtain great savings in long-distance charges. There has always been a considerable gap in information access between the

A LONG-DISTANCE CONVERSATION

Imagine a room in Denver, sometime in the future, where two people are sitting on a couch, having an intense conversation. However, not all is as it seems. One of the people is not really there, but has been brought to the room electronically, so that her image is there but not her person. She is actually in Karuizawa, Japan, where the other person's image has also been brought to her living room. Their images have been transported to each other's location in three dimensions, via holography, and they can make eye contact while carrying on the conversation.

With such technology, it may not be necessary for people to travel as much as they used to, since the experience of visiting someone, complete with body language, can be achieved electronically. The same can be said for other kinds of experiences, such as going on drives and visiting museums. In the future, therefore, communication may well increasingly be substituted for transportation. There are limits to such substitutions, however; for example, few people would want to go on an electronic date!

largest cities and peripheral areas; from available evidence, this gap is widening, not getting smaller.

Finally, the social consequences of the communications revolution will be enormous, though they are far from fully understood. There is little question that the greatly enlarged scale of new communications systems will shape the way we think and behave as individuals and members of communities. In the past few thousand years people moved from traditional societies, where limited information was transmitted orally from generation to generation, to modern print societies, which fostered the growth of science, literature, and technology. Now we are well on the way into the era of the electronic media, which (unlike the print medium) can simultaneously provide visual information, animation, video, and sound. The written word emphasized the importance of, and the sharing of, ideas, while electronic media tend to emphasize feeling, mood, and appearance; they foster the sharing of experience and emotional sensory involvement. This shared experience can easily cross national boundaries, religious affiliations, and other traditional group associations, for with electronic media distance means little. Thus they can extend our senses and give us what seems to be direct experiences of events on a global scale. That capability, in turn, almost certainly will diminish the role of the state, and even of the family, in shaping cultural norms. What, exactly, the long-term impacts of these possibilities will be nobody can say, but they certainly make stimulating topics for classroom discussion.

Summary

Transportation allows areal specialization to take place. The better the transportation, the less the friction of distance and the greater the scale of this areal specialization. When areal specialization occurs, those places that can produce a good or service most efficiently concentrate on that kind of output, and with trade, everyone can enjoy a higher standard of living. Today's world scale of areal specialization depends on low transportation costs; if these costs were to rise significantly, there would have to be a return to something closer to areal self-sufficiency.

Americans and Canadians spend far more on passenger than on freight transportation, chiefly because the very expensive private automobile dominates the movement of people. Truckers get the great bulk of the freight dollar, distantly followed by the railroads. Railroads, however, carry more intercity freight ton-kilometers than any other form of transportation.

Since the late 1950s, the geographic impacts of highway transport have been strengthened in the United States by the construction of the I-system, which is vital to the movement of people and goods both within and between cities. This system has had a great impact on the geography of our cities, acting as a skeleton around which recent urban growth has taken place. Generally, these freeways have encouraged urban sprawl—the movement of population and economic activities away from the old centers and toward the suburban sections of the I-system. Airlines, railroads, pipelines, and waterways have also influenced the location of certain businesses, but their locational impacts have been far less widespread than those of the truck and private automobile.

Communication is the movement of messages or ideas between places. For centuries, communication required transportation and was slow, but since the 1840s, there have been a series of revolutionary innovations. The telegraph and telephone favored the largest cities, and the most advanced of the recent innovations also favor such locations. The mass communications media of the 19th and 20th centuries, on the other hand, fostered cultural uniformity.

Current developments, as illustrated by the rapidly expanding Internet, can be seen as just the latest of a long-standing series of changes in electronic communications. However, they are also revolutionary, for the mating of advanced transmission hardware to the computer can combine the features of all the previous electronic media to bring to people sensory experiences from all over the world that are reminiscent of the spontaneity of traditional oral societies As such, they are almost certain to lead to enormous and far-reaching, although as yet undefined, social changes. In education, discussions about virtual classrooms and virtual universities only hint at some other possibilities springing from these developments. The communications revolution already has led to substantial economic changes, but the potentials for further developments along these lines seem unlimited. Although the impacts of these changes already are felt in most corners of the world, the largest cities in the more advanced countries consistently have the best access to the newest communications technology, and as such they will be the geographic locus of most of the related economic growth.

Key Words

Amtrak	intermodal traffic
areal specialization	land bridge services
circumferential freeways	light rail
deregulation	St. Lawrence Seaway
friction of distance	smart buildings
heavy rail	teleport
	Via Rail

Gaining Insights

1. What is areal specialization? What role does transportation play in areal specialization?
2. What form of transportation dominates expenditures for moving passengers in the United States and Canada? What is the leading form of public passenger transportation in most cities today? Why?
3. Describe several impacts of the interstate highway system on the geography of the United States. When did the system originate?

4. How have trucks contributed to the increasing suburbanization of the United States and Canada?

5. What kinds of freight commodities are most effectively handled by railroads? By inland water transportation?

6. Why are the United States and Canada well-suited to the use of air transportation?

7. What were some impacts of the Airline Deregulation Act on the geography of air transportation?

8. What are the two most important inland waterways in the United States? Why have they been experiencing very different traffic trends?

9. What is the St. Lawrence Seaway? When was it finished, and why has its impact since then been far less than was anticipated?

10. What is communication? Discuss communication as a powerful force for cultural uniformity.

11. What are some future cultural impacts that you expect from the communications revolution? How about economic impacts? Why?

Selected References

Brunn, Stanley D. and Thomas R. Leinbach, eds. *Collapsing Space & Time: Geographic Aspects of Communication and Information.* London: Harper Collins Academic, 1991.

Hanson, Susan, ed. *The Geography of Urban Transportation.* New York: Guilford Press, 1986.

Hepworth, Mark. *Geography of the Information Economy.* New York: The Guilford Press, 1990.

Lewis, Pierce. "America between the Wars: The Engineering of a New Geography." In *North America: The Historical Geography of a Changing Continent,* edited by Robert T. Mitchell and Paul A. Groves. Savage, Md.: Rowman and Littlefield, 1990.

Mayer, Harold M. "Cities: Transportation and Internal Circulation." *Journal of Geography* 68 (1969): 390–408.

Meyer, David R. "The National Integration of Regional Economies, 1860–1920." In *North America: The Historical Geography of a Changing Continent,* edited by Robert T. Mitchell and Paul A. Groves. Savage, Md.: Rowman and Littlefield, 1990.

Muller, Peter O. "Transportation and Urban Growth: The Shaping of the American Metropolis," *Focus* 36 (1986): 8–17.

O'Neill, Michael J. *The Roar of the Crowd.* New York: Times Books, 1993.

Richards, Curtis W. and Michael L. Thaller. "United States Railway Traffic: An Update," *The Professional Geographer* 30 (1978): 250–55.

Sealy, Kenneth R. *The Geography of Air Transport.* Chicago: Aldine, 1966.

Sui, Daniel Z. and Robert S. Bednarg. "The Message is the Medium: Geographic Education in the Age of the Internet," *Journal of Geography* 98 (1999): 93–99.

Vance, James E. "Revolution in American Space Since 1945, and a Canadian Contrast." In *North America: The Historical Geography of a Changing Continent,* edited by Robert T. Mitchell and Paul A. Groves. Savage, Md.: Rowman and Littlefield, 1990.

Zelinsky, Wilbur. *The Cultural Geography of the United States,* rev. ed. Englewood Cliffs, N.J.: Prentice-Hall, 1992.

CHAPTER

9

Cities

One of America's most stunning and enduring contributions to architecture and urban form is the skyscraper. Even the use of the word *skyscraper* to denote a many-storied building is an American coinage. The skyscraper first took shape in Chicago and New York City late in the 19th century, and although it has now spread around the world, United States and Canada remain preeminent in its use, containing 100 of the world's 120 tallest buildings.

For much of the 20th century, it appeared that American cities were competing to build the tallest structures they could. An urbanizing population, the concentration of business functions at the city core, and the high price of land and limited space in the central business district made it desirable to build upward. Skyscrapers became a source of pride, a focal point for city residents. They brought a certain distinctiveness to a city, one that could be converted into business prestige and tourist interest. Very tall buildings were erected in the cities of the Northeast and Midwest, culminating in Chicago's Sears Tower (Figure 9.1) and the twin towers of the World Trade Center in New York, all more than 100 stories tall.

Southern and western American cities have a lower profile than New York or Chicago. Although most of these cities have buildings of 60 or 70 stories, they have far fewer of them. These cities have realized their greatest growth in more recent years, when there has been a tendency to decentralize. Suburban land is relatively inexpensive; space at the core of most southern and western cities is not as lim-

ited as in northeastern and midwestern ones; and office functions and the workforce need not be concentrated in a small area. When skyscrapers are built in an Atlanta, Dallas, Houston, Los Angeles, or Seattle, they rarely exceed 70 stories. The contrast in skylines between Los Angeles and New York, the two largest American metropolises, is a telling reminder of the different history and geography of these two cities (Figure 9.2).

(a)

(b)

Figure 9.2
Compare the skylines of Los Angeles (a) and New York City (b). The latter remains the skyscraper capital of the world, with about 150 buildings taller than 150 meters (492 ft). The World Trade Center is 26 meters (86 ft) shorter than the Sears Tower in Chicago. The Empire State Building, visible in the background, was, for over 40 years, the world's tallest building.

Figure 9.1
The skyline of Chicago. At 443 meters (1454 ft), the Sears Tower, completed in 1974, is the world's second tallest building. The tallest self-supporting *structure* in the world is the CN Tower in Toronto, at 555 meters (1821 ft).

In this chapter, our first objective is to consider the major factors responsible for the size and location of American and Canadian cities. The second goal is to identify the land use patterns within those urban areas. We shall see that the cities have certain similarities and a number of striking differences.

Functions of Urban Areas

Our human support systems are based on the flow of information, goods and services, and cooperation among people who are located at convenient places relative to one another. Unless individuals can produce all that they need themselves, and relatively few can, they must depend on shipments of food and supplies to their home place or convenient outlet centers. We establish stores, places of worship, repair centers, and production sites as close to home places as is possible and reasonable. The result is the establishment of towns. These may grow to the size of a New York metropolitan area (about 20 million people today).

Whether they are villages, towns, or cities, these places exist for the efficient performance of functions required by society, functions that cannot be adequately carried out in dispersed locations. They reflect the saving of time, energy, and money that the clustering of people and activities implies. The more accessible the producer to the consumer, the worker to the workplace, the citizen to the city hall, the worshiper to the church, or the lawyer or doctor to the client, the more efficient is the performance of their separate activities, and the more effective is the integration of urban functions.

Urban areas provide all or some of the following types of functions: retailing, wholesaling, manufacturing, business service, entertainment, religious service, political and official administration, military defensive needs, social service, public service (including sanitation and police), transportation and communication service, meeting place activity, visitor service, and places to live. Because not all urban functions and people can be located at a single point, cities themselves must take up space, and land uses and populations must have room. The more limited space is, the greater is the tendency to build up rather than out. Because connections to other places are essential, the nature of the transportation system will have an enormous bearing on the number of services that can be performed and the efficiency with which they can be carried out.

Some Definitions

Urban areas are not of a single type, structure, or size. What they have in common is that they are nucleated, nonagricultural settlements. At one end of the size scale, urban areas are small towns with perhaps a single main street of shops; at the opposite end, they are complex, multifunctional metropolitan areas or supercities. The word *urban* often is used in place of such terms as *town, city, suburb,* and *metropolitan area,* but it is a general term, and it is not used to spec-

ify a particular type of settlement. Although the terms designating the different types of urban settlements, such as *city,* are employed in common speech, they are not uniformly applied by all users. What is recognized as a city by a resident of rural Vermont or West Virginia might not at all be afforded that name and status by an inhabitant of California or New Jersey. It is necessary in this chapter to agree on the meanings of terms commonly employed but varyingly interpreted.

The words **city** and **town** denote nucleated settlements, multifunctional in character, including an established central business district and both residential and nonresidential land uses. Towns are smaller and less functionally complex than cities, but they still have a nuclear business concentration. **Suburb** denotes a subsidiary area, a functionally specialized segment of a large urban complex. It may be dominantly or exclusively residential, industrial, or commercial, but by the specialization of its land uses and functions, a suburb depends on urban areas outside of its boundaries. Suburbs, however, can be independent political entities. For large cities having many suburbs, it is common to call that part of the urban area contained within the official boundaries of the main city around which the suburbs have been built the **central city** (Figure 9.3).

The **urbanized area** refers to a continuously built-up landscape defined by building and population densities with no reference to political boundaries. It may be viewed as the physical city and may contain a central city and many contiguous cities, towns, suburbs, and other urban tracts. A **metropolitan area,** on the other hand, refers to a large-scale functional entity, perhaps containing several urbanized areas, discontinuously built-up but nonetheless operating as an integrated economic whole.

Location of Urban Settlements

Cities or metropolitan areas are functionally connected to other urban and rural areas. In fact, the reason for the existence of a city is not only to provide goods and services for itself, but also to furnish these things for others outside of its boundaries. People of the city consume food, process materials, and accumulate and dispense goods and services, but they must rely on outside areas for supplies and as markets for their activities. The Chicago metropolitan area houses about 8 million people who must be fed every day. An enormous amount of food must arrive in Chicago daily to supply these people. Many of the people of Chicago work in manufacturing trades, the product of which is sold in Chicago and other parts of the United States, Canada, and the world. To perform adequately the tasks that support it and to add new functions as demanded by the larger economy, cities such as Chicago must be efficiently located. That efficiency may be marked by centrality to the area served. It may derive from the physical characteristics of its site. Or placement may be related to the resources, productive regions, and transportation network of

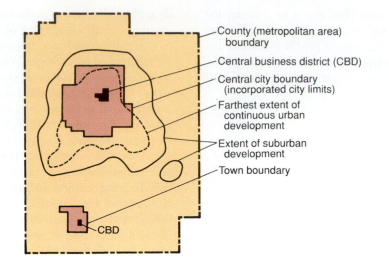

Legend:
- County (metropolitan area) boundary
- Central business district (CBD)
- Central city boundary (incorporated city limits)
- Farthest extent of continuous urban development
- Extent of suburban development
- Town boundary

CBD

Figure 9.3

A hypothetical spatial arrangement of urban units within a metropolitan area. Sometimes official limits of the central city are extensive and contain areas commonly thought of as suburban. Older eastern U.S. cities more often have restricted limits and contain only part of the high-density land uses associated with them.

DEFINITION OF *METROPOLITAN* IN THE UNITED STATES

Definitions of various types of urban areas must be clear if proper accounting is to be made by governmental authorities. The U.S. Bureau of the Census has refined and redefined the concept of "metropolitan" from time to time to summarize the realities of the changing population, physical size, and functions of urban regions.

Until 1983, the *Standard Metropolitan Statistical Area* (SMSA) was recognized. It was made up of one or more functionally integrated counties focusing upon a central city of at least 50,000 inhabitants. Now, the minimum size requirement for central cities has been dropped, and central city status is determined by other qualities—whether, for example, a city is an employment center surrounded by bedroom-community-type suburbs.

With this change, the number of central cities (and metropolitan areas) automatically increased. In Canada, the Census Metropolitan Area (CMA) is equivalent to the SMSA, except that whole counties are not included at the outer reaches of the commuting range.

In the mid-1980s, old and new metropolitan areas were redefined into *Metropolitan Statistical Areas* (MSAs are economically integrated urbanized areas in one or more contiguous counties), *Primary Metropolitan Statistical Areas* (PMSAs are those counties that are part of MSAs that have less than 50% resident workers working in a different county), and *Consolidated Metropolitan Statistical Areas* (an MSA becomes a CMSA if it contains 1 million or more people and is composed of PSMAs).

the country, so that the effective performance of a wide array of activities is possible.

In discussing urban settlement location, we frequently differentiate between site and situation. The **site** is the exact location of the settlement and can be described in terms of latitude and longitude or physical characteristics. For example, the site of Philadelphia is an area bordering and west of the Delaware River north of the intersection with the Schuylkill (pronounced *skool-kill*) River in south-

east Pennsylvania. The description can be more or less exhaustive, depending on its purpose. In the Philadelphia case, the fact that the city is partly on the Atlantic Coastal Plain, partly on a piedmont (foothills), and is served by navigable rivers is important if one is interested in the development of the city during the Industrial Revolution. Water transportation was an important localizing factor when the major U.S. and Canadian cities were established (Figure 9.4).

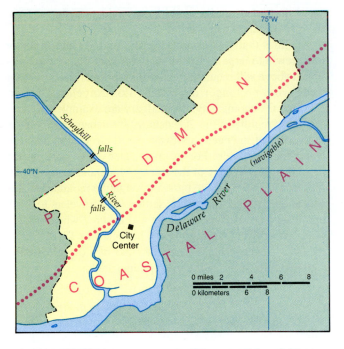

Figure 9.4
The site of Philadelphia.

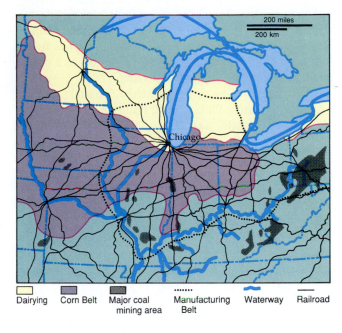

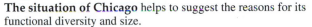

Figure 9.5
The situation of Chicago helps to suggest the reasons for its functional diversity and size.

If site suggests absolute location, **situation** indicates relative location. The relative location places a settlement in relation to the physical and human characteristics of the surrounding areas. It is important to know what kinds of possibilities and activities exist in the area near a settlement, such as the distribution of raw materials, market areas, agricultural regions, mountains, and oceans. The site of central Chicago is 45°52′ North, 87°40′ West, on a lake plain, but more important is its situation close to the deepest penetration of the Great Lakes system into the interior of the country, astride the Great Lakes–Mississippi waterways, and near the western margin of the Manufacturing Belt, the northern boundary of the Corn Belt, and the southeastern reaches of a major dairy region. References to railroads, coal deposits, and ore fields would amplify its situational characteristics. From this description of Chicago's situation, implications relating to market, raw materials, and transportation centrality can be drawn (Figure 9.5).

The site or situation that originally gave rise to an urban unit may not long remain the essential ingredient for its growth and development. Pittsburgh is found where the Monongahela and Allegheny Rivers join to form the Ohio River, a perfect location in 1764 for a town to be founded near a military fort that offered protection to the rugged western Pennsylvania area against attack by American Indians or the French. After the Revolutionary War, Pittsburgh became a starting and supply point for pioneers traveling west. Towns, originally successful for whatever reason,

may by their success attract people and activities totally unrelated to the initial localizing forces. By what has been called a process of "circular and cumulative causation," a successful place may acquire new populations and functions attracted by existing markets, labor force, and urban facilities. In the 1800s, connecting Pittsburgh by road and rail to the eastern cities became more important than retaining the fort, and during this period, the coal found in the hills and valleys in and around Pittsburgh became the base for glass and iron industries. Not only did Pittsburgh develop into a huge iron and steel making center, but it also became a major market for all manner of goods and services.

In the United States and Canada, the early settlements were either agricultural service centers or ports, located along ocean coasts or along rivers and lakes. In the 1600s and 1700s, the most important form of transit and trade was by water. Early settlers naturally sought places where they could dock, load and unload goods, protect shipping from coastal storms, be certain of a supply of drinking water, have enough relatively flat land so that houses, commercial warehouses, and roads could be built, and be assured of a hinterland that could supply the settlement with food and other necessities. A **hinterland** is the area surrounding a city that directly serves and is served by a city. Today, that concept is not as easily delineated on a map as it was in the early days of settlement. Then, one could draw a line around the agricultural area of southeastern Pennsylvania

Cities

and southern New Jersey to see that Philadelphia had a hinterland rich in agricultural potential. Today, Philadelphia, like all American cities, receives its food supplies from a vast number of sites throughout the world, but to a degree, it still depends on its original hinterland.

It is interesting to note that the "coastal" cities of New York, Boston, Philadelphia, Baltimore, New Orleans, Vancouver, Seattle, San Diego, Miami, Montréal, and Washington, D.C., are not really directly on the coast. They either are protected by islands or are sufficiently far up a river or bay so as not to be vulnerable to coastal storms and to be nearer the heart of their agricultural hinterland. Note that San Francisco's harbor faces the bay rather than the ocean. Such inland cities as Chicago, Detroit, Kansas City, Toronto, St. Louis, Pittsburgh, Minneapolis, and Cincinnati are sited along rivers or lakes at key locations where inexpensive water transportation en-

sured that goods easily could be brought to and sent from the city (Figure 9.6).

Cities that were founded in the mid- to late-1800s or that grew into large metropolitan areas later than those mentioned above still required fresh drinking water, but rivers for trade were no longer necessary. Los Angeles, Denver, Phoenix, Oklahoma City, and Calgary are examples of cities that grew rapidly in the 20th century by depending more on railroads and interstate highways for their success than on water transportation. For example, Denver is sited on the high plains at the edge of the Rocky Mountains at the beginning of a route leading to passes through the mountains. This was an ideal point for miners, settlers, and travelers to stock up on supplies. Denver's success was ensured when the railroad arrived from Chicago (via Cheyenne) in 1870. Unlike the cities sited in the past, the new cities of today are usually suburban extensions of the metropolitan area near freeways.

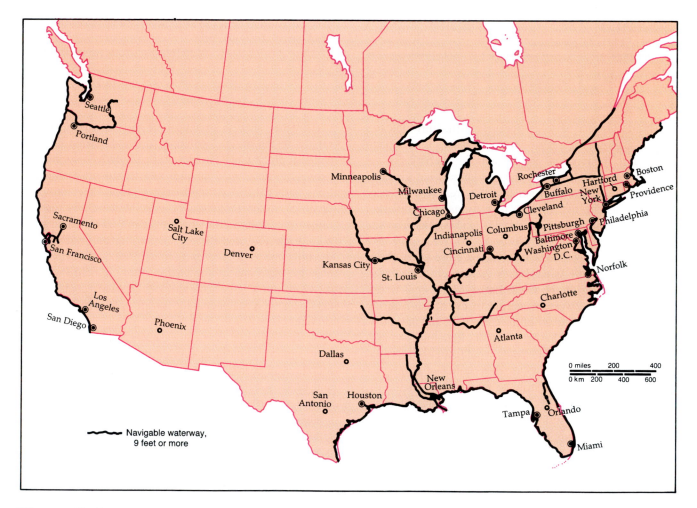

Figure 9.6

The association of major metropolitan areas and navigable water. Before the advent of railroads in the middle of the 19th century, all major cities were associated with waterways.

Systems of Urban Settlements

The various functions that an individual urban area performs are reflected not only in its size but also in its relationship with other urban units in the larger system of which it is part. As Denver grew, it became the focal point of activities that tied it to Colorado Springs, Boulder, and many smaller towns along the eastern edge of the Rocky Mountains, as well as to mining towns, such as Leadville, in the Rockies. But Denver is also strongly tied economically to Chicago and Kansas City to the east and to Salt Lake City and San Francisco to the west. Denver is part of an interdependent urban system whose prosperity or decline will be reflected in the city itself. New buildings are erected in times of economic growth but stand vacant when any major part of a city's network suffers a downturn.

Urban Hierarchy

Perhaps the most effective way to recognize how systems of cities are organized is to consider the **urban hierarchy.** Cities can be divided into size classes on the basis of the numbers and kinds of services each city or metropolitan area provides. The hierarchy is then like a pyramid: the few large and complex cities are at the top, and the many smaller cities are at the bottom. There are always more smaller cities than larger ones. One can envisage, say, a seven-level hierarchy where the complexity of cities increases as one rises in the pyramid. In the United States, New York is clearly at the top of the pyramid. It is a national city of great size and influence. It faces Europe and interacts financially on a daily basis with every part of America and all parts of the world. Just below New York in the pyramid are the great regional capitals of the Midwest and West, Chicago and Los Angeles. They compete with New York but also look to New York for leadership in many types of activities.

When a location dimension is added to the hierarchy, it becomes clear that a spatial system of metropolitan centers, large cities, small cities, and towns exists. Just below Chicago and Los Angeles are the subregional capitals: Boston in the Northeast, Philadelphia in the mid-Atlantic region, Atlanta in the Southeast, Miami in the Caribbean, Detroit in the eastern Midwest, Minneapolis in the northern Midwest, Kansas City in the western Midwest, Dallas in the Southwest, Denver in the mountain states, San Francisco in the central West, and Seattle in the Northwest. The Canadian equivalents are Toronto at the top of the pyramid, followed by Montréal as the eastern (Québec) capital, and Vancouver in the West. The regional capitals include Halifax in the Maritime East, Winnipeg in the Midwest, and Calgary in the Rocky Mountains.

Goods, services, communication lines, and people flow up and down the hierarchy. The few high-level metropolitan areas provide the most specialized functions, especially financial services, for large regions, while the smaller cities provide fewer services and serve smaller regions. Together, all cities at all levels in the hierarchy constitute an urban system. We have mentioned the top three levels in the hierarchy. The next four levels include all of the remaining cities of the United States and Canada (Figure 9.7).

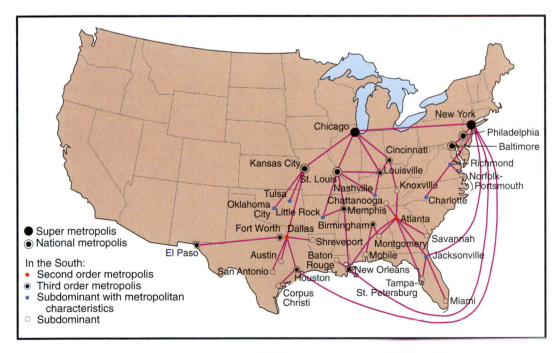

Figure 9.7
Hierarchical relationships among southern U.S. cities.

Urban Influence Zones

A small city may influence a local region of, say, 50 square kilometers (19.3 sq mi) if, for example, its newspaper is delivered to that region. Beyond that area, another city may be the dominant influence. **Urban influence zones** are the areas outside of a city that are affected by it. As the distance away from a city increases, its influence on the surrounding countryside decreases. The sphere of influence of an urban unit is usually proportional to its size.

A large city located 160 kilometers (100 mi) from a small city may influence that city and other small ones through its banking services, TV station, and large shopping malls. There is an overlapping hierarchical arrangement, and the influence of the largest cities is felt over the widest areas.

Intricate relationships and hierarchies are common. Consider Fargo, North Dakota, which for local market purposes dominates the rural area immediately surrounding it. However, Fargo is influenced by political decisions made in the state capital, Bismarck. For a variety of cultural, commercial, and banking activities, Fargo is influenced by Minneapolis. A center of wheat production, Fargo is subordinate to the grain market in Chicago. The pervasive agricultural and other political controls exerted from Washington, D.C., on Fargo indicate how large and complex are the urban zones of influence.

Perhaps the most used indicator of influence zones is the location of daily newspaper deliveries. It represents only one level of influence, but it is an important indicator of the places where the day-to-day news of a particular city is of interest to a wider audience than that in the city itself. Of considerable interest to commercial concerns is that it also represents, to a large extent, the television market area. It may be the sports teams, the social news, the financial news, or cultural activities that are of main interest to the surrounding populace. Of course, most newspapers reinforce their influence by printing news of the areas in which they make their daily deliveries. Figure 9.8 shows the

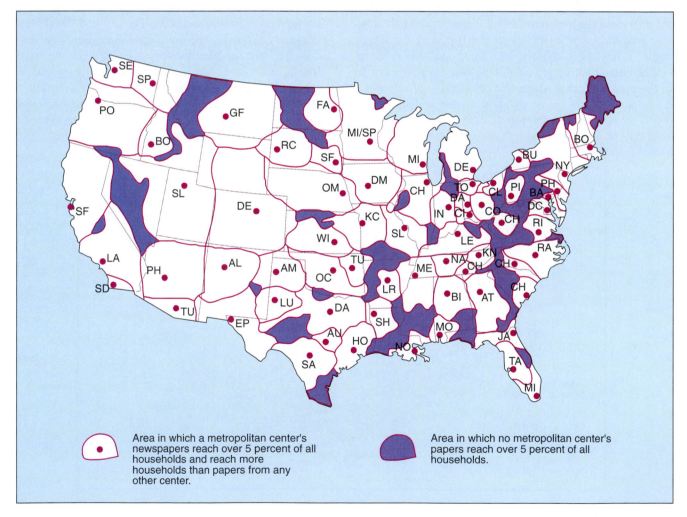

Area in which a metropolitan center's newspapers reach over 5 percent of all households and reach more households than papers from any other center.

Area in which no metropolitan center's papers reach over 5 percent of all households.

Figure 9.8
Metropolitan newspaper influence zones.

newspaper influence zones of the major U.S. daily newspapers. Although the Denver and Salt Lake City influence zones are the largest in the United States, the greatest circulation of newspapers is in and around the largest cities, such as New York, Los Angeles, and Chicago. Note that not all regions are under the influence of a metropolitan area.

Other indicators that show influence at smaller scales of resolution are the spokes representing connecting flight lines of major hub airports, the Federal Reserve Bank districts, and **migration fields.** Migration fields show from which area most internal migrants originate. Note in Figure 9.9 that Seattle's influence spreads across the northern tier of western states, and Chicago's influence extends deeply into the South. Several coastal areas along the Atlantic are more closely tied to the powerful influence of New York than to nearby metropolitan areas. The major military base areas along the Gulf Coast in northern Florida and Alabama are more closely tied to the Washington, D.C., area than to Atlanta, Miami, or New Orleans.

Economic Base

When one or more urban settlements within a well-linked system increase their productivity, perhaps because of an increase in demand for the special goods or services that they produce, all members of the system are likely to bene-

fit. For example, if the demand for computers increases, the cities and towns of Silicon Valley, California (San Jose, Cupertino, Sunnyvale, Mountain View, Santa Clara, and so on), will benefit, but so too will San Francisco, where much of the financial work for the increased demand will be done. The concept of the **economic base** shows how settlements are affected by changes in economic conditions.

Part of the employed population of an urban place is engaged in the production of goods or the performance of services for areas and people outside the city itself. They are workers engaged in "export" activities, whose efforts result in money flowing into the community. Collectively, they constitute the **basic sector** of the city's total economic structure. Other workers support themselves by producing things for residents of the urban unit itself. Their efforts, necessary to the well-being and the successful operation of the city, do not generate new money for it but comprise a **service,** or **nonbasic sector** of its economy. These people are responsible for the internal functioning of the urban unit. They are crucial to the continued operation of its stores, professional offices, city government, and local transit and school systems. The computer workers in Silicon Valley are responsible for bringing new money into their region; thus, they are basic workers. Those responsible for teaching the children of the employees of the computer companies are nonbasic workers. The community must have both types of workers to function.

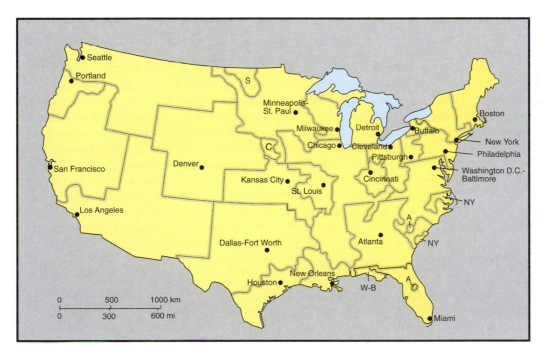

Figure **9.9**

Metropolitan migration fields, 1965–1970. Each of the 24 migration fields is focused on a particular metropolitan center. The field influences the characteristics of the center's population. Chicago, for example, receives many southern migrants, while Seattle does not. A denotes Atlanta; C Chicago; NY New York City; S Seattle; and W-B Washington, D.C.-Baltimore.

The total economic structure of a city equals the sum of its basic and nonbasic activities. In actuality, it is the rare urbanite who can be classified as belonging entirely to one sector or another. For example, not all the computers that are produced in Silicon Valley are sold outside of its borders. Some part of the work of most people involves financial interaction with residents of other areas. Doctors, for example, may have mainly local patients and thus are members of the nonbasic sector, but the moment they provide a service to someone from outside the community, they bring new money into the city and become part of the basic sector. Given the size of the San Jose area, one might believe that one can obtain every kind of good and service there, but this is not the case. If, for example, a small child needs a special type of operation, the family might take the child to San Francisco. The doctors who take care of the youngster are responsible for bringing new money into San Francisco; therefore, they are basic workers when they tend the child but nonbasic workers when they provide services for people from San Francisco.

Most cities perform many export functions, and the larger the urban unit, the more multifunctional it becomes. Nonetheless, even in cities with a diversified economic base, one or a very small number of export activities tends to dominate the community and to identify its role within a system of cities. Figure 9.10 indicates the functional specializations of large U.S. cities.

Recalling that most cities are multifunctional, in Figure 9.10 note that the cities that are primarily manufacturing centers are in the Northeast and Midwest. Here are found the large cities of Detroit, Milwaukee, Cincinnati, Buffalo, and Cleveland, all major manufacturing centers. In fact, the only city outside of the region is San Jose.

The transportation centers are either ports, such as San Francisco and New Orleans, or important air and rail transfer centers, such as Chicago, New York, and Atlanta.

All cities are retail centers, but tourist and border cities, such as El Paso and San Diego, provide far more retail services than most cities. This is due to the demand for American goods by Mexican residents of the large adjoining cities of Ciudad Juárez and Tijuana, respectively.

New York is the major financial center of the United States. If the map showed Canadian cities as well, Toronto would stand out as a major financial center. Large insurance companies are found in Hartford, Newark, Columbus, and Omaha. Dallas is the financial center of the Southwest, and Miami is the financial center for the Caribbean region.

Public administration means government employment. As the national capital, Washington, D.C., is the most specialized large city in the United States. Ottawa, a much smaller city, plays the same role for Canada. State capitals that have rather few other basic functions, such as Harrisburg, Pennsylvania, and Sacramento, California, also specialize in public administration. Since military personnel or people who work for the military are included in this category, cities such as San Diego and San Antonio stand out.

The most diversified cities are usually the largest. They became large mainly because they specialized in a variety of activities ranging from manufacturing to financial services.

In much the same way as settlements grow in size and complexity, so do they decline. When the demand for the goods and services of an urban unit falls, fewer workers are needed, and thus both the basic and service components of a settlement system are affected. There is, however, a resistance to decline that impedes the process and delays its impact. Whereas migrants can quickly fill the need for more workers, under conditions of decline those who have developed roots in the community are hesitant or may be financially unable to move to another locale.

Growth of Metropolitan Regions

In 1996, 47 metropolitan regions in the United States and 4 in Canada had a population of more than 1 million (Tables 9.1 and 9.2). In 1960, these figures were 24 and 2. Not only has there been a steady growth of the population of the two countries, but metropolitan area population as a proportion of the total population has increased from 62% to 80% since 1960. The growth has not been even over the entire region, however. If we look at recent increases in population in the United States, it is clear that there is in fact a Sun Belt/Rust Belt phenomenon (Figure 9.11). The attraction to warmth, newness, and job opportunities, the development of retirement communities, and large Hispanic migrations have been responsible for the greatest recent growth in the Sun Belt. The heavy manufacturing areas from Pittsburgh to Detroit have lost population. In the 20- to 30-year period leading up to 1980, the greatest growth was to the South and West, mainly stimulated by the development of petroleum, aerospace, and defense industries and the huge government outlays in those areas. In the last 40 years, the growth of air transportation and the interstate highway system have made all metropolitan areas accessible to each other. At one time, Denver, Salt Lake City, Seattle, Calgary, and Vancouver were thought to be outposts of civilization.

In a number of cases, metropolitan areas have grown together into great **conurbations,** or urban clusters. The major urban cluster, and the one called *the* **Megalopolis** of America, is the continuous urban string that stretches from north of Boston (southern New Hampshire) to the many suburban towns of northern Virginia and includes Boston, Providence, Hartford, New York, Newark, Trenton, Philadelphia, Wilmington, Baltimore, and Washington, D.C. There is limited open space in this

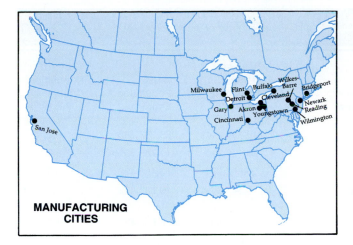

MANUFACTURING CITIES

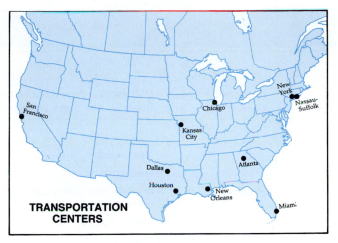

TRANSPORTATION CENTERS

RETAIL CENTERS

FINANCE, INSURANCE, AND REAL ESTATE

PUBLIC ADMINISTRATION (Including Military)

MOST DIVERSIFIED CITIES

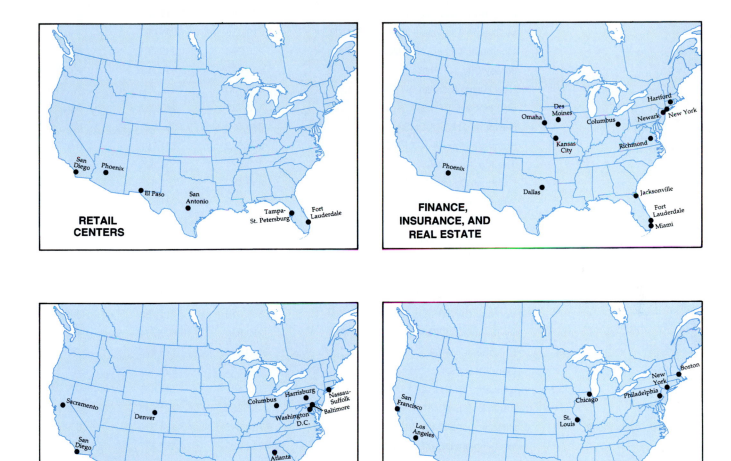

Figure **9.10**

Functional specialization of U.S. metropolitan areas. Five categories of employment were selected to show patterns of specialization for some U.S. metropolitan areas. A sixth category, "Most Diversified," represents those cities that have a generally balanced employment distribution. Note that the most diversified urban areas tend to be the largest.

Cities

Table 9.1
Population of Metropolitan Centers and Metropolitan Areas in the United States, 1996

Rank	Metropolitan Center	Metropolitan Area Includes Center, Suburbs, and Areas Listed Below	Revised 4/1/90 Population Census	7/1/96 Population Estimate	% of Population Change 1990–1996
		United States	248,718,301	265,283,783	6.7
		Metropolitan	198,170,228	211,785,351	6.9
		Nonmetropolitan	50,548,073	53,498,432	5.8
1.	New York	Northern New Jersey-Long Island, NY-NJ-CT-PA CMS	19,549,649	19,938,492	2.0
		New York, NY PMSA	8,546,846	8,643,437	1.1
		Nassau-Suffolk, NY PMSA	2,609,212	2,660,285	2.0
		Newark, NJ PMSA	1,915,694	1,940,470	1.3
		Bergen-Passaic, NJ PMSA	1,278,682	1,311,331	2.6
		Middlesex-Somerset-Hunterdon, NJ PMSA	1,019,858	1,091,097	7.0
		Monmouth-Ocean, NJ PMSA	986,296	1,065,284	8.0
		Jersey City, NJ PMSA	553,099	550,789	−.4
		New Haven-Meriden, CT PMSA	530,180	523,724	−1.2
		Bridgeport, CT PMSA	443,722	443,637	−0.0
		Newburgh, NY-PA PMSA	335,613	362,561	8.0
		Stamford-Norwalk, CT PMSA	329,935	331,767	0.6
		Trenton, NJ PMSA	325,824	330,226	1.4
		Dutchess County, NY PMSA	259,462	262,675	1.2
		Waterbury, CT PMSA	221,629	221,894	0.1
		Danbury, CT PMSA	193,597	199,315	3.0
2.	Los Angeles	Riverside-Orange County, CA CMSA	14,531,529	15,495,155	6.6
		Los Angeles-Long Beach, CA PMSA	8,863,052	9,127,751	3.0
		Riverside-San Bernardino, CA PMSA	2,588,793	3,015,783	16.5
		Orange County, CA PMSA	2,410,668	2,636,888	9.4
		Ventura, CA PMSA	669,016	714,733	6.8
3.	Chicago	Gary-Kenosha, IL-IN-WI CMSA	8,239,820	8,599,774	4.4
		Chicago, IL PMSA	7,410,858	7,733,876	4.4
		Gary, IN PMSA	604,526	622,303	2.9
		Kenosha, WI PMSA	128,181	141,646	10.5
		Kankakee, IL PMSA	96,255	101,949	5.9
4.	Washington	Baltimore, DC-MD-VA-WV CMSA	6,726,395	7,164,519	6.5
		Washington, DC-MD-VA-WV PMSA	4,222,830	4,563,123	8.1
		Baltimore, MD PMSA	2,382,172	2,474,118	3.9
		Hagerstown, MD PMSA	121,393	127,278	4.8
5.	San Francisco	Oakland-San Jose, CA CMSA	6,249,881	6,605,428	5.7
		Oakland, CA PMSA	2,080,434	2,209,629	6.2
		San Francisco, CA PMSA	1,603,678	1,655,454	3.2
		San Jose, CA PMSA	1,497,577	1,599,604	6.8
		Vallejo-Fairfield-Napa, CA PMSA	450,236	482,048	7.1
		Santa Rosa, CA PMSA	388,222	420,872	8.4
		Santa Cruz-Watsonville, CA PMSA	229,734	237,821	3.5

Rank	Metropolitan Center	Metropolitan Area Includes Center, Suburbs, and Areas Listed Below	Revised 4/1/90 Population Census	7/1/96 Population Estimate	% of Population Change 1990–1996
6.	Philadelphia	Wilmington-Atlantic City, PA-NJ-DE-MD CMSA	5,893,019	5,973,463	1.4
		Philadelphia, PA-NJ PMSA	4,922,257	4,952,929	0.6
		Wilmington-Newark, DE-MD PMSA	513,293	550,892	7.3
		Atlantic-Cape May, NJ PMSA	319,416	333,699	4.5
		Vineland-Millville-Bridgeton, NJ PMSA	138,053	135,943	−1.5
7.	Boston	Worcester-Lawrence, MA-NH-ME-CT CMSA	5,455,403	5,563,475	2.0
		Boston, MA-NH PMSA	3,227,707	3,263,060	1.1
		Worcester, MA-CT PMSA	478,384	485,229	1.4
		Lawrence, MA-NH PMSA	353,232	372,693	5.5
		Lowell, MA-NH PMSA	280,578	290,753	3.6
		Brockton, MA PMSA	236,409	246,082	4.1
		Portsmouth-Rochester, NH-ME PMSA	223,271	230,625	3.3
		Manchester, NH PMSA	173,783	182,173	4.8
		Nashua, NH PMSA	168,233	178,335	6.0
		New Bedford, MA PMSA	175,641	175,090	−.3
		Fitchburg-Leominster, MA PMSA	138,165	139,435	.9
8.	Detroit	Ann Arbor-Flint, MI CMSA	5,187,171	5,284,171	1.9
		Detroit, MI PMSA	4,266,654	4,318,145	1.2
		Ann Arbor, MI PMSA	490,058	529,898	8.0
		Flint, MI PMSA	430,459	436,128	1.3
9.	Dallas	Fort Worth, TX CMSA	4,037,282	4,574,561	13.3
		Dallas, TX PMSA	2,676,248	3,047,983	13.9
		Fort Worth-Arlington, TX PMSA	1,361,034	1,526,578	12.2
10.	Houston	Galveston-Brazoria, TX CMSA	3,731,029	4,253,428	14.0
		Houston, TX PMSA	3,321,926	3,791,921	14.1
		Galveston-Texas City, TX PMSA	217,396	240,653	10.7
		Brazoria, TX PMSA	191,707	220,854	15.2
11.	Atlanta, GA	MSA	2,959,500	3,541,230	19.7
12.	Miami	Fort Lauderdale, FL CMSA	3,192,725	3,514,403	10.1
		Miami, FL PMSA	1,937,194	2,076,175	7.2
		Fort Lauderdale, FL PMSA	1,255,531	1,438,228	14.6
13.	Seattle	Tacoma-Bremerton, WA CMSA	2,970,300	3,320,829	11.8
		Seattle-Bellevue-Everett, WA PMSA	2,033,128	2,234,707	9.9
		Tacoma, WA PMSA	586,203	657,272	12.1
		Bremerton, WA PMSA	189,731	231,741	22.1
		Olympia, WA PMSA	161,238	197,109	22.2
14.	Cleveland	Akron, OH CMSA	2,859,644	2,913,430	1.9
		Cleveland-Lorain-Elyria, OH PMSA	2,202,069	2,233,288	1.4
		Akron, OH PMSA	657,575	680,142	3.4

continued

Rank	Metropolitan Center	Metropolitan Area Includes Center, Suburbs, and Areas Listed Below	Revised 4/1/90 Population Census	7/1/96 Population Estimate	% of Population Change 1990–1996
15.	Minneapolis	St. Paul, MN-WI MSA	2,538,776	2,765,116	8.9
16.	Phoenix	Mesa, AZ MSA	2,238,498	2,746,703	22.7
17.	San Diego, CA	MSA	2,498,016	2,655,463	6.3
18.	St. Louis, MO	IL MSA	2,492,348	2,548,238	2.2
19.	Pittsburgh, PA	MSA	2,394,811	2,379,411	−16
20.	Denver	Boulder-Greeley, CO CMSA	1,980,140	2,277,401	15.0
		Denver, CO PMSA	1,622,980	1,866,978	15.0
		Boulder-Longmont, CO PMSA	225,339	258,234	14.6
		Greeley, CO PMSA	131,821	152,189	15.5
21.	Tampa	St. Petersburg-Clearwater, FL MSA	2,067,959	2,199,231	6.3
22.	Portland	Salem, OR-WA CMSA	1,793,476	2,078,357	15.9
		Portland-Vancouver, OR-WA PMSA	1,515,452	1,758,937	16.1
		Salem, OR PMSA	278,024	319,420	14.9
23.	Cincinnati	Hamilton, OH-KY-IN CMSA	1,817,569	1,920,931	5.7
		Cincinnati, OH-KY-IN PMSA	1,526,090	1,597,352	4.7
		Hamilton-Middletown, OH PMSA	291,479	323,579	11.0
24.	Kansas City, MO	KS MSA	1,582,874	1,690,343	6.8
25.	Milwaukee	Racine, WI CMSA	1,607,183	1,642,658	2.2
		Milwaukee-Waukesha, WI PMSA	1,432,149	1,457,655	1.8
		Racine, WI PMSA	175,034	185,003	5.7
26.	Sacramento	Yolo, CA CMSA	1,481,220	1,632,133	10.2
		Sacramento, CA PMSA	1,340,010	1,482,208	10.6
		Yolo, CA PMSA	141,210	149,925	6.2
27.	Norfolk	Virginia Beach-Newport News, VA-NC MSA	1,444,710	1,540,252	6.6
28.	Indianapolis, IN	MSA	1,380,491	1,492,297	8.1
29.	San Antonio, TX	MSA	1,324,749	1,490,111	12.5
30.	Columbus, OH	MSA	1,345,450	1,447,646	7.6
31.	Orlando, FL	MSA	1,224,844	1,417,291	15.7
32.	Charlotte	Gastonia-Rock Hill, NC-SC MSA	1,162,140	1,321,068	13.7
33.	New Orleans, LA	MSA	1,285,262	1,312,890	2.1
34.	Salt Lake City	Ogden, UT MSA	1,072,227	1,217,842	13.6
35.	Las Vegas, NV	AZ MSA	852,646	1,201,073	40.9
36.	Buffalo	Niagara Falls, NY MSA	1,189,340	1,175,240	−1.2
37.	Hartford, CT	MSA	1,157,585	1,144,574	−1.1
38.	Greensboro, NC	MSA	1,050,304	1,141,238	8.7
39.	Providence, MA	MSA	1,134,350	1,124,044	−.9
40.	Nashville, TN	MSA	985,026	1,117,178	13.4
41.	Rochester, NY	MSA	1,062,470	1,088,037	2.4
42.	Memphis, TN	AR-MS MSA	1,007,306	1,078,151	7.0
43.	Austin	San Marcos, TX MSA	846,227	1,041,330	23.1
44.	Oklahoma City, OK	MSA	958,839	1,026,657	7.1
45.	Raleigh, NC	MSA	858,485	1,025,253	19.4
46.	Grand Rapids, MI	MSA	937,891	1,015,099	8.2
47.	Jacksonville, FL	MSA	906,727	1,008,633	11.2

Table 9.2
Estimated Canadian Metropolitan Area Population, 1996

1. Toronto, Ontario	4,264,000
2. Montréal, Québec	3,327,000
3. Vancouver, British Columbia	1,832,000
4. Ottawa-Hull, Ontario-Québec	1,010,000
5. Edmonton, Alberta	863,000
6. Calgary, Alberta	822,000
7. Québec City, Québec	672,000
8. Winnipeg, Manitoba	667,000
9. Hamilton, Ontario	624,000
10. London, Ontario	399,000
11. Kitchener, Ontario	383,000
12. St. Catherines-Niagara, Ontario	372,000
13. Halifax, Nova Scotia	333,000
14. Victoria, British Columbia	304,000
15. Windsor, Ontario	279,000
16. Oshawa, Ontario	269,000
17. Saskatoon, Saskatchewan	219,000

600-km (360-mi) swath of residential, commercial, and industrial districts. Other American conurbations include (Figure 9.12):

- The southern Lake Michigan region stretching from north of Milwaukee through Chicago and across northwestern Indiana and southwest Michigan
- The San Francisco Bay area, which includes a continuum of cities and towns starting with Richmond and Berkeley in the northwest and continuing in a clockwise direction around the bay embracing Oakland, the Silicon Valley to San Francisco, and the suburban towns north of the Golden Gate Bridge
- Los Angeles, including the myriad suburbs and cities extending from Santa Barbara to Ventura, Los Angeles, Long Beach, San Diego, and even farther south to include more than 1 million people living in Tijuana, Mexico
- The spreading urban agglomeration in and around Dallas and Fort Worth

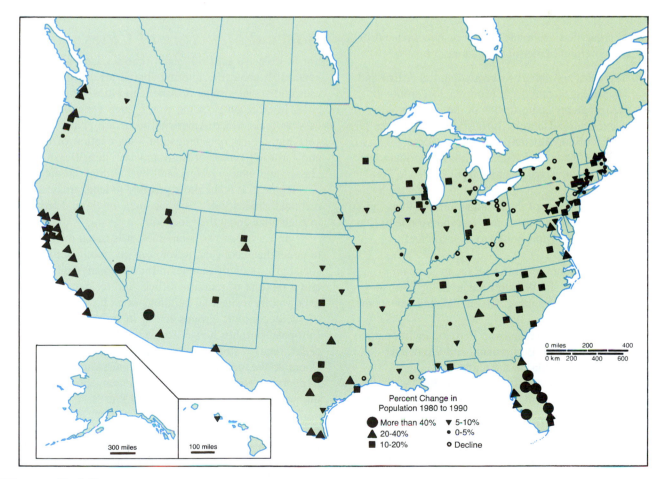

Figure 9.11
The pattern of metropolitan growth and decline in the United States, 1980–1990.

Source: Data from U.S. Bureau of the Census.

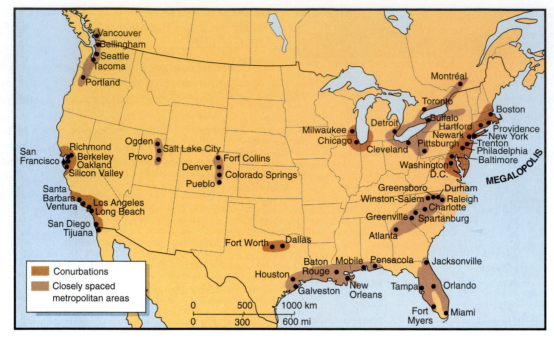

Figure 9.12
Megalopolis and other conurbations and areas that are made up of unconnected but closely spaced metropolitan districts.

The areas that are made up of an unconnected but closely spaced group of metropolitan areas are:

- The cities and suburbs surrounding Lakes Erie and Ontario, stretching from Montréal to Toronto to Detroit along the north side of the lakes to Cleveland, Pittsburgh, Buffalo, and through central New York on the south side of the lakes
- The Gulf Coast and Atlantic rims of Florida containing the coastal cities from Tampa to Fort Myers and Jacksonville to Miami, connected by the cities and towns of the Orlando metropolitan region
- The Gulf Coast from Pensacola and Mobile to New Orleans and Baton Rouge to Houston and Galveston
- The Piedmont region from Raleigh-Durham, Greensboro, and Winston-Salem in the north to Charlotte, Spartanburg, Greenville, and Atlanta in the south
- The Northwest region from Portland in the south to Tacoma, Seattle, Bellingham, and Vancouver (British Columbia) in the north
- The line of cities at the front of the Rocky Mountains centering on Denver, from Fort Collins in the north to Colorado Springs and Pueblo in the south
- The cities on the eastern side of the Great Salt Lake, including Ogden, Salt Lake City, and Provo

Internal Structure of Cities

For at least the past 80 years, scholars have attempted to generalize about the internal spatial arrangement of land uses and social groups in American cities. Compared with cities in other parts of the world, such generalizations are easy to make since American cities are characterized by very high levels of segregation—segregation of land uses and building types, of high and low densities, of social classes and ethnic groups, and by building age and era of development. It is therefore possible to label areas as "commercial," "industrial," "low income," or "high density" and to generalize about their location.

In European or Asian cities, it is much more difficult to generalize about the location of economic activities and social groups. For example, in Europe, many buildings contain a mixture of retailing, office, and residential uses, and even wholly residential buildings often have a wide variety of socioeconomic groups living in them. The traditional European burgher house, for example, had storage in the cellar, retail at street level, prosperous residents on the two floors above, and lower-income renters on the top floors. Even today, such mixing persists in many cities, even those that were totally rebuilt after World War II.

In most of Asia, there is no zoning, and small-scale industrial activities commonly operate in residential areas. Even in Japan, a house may contain several people doing piecework for a local industry. In Europe and Japan, neighborhoods have been built and rebuilt over time, and a wide variety of building types from various eras are often mixed together on the same street. In the United States and Canada, such mixing is much rarer and is often viewed as a temporary condition because areas are in transition toward total redevelopment. Perhaps the only exception to this in a large city is Houston, where there are no zoning regulations.

Segregation in American Cities

Why are American cities so segregated? The answer to this question is long and complicated and involves everything from cultural biases against urban life, the cultural diversity of a land of immigrants, and the nature of property markets in a capitalist system unfettered by traditional ways of doing things. The value system in America, as opposed to that in Italy, for example, allows for the destruction of old buildings when they no longer serve a useful economic purpose. Perhaps most of all, however, segregation has resulted from a combination of planning and zoning ideologies and affluence. Throughout most of the 20th century, American planning policies called for segregating people and land uses and, unlike in some other places where such policies also were present, the money was there to continuously shape and reshape urban form in line with these policies.

The basic idea was this: good rational planning will result in efficient, well-organized urban areas with everything in its proper place. Cities were zoned so as to have central business districts, industrial areas, warehousing areas, high-density residential areas, areas of single-family homes, areas of very large single-family homes, commercial strips, and the like. For most of the 20th century, mixed uses were taboo. Living above shops was discouraged, and owners of houses in areas zoned for industry were denied mortgages and home improvement loans. Corner stores and repair businesses were gradually eliminated from residential areas, just as houses were eliminated from commercial strips. People, too, were segregated. Lenders denied money to those who wished to live in ethnically mixed areas, and even government agencies encouraged racial segregation. Suburban governments disallowed apartment buildings, and many areas were zoned exclusively for large, expensive single-family homes. Governments constructed low-income public housing in only those areas with limited political resistance, usually areas that were already poor and disorganized. And so it was that the American city, more so than cities in any other part of the world, became neatly packaged and segregated into areas and districts of similar character and appearance. Descriptive words such as *downtown, inner city, commercial strip,* and *residential suburb* conjured up images of places with uniform character. The messy mixing of traditional cities was not valued or tolerated in the modern cities of America.

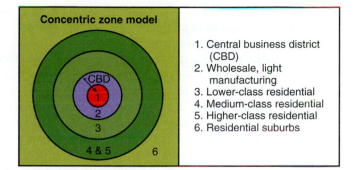

Figure 9.13
The concentric zone model.

Models of City Structure

The previous discussion is necessary for the understanding of both the American city and our models and generalizations about it. We not only have attempted to create "order" over time through various governmental and economic policies, but we also have needed to see order so that we might explain the city by describing the spatial arrangement of its component parts. The very fact that we choose to create land use models of American cities tells as much about us as it does about the cities we describe. Such efforts would seem much less useful to the French or Japanese. It is only in this context, for example, that our concentric zone and sector models make sense. It is important to recognize that the patterns we observe have resulted from particular cultural requirements and specific political and economic policies. It is also important to recognize that for whatever cultural reasons, we need to see such order. We are not comfortable with chaos. With these caveats in mind, let us begin to review some models that have been created to facilitate our search for order in the metropolis.

Concentric Zone Model

The oldest popular model or generalization of the internal structure of the American city is the **concentric zone model** (Figure 9.13). It was developed by sociologists at the University of Chicago during the 1920s to describe and explain the rapidly changing "social ecology" of the city. Over the years, it has been modified and refined by a wide variety of social scientists and has been merged with other models, such as various models of stages of housing decline and racial change. The model postulates a series of concentric rings of decreasing land value and land use intensity around a strong downtown or **central business district.**

At the very heart of the city is a **peak land value intersection (PLVI)** where transit lines and major highways come together to form a place of maximum accessibility and visibility. Here, under the bid-rent model, people will bid the most for a location, and so highly profitable activities, such as tall office buildings and big department stores, dominate (Figure 9.14). Zoning has institutionalized the downtown in most cities because it was only there (at least until recently) that skyscrapers were permitted and the infrastructure to service them was provided.

Just beyond the central business district, the model postulates a **zone in transition** characterized by marginal land uses and prevalent social and economic decay and disorganization. Such areas are popularly known as *skid rows.* Few cities outside of America have skid rows, but they loom large in the images of our cities. Often the problem is that skid row areas are underbuilt compared to what is allowable under downtown zoning. For example, there is little motivation to keep a two-story building in repair when its site allows for the construction of a 30-story structure. Many owners simply "milk" older buildings by using them as pool halls, or fleabag hotels while awaiting an offer from a developer wanting to build a skyscraper. Usually the offer takes decades to arrive, so slum conditions appear. The situation is aggravated by the construction of large towers in other parts of the downtown, since skyscrapers tend to pull activities out of the smaller, older buildings and make them redundant. In many cases, government policies make it difficult to maintain older structures because if significant repairs are carried out or if building use is changed, the buildings must be brought up to all current codes (fire, seismic, electricity, plumbing, and so on), an economic impossibility for the owner of a small building in a marginal area. While some skid rows still exist, in recent years, many have become historic districts ("Olde Townes") or have been cleared for waterfront parks or sports arenas.

The zone in transition is also transitional between light industry, wholesaling, and warehousing and the heavier industry beyond. The ring just beyond skid row, for example, contains a variety of city-serving, light industrial activities (for example, office equipment repair, janitorial services, and printing and publishing), as well as some labor-intensive industries that need high levels of communication and can use urban loft space in older buildings. Although it is gradually decentralizing, the garment industry is the classic example of such an activity. In some cities, large food markets added a lively dimension to the district.

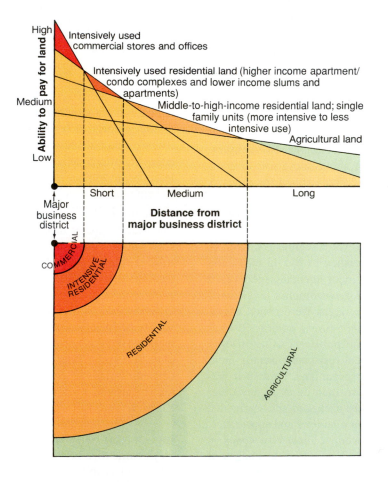

Figure 9.14

Generalized urban land use patterns. The model depicts the location of various land uses in an idealized city where the highest bidder gets the most accessible land.

Heavy industry and transportation facilities, such as ports and railroad terminals, tend to occupy the next concentric ring of land uses. These space-extensive activities cannot pay as much for land as office and retail activities but, at one time anyway, they were profitable enough to bid for a reasonably central location. In recent years, many of these industrial zones have been abandoned as older heavy industry has died or moved out of the city. Vast tracts of empty, decaying industrial buildings surround the cores of the typical older American city. In most other parts of the world, including much of Europe as well as in developing countries, heavy industry was initially located farther away from the central core since industry arrived after the cities were well established. Many American cities are now faced with the irony that hundreds of acres of theoretically valuable central city land are virtually empty. There has been talk of enterprise zones, subsidized housing, and other new uses, but many areas are contaminated by toxic wastes as well as evil reputations. It is sometimes easier to just ignore them and move on.

The residential city begins only in the next ring with the peculiarly American assumption that no one, other than perhaps a few poor denizens of skid row, lives in the center of the city. Indeed, graphs of declining population density away from the American city center typically show a low point at the PLVI with the trend line rising to a maximum height at a **density rim** just beyond the central industrial zone (Figure 9.15).

In the classic concentric zone model, the inner ring of housing is predominantly high-density and low-income. It is the classic and stereotypical "inner city." The houses and apartment buildings tend to be older and poorly maintained, and many suffer from overcrowding as too many people have occupied the area to be close to employment in the central city while minimizing transportation costs. This is also the zone of entry for new immigrants to the city, whether they be from foreign countries or rural parts of the United States. Ethnic diversity abounds as new arrivals make ends meet by sharing flats while learning the ropes and seeking meaningful employment. Initially, the model was based on ecological notions of invasion and succession, whereby the poorest, most recent immigrants would cram into the spaces left behind by those gradually moving up the ladder to newer housing. Very often, however, ethnic districts were created by overt discrimination such that blacks, Chinese, or Irish were not allowed to live any place else. Overcrowding resulted from the inability to search for housing in "white" areas, as well as from poverty.

In recent years, the inner city has changed drastically in most American cities. In some cases, it has thinned out as no new immigrants have arrived to replace those moving up and out. Populations have plummeted to a fraction of former times, leaving only a few crack houses among the empty fields. Here, as in the industrial wastelands, the bid-rent model seems to fail since no one is bidding for the space. In other cases, inner-city districts have been discovered and gentrified by downtown office workers seeking a combination of a convenient location and potentially attractive historic architecture (see "Gentrification"). This is particularly true in cities with attractive physical sites where the older residential areas occupy hills with views of an emerging skyline. The recent acceptance of high-rise condominium and apartment living in American cities has facilitated the recapturing of many areas by the upper and middle classes. In still other cases, the immigrants are still coming, and the once Irish or German neighborhoods are now thriving Vietnamese or Mexican communities. It is becoming increasingly difficult to generalize about the typical American inner-city neighborhood.

In the remaining residential rings (the specific number varies with the purpose of the model), population density decreases and social status increases as we travel from the inner rings outward toward suburbia. In the classic model, the rich could live farther out on cheaper land because they

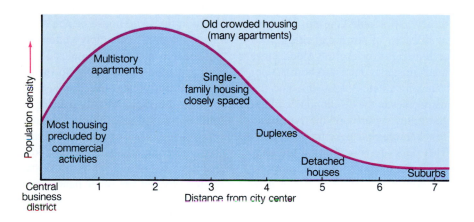

Figure **9.15**
The density rim. As distance from the area of multistory apartment buildings increases, the population density declines.

Cities

GENTRIFICATION

As local, state, and federal authorities attempt to find ways to revitalize the ailing older American cities, a potentially significant process is occurring: **gentrification,** or the movement of middle-class people to deteriorated portions of the inner city. This movement does not yet counterbalance the exodus to the suburbs that has taken place since the end of World War II, but it marks an interesting reversal to what had seemed to be the inevitable decline of the cities. According to one estimate, 70% of all sizable American cities are experiencing a significant renewal of deteriorated areas. Gentrification is especially noticeable in the major cities of the North and the East, from Boston down the Atlantic Coast to Charleston, South Carolina, and Savannah, Georgia.

During the early years of suburbanization, there was also a major movement of low-income minorities into the central cities. The migration stream has now slowed to a trickle, and attention is centered on the movement of young, affluent, taxpaying professionals into the neighborhoods close to the city center. These formerly depressed areas are being rehabilitated and made attractive by the new residents. The prices of houses and apartments in former slums have soared. New, upscale restaurants and specialty shops open daily.

The reasons for the upsurge of interest in urban housing reflect to some extent the recent changes in American family structure and the employment structure of central business districts. Just 30 years ago, the suburbs, with their green spaces, were a powerful attraction for young married couples who considered single-family houses with ample yards ideal places to raise children. Now that the proportion of single people (whether never or formerly married) and childless couples in the American population has increased, the attraction of nearby jobs and social and recreational activities close to the city center has become an important residential location factor. Many of the new jobs in the central business district are designed for professionals in the fields of banking, insurance, and financial services. These jobs are replacing the manufacturing jobs of an earlier period.

One negative feature of the gentrification phenomenon has been the displacement of local residents. As rents or housing prices have risen, low-income residents have been forced out. Overall, however, gentrification has been a positive force in the renewal of some of the depressed housing areas in neighborhoods surrounding the central business district.

---

could afford higher transportation costs. Thus, large houses on sprawling lots were possible in the outer rings. In addition, many areas were zoned so that only expensive houses on large lots were allowed. In the classic model, the typical (white) American would gradually work his or her way up and out from the inner rings to the outer ones à la Horatio Alger. This has led to a reliance on housing that has "filtered down" from middle class to working class to poor over and above the construction of new, affordable housing. This policy exaggerates the zonal character of American cities, since the poor have no choice but to move into the older, inner rings of the city. This stands in stark contrast to both the peripheral working-class housing estates of Europe and the squatter settlements of many cities in the developing world.

In recent years, the outer suburbs of the American city have become far less remote and peripheral as beltways and suburban office parks have encircled the city. Protection from nonconforming land uses must now be more a matter of political power and less a matter of geographic isolation. In many cities, suburban residential space has all but been squeezed out by shopping malls, industrial and office parks, airports, freeways, and new versions of the commercial strip.

Although the concentric zone model was postulated in the 1920s and must be continuously revised to be at all helpful, some aspects of the American city can still best be described through the use of concentric zones. Family status and life cycle, for example, still vary noticeably between the inner and outer city. The inner rings tend to be occupied by a high percentage of older people (retired "empty nesters," single widows and widowers). Many have lived in the same house or neighborhood for much of their adult life and have simply gotten older on site. Others have moved into apartment complexes built specifically for elderly populations, often close to medical and social services. Government policies have sometimes played a role here as high- and mid-rise elderly housing projects have often been built in renewal programs in and around downtown.

The inner rings are also characterized by large numbers of young adults living around universities or in apartments close to downtown and the bar scene. The outer rings of the city, on the other hand, tend to have more middle-aged families with children since a single-family house on a quiet street in a good school district is an American requirement for proper family life. Like all generalizations, the one about family status is constantly changing and in need of revision. As young families move into older, gentrified central-city neighborhoods and as massive "singles" apartment complexes are built around suburban malls, the old models must be redefined.

An important modification of the concentric zone model that takes into account the rise of beltway shopping malls and office parks involves the inclusion of a bid-rent curve that slopes downward from the central business district until it reaches the outer ring of the city and then climbs upward at a steep angle. Here, land is far more expensive than anywhere else except the downtown core, having been bid up by developers of suburban office parks and shopping malls. Large suburban nodes may generate their own sets of concentric zones as discount stores, automobile showrooms, and suburban apartment complexes compete for space close to the mall.

With constant revision and modification, the concentric zone model continues to provide a useful framework for generalizing about the spatial organization of American cities. It need not stand alone, however, since other models have been added to the stable over the years. Chief among them is the sector model.

Sector Model

The concentric zone model best describes older, compact cities with limited transportation options and short commuting distances. Indeed, the model requires that transportation be equally available in all directions away from the central business district—a situation that better describes the city of carriages and omnibuses than the more recent city of commuter railroads and freeways. Similarly, the model requires that physical amenities and natural corridors and obstacles, such as rivers and mountains, play no role in shaping the city. As early as 1939, when the **sector model** was put forward, it was realized that neither of these assumptions was always met (Figure 9.16). While the concentric zone model could be modified to some degree by allowing bulges where transportation was best and indentations where obstacles were insurmountable, the modification options were limited.

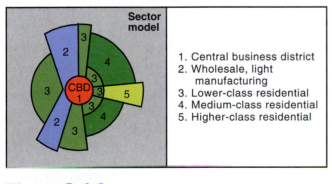

Figure 9.16
The sector model.

The sector model was developed to fully incorporate the fact that major highways, rail lines, beachfronts, commercial strips, and other linear features greatly affect urban form. It also incorporated the idea that people tend to move both inward and outward along specific corridors, such as freeways, thus leading to sectors of ethnicity and class. For example, once the "good side of town" has been established, it tends to perpetuate itself as high-class developments are built nearby but farther out. The same is true of industrial districts and the like. The sector model suggests that the city is best depicted by a number of unlike sectors emanating from the central core. They need not be the same length or width. Some sectors describe land uses, such as industrial zones or commercial strips, while others describe levels of affluence and ethnicity. Once again, Chicago provided the initial empirical example with such sectors as the Gold Coast affluent sector northward along the lakefront, as well as the Black Belt extending southward toward the industries of Calumet City.

The sector model has been fleshed out over the years with studies of how residents perceive the city. Most people, for example, are familiar with the downtown and the suburban parts of their own sector. Shopping and social contacts tend to take place in the sector of familiarity, and most moves are within the sector to allow the maintenance of social linkages. Rarely do people move through the downtown to an across-town location, since a variety of apartments, older housing, and newer housing should exist in their own sector.

The sector model is perhaps the best generalization for describing ethnic and racial patterns, especially in the post–civil rights era. At one time, minorities could be kept in neatly bounded inner-city areas through the use of blatantly discriminatory techniques. With the expansion of the minority middle class and the enforcement of open housing laws in recent decades, however, largely minority sectors have appeared that extend well into peripheral suburban locations. Blacks, Hispanics, and other minorities can thus obtain the suburban American dream while remaining close to friends and attractions within a familiar sector. Often densities are quite low at the suburban end of the sector compared to similar locations in predominantly white sectors since there may be fewer high-density developments, such as office parks and related apartment complexes.

Most recently, the sector model has allowed for increasing complications beyond the idea of relatively homogeneous sectors of land use and social class. For example, many cities have become lopsided as one or more sides of town expand rapidly while the other sectors remain stable. Often, the rapidly growing sectors are much more socially diverse than the stable ones, since movement outward creates excess vacancies for marginal populations and businesses to fill. Thus rich, mobile people and poor, ethnic minorities may gradually occupy the same rapidly growing sector, while working-class sectors cease to grow or

Cities

Many geography majors end up working in city planning after they graduate. What is city planning? What kinds of jobs do planners perform?

In the broadest sense, city planning is the profession that protects the quality of the urban environment. People who work in this field may be concerned with estimating the noise impacts of proposed new airports or sports stadiums, studying alternative routes for new roads to minimize their impacts on forest or other valued natural areas, or helping merchants design an attractive architectural theme for revitalizing a downtown. Others may work with builders to make sure that homes and businesses meet community requirements, such as proper building setbacks from streets and the required number of parking lots per business.

Many states require cities to make *comprehensive plans* or *general plans.* These are documents, with texts and maps, that outline the goals and objectives of a city, especially for future urban development. They deal with a wide range of topics, such as desired residential densities, the general locations for broad categories of land uses, goals for the amount of park land, and standards for traffic flow. Planners gather information for these plans, conduct public meetings to get community input, do technical analyses, and make draft plans for consideration (and eventually for adoption) by elected decision makers. People who deal with such topics are often called *long-range planners.* They may also work on other tasks, such as plans to (1) incorporate more affordable housing into the community, (2) improve the local bus system, (3) increase the area in open space devoted to wildlife protection, or (4) make new regulations about the location and design of billboards or signs for businesses.

Once a comprehensive plan is adopted, ways must be found to reach the goals identified. The implementation part of planning is often called *current planning* or *zoning administration.* Implementation usually occurs via *zoning,* where certain areas are designated for particular uses (such as single-family homes or commercial development) and incompatible uses (such as factories in residential areas) are not allowed. Zoning also establishes standards, such as setbacks and height limits for buildings, for development. All states allow zoning.

As can be imagined, citizens who find that zoning does not allow them to develop their lands as they wish often object vehemently to such restrictions. However, this is a case where one person's freedom to develop his or her property as they please can infringe on the quality of life of others, and the courts have generally upheld the rights of cities to adopt land use regulations through zoning.

Clearly, city planners do a wide variety of jobs. People who want to work in urban planning probably would do best if (1) they have a good general education, as well as (2) training in a specialty area (such as environmental geography or design), (3) work well with people, and (4) have a whole lot of common sense.

change. In growing sectors, there is also rapid outward movement of businesses as owners seek locations in prosperous suburban malls well away from the inner city, while older commercial strips and corner stores may remain in the stable, working-class sector.

Multiple Nuclei Model

The sector and concentric zone models are both based on the premise of gradual, piecemeal growth around a consensus downtown core. In such a scenario, the existing city with its transportation lines and established neighborhoods suggests the kinds of additions that are appropriate in each ring or sector. Over the past few decades, there have been several important changes in the way cities grow and change. Foremost among them is the scale of new development. No longer do builders buy a few lots and construct a handful of houses. Today, developers may put up as many as 20,000 homes in one relatively self-contained community. Office and industrial parks with as much space as downtown can be built all at once in suburban locations near new regional airports, artificial lakes, and retirement cities. Such developments are often completely independent of existing ring and sector considerations. They are large enough to create their own realities.

The trend toward massive, unrelated development was recognized as early as the 1940s, when the **multiple nuclei model** was put forward (see Figure 9.17). This model suggests that large complexes, such as airports or freeway interchanges, attract and repel land uses independently of their relationship to downtown. For example, an airport might attract hotels, warehousing, and certain types of office space, but it might repel shopping centers, luxury housing, and mental hospitals. Cities with physically diverse sites are especially well described by the multiple nuclei model because view lots on mountains, waterfront lots around lakes, and other physical features may play a far more important role in shaping the social geography of the city than location relative to downtown and transportation lines.

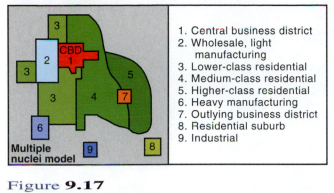

Figure 9.17
The multiple nuclei model.

Legend for figure:
1. Central business district
2. Wholesale, light manufacturing
3. Lower-class residential
4. Medium-class residential
5. Higher-class residential
6. Heavy manufacturing
7. Outlying business district
8. Residential suburb
9. Industrial

The new metropolitan areas of the South and West are perhaps best described by the multiple nuclei model. They are more likely to locate close to mountain, desert, or waterfront amenities, and they are more likely to have been built in large, independent, automobile-related sections. In metropolitan areas such as Orange County, California, little may be recognizable as traditional American city structure complete with downtown and suburbs. The recent phenomena of *edge cities* epitomizes the trend toward the multiple nuclei metropolitan areas. Edge cities are massive complexes usually built by a single developer either at an existing transportation node within an urban area or in a remote location on the edge of one. Some have suggested that these are the cities of tomorrow and that older central cities will eventually fade away. Others have maintained that they are aesthetic and environmental disasters that will never achieve true urbanity. There are probably many examples of each type.

Combination Models

Most attempts to generalize about American cities today involve the use of a combination of rings, sectors, and multiple nuclei. Some cities, for example, may still be dominated by earlier rings of development with only a few major sectors or competing nuclei. Others may be characterized by uneven sectoral growth with only a few remaining concentric zonal attributes. Still others, usually newer cities, may appear to be little more than a random collection of multiple nuclei. In most cases, however, particular social, demographic, and land use dimensions can be adequately described through the use of one or more of the classic spatial models.

Geographic Variations in the Application of Models

Models were developed so that students of the city could recognize the similarities and predictable regularities among cities to better understand the processes shaping them. A generalized model allows confusing idiosyncrasies to be pushed aside to facilitate the study of constant under-

lying forces. We have accomplished this. We know a great deal about the essential economic, political, and social forces that have shaped the city, and we have been able to study them in general, as though they operate everywhere in a similar fashion. Perhaps it is time to recognize that the models we have discussed not only need to be modified over time and brought up to date periodically, but also modified over space so as to better fit particular regional and subregional characteristics. Not all American cities are the same. It will be some time before Boston is confused with Phoenix.

An infinite number of variables serve to make cities unique places—architectural traditions, industrial heritage, transportation facilities, population characteristics, eating and drinking habits, climate, and the like. We will discuss only a few of them here, but we hope that this will be sufficient to make the point that models can be adjusted to apply differentially to cities of different types. Five variables are emphasized here: era of formation and development, physical site characteristics, past and present economic function, racial and ethnic distributions, and political policies. These five variables can be used alone or in combination to create cities of differing types. For example, we could argue that an older city with a constricted physical site that functions as a major financial center with a strong tradition of planning will differ from a young, sprawling city that functions as an unplanned tourist center in regular and predictable ways. Let us examine the variables in sequence.

Era of Formation and Development

Even though the United States and Canada are sometimes known as part of the New World, their cities vary tremendously in age. Boston and Québec were thriving places before 1700, while Miami and Phoenix were barely functioning in the early decades of the 20th century. To simplify matters, we can identify just three types of city eras: pre-1850, 1850–1950, and post-1950.

Cities growing up during the first era were built either with narrow, circuitous streets and tiny building lots (Boston is an example) or in a grid pattern (Philadelphia). Before the transportation revolutions of the mid-19th century, they were **walking cities** in which everything was located within a mile or two of everything else. Consequently, they were built up in a very dense fashion with little wasted space. Although there were specialized districts, such as financial centers and waterfronts, by the early 1800s, they were small, compact, and close together. It was an easy walk from downtown to a fashionable residential suburb. Since space was valuable, buildings tended to be solid, well-constructed, and relatively large (at least three or four stories). In addition, architectural aesthetics and consistency were important, with classical (Georgian) detailing and harmonious design predominating. As in Europe, heavy industry and railroad facilities came only after these cities were well-established, and so the "industrial ring" seems relatively remote from the city center.

Figure **9.18**
Boston: New skyscrapers and historic neighborhoods around Boston Common.

To generalize, we can assign those cities that were well-established urban places before roughly 1850 into this first category. These would include Boston, New York, Philadelphia, Québec, Montréal, Savannah, Cincinnati, and Charleston, South Carolina, among a great many others large and small. These cities would have relatively large central (downtown) zones and a tradition of high-density, central-city living (Figure 9.18). More recent sectoral and nuclei developments would tend to begin farther out, away from the historic and often protected central neighborhoods. Geometric aspects of the spatial models can be modified accordingly.

The cities that emerged after 1850 tended to have a very different morphological pattern because they often grew up around heavy industry and transportation facilities. Instead of a traditional mixed-use central area, these cities had purpose-built central business districts with little or no housing. Railroad tracks, steel mills, oil refineries, port facilities, and a variety of warehouses closely abutted the specialized downtowns. Industry itself was an unattractive feature, and so elites quickly moved to remote suburbs as they expanded along trolley lines, leaving the inner rings for working-class tenements. The classic models were developed for Chicago in the 1920s and 1930s. Cleveland, Detroit, Buffalo, Milwaukee, Akron, and Kansas City would join Chicago as the cities that best fit these models (Figure 9.19).

Finally, there are the newer cities that have grown up since World War II. Many of them have never developed a strong central core but rather have relied on a scattered, multiple nuclei pattern of activity from the beginning. Such cities may have a few downtown skyscrapers, but they are often surrounded by parking lots and remnant single-family

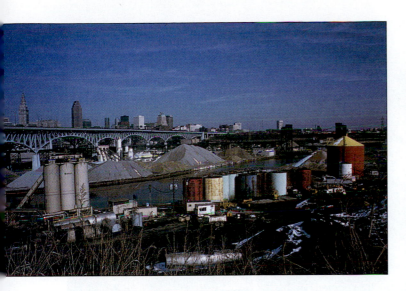

Figure 9.19
Central Cleveland's skyline emerging from heavy industry.

Figure 9.20
San Diego, with the San Diego Yacht Club in the foreground.

houses rather than more traditional urban landscapes. Since the central city never had a high-density residential population living in tenements and apartments, it often has little or no density gradient. Large apartment complexes are more likely to be located around suburban malls than in inner-city neighborhoods. In addition, these cities grew up around the automobile and major highways rather than water and rail connections, and so the central rings are likely to be devoid of the kinds of heavy industry that developed around older ports and rail yards. Phoenix, Dallas, Miami, San Diego, and Albuquerque illustrate this type of city (Figure 9.20).

Physical Site Characteristics

To avoid the bugaboo of environmental determinism, most geographers have tended to downplay the role of physical site in the creation of models of city structure. "People can do whatever they want" replaced "People had to do this because of the river" as the basic assumption of city form. Indeed, during the late 19th and early 20th centuries, many rivers were made to change course and many hills were sluiced into submission as people built the kinds of cities they wanted. Nevertheless, it seems unwise at this point to eliminate completely the role of the physical environment as an influence on urban form.

At the simplest level, we can at least divide cities into those with complex and variegated physical sites that tend to influence the basic shapes put forward in our models and those that occupy uncomplicated sites where the human forces shaping our cities are free to reign unfettered by the occasional canyon or mountain range. Would Nob Hill in San Francisco have remained an elite area if it had been possible for horses to pull produce wagons and other dé-

Figure 9.21
San Francisco has many close-in luxury apartment districts, partly because topography affords both scenic views and protection from through traffic.

classé materials up its steep streets (Figure 9.21)? Would it have become a part of the adjacent skid row? Would Beverly Hills be what it is if it had the flat topography of Watts? Would the Gold Coast in Chicago exist without the coast? In many cities, such as San Francisco, San Diego,

Figure 9.22
A bird's-eye view of Toronto showing clusters of tall buildings in the central business district and at important intersections.

Seattle, and Vancouver, the physical sites add a tremendous complexity to what might otherwise be generalized patterns. In many other cities, of course, there are no such complexities. Perhaps our models can be modified to differentiate such places.

Past and Present Economic Function

American and Canadian cities vary tremendously in basic employment, and these variations rather obviously show up in their internal structure and morphology. Cities that are predominantly office centers, such as those specializing in banking, insurance, and government, are likely to have a strong downtown, as well as a number of competing office nodes. Cities such as New York, Boston, Chicago, San Francisco, Toronto, and Washington, D.C., have downtowns bristling with office towers, and downtown employment levels tend to be high and rising (although a decreasing percentage of total metropolitan employment) (Figure 9.22). Usually, there are many spin-offs

from this employment, including convention centers, hotels and restaurants, downtown condominium housing, and the like. In these cities, downtown as a concept and as a place is here to stay. In several, office development has facilitated the rise of new subdowntowns, such as Bellevue in Seattle or Century City in Los Angeles.

In cities that have relied more on heavy industry, transportation, mining, and other economic activities, the trends may be quite different. Detroit, Buffalo, Memphis, and Birmingham have much weaker demand for downtown office space and related developments. Although such contrasts may diminish over long periods of time, the economic functions of cities do not change overnight. While San Francisco has placed a cap on the amount of office space that can be built in it in any year, many other cities are searching for developers who will build downtown (Figure 9.23).

Industrial cities, rather obviously, tend to have large industrial rings and sectors, as well as related zones of

Figure 9.23
Vacant lots in a formerly commercial area of St. Louis.

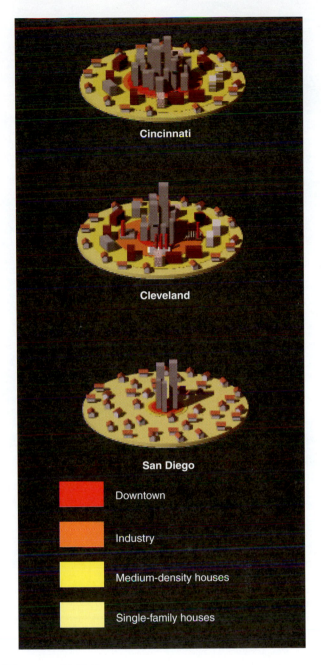

Figure 9.24
The concentric zone model is altered to reflect the most important land use zone. Cincinnati is a major regional center with an extensive business zone, Cleveland is a major industrial city, and San Diego is a military and tourist city.

working-class housing (Figure 9.24). Tourist cities tend to have beaches, marinas, forests, ski runs, and other features not mentioned in the classic models of city structure. The demand for housing, shopping, hotels, and other facilities is usually related to the present and historic distribution of rings and sectors of amenities and unattractive features. Our models can be modified to apply specifically to certain types of cities classified by basic employment.

Racial and Ethnic Distributions

The United States and Canada have always been lands of immigrants, and cities have always taken more than their fair share of this immigration. In recent years, however, divergence among various cities has been increasing as some continue to take massive numbers of immigrants, while others receive few or none. In addition, some cities are getting only immigrants from one general source area (such as Latin America), while others are accepting immigrants from all over the world. Such differences obviously affect the social and internal geographies of cities.

Cities may be classified in a variety of ways with regard to ethnic composition. For example, we can differentiate between Northwestern European stock white (Salt Lake City), Northwestern European stock white and ethnic white (Boston), Northwestern European stock white and black (Birmingham), Northwestern European stock white and Hispanic (San Antonio), diverse white and Asian (San

Francisco), and total diversity (Los Angeles). In cities that are basically Northwestern European stock white, the social geography may include only subtle variations from zone to zone and sector to sector. In white and black cities, there may be enormous contrasts between vast and obviously different parts of the city. In cities with a diverse set of new immigrants, the social geography may be extremely complex with many different social worlds competing for space

253

Figure 9.25
Abandoned houses in central Detroit.

in small areas. It would be very difficult for one model to describe both Birmingham and San Francisco.

Because of past patterns of discrimination, cities that have long had giant ghettos or barrios tend to be far more decrepit than those that have not. Seattle, Portland, and San Diego have nothing nearly comparable to central Detroit or South Chicago, even though they contain distinctive black communities (Figure 9.25).

Political Policies and Patterns

American and Canadian cities have been shaped by political decisions, as well as by economic and social forces. For example, while some cities instituted strict height limits during the first half of the 20th century, others encouraged the construction of monumental skyscrapers. While some

cities discouraged the development of polluting industries in central areas, others adopted an "anything goes" attitude. Some cities built thousands of units of low-income housing in central areas, thus institutionalizing a low-income ghetto, while others built none. A few cities constructed dense networks of mass transportation, while others relied on broad boulevards and seas of parking lots. A variety of political policies has served to differentiate cities over the years.

Cities that serve as political capitals often have sacred centers protected to at least some degree from economic demands. Central Washington, D.C., with its focus on a vast, grassy mall rather than a set of office buildings, is a case in point. Many state capitals, such as Columbus, Ohio, have at their PLVIs a state house surrounded by acres of lawn, as well as 40-story skyscrapers. In recent years, government redevelopment projects have served to create immense differences between cities. The formerly industrial waterfront of Baltimore, for example, has been transformed into a recreational playground of cafes and tourist attractions as a result of urban renewal (Figure 9.26). The inner ring has certainly changed and no longer conforms to much that is in the classic models.

Perhaps the primary way that cities vary systematically as a result of political policies is in **political fragmentation.** Any model attempting to describe the spatial organization of the city would do well to differentiate between those that have either some form of metropolitan government or have been able to annex territory easily (such as Toronto, Houston, Phoenix, Oklahoma City, and San Diego) and those that are surrounded by politically independent suburbs. In the latter cases, the zonal character of the city is often enhanced by the fact that the suburbs have practiced exclusionary zoning and disallowed low-income housing of any kind. These suburbs also usually have separate school districts and have thus become refuges for those fleeing the diverse city systems. In politically fragmented urban areas, a distinct line often separates the social world of the city and those of the relatively idyllic suburbs. The central city becomes the inner city. In metropolitan areas in which the central city has annexed widely, the contrast between the inner and outer zones is likely to be less pronounced.

Summary

Cities that grow beyond their village origins take on functions that tie them to the countryside and to a larger system of cities. Site and situation tell us much about the early origins and functional specialization of American and Canadian cities. The economic base, composed of both basic and nonbasic activities, is a key to understanding how cities grow and decline. Cities form into hierarchical systems in which there is a great deal of interaction and where some cities contain other cities in their influence zones. A large conurbation called Megalopolis has developed on the East Coast. There are, however, a number of other clusters in many regions throughout the United States and southern Canada.

Geographers have created a number of models that help us to understand the incredible diversity of land uses and social groups within the city. The concentric zone model, the sector model, and the multiple nuclei model, as well as combinations of these, enable us to analyze the spatial structures of cities and to better understand the processes that shape them. It is wise to realize, however, that American and Canadian cities still differ from each other in various systematic ways. Older cities vary from newer ones, ethnically diverse cities vary from homogeneous ones, and politically fragmented cities are likely to take a slightly different shape than those in which one government entity prevails. Our models not only describe cities in general in order

Figure **9.26**
Baltimore's restored waterfront, Harbor Place.

to highlight process, but they also demonstrate that processes vary systematically over space and time.

Key Words

basic sector	peak land value intersection (PLVI)
central business district	political fragmentation
central city	sector model
city	service or nonbasic sector
concentric zone model	site
conurbation	situation
density rim	suburb
economic base	town
gentrification	urban hierarchy
hinterland	urban influence zones
Megalopolis	urbanized area
metropolitan area	walking cities
migration field	zone in transition
multiple nuclei model	

Gaining Insights

1. Consider the city or town in which you live or attend college or with which you are most familiar. In a paragraph, discuss that community's site and situation. Point out the connection, if any, between its site and situation and the basic functions that it earlier performed or now performs.

2. Again, focusing on the community named above, where is it in the urban hierarchy? Which cities are above or below it? Draw a sketch map of the community's influence zone. What cities and towns are in its influence zone?

3. Given your community's predominant basic activities, what are the possibilities for urban growth and expansion in the next ten years?

Cities

4. Why do we create models or generalizations to describe land use in cities? Does this help us to understand how cities evolve?

5. What are the three basic models of American city structure? What are the assumptions and expectations that each of these models are based upon?

6. Do you think that models developed at one point in time (say, the 1920s) can be used at another (say, in the year 2000 or later)?

7. As some cities prosper and others decline, do you believe that American and Canadian cities will remain similar or change in form and structure? Explain your answer.

8. How do physical site and political policies affect land use in the inner and outer rings of the city?

Selected References

Association of American Geographers Comparative Metropolitan Analysis Project. *A Comparative Atlas of America's Great Cities: Twenty Metropolitan Regions.* Vol. 3. Minneapolis: University of Minnesota Press, 1976.

Bernich, Michael and Robert Cervero. *Transit Villages in the 21st Century.* New York: McGraw-Hill, 1997.

Bourne, Larry, ed. *Internal Structure of the City: Readings on Urban Form, Growth and Policy.* New York: Oxford University Press, 1982.

Christian, Charles M., and Robert A. Harper, eds. *Modern Metropolitan Systems.* Columbus, Ohio: Charles E. Merrill, 1982.

Davis, Mike. *City of Quartz: Excavating the Future in Los Angeles.* New York: Verso, 1990.

Fishman, Robert. *Bourgeois Utopias: The Rise and Fall of Suburbia.* New York: Basic Books, 1987.

Ford, Larry R. *Cities and Buildings: Skyscrapers, Skid Rows, and Suburbs.* Baltimore: The Johns Hopkins University Press, 1994.

Garreau, Joel. *Edge City: Life on the New Frontier.* New York: Doubleday, 1991.

Goldberg, Michael A., and John Mercer. *The Myth of the North American City: Continentalism Challenged.* Vancouver: University of British Columbia Press, 1986.

Hall, Peter. *Cities of Tomorrow.* Cambridge, MA: Basil Blackwell, Inc., 1990.

Hannigan, John. *Fantasy City.* New York: Routledge, 1998.

Hart, John Fraser, ed. *Our Changing Cities.* Baltimore: The Johns Hopkins University Press, 1991.

Hartshorn, Truman A. *Interpreting the City: An Urban Geography,* 2nd rev. ed. New York: John Wiley & Sons, 1990.

Jakle, John and David Wilson. *Derelict Landscapes: The Wasting of America's Built Environment.* Savage, Md.: Rowman & Littlefield, 1992.

Johnston, R. J. *The American Urban System: A Geographical Perspective.* New York: St. Martin's Press, 1982.

Kariel, Herbert G. "Canadian Urban Hierarchies," *Canadian Geographer* 28, 4(1984): 383–90.

Langdon, Philip. *A Better Place to Live: Reshaping the American Suburb.* Amherst, MA: University of Massachusetts Press, 1994.

Liebs, Chester. *Main Street to Miracle Mile: American Roadside Architecture.* Boston: Little, Brown, 1985.

Mayer, Harold, and Richard Wade. *Chicago: Growth of a Metropolis.* Chicago: University of Chicago Press, 1969.

Osofsky, Gilbert. *Harlem: The Making of a Ghetto.* New York: Harper & Row, 1966.

Palm, Risa. *The Geography of American Cities.* New York: Oxford University Press, 1981.

Pred, Allan. *City-Systems in Advanced Economies.* New York: John Wiley & Sons, 1977.

Relph, E. C. *The Modern Urban Landscape.* Baltimore: The Johns Hopkins University Press, 1987.

Scott, Allen J. *Metropolis: From the Division of Labor to Urban Form.* Berkeley: University of California Press, 1988.

Sharp, William, and Leonard Wollock, eds. *Visions of the Modern City: Essays in History, Art, and Literature.* Baltimore: The Johns Hopkins University Press, 1987.

Smith, Neil, and Peter Williams, eds. *Gentrification of the City.* Winchester, Mass.: Allen and Unwin, 1986.

Teaford, Jon. *The Rough Road to Renaissance: Urban Revitalization in America, 1940–1985.* Baltimore: The Johns Hopkins University Press, 1990.

Ward, David. *Cities and Immigrants: A Geography of Change in Nineteenth Century America.* New York: Oxford University Press, 1971.

Wright, Gwendolyn. *Building the Dream: A Social History of Housing in America.* New York: Pantheon Books, 1981.

Yeates, Maurice. *The North American City,* 4th ed. New York: Harper & Row, 1990.

CHAPTER

10

Neighborhoods

Sometime early in the 21st century, Los Angeles is expected to bypass New York to become the United States' most populous metropolitan area. In 2000, greater Los Angeles had about 16 million people, compared to perhaps 20 million in and around New York. The former, however, has been growing at roughly three times the rate of New York. This growth is fueled, especially, by an enormous influx of immigrants.

Several years ago, one of the country's leading weekly news magazines labeled Los Angeles America's "new melting pot." In recent decades, the metropolitan area (which consists of five counties) has become by far the single greatest destination for immigrants, a title it surely holds not just for the country but for the world. Most immigrants today are coming from Latin America and Asia, but the list of countries furnishing immigrants is large. One result, and one that can make a teacher's job very difficult, is that Los Angeles school districts today have native speakers of about 80 distinct foreign languages.

From some individual countries, the flow of immigrants has been enormous. It has often been noted that if just the Mexican-Americans were counted, Los Angeles would be the world's second largest Mexican city (after only Mexico City). Similarly, the metropolis contains the second largest agglomeration of Koreans anywhere. But these facts are just a hint of the region's great diversity.

Many recent immigrants have segregated themselves voluntarily into ethnic communities, or neighborhoods. The result includes such well-known ethnic enclaves as the East Los Angeles Barrio (Hispanics), Koreatown, Little Tokyo, and Little Saigon. Yet these are only the more famous areas. There are dozens of other enclaves of immigrants from Armenia, Russia, Cambodia, Iran, the Philippines, Arab-speaking countries, China, Central America, and many other places. Adding to the complexity, there are other, often older, distinctive areas such as black ghettos, places with concentrations of gays, and neighborhoods known for wealth. In fact the metro area is a mosaic of ethnic and socioeconomic neighborhoods that would take a person a whole lifetime to study in detail. Greater Los Angeles is now America's, and probably the world's, most diverse large metropolis.

Within less than a mile in Los Angeles, or any other metropolitan area of the country, neighborhoods of greatly different character are evident. Cities of North America are mosaics; neighborhoods of remarkably different character exist side-by-side. Rarely is a neighborhood completely homogeneous. A Korean neighborhood is not all Korean (more Hispanics live in Koreatown in Los Angeles than Koreans); not all residents of an elite neighborhood are wealthy. Enough residents will be alike, however, that they can typify the district.

In this chapter, we describe the differences that exist from small area to small area. In particular, we emphasize the location of neighborhoods in urban, suburban, and rural areas. In many respects, the material of this chapter is an extension of the discussion of cities in the last chapter. The location of many of the neighborhoods fits nicely into the various rings and sectors discussed in Chapter 9.

Criteria for Delimiting Regions

Regions are difficult to define and delimit, yet we must have them to make sense out of complexity. In this chapter, we divide America into regions. Our purpose is to consider the very small regions, perhaps only of neighborhood size, that help us to see the pattern of settlement from a social perspective. There are sufficient differences in the cultural development of Canada (discussed in detail in Chapter 14) to require that we append a section on Canadian neighborhoods to the discussion of American regions.

We identify groups of people living in an area who can be considered as homogenous in terms of the kinds of jobs they hold, their incomes, and their demographic characteristics. By so doing, we find repetitive patterns of settlement from one metropolitan area to another. For example, we note that the rolling hills or lakeshores close to large cities contain the homes of a group of people who can be described as the **elite,** that is, the educated, affluent executives, the country club set, whose head of household is usually male, married, middle-aged or older, and whose children may be tots or young marrieds (Figure 10.1). In terms of our discussion in Chapter 9 of models of city structure, this group lives in the outer ring of the concentric zone model.

Social Areas of Cities

The more complex cities are economically and socially, the stronger is the tendency for their residents to segregate themselves based on *social status, family status,* and *ethnicity.* In a large metropolitan region, this territorial behavior may be a defense against the unknown or the unwanted, a desire to be among similar kinds of people, a response to income constraints, or a result of social and institutional barriers. Most people feel more secure when they are near those with whom they can easily identify. In traditional societies, these groups are the families and tribes. In modern society, people group according to income or occupation (social status), stage in the life cycle (family status), and language or race (ethnic characteristics). Many of these groupings are fostered by the size and value of the available housing. Land developers, especially in cities, produce homes of similar quality in specific areas. Of course, as time elapses, the quality of houses changes, and new groups may replace old groups. In any case, neighborhoods of similar social characteristics evolve.

Social Status

The social status of an individual or a family is determined by income, education, occupation, and home value. In the United States, high income, a college education, a

Figure 10.1
A house in a typical elite community outside of Cincinnati, Ohio.

professional or managerial position, and high home value constitute high status. High home value can mean an expensive rented apartment as well as a large house with extensive grounds.

A good housing indicator of social status is persons per room. A low number of persons per room tends to indicate high status. Low status characterizes people with low-income jobs living in low-value housing. There are many levels of status, and people tend to filter out into neighborhoods where most of the heads of households are of similar rank.

In most cities, people of similar social status are grouped in sectors whose points are in the innermost urban residential areas. If the number of people within a given social group increases, they tend to move away from the central city along an arterial connecting them with the old neighborhood. Major transport routes leading to the city center are the usual migration routes out from the center. Social-status patterning agrees with the sector model described in Chapter 9.

Family Status

As the distance from the center of the city increases, the average age of the head of the household declines or the size of the family increases or both. Within a particular sector—say, that of high status—older people whose children do not live with them or young professionals without children tend to live close to the city center. Between these are the older families who lived at the outskirts of the city in an earlier period. The young families seek space for child rearing, and older people covet the accessibility to the cultural and business life of the city. Where inner-city life is unpleasant, older people tend to migrate to the suburbs or retirement communities. Within the lower-status sectors, the same pattern tends to emerge.

Transients and single people are housed in the inner city, and families, if they find it possible or desirable, live farther from the center. The arrangement that emerges is a concentric-circle patterning according to family status. In general, inner-city areas house older people, and outer areas house younger people.

Ethnicity

For some groups, ethnicity is a more important residential locational determinant than social or family status. Areas of homogeneous ethnic identification appear in the social geography of cities as separate clusters or nuclei, reminiscent of the multiple-nuclei concept of urban structure. For some ethnic groups, cultural segregation is both sought and vigorously defended, even in the face of pressures for neighborhood change exerted by potential competitors for housing space. The durability of Little Italys and Chinatowns and of Polish, Greek, Armenian, and other ethnic neighborhoods in many American cities is evidence of the persistence of self-maintained segregation.

Certain ethnic or racial groups, especially blacks, were at one time segregated in nuclear communities. Nearly every city has one or more black areas, which in many respects may be considered cities within a city, with their own self-contained social geographies of social status, income, and housing quality. The barriers to movement outside the area historically were high.

Figure 10.2 presents a summary model of these social geographic patterns. Of the three components, family status has undergone the most widespread change in recent years. Today, the suburbs house large numbers of singles and childless couples, and areas near the central business district have become popular with young professionals. Much of this is a result of changes in family structure and the advent of large numbers of new jobs for professionals in the suburbs and the central business districts, but not in between.

The socioeconomic spatial pattern that emerges reflects the **segmentation** of America. Segments are roughly homogeneous areas based on social status, family status, and ethnicity. The marketing research firm Claritas has divided the United States into segments based on these using the following actual variables: household density per square mile, area type (city, suburb, town, farm), degree of ethnic diversity, family type (married with children, single, and so on), predominant age group, extent of education, type of employment (white-collar, blue-collar), housing type (single unit, multiple unit), and neighborhood quality (home upkeep, and so forth). The Claritas system is particularly useful because it allows us to develop a view of the United States that is not so general as to be vague or so detailed as to be tedious. In this chapter, we use both the information provided by the Claritas Corporation and the services of the University of Kansas Geographic Bureau to identify the location of 12 major groups of Americans. Depending on the degree of heterogeneity within a group, each is subdivided into subgroups. To a large extent, we depend on the descriptions of the various segmented subgroups provided by Michael J. Weiss in his book *The Clustering of America*. In that book and in the Claritas system, 40 "clusters" are identified. Catchy names are given to the subgroups, such as blue blood estates, urban gold coast, gray power, coalburg and corntown, and backcountry folks. We use a somewhat more general classification with names that are more closely associated with standard socioeconomic terminology.

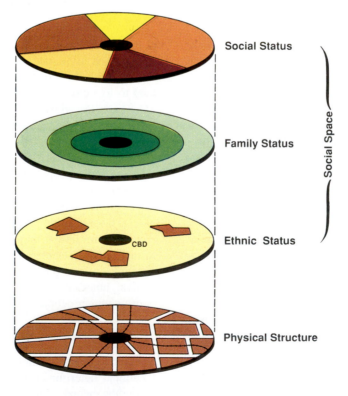

Figure **10.2**
The social space of American and Canadian cities.

Social Segmentation

In this section, we describe in modest detail the nature of each of 12 segmented groups. The reader must recognize that 12 groups represent a generalization and are made up of many subgroups, some of which are identified below (Table 10.1). We use a classification scheme that took great pains to ensure that groups would be as homogeneous as possible. The groups vary in size from 5% to 13% of all American households. The sections are ordered by their location relative to the centers of the large metropolitan areas (Figure 10.3). The more urban groups are described first, followed by groups that tend to live in the suburbs, then by small industrial-town groups and various rural groups. Recall the discussion in Chapter 9 of socioeconomic status, life-cycle status, and ethnic status as the underlying social-ecological forces responsible for the residential spatial patterns of American cities.

Upscale Urban

All four subgroups in the **upscale urban** category can afford expensive urban living. The first socioeconomic subgroup is the *urban high-rise dwellers* (Figure 10.4). These are made up of singles and couples, often elderly widows and widowers. The younger among them are white-collar professionals, usually without children, who enjoy the accessibility urban living affords. Every major metropolitan area contains these high-rise structures that advertise such amenities as a pool, security guard, and a gym. This subgroup is not to be confused with one of the elite groups mentioned later, although some of this group who live in the penthouses of high-rises might belong to the elite group. This subgroup, in general, is solidly in the upper middle class.

The second subgroup includes *the avant-garde* artists, intellectuals, well-to-do homosexuals, and others associated with the Bohemian population. They live in comfortable apartments near the city center where they can meet people

---

Table 10.1
Social Segmentation by Distance from the Central Business District

Central Business District

Upscale Urban
- urban high-rise dwellers
- the avant-garde
- prosperous black families
- singles in upscale urban condominiums

  Urban Poor
- unemployed heavy-industrial workers
- southern working-class blacks
- poor Hispanics
- poor blacks

    Working-Class Urban
- immigrant neighborhoods
- white industrial workers
- black workers
- working-class singles

      Elderly Suburban
- retirement communities
- old planned communities
- blue-collar elderly

        Upscale Suburban
- well-established, single-family-home suburbs
- multiethnic, high-income suburb
- young professionals

          Elite
- blue bloods
- urbane
- new rich

            Middle-Class Suburbs
- white-collar professionals
- highest-paid blue-collar workers

              Blue-Collar in Suburbs and Small Cities

                Young Mobiles
- boom towns
- college towns

                  Small Towners
- rustic communities
- mining, lumbering, fishing towns
- mill towns

                    Farm Business-Related Towns

                      Rural poor
- rural poor, white workers
- rural blacks

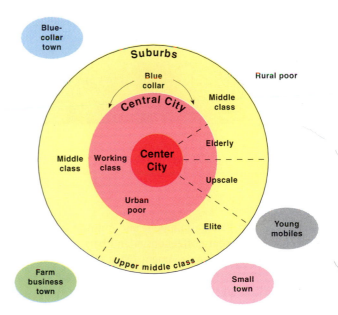

Figure 10.3

The location of the various neighborhood segments with regard to the metropolitan area center, the central city, the suburbs, and small town and rural areas.

Figure 10.4
An upscale urban high-rise in Miami, Florida.

Figure 10.5
Meeting place of the avant-garde.

easily, take part in entertainment and cultural activities in the city, and frequent coffeehouses and bars (Figure 10.5). An urban university environment often stimulates the development of such neighborhoods. Well-known locales for the avant-garde are Greenwich Village in New York City, Cambridge in the Boston area, Haight-Ashbury in San Francisco, Lincoln Park in Chicago, and Georgetown in Washington, D.C.

Tucked away in the inner city, apart from the poverty and crime of the poor, are neighborhoods containing *pros-*

perous black families. These families keep up the fine, closely spaced houses that once were the homes of well-to-do whites in what were suburban areas at the turn of the century. The heads of households, sometimes millionaires, are prosperous businesspeople, landowners, and others who have a vested interest in the black community. Such neighborhoods exist in the well-known black area of Harlem in New York. Other examples include the Mount Airy section of Philadelphia, Cranwood in Cleveland, and Seven Oaks in Detroit.

The fourth subgroup, the largest in this category, contains primarily *singles living in upscale urban condominiums,* garden apartments, and bungalows. More often than not, these neighborhoods are located toward the outer portion of the inner city. A large portion are found in rapidly growing metropolitan areas. Singles, new to the area, find a welcome haven among other recent migrants in such neighborhoods. Many of the subgroup members conserve their financial resources by living together and sharing the rent. Together, they benefit from the amenities of upper middle-class living. Englewood in Denver, University Town Center in San Diego, and Park Place in Houston are examples of this type of neighborhood.

Urban Poor

Four subgroups make up the **urban poor.** If the subgroup is white, its members tend to be *unemployed heavy-industrial workers.* As industry declined, so too did the neighborhoods of the workers. Those who remain depend for money on odd jobs, their families, or unemployment compensation. The subgroup made up of *southern working-class blacks* lives in old neighborhoods in the southern cities where jobs were once held in labor and service work. They live in ramshackle houses, try to keep their families together, and given the circumstances, send a surprising number of their children to college. The subgroup made up mainly of *poor Hispanics* often lives in large barrios of recent immigrants from Mexico, Guatemala, El Salvador, or Puerto Rico. Many came from poor, rural areas of their native country. They have few skills and are relegated to unskilled labor jobs, such as maids and gardeners. East Los Angeles is America's largest barrio, but Spanish Harlem in New York City, El Paso and San Antonio in Texas, and Miami also contain large Hispanic neighborhoods. If the subgroup is made up primarily of *poor, inner-city blacks,* then the neighborhoods are the run-down, drug- and crime-infested environments well publicized by the media. Every American city has a ghetto that contains people, usually singles and single-mother families, who depend on public assistance for their survival (Figure 10.6). They live in broken-down row houses and apartments and in inadequately serviced, government-built high-rises.

Working-Class Urban

Four subgroups make up the **working-class urban** category. The first subgroup makes up the large *immigrant neighborhoods* in the largest cities and in coastal cities, such as Miami. Today, instead of being dominated by Italians, Poles, and Russians, these old, densely settled housing districts are, for example, home to Haitians in Miami, Hispanics in Houston, Cambodians in Long Beach, and Koreans in Chicago (Figure 10.7). This subgroup often dwells in a neighborhood of apartment houses or multifamily complexes. Family income is relatively high (many workers to a household) but individually relatively low. Many of the more educated of this group are moving into white-collar occupations.

The second subgroup are the primarily *white industrial workers* who had been working in textile mills for several generations before the plants began to close. Others work in metals plants or in the fishing industry. Many are members of labor unions. There is a strong English and Irish makeup to the population, although Italians and Portuguese have their own, similar neighborhoods. Many live in row houses typical of old northeastern and New England industrial towns. Revere, Massachusetts, is a good example of this type of neighborhood. It is an old city with a long tradition of producing high-quality cooking and eating utensils.

The third group consists of those *black workers* who live in formerly densely settled white areas of the central city. These are families, with many singles, struggling to raise themselves from poverty. They are engaged in blue-collar and service occupations. Newburg in Cleveland; East Orange, New Jersey; and the Anacostia section in the Washington, D.C. area are examples.

The fourth group are those who live in and around the upscale, avant-garde group mentioned previously: these

Figure 10.6
A poor inner-city black area in New York City.

are chiefly *working-class singles* who hold blue-collar and service jobs. They usually live in apartment houses and generally have low incomes. Many of the cities in the West and Southeast have enclaves where this group predominates, mainly because the singles are in transition, having lived in the district for relatively short periods of time. Ocean Beach in San Diego, Venice Beach in Los Angeles, and North Austin, Texas, are examples of this type of neighborhood.

Neighborhoods

Figure 10.7
An Asian immigrant neighborhood in the Flushing, Queens, area of New York City.

Elderly Suburban

In the older suburbs such as Levittown, Pennsylvania (outside of Philadelphia), and Cicero, Illinois (in suburban Chicago), and retirement communities, such as Sun City, Arizona (near Phoenix), and Clearwater, Florida (outside of Tampa), are many older Americans who give a distinctive

Figure 10.8
A retirement community in Fort Myers, Florida.

flavor to their communities. Three subgroups constitute the **elderly suburban** group.

The first includes a range of well-to-do elderly who can afford to retire to well-kept communities, such as Laguna Hills, California; Sun City, Arizona; or Sarasota, Florida (Figure 10.8). Their retirement incomes are sufficient to enable them to maintain a high standard of living. Their children reside elsewhere. Many activities are provided by the towns or the *retirement community* leaders to keep the residents busy. Golf courses and interest group meetings abound.

As many of the young left the older suburbs, the population aged in place. This second group includes those who, after World War II, were among the first to move to the now *old planned communities* that were developed at that time. Levittowns were built in Pennsylvania, New Jersey, and New York (Figure 10.9). A number of similar suburban towns also were built then. Most of these communities now house elderly who have long since remodeled their houses to suit their family needs. The original repetitive patterns of similar house types are no longer identifiable.

The third group of suburban elderly also moved to the suburbs many years ago, raised their families, and now see their children and grandchildren on weekends and holidays. These elderly can be differentiated from the others because they are *blue-collar elderly* with a strong ethnic flavor. Eastern Europeans who worked in the factories, joined labor unions, and now live on pensions are typical of this subgroup. Unfortunately, as the population of these suburbs contracts, the areas become run-down. Examples include Wyandotte in the Detroit area, which is in the automobile-producing region, and Sparrow's Point in the Baltimore area and Carnegie in the Pittsburgh region, which are major steel centers.

Figure 10.9

Levittown, New York, in 1949, an early example of a planned community. After World War II, William J. Levitt built 17,447 houses on what had been farmland on Long Island. The mass-produced houses all had the same floor plan and were sold at prices below the current market level.

Figure 10.10

The well-established, single-family neighborhood of Shipley in Anne Arundel County, Maryland. These houses have approximately twice the square footage as the Levittown homes shown in Figure 10.9.

HOUSING THE POOR

Large sections of metropolitan areas are densely settled with poor people who depend on public assistance. In the very large cities, such as New York and Chicago, many live in large public housing projects. Such well-known developments as the Robert Taylor Homes in Chicago or the now demolished Pruitt-Igoe project in St. Louis were intended to give the poor the opportunity to have a decent residence.

Unfortunately, many of the huge apartment houses have deteriorated badly because of lack of upkeep or misuse by residents. These projects, generally not built in the suburbs, have created huge neighborhoods of poverty-stricken people in the inner cities of America.

Upscale Suburban

The **upscale suburban** category can be divided into three subgroups. The first includes those who live in the *well-established, single-family-home suburbs* (Figure 10.10). The heads of households tend to be older, that is, 45 years of age and up. As a result, many families have no children living at home. Families are small but houses are large. These suburbanites are comfortable, with high incomes and high-paying white-collar jobs. Their homes usually have pools and sprawling patios and the latest electronic gadgetry. Around each of the major cities is usually more than one of these neighborhoods; some are made up of reasonably new homes. The homes in the older suburbs in this category have been upgraded to

Neighborhoods

Associated with the upscale suburban segment is a new type of community that appeared in America in the late 1970s, blossomed in the 1980s, and now flourishes in urban areas of the South and West. One might identify as its ancestors the walled towns of medieval Europe, for the distinguishing characteristic of the *gated community* is the residents' desire for security.

From Miami to Houston to Los Angeles, people have fled the suburbs of the 1950s to reside in master-planned, upscale developments surrounded by walls or fences. Entry to these communities within communities is restricted; gates are controlled by guards or accessible only by computer key-card or telephone. Some of the communities hire private security forces to patrol the streets. Surveillance systems monitor common recreational areas, such as community swimming pools, tennis courts, and health clubs. Houses are commonly equipped with security systems.

The typical gated community has been built by a developer following a master plan. To preserve the upscale nature of the development and protect land values, self-governing community associations enact what are called *CC&Rs*—covenants, conditions, and restrictions. Pervasive and detailed, they specify such things as the size, construction, and color of walls and fences; the size and permitted uses of rear and side yards; and the design of exterior lights and mailboxes. Some go so far as to tell residents what trees they can plant, what pets they may raise, and where they may park their boats or Winnebagos.

The singles, families, and retirees residing in the gated community typically have high average annual incomes. Troubled by the high crime rates, drug abuse, gangs, and drive-by shootings that characterize many urban areas, they seek safety within their walled enclaves. Even Eden had its serpent, however, and it is not possible to wall out all crime. Incidents of robbery, burglary, child abuse, rape, arson, and murder occur even in the poshest and "safest" gated communities, perpetrated by the residents themselves.

---

include landscaped front lawns and well-kept brick, stucco, or stone facades.

A second subgroup includes a relatively new feature of American suburban life, the *multiethnic, high-income suburb.* Educated Hispanics, Asians, and blacks who are professionals (technicians, lawyers, labor union officials, and high-level government employees) have joined with the white majority in these suburbs. Successful business-people are also among the residents. This type of neighborhood tends to be located outside of America's largest cities. Examples are Flushing (near New York City), parts of Skokie (outside of Chicago), and Rancho Park (near Los Angeles).

The third subgroup are the *young professionals* who prefer to live in fancy suburban townhouses, new apartments, and condominiums. These are skilled professionals, often in two-earner households with few or no children. They may be considered in the pre-family stage of the life cycle. Such neighborhoods include Greenbelt on the edge of both Baltimore and Washington and Glendale just outside of Denver.

Elite

The *elite* classification consists of three subgroups. The first is made up of America's millionaires and billionaires who live on estates in the suburbs of the principal cities. About 1% of the American population can be considered *blue bloods.* Very often their wealth is based on industrial success of parents, grandparents, or even great grandparents. Those who maintained the family fortune usually did so by continuing to engage in financial, industrial, or service activities (such as owning television stations, newspapers, and stock in established industries). Good examples of communities of the elite are Scarsdale (outside of New York City), Kenilworth (just north of Chicago), and the horse-farm area of Rancho Santa Fe near San Diego (Figure 10.11).

The second subgroup includes those who live in townhouses, condominiums, and apartments in America's largest cities. They have an *urbane* life-style that includes attending the latest stage shows and opera productions and are active in the local theater and museum guilds. Very often they engage in charitable activities. They may live in the Rittenhouse Square area of Philadelphia, on Nob Hill in San Francisco, or in the midtown east side of New York along Park Avenue or Beekman Place (Figure 10.12).

A third subgroup, the *new rich,* includes those who have built up their fortunes in the world of commerce and finance. These wealthy families with their pools and tennis courts live in metropolitan bedroom suburbs, such as Lexington (outside of Boston), La Jolla and Bel Air in California, and some of the newer posh suburbs represented by Paradise Valley in the Phoenix area and along the mainline in suburban Philadelphia.

Figure **10.11**
Cornelius Vanderbilt's "The Breakers" in Newport, Rhode Island, a luxurious summer resort.

Figure **10.12**
The urbane attending the Metropolitan Opera in Lincoln Center, New York City.

Middle-Class Suburbs

Middle-class suburbs contain the child-raising families who live in comfortable single-family homes in the suburbs. They are financially burdened by home upkeep and children's education costs. The two subgroups in the middle-class suburbs can be differentiated only by the occupation of the head of household. The first is made up largely of *white-collar professionals* and well-paid civil servants, while the second group contains the *highest-paid blue-collar workers*. These are the most skilled workers or those who by seniority have gained a measure of financial independence. Every metropolitan area is ringed with these vast suburbs, although every now and again one of the suburban subgroup types mentioned before will break the monotony of middle-class areas (Figure 10.13).

Blue-Collar in Suburbs and Small Cities

There are no distinguishable subgroups of **blue-collar in suburbs and small cities,** a category made up of well-paid machinery operators, steelworkers, automobile workers, company service representatives, mechanics, plumbers, secretaries, and nurses. Nearly all have high school educations and good blue-collar jobs. They live in single-family

Figure **10.13**
A middle-class suburban neighborhood outside of Philadelphia, Pennsylvania.

houses, with the younger families busy raising children. Sometimes the communities are in the suburbs, such as Magnolia in the Houston area, and sometimes in small cities, such as Decatur, Illinois.

Neighborhoods

Young Mobiles

Young mobiles has two discernible subgroups living in a variety of neighborhoods in cities and towns. Both subgroups consist of young people with modest incomes living in small households. They have strong possibilities for higher incomes and are likely to be living in another metropolitan area in the near future. Very often, they are living away from their families, which tend to be members of higher status groups.

The wealthier of the group are attracted to *boom towns,* where possibilities for advancement are strong. They may be working in high-tech industries or in flourishing resort areas. Plainsboro, near Princeton University, along the high-tech Route 1 in New Jersey, and Corrales in Albuquerque, New Mexico, are examples of this phenomenon (Figure 10.14). Residents usually stay long enough to gain the experience needed to move up to a better paying job. The less wealthy have less education and work at blue-collar jobs, but seek the chance to improve their well-being by heading for a boom town. Very often, they are people from urban areas who seek the independence and outdoor life that exist in and around the prosperous towns in the Rocky Mountains region, such as Pocatello, Idaho; Aspen, Colorado; and Billings, Montana; and in Washington and Oregon.

The second subgroup lives in the *college towns* of America, which count among their residents undergraduates, graduate students, faculty, and service personnel. Very often, these towns are centers of modest wealth in farm areas far from major metropolitan areas. Most of the students expect to have a higher standard of living in the future. They consider themselves as temporary residents. Good examples are Champaign-Urbana, Illinois; State College, Pennsylvania; and College Station, Texas. Every state has at least one large state university whose presence has created the college town environment.

Small Towners

Small towners live in communities that tend to be microcosms of the larger cities, although they lack some of the groups (or neighborhoods) mentioned earlier. In this category, we note just those small town neighborhoods that have no urban counterpart.

There are two major types of small towns: the non-farm type that we treat under this heading and the farm business-oriented towns that are home to the next group. Most small towners live in neighborhoods made up of the types in the nine categories mentioned previously.

The first subgroup includes those who live in *rustic communities* with pleasant, small cottages located along the coast, in the mountains, or near lakes. Some reside in what were once summer homes for the affluent who decided, with improved transportation and added retail stores, to make these their permanent homes. Others came to these areas to serve the residents and enjoy the small town envi-

ronment. Laconia and Wolfboro in New Hampshire and Mountain Home in Arkansas are surrounded by excellent fishing and boating lakes; residents of Cape May in New Jersey and Pismo Beach in California enjoy small town living in coastal environments.

The second subgroup includes those living in towns associated with *mining, lumbering, and fishing* activities. These are usually run-down, struggling towns where incomes are as variable as the abundance or price of the raw material that is extracted. Wilkes-Barre, Pennsylvania (coal); Butte, Montana (copper); Hoquiam, Washington (lumbering); and Thibodaux, Louisiana (shrimp fishing); are just a few examples.

The third subgroup lives in the hundreds of textile and lumber *mill towns* stretching across the South, from southern Virginia to Louisiana (Figure 10.15). These are working-class towns that have attracted industry over the years with low, nonunion wages. As wages have risen, however, many of these towns have lost out to foreign countries with lower wages. Thus, these towns often have a shabby appearance.

Although not numerous enough to consider as a fourth subgroup, many small towns near military bases cater to the needs of military personnel.

Farm Business-Related Towns

One could divide **farm business-related towns** into subgroups according to the area in which they predominate or

Figure 10.14
An industrial plant in the high-tech boom town of Cupertino, in the Silicon Valley of California.

the kind of agriculture that is practiced nearby, but they all have one thing in common: they serve the nearby farm community with wholesale and retail supplies. The goods can be tractors or feed and seed, not to mention the usual supplies for a farm population, such as food and places to gather (community centers, cafes, and churches). The communities have some imposing old Victorian houses that are the homes of the well-to-do, but most residents live in modest, wood two-story houses just off of the main street or in newer ranch-style houses at the edge of town (Figure 10.16). If the farm area that surrounds the town is prosperous, the town tends to be, too.

Another characteristic of these towns is that many of their young people have migrated. As farm sizes have

CHANGES IN FARM TOWNS

As we indicated earlier, the modernization of agriculture and the resultant enlargement of farms reduced the number of people engaged in farm-related activities. Not only did the farmers move elsewhere, but so too did the small-town merchants who supplied the farmers with machinery and retail goods. Those who remain in the agricultural areas either live on modest means or are well-to-do, land-holding farmers and businesspeople. The smaller population has forced school districts to consolidate into large regional districts for the sake of efficiency; many children in these towns now commute long distances to school. To a large extent, farm towns close to cities have become commuter towns.

Those who leave, if they are elderly, seek places where living costs are low. They are attracted to those areas of the Sun Belt where rents are relatively low and heating costs are minimal. The younger, displaced farm population is forced to move to the cities, where job opportunities are more plentiful. Those who have no more than a high school education tend to move to small towns, where they work in either service or blue-collar factory jobs. The educated tend to join the other young singles and families in the various urban and suburban neighborhoods discussed above.

In recent years, a new countervailing force appears to be reviving some small towns. Emerging in small towns are companies that depend on high technology, such as facsimile machines, computers, and phones, but do not need to be near centers of commerce. Many small business owners are drawn to small towns as an alternative to high-rent, high-crime urban areas.

Figure **10.15**
A pulpwood mill in a small town in northern Florida near Apalachicola.

Figure **10.16**
A residential street in a small town in Missouri.

Neighborhoods

increased, the population needing nearby services has decreased. In many of these towns, a house that was once the residence of a farm business merchant is now inhabited by a commuter who works in a nearby city or town. Oftentimes, the farm town's central business district contains many vacant stores. Traveling the nearby high-speed highways, most farm town residents shop in the malls on the outskirts of the cities that were at one time relatively inaccessible. Farm business-related towns in this category are found everywhere the farm economy dominates, except in the South.

Rural Poor

A number of communities are related to the nearby industrial or farm economy but are not members of the types of small towns just discussed. Found primarily in the rural South and Southwest, these are **rural poor** white, black, and American Indian communities. Three subgroups make up this category.

The first subgroup includes those who live in towns that are peripheral to the mainstream of American economic life. These are the *rural poor, white workers* who live in small communities in the Appalachians from Maine to Georgia, in the Ozarks, and across the rural South. Little has changed in these communities over the years. Workers may be dairy, chicken, or other small farmers; lumberjacks; or odd-job holders. The communities have a distinct backcountry, tattered appearance. Often, whites live on one side of the railroad tracks, blacks on the other. The whites in these communities tend to be better off than their black counterparts. The region where this description holds best is often called America's Bible Belt.

The second subgroup is, again, primarily southern. It includes the *rural blacks* living in the small towns mentioned in the preceding paragraph, as well as residents of completely black towns in the low country of the western portion of Mississippi and the coastal Carolina region (Figure 10.17). They work on farms owned by whites, are poorly educated, and live in dilapidated shanties that often lack indoor plumbing. Parts of these towns have a distinct Third World flavor.

The third subgroup is a catchall mainly designed to include those living in the truly *remote* sections of the country. It encompasses all those rural dwellers not already accounted for. They include whites living in the hollows of West Virginia and Tennessee; American Indians in Arizona and New Mexico; Hispanics in south Texas, New Mexico, and Arizona; and various groups living in the most remote settlements of North Dakota, South Dakota, and northern Minnesota. They may be farm laborers, local service workers, itinerant lumberjacks, or, in the case of those living in the hollows, subsistence farmers.

Case Studies

In this section, we briefly describe the residential patterns of the segmented population in two American metropolitan

Figure 10.17
A house in a rural black community near Plains, Georgia.

areas: Philadelphia and San Diego. Each represents an entirely different urban milieu. Philadelphia is an old, industrial city whose greatest growth occurred before World War II, while San Diego is a Sun Belt city that experienced a great deal of growth in the past few decades. Philadelphia represents a city with a diversified economy where growth is evident in the central business district and in the new suburbs, but where there is a great deal of urban decay, especially within the city limits. San Diego, on the other hand, is a tourist center, but its size is more a function of the military bases and of defense-related high-tech industries that are scattered about the city and suburbs.

By American standards, the two cities could not be more different, yet we will see that the neighborhood districts are roughly similar in socioeconomic qualities.

Two points are worth noting: with the exception of the small town and rural categories, American cities, new and old, contain all of the groups and subgroups already described; and given topographic considerations, the locations of the various groups differ only slightly from city to city.

Philadelphia

The center of Philadelphia is just to the west of the Delaware River (Figure 10.18). Directly across the river is

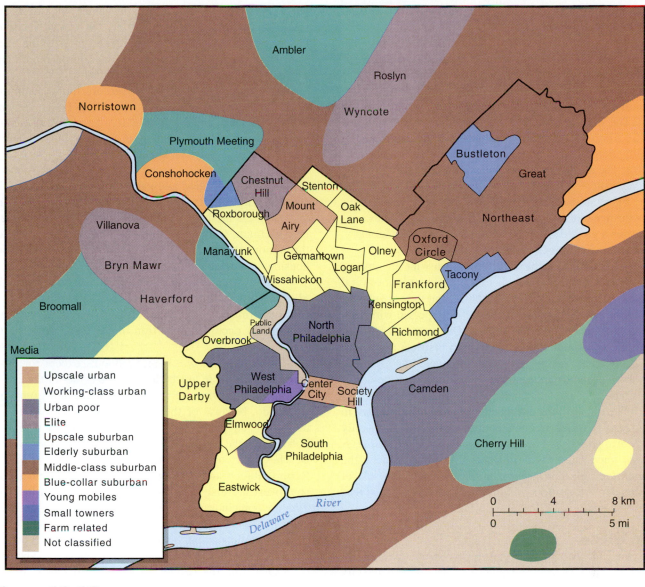

Figure 10.18
The segmentation of Philadelphia.

Legend:
- Upscale urban
- Working-class urban
- Urban poor
- Elite
- Upscale suburban
- Elderly suburban
- Middle-class suburban
- Blue-collar suburban
- Young mobiles
- Small towners
- Farm related
- Not classified

Camden, New Jersey. Bisecting Philadelphia and extending from northeast to the southwest is the boundary between the flat Atlantic Coastal Plain and the rolling hills of the Piedmont district. Unimpeded by mountains or oceans, the pattern of social groups roughly forms concentric circles moving outward from the center of the city in the order of the descriptions of the segmented groupings presented before.

Surrounding the center of the city is an *upscale urban area.* This includes the gentrified Society Hill section close to the river, the fancy high-rise apartment houses around Rittenhouse Square and along Benjamin Franklin Boulevard, and the townhouses of the well to-do and avant-garde along Spruce and Pine Streets.

Farther from the center and extending in all directions are the *urban poor.* North, West, and South Philadel-

phia (northern section) and Camden, New Jersey, contain a huge population of poor, mainly blacks and Hispanics, who live in the working-class row houses built at the turn of the century for the thousands of migrants from Europe.

Even farther from the center, in districts within the city limits and stretching into the northwest suburbs, is a large population of *working-class urban.* Perhaps the best known of these districts is the largely Italian-American area of South Philadelphia. Outside of the city limits, representative *blue-collar suburban* districts are Conshohocken and Norristown.

A varied pattern begins beyond and around the working-class urban areas. There are wedges of *middle-class suburbs* in northeast Philadelphia (Oxford Circle and beyond), *elderly suburban* west of the International Airport and in

Neighborhoods

places around Bustleton in the Great Northeast, and *upscale urban* in Mount Airy.

Two groups dominate the pattern farther out from the center. The first large area surrounding the city contains *upscale suburban* groups. These are the well-to-do suburbs of the city, such as Media, Broomall, Plymouth Meeting, Ambler, and Cherry Hill. The second is the location of the *elite*. The well-known Mainline (named after the main line of the historic Pennsylvania Railroad) contains such wealthy neighborhoods as Bryn Mawr, Haverford, and Villanova. The northern suburbs include among others, Roslyn and Wyncote.

Surrounding these groups and forming a belt of their own farther out are the *middle-class suburbs.* These account for a great deal of the area of Chester, Montgomery, Bucks, Camden, and Gloucester Counties.

At the edge of the metropolitan district are several other *blue-collar suburban* districts. These are more rural in nature than the middle-class suburbs, but they are closely tied to the urban economy. They are interspersed with such towns as Coatesville, Pottstown, and Quakertown.

In and around the previous two districts are *young mobiles* who are taking advantage of developing high-tech industries primarily in the rolling countryside of the northern suburbs.

The rural groups described earlier are concentrated in New Jersey, which is connected to Philadelphia by only three bridges. There are *small towners, farm business-related towns,* and even *rural poor.* Scattered among the *blue-collar suburban* groups on the Pennsylvania side of the border are some *small towners* and *farm business* neighborhoods.

San Diego

The center of the city is a short distance to the east of San Diego Bay, across from the peninsula that contains the city of Coronado (Figure 10.19). The Pacific Ocean and topography, along with a large federal presence, disturb the consistent concentric circle pattern noted in Philadelphia. San Diego is built on low hills, marine terraces, and the valleys between them. Because the climate is dry, the hills (called mesas) are not rounded as in Philadelphia. Precipitous boundaries often exist between neighborhoods, which tend to be located on the mesas, while freeways and commercial developments occupy the narrow valleys between the mesas. To the east, rugged mountains limit urban settlement, as does the ocean to the west. To the south is the international border that separates San Diego from Tijuana, Mexico. Large naval bases line San Diego Bay, and the huge Miramar Air Station is in the north central part of the city.

In and around the downtown are new high-rise apartment houses and mid-rise condominiums characteristic of *upscale urban* neighborhoods. Just to the northeast of downtown is the Hillcrest district, a center of the *avant-garde.* Along the coast, close to the center, are singles and small-family *upscale urban* districts (Mission Beach and Pacific Beach). On mesas both north and south of the Interstate Highway 8 freeway are such upscale neighborhoods as Kensington and Del Cerro.

The *urban poor* of San Diego are found south and east of the center in such neighborhoods as Logan Heights, East San Diego, and National City. These areas contain mostly Hispanics but, especially in East San Diego, there is a large area of mainly black residents.

The *working-class urban* neighborhoods are found in the north in Serra Mesa and parts of Clairemont and in the south in National City, Chula Vista, and San Ysidro. Living in these areas are large numbers of immigrants from Mexico, Vietnam, Cambodia, and the Philippines. Scattered throughout these areas are well-defined neighborhoods housing navy-related families.

Many of the elderly of San Diego can be found in the *elite* (La Jolla) and *upscale suburban* districts, such as Coronado (many retired naval officers). A large retirement community occupies a portion of the Rancho Bernardo area.

Upscale suburban neighborhoods, except for those along the ocean, are much closer to the center of the city in San Diego than they are in Philadelphia. Point Loma and Coronado, because of their former inaccessibility to the downtown, are really suburban communities. The neighborhoods along the north coast are also in this category. Solana Beach, Cardiff, and Encinitas are examples. Further inland, Scripps Ranch and Poway in the east and Jamul in the southwest are also *upscale suburban* communities. The *elite* areas include fancy ocean suburbs, such as La Jolla and Del Mar. A particularly noteworthy area, Rancho Santa Fe, in the rolling countryside northeast of La Jolla, contains large estates with stables, tennis courts, and golf courses.

Out from the center, from south to north, are the *middle-class suburbs* of Chula Vista, La Mesa, Spring Valley, the College Area, and Tierrasanta.

The *blue-collar suburban* districts are well represented in many neighborhoods of the suburban cities of El Cajon and Santee and in much of the small city of Escondido to the north (outside the map area).

The *young mobiles* are found in large numbers in the University City and Golden Triangle areas. Many young, well-educated singles live in fancy condominiums and work either in the nearby high-tech biological laboratories or in companies doing research for defense-related establishments.

Settlement Restructuring

The neighborhoods of America are constantly changing. Some types are flourishing, while others are depressed. Districts that have become poorer have lost population, while the economically successful areas attract new

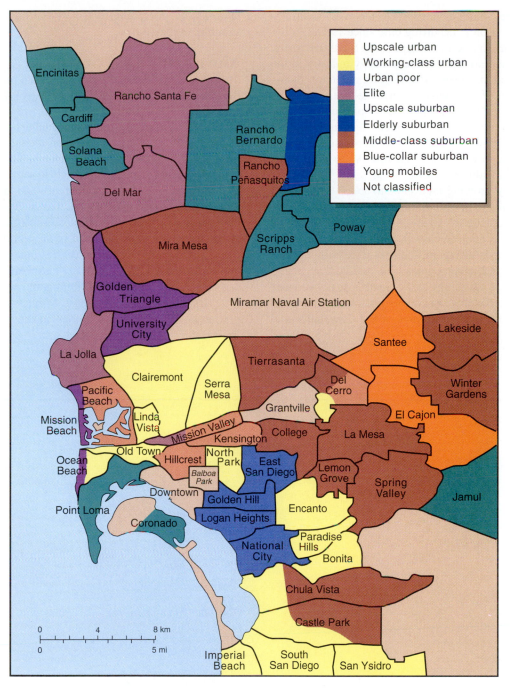

Figure 10.19
The segmentation of San Diego.

residents. Where growth is most pronounced, the segmentation process is most evident. There has been a remarkable movement of people around the country. In this section, we attempt to identify the forces at work in the society that are responsible for the decline and expansion of neighborhoods. In addition, we indicate where in the country the neighborhood changes are most recent and most pronounced.

Rust Belt to Sun Belt Movement

Perhaps the most pronounced movement of population in the last 40 years has been a migration from the industrial cities of the North to the South and the West. As noted in Chapter 3, the movement west began soon after the country was founded. Americans have always been curious about the West, and in times when opportunity for higher incomes prevailed there, large migrations followed. The movement to the South, as opposed to the West, is a relatively new phenomenon, occurring only since 1970.

The reasons for the movement to the South are many, but most important are the decline in heavy industry in the North and the rise of newer, smaller, and nonunion industries in the South. Wages were lower there, but so too was the cost of living. Housing prices and upkeep and transportation costs were lower in the South than in the North. Southern industries were able to attract labor and keep their costs low. Tax and environmental laws were also much more favorable to industries in the South.

The rapidly growing metropolitan areas of the South began in the 1960s to be patterned like other major cities in the United States (Figure 10.20). Upscale neighborhoods containing high-income executives developed in the suburban rolling hill lands where golf courses were built, such as in the Charlotte and Atlanta areas. The middle class found new suburban-type dwellings in the South, usually roomier than their northern or western counterparts, while the urban poor remained in the central cities of the Clevelands and Detroits, basically unaffected by the huge change in the structure of their cities.

Movement to the Suburbs

The low-income inner cities of America's metropolitan areas have been the reservoir for the peopling of the suburbs. Ever since the end of World War II, as incomes rose, Americans have sought greener pastures in the suburbs. This movement to the suburbs has not ceased, even though the flow diminishes during downturns in the economy. The usual pattern is for a particular group of immigrants to settle first in the inner city and then as soon as possible move to the suburbs, leaving behind a remnant population, that is, the old and the very poor of their group. Oftentimes, an entirely new ethnic group, usually from another part of the world, will move into the vacated residences. Today, the inner-city immigrant groups are mainly from Asia and Latin America; during the early part of the century, most immigrants were Europeans.

Blacks, however, have had a somewhat different migration history. They, too, settled in the inner portions of

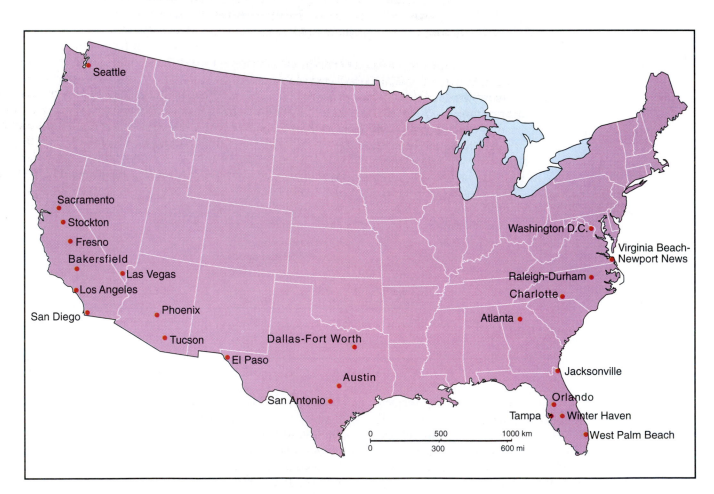

Figure 10.20
Metropolitan Statistical Areas that grew by more than 20% from 1980–1990.

the large cities when they migrated from agricultural areas of the South to the North, but only in recent decades have conditions permitted their movement to suburban locales. Even with increasing incomes, blacks were at one time denied entry to the suburbs. Within any metropolitan area are inclusive and exclusive suburbs, and the latter once used many different techniques to limit the numbers of minorities in them. In recent years, large numbers of blacks have moved to the suburbs. In many instances, they have not been replaced by new ethnic groups in the inner city. The remaining blacks are often left in a dysfunctional urban environment that fails to attract new migrants. Drugs, crime, poor schools, and inferior governmental services permeate large parts of the inner city, making it the least desirable place to live.

Waning of Organized Labor

Union power peaked in the 1950s, when one-third of the labor force belonged to unions (the peak in numbers of members was in 1980), such as the United Automobile Workers, United Steelworkers, Teamsters, United Mine Workers, and Ladies Garment Workers (Figure 10.21). Often waging an uphill battle against the large employers, unions were eventually able to increase wages and therefore enhance the living conditions of workers. Housing quality in neighborhoods of working-class urban residents was continuously being upgraded. No other country had such prosperous working-class neighborhoods. Union workers entered and swelled the middle-class. Many

moved to and helped expand middle-class suburban neighborhoods.

The great success of American labor under such notable leaders as Walter Reuther and John L. Lewis has been undermined in recent years by industrial automation, downsizing, and the rise of the nonunion shop, particularly in areas away from the union power bases, such as Pittsburgh and Detroit. The nonunion areas of the South gained at the expense of the union workers of the North. The major Japanese carmakers located some of their plants in Kentucky and Tennessee, away from strong labor but still close to many of the parts suppliers. Now that many large firms are reducing their size, union labor is negatively affected. The current mode of having many small firms do some of the work that was done by a formerly large-scale integrated giant like U.S. Steel undermines the labor movement because small firms are less susceptible to strikes and labor stoppages that could threaten to wreak havoc on the nation's steel or transportation industry. The areas of the city that housed many union laborers are now the homes of new, nonunion ethnic groups. Many union people have retired and moved to areas of the country where their pensions will keep them at a modest, but reasonable, standard of living.

Government-Supported Industrial Locations

During World War II, the U.S. government supported a military buildup that generated contracts worth many

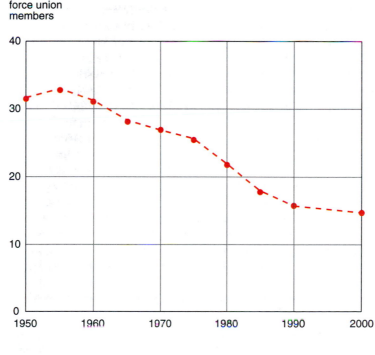

Figure 10.21
Union membership in the United States, 1950–2000.

millions of dollars for the development of war material. Since that time, additional wars, including the cold war, kept the defense industries flourishing. Only in recent years, with the decline and breakup of the Soviet Union, has defense spending been cut back.

The government contracts were not distributed evenly around the country (Figure 10.22). Through the political process and with an eye on costs, the government helped to enhance the industrial capacity of such areas as Texas and California. These states and others, such as Washington and Michigan, became the major builders of planes, tanks, ships, and ordnance. As a result, large numbers of people migrated to these states to take advantage of economic opportunities. Jobs related to government expenditures have generally produced a large middle class of engineers, technicians, office workers, and skilled laborers. For example, in southern California, large parts of Los Angeles, Orange, Riverside, and San Diego counties are home to vast suburban developments of single-family homes for these workers. Other areas benefited greatly from defense dollars. Space exploration expenditures helped build Houston into a large metropolitan area. In addition, the Huntsville, Alabama, and Titusville, Florida, areas grew dramatically due to space-related expenditures.

Boom and Bust Towns

The oil crises of the 1970s, perpetrated by wars and unrest in the Middle East, also affected the movement of the labor force. Oil shale research and development efforts drew many people to the Green River area of Utah, Wyoming, and Colorado. Parts of Texas and Louisiana benefited as once-abandoned oil areas were further developed by secondary recovery techniques. The ultimate boom area was in Alaska, where North Slope oil and the pipeline to transport it were developed. Neighborhoods of blue-collar workers grew up in suburbs and small towns in all of these areas. Later, the oil shale industry declined, and many former new towns in Wyoming and western Colorado have disappeared.

Today, the boom towns are associated with resort areas (Aspen and Vail, Colorado; Brian Head, Utah; Stowe, Vermont; Maui, Hawaii), high-tech industry (western suburbs of

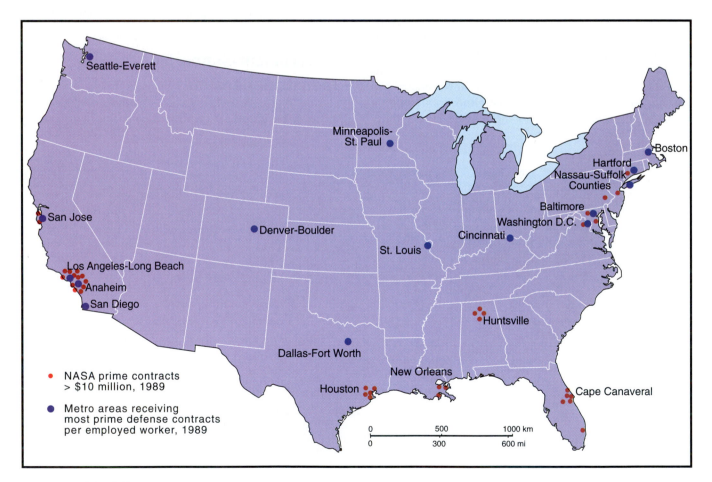

Figure 10.22
The location of prime contracts for 1989 by the Department of Defense and National Aeronautics and Space Administration (NASA).

Sources: From R. D. Atkinson, *Economic Geography,* 69(2):119 (1993); and from A. T. Scott, *Economic Geography* 69(2):146 (1993).

Chicago; Research Triangle, North Carolina), and relocation beneficiaries (people who cash in on their California properties and head for lower cost, less crowded areas in Washington, Oregon, and Idaho). This movement has been enhanced by the decline in defense-related industries, which in the early 1990s caused many Californians to seek new job opportunities in the Pacific Northwest, Idaho, and Utah. One city that has benefited from a combination of all of these things, but most of all from the expansion of gambling and entertainment activity, is Las Vegas (discussed below).

Rise of Retirement Centers

Segmentation has fostered the spatial separation of older from younger Americans. The idea of a *retirement community* in a pleasant, low-cost environment, away from crying children and the economic problems of the industrialized Northeast, has caught on to a great extent. Florida (especially the Gulf Coast), Arizona, Arkansas, Kentucky, and southern California are the prime locations for retirement communities (Figure 10.23). Sun City, near Phoenix, and

Leisure World, near Silver Spring, Maryland, are examples of planned developments that have age and no-children restrictions.

There are, in addition, a growing number of *retirement facilities* in cities, towns, and suburbs all across the country. These may be clusters of apartments, condominiums, or houses (or sometimes a mixture of all three); many have nursing homes or other health-care facilities on the premises. Often, they provide such basic services as banks, grocery stores, hair salons, libraries, chapels, and swimming pools, so that residents need not venture out into the larger community. These "golden-age ghettos," then, foster the separation of the old from the young within a given town or city.

Many retirees who can afford it and yearn to travel have bought recreational vehicles (RVs). With these, many escape the cold winters of the North and "camp out" for the season in huge RV parks in places like Yuma, Arizona, and Bradenton, Florida. These parks, then, also function as retirement communities.

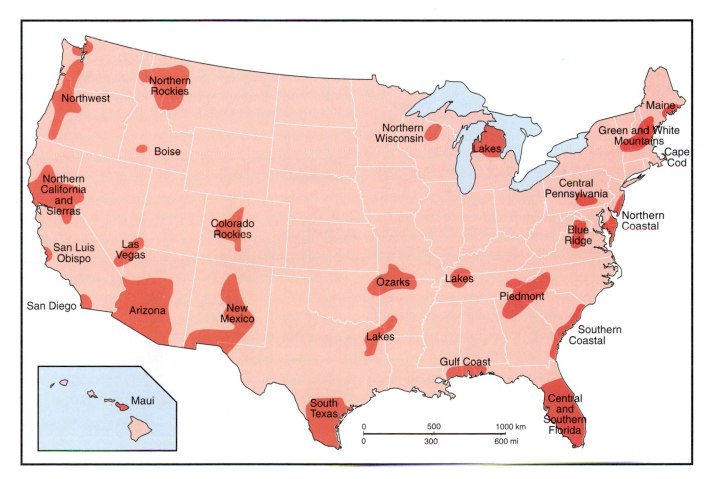

Figure 10.23

The location of the leading retirement centers as determined by R. Boyer and D. Savageau in *Places Rated Retirement Guide* (Chicago: Rand McNally, 1984).

Neighborhoods

Entertainment and Gambling Centers

With increasing income and mobility, Americans flocked to places like Las Vegas, Nevada, and Disneyland in Anaheim, California, to vacation. This fostered the initial growth of nearby areas in, respectively, Nevada and Orange County, California. But the rise of Walt Disney World and other theme parks in the Orlando, Florida, area as an entertainment center is unequaled. Planes, trains, and cars arrive daily with passengers from all over the United States and the world to enjoy the activities at such entertainment centers. The Orlando area has grown from a small agricultural town to a large city in very short order. Examples of other centers that have grown because of gambling and theme parks are Las Vegas (now a very large city), Reno, and Laughlin, Nevada; and Williamsburg, Virginia. The enormous number of service workers needed in the entertainment industry has resulted in huge complexes of apartments, condominiums, and trailer parks for the many working-class singles.

Homogenization of the United States

Outlined above are the reasons for and the spatial outcomes of the segmentation of the population of the United States. Every new immigrant and every change in industrial structure accentuates social heterogeneity. During the period of relatively greatest immigration, the early part of the 20th century, distinct ethnic neighborhoods were established, thereby increasing social heterogeneity. The industrial restructuring of the late 1980s and 1990s, together with a large immigrant population, likewise led to increased segmentation.

At the same time as forces are at work for greater heterogeneity and neighborhood segmentation, so too are strong counterforces that promote a homogeneous society. To bring balance to our discussion of the creation of distinct neighborhoods, in this section, we briefly describe the nature of these countervailing forces. They are not strong enough to erase the boundaries between neighborhoods, but they do limit the degree of separateness of neighborhoods. Were it not for the forces of homogeneity, the country would be much more segregated into socioeconomic classes and racial and ethnic groups than it is.

Language

By far the strongest force for homogeneity in the American population is the English language. In the great immigration period at the turn of the 20th century, most new residents strove to learn English as quickly as possible. They wanted to be "American" so they could join the mainstream of the society. They recognized the importance of learning how business and commerce operated in their new country. Although some recent immigrants appear less enthusiastic about engaging in the same rapid learning process, most want to learn the English language so that they can improve their circumstances.

The English language is the most powerful force in America for people to understand and influence one another. It allows for the smooth functioning of the myriad elements of which the society is composed.

Television

A second powerful force for homogenization is television. Although there is a great deal of choice in TV viewing (and the options are growing), the majority of people watch the same set of three or four channels. The huge television networks, realizing that they must appeal to a large audience if they are to survive economically, are satisfied only when 20 million or more (out of 100 million) households watch their programs. And indeed, on most weeknights, about 80% of prime-time viewers are tuned to the three or four large networks. American viewers watch television for an average of three hours and 46 minutes a day. Since the fare is roughly the same on the large networks, one can argue that most Americans have the same viewing experiences, watch similar news broadcasts, and are influenced by the same kind of reporting of news, sports, and social events. Television is the only medium that influences virtually all of the people.

Mass Advertising

Mass-market advertising is a third powerful force for homogeneity. The ads on TV and in the major magazines are paid for by a relatively few mass-market advertisers. Automobile, beer, snack food, cosmetics, cereal, and soap companies dominate the media (Figure 10.24), with the result that

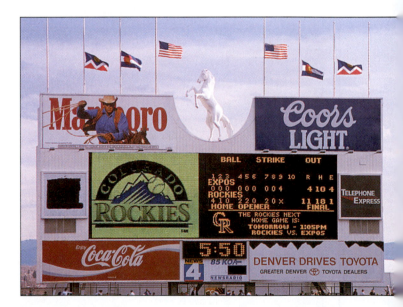

Figure 10.24

Scoreboard advertising at a Colorado Rockies baseball game in Denver. In 1990, the leading mass-market advertisers were Procter and Gamble, Philip Morris, Sears Roebuck, and General Motors. Each spent more than $1.5 billion on advertising.

the tastes and preferences for material goods of Americans are similar. People are bombarded with the same commercial advertisements time and time again, so it is no surprise that their buying habits tend to be similar.

Business and Industrial Structure

The fourth powerful force for homogeneity is the way businesses organize themselves. For example, if banks are only open during the daylight hours, many other businesses and industries are forced to follow suit. Businesses form networks to buy and sell goods and services. Banks are the glue that holds the system together. Thus, most workers in the country start their workday at 8 or 9 A.M. and finish at 4, 5, or 6 P.M. Not only do most working Americans have the same commuting patterns, but they also have the same vacation periods, weekends for relaxation, and evenings to watch television.

The organizational effects of businesses, however, must be viewed in larger, structural terms than daily movement patterns. In the 1950s, the proportion of the work force in manufacturing was greater than it is today. Restructuring has sent many manufacturing jobs overseas, replaced by jobs in service and high-tech industries. White-collar workers are more in evidence today than in the 1950s, when there was more of a balance between the number of white- and blue-collar workers. Thus, the country is becoming more homogeneous with regard to the kinds of jobs that are available. In addition, the franchise system ensures that the same types of restaurants and shops exist everywhere in the country. Shops in malls tend to be the same from place to place.

Even the large mail-order houses contribute substantially to homogeneity in tastes and preferences.

Educational System

The American system of precollege education is a fifth homogenizing force. Although there can be no single system of education in a country where public schools are directed by local school boards, the public schools have traditionally imbued students with the principles upon which the country operates. Schools are expected to instill society's goals, traditions, and history. Children study the U.S. Constitution and the Bill of Rights and learn about the structure of government (Figure 10.25). In this atmosphere, American students are taught to respect peoples of differing races and ethnicities. Furthermore, much emphasis is placed on instruction in the English language. In these ways, the schools represent a strong agent for the people to gain a sense of being Americans; they respect heterogeneity but promote homogeneity.

Canadian Neighborhood Differences

Within any seemingly homogeneous culture realm there are subtle but significant differences. Although the economies of the United States and Canada are similar, segmentation expresses itself in the two countries somewhat differently. The Canadian city, for example, is more compact than its American counterpart of equal population size, with a higher density of buildings and people and a lesser degree of suburbanization of population and functions.

Figure 10.25
School children in New Brighton, Minnesota, reciting the pledge of allegiance.

Neighborhoods

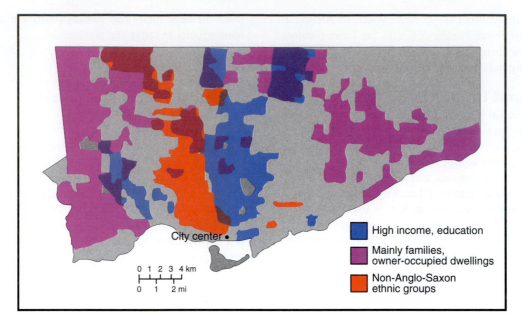

Figure 10.26
Social areas of Toronto.

Source: Redrawn from I. Fabbro, "The Social Geography of Metropolitan Toronto: A Factorial Ecology Approach." Department of Geography, York University, Toronto, unpublished research paper, 1986.

A larger proportion of the wealthy live within central cities in Canada (Figure 10.26). Space-saving, multiple-family housing units are more the rule, so a greater proportion of middle-income groups live in more urban-like surroundings. The Canadian city is better served by and more dependent on mass transportation than is the U.S. city. Since Canadian metropolitan areas have only one-quarter the number of expressway lanes per capita as U.S. metropolitan areas, suburbanization of peoples and functions is less extensive north of the border. For example, Vancouver and Winnipeg rejected the notion of building expressways into the central city. Because there has been much less flight to the suburbs by middle-income Canadians, neighborhoods in Canada show greater social stability, more retention of shopping facilities, and more employment opportunities and amenities than their U.S. counterparts.

The ethnic makeup of Canada differs considerably from that of the United States (see Chapter 14 for a full discussion of these differences). Not only is there the large French-speaking population of Québec (one of four Canadians), but Canadian communities have a higher proportion of foreign-born residents. U.S. central cities, however, exhibit far greater internal distinctions in race, income, and neighborhood types and more pronounced contrasts between central city and suburban residents. Canada contains no large black or Hispanic population. Canadian foreign groups tend to be British (English, Scots, Welsh) or Eastern and Southern European (Polish, Ukrainian, Italian, Greek), although Chinese occupy large areas of Vancouver and Toronto. Toronto is a veritable polyglot city, with neighborhoods as diverse as those housing West Indians and Italians side-by-side.

Canada favors local control of neighborhoods within fairly strong metropolitan governments. In particular, it does not have the problem of rivalry between central cities and well-defined "outer cities" of suburbia that has so spread and fragmented U.S. metropolitan complexes. The elite neighborhoods like Rosedale in Toronto and Outremont in Montréal are made up of single-family housing close to the center, which sends up land values in neighborhoods that surround them. The blue bloods live in these neighborhoods, but they may also have estates at the edge of the large cities.

The urban poor, rather than being made up of blacks and Hispanics as they are in the United States, are usually new immigrants. Canada has no rural black communities. Colleges and universities are usually located within the large cities; college towns are not in evidence. Young professionals usually seek residences in upscale urban neighborhoods; there are relatively few upscale suburban neighborhoods. The eastern provinces have a preponderance of mining, lumbering, and fishing towns, while many of the farm business-related towns of the Prairie Provinces house Eastern European immigrants.

Summary

Segmentation, although an important subject in the field of sociology, has been only lightly treated in the geographic literature. Our conviction is that geographically, the United

States and Canada are better understood by recognizing that different "kinds" of people occupy different kinds of locations.

There is no such thing as a balance between homogeneity and heterogeneity. The United States and Canada have always had a segmented neighborhood structure representing different sets of socioeconomic factors. In some periods, however, when the countries are undergoing either a social or economic transformation, the tendency is for segmentation to increase.

Any view of segmentation must be accompanied by an understanding of the geographic scale of analysis. A large region, like a state or province, is not itself a segmented portion of the country. It is diverse in every respect. Large regions have certain general characteristics that can be described. The smallest homogeneous group is, of course, the individual or the family unit. But as we have seen in this chapter, people and family units tend to reside in similar areas called neighborhoods. While neighborhoods are not consistently homogeneous, they have evident and identifiable similarities. It is at the neighborhood level that segmentation occurs.

Key Words

blue-collar in suburbs and small cities	small towners
elderly suburban	upscale suburban
elite	upscale urban
farm business-related towns	urban poor
middle-class suburbs	working-class urban
rural poor	young mobiles
segmentation	

Gaining Insights

1. In what ways do social status, family status, and ethnicity affect the residential choices of households?
2. What are the expected distributional patterns of the various segmented groups?
3. Create a map of the segmentation of your community.
4. What are the characteristic differences between upscale urban and upscale suburban neighborhoods? Between urban poor and rural poor? Between working-class urban and blue-collar suburban?
5. Why are singles attracted to upscale urban condominiums?
6. What are some major differences in the segmented patterns of Philadelphia and San Diego? What factors contribute to the differences?
7. What areas of the country particularly benefited from government expenditures? With the cutback in defense spending, what will happen to those areas that gained from the armaments industry?
8. What is the locational pattern among the fastest growing cities? Why is it this way?
9. In what ways does the Canadian city differ from its U.S. counterpart?
10. Do you foresee in the future a relaxation of the fairly rigid segmentation witnessed in the United States? If so, why? If not, why not?

Selected References

Atkinson, Robert D. "Defense Spending Cuts and Regional Economic Impact: An Overview," *Economic Geography* 69 (1993): 107–22.

Baine, R. P. and A. L. McMurray. *Toronto: An Urban Study*. Toronto: Clark, Irwin, 1977.

Berry, B. J. L. and J. D. Kasarda. *Contemporary Urban Ecology*. New York: Macmillan, 1977.

Bourne, L. S. *The Geography of Housing*. London: Arnold, 1981.

Brown, M. A. "A Typology of Suburbs and Its Public Policy Implications," *Urban Geography* 2 (1981): 288–310.

Davies, R. B. and A. R. Pickles. "A Panel Study of Life-Cycle Effects in Residential Mobility," *Geographical Analysis* 17 (1985): 199–216.

Golant, S. M. *A Place to Grow Old: The Meaning of Environment in Old Age*. New York: Columbia University Press, 1984.

Goldberg, M. A. and J. Mercer. *The Myth of the North American City: Continentalism Challenged*. Vancouver: University of British Columbia Press, 1986.

Gruen, N. J. "Sociological and Cultural Variables in Housing Theory," *Annals of Regional Science* 18 (1984): 1–10.

Harris, R. "Home Ownership and Class in Modern Canada," *International Journal of Urban and Regional Research* 10 (1986): 67–86.

Herbert, D. T. and R. J. Johnston, eds. *Social Areas in Cities*. 2 vols. New York: John Wiley & Sons, 1976.

Hoyt, Homer. *The Structure and Growth of Residential Neighborhoods in American Cities*. Washington D.C.: Government Printing Office, 1939.

Jackson, K. T. *Crabgrass Frontier: The Suburbanization of the United States*. New York: Oxford University Press, 1985.

Kemp, K. A. "Race, Ethnicity, Class and Urban Spatial Conflict: Chicago as a Crucial Test Case," *Urban Studies* 23 (1986): 197–208.

Lake, R. W. *The New Suburbanites: Race and Housing in the Suburbs*. New Brunswick, N.J.: Rutgers University, Center for Urban Policy Research, 1981.

Lauria, M. and L. Knopp. "Toward an Analysis of the Role of Gay Communities in the Urban Renaissance," *Urban Geography* 6 (1985): 152–69.

Ley, D. *A Social Geography of the City*. New York: Harper & Row, 1983.

McCarthy, K. F. "The Household Life-Cycle and Housing Choices," *Papers of the Regional Science Association* 37 (1976): 55–80.

Maldonado, L. and J. Moore, eds. *Urban Ethnicity in the United States: New Immigrants and Old Minorities.* Beverly Hills, Calif.: Sage, 1985.

Massey, D. *Spatial Divisions of Labor: Social Structures and the Geography of Production.* New York: Methuen, 1984.

Morrill, R. L. and E. H. Wohlenberg. *The Geography of Poverty in the United States.* New York: McGraw-Hill, 1971.

Morrow-Jones, H. A. "The Geography of Housing: Elderly and Female Households," *Urban Geography* 7 (1986): 263–69.

Murdie, R. A. "The Social Geography of the City: Theoretical and Empirical Background." In *Internal Structure of the City,* edited by L. S. Bourne. New York: Oxford University Press, 1971.

Palm, R. "Ethnic Segmentation of Real Estate Agent Practice in the Urban Housing Market," *Annals of the Association of American Geographers* 75 (1985): 58–68.

Rees, P. "Concepts of Social Space: Toward an Urban Social Geography." In *Geographical Perspectives on Urban Systems,* edited by B. J. L. Berry and F. Horton. Englewood Cliffs, N.J.: Prentice-Hall, 1970.

Rose, Harold. *The Black Ghetto: A Spatial Behavioral Perspective.* New York: McGraw-Hill, 1971.

Shevky, E. and W. Bell. *Social Area Analysis: Theory, Illustrative Applications and Computational Procedures.* Stanford, Calif.: Stanford University Press, 1955.

Weiss, Michael J. *The Clustered World.* New York: Little Brown & Company, 2000.

———. *The Clustering of America.* New York: Harper & Row, 1988.

CHAPTER

11

Recreational Resources

The nearly 20 million square kilometers (7.5 million sq mi) of the United States and Canada contain a surprisingly large portion of the recreational resources of the world. Landscapes ranging from the great mountains of the West to the vast plains in the center, vegetation from the redwoods of California to the Everglades of Florida, climates from one of the wettest places on the globe (Hawaii) to one of the hottest and driest (Death Valley), and cities from New York to Anchorage and Honolulu provide an incredible variety of recreational resources.

The standard breakdown of time in our 24-hour day is a third for work, a third for sleep, and a third for play. **Recreation,** defined as a means of refreshing or entertaining oneself after work by some pleasurable activity, is also viewed by many Americans as a natural right, one cited in the Declaration of Independence: "life, liberty and the pursuit of happiness." That many choose to use their one-third play time for more work or activities that are far from pleasurable does not negate the statement that much of our lifetime in America focuses on recreation. It can be active sports on a playground, body-building and spirit-raising activities like jogging and cross-country skiing, travel for ed-

ucation or pure fun, or sedentary activities, such as reading and watching television.

Recreational activities have important economic implications for the individual, who spends money to travel, buys sports equipment, and visits recreational areas; for travel companies, manufacturers of sports equipment, and owners and operators of private recreational areas; and for local, state, and federal governments, which spend a significant portion of their budgets to provide recreational facilities. Tourism is now the world's largest industry, accounting for over $2.1 trillion annually. In the United States, where only a few years ago, a limited number of states had billion-dollar tourist businesses, a number of cities now have tourist industries of the same magnitude. In addition, it has been estimated that the U.S. National Park Service maintains the most extensive system of travel attractions in the world (Figure 11.1).

Demand for Recreational Resources

Many factors affect the demand for and supply of recreational resources. In this section, we explore the demand side of the equation; the supply and use of recreational

Figure 11.1
The Grand Canyon in Arizona, one of the earth's most dramatic physical features. It is one of 51 national parks in the United States, which are part of the 367-unit National Park System.

resources are discussed later in the chapter. In the United States and Canada, age, income, amount of leisure time, and other variables help determine the kinds of activities we pursue and how often and where we engage in them.

Population

Most forms of recreation have been growing more or less consistently for decades. Part of this growth can be tied very simply to population, although many factors related to population can influence the popularity of a particular recreational form or site. Taking the growth in visits to Yellowstone National Park in Wyoming as an example, one would expect the number of park visitors to have more than doubled between 1930 and 1998 because the population of the United States also more than doubled during that period. Actually, visitors to Yellowstone increased more than 13 times during those years (Figure 11.2). The growth includes a sizable increase in foreign visitors, who were present in 1930 only in very small numbers. Also, the population of the western United States, including the area in the states surrounding Yellowstone, has been growing faster than that of the country as a whole during the entire period.

The populations of the United States and Canada show no signs of stabilizing, which indicates that the demand for recreational resources will continue to grow. This may be good news for those whose livelihoods depend on it, but overcrowding and the resulting wear and tear on recreational resources will also increase.

Urbanization

In the United States and Canada, the movement from rural to urban areas has been as consistent and long-lived as population growth. While there are many implications of this movement, it affects recreational activities directly: most Americans desire to escape from the city whenever they can. Such was not the case when the population was mostly rural, as recently as the early part of this century. People living on farms or in small towns do not have as much desire to participate in most forms of outdoor recreation as do their urban counterparts. Forest, stream, and open space are daily companions, and the pressure from noise, pollution, and crowding is much diminished or absent. Having jobs that often are more physical and incomes that probably are lower than their urban counterparts also affects the desire and economic ability of nonurban people to participate in many forms of recreation.

Occupation

The move to urban areas has resulted in less physically demanding jobs and, therefore, more need for physical forms of recreation. In the last few decades, many factory jobs have been replaced by work in service, managerial, and professional industries, which require less physical activity. These jobs, moreover, are often unchallenging, repetitive, and stressful. Recreation, then, is used to fill the need for physical activity, as shown by the tremendous increase in health club memberships and involvement in walking, jogging, cycling, and other sports (Figure 11.3).

Income

In the long run, the income of the average American has soared since the founding of the country. This extra income can go to "luxuries" like recreation.

Income influences recreation in a number of ways. There is nearly always a travel component for recreation, which in the United States usually requires an automobile.

Figure 11.3
Seldom do joggers appear in numbers like this, the annual Bay to Breakers Run in San Francisco, but the race does indicate the popularity of recreation, with fitness and health as a primary focus.

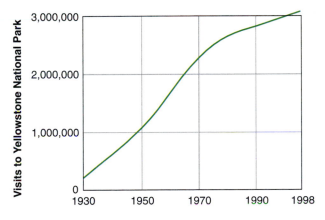

Figure 11.2
Although the population of the United States more than doubled from its 1930 total of 123 million, the number of visitors to Yellowstone National Park has increased 13 times.

Recreational Resources

Also, recreational equipment, while not always essential, is nevertheless figured as part of the cost of participating. With $500 sleeping bags and $1000 mountain bikes, the cost can be steep indeed (Figure 11.4). Finally, there is a direct correlation between income and time for recreation.

An important influence of economics on recreation is not just the average income but the increasing number of people in the upper-income groups. Studies of recreational land use, for example, show that upper-income groups visit national parks and wilderness areas more and participate much more heavily in "high-tech" outdoor recreation, such as skiing, than do lower-income groups.

Leisure Time

Until quite recently, the amount of leisure time available for each working American or Canadian had steadily increased. This increase came from fewer average hours worked per week, fewer years spent in the work force, more holidays, and a greater number of paid vacations. Much of the increased amount of leisure time has been spent close to home (watching television, doing yard work, exercising, and so on), but the larger blocks of time— three-day weekends and long vacations—have had a major impact on travel for recreation. Visits to national parks and national forests increase because of greater leisure time, but so does participation in such sports as backpacking, skiing, and mountaineering, which seldom can be carried out close to home.

The increase in work hours during the last decade may indicate that the current generation, contrasted with those that preceded it, prefers higher income to increased leisure time. It is still too early to predict the ramifications of this. It is possible the "close to home" forms of recreation will benefit, while those that require extensive travel will decline.

Mobility

The near universality of car ownership, the building of interstate highways, and the relative cheapness of motor fuel and airline fares have had a great impact on recreational travel. All the major forms of outdoor recreation, from pleasure driving to river running, benefit from the ultimate freedom afforded by the private automobile. The impact on the large recreational areas, such as national parks and forests, and on "recreational states," such as Florida, California, and Hawaii, has been profound. Increased mobility has also strongly influenced the forms of recreation Americans pursue.

The probable decline of auto travel in the future as fuel and car prices inevitably escalate may be balanced by increased investment in public transportation. There is the possibility, for example, that the traveler to Yosemite Park in California might someday *have* to tour the valley by public transportation to relieve overcrowding (Figure 11.5).

Age

One of the most important determinants of the preferred forms of recreation is age. After all the attention given youth during the 1960s as the baby boomers hit their teens, the concern over the next decade or two probably will be with the recreational preferences of an aging population. The percent of the population over 60 will continue to grow for the foreseeable future, and the impact will be mixed. Older people have significantly more leisure time if they are retired, and they engage in some forms of recreation, such as travel and walking for pleasure, more than the general public. However, older people tend to participate less in most forms of outdoor recreation, one reason why many ski areas offer free lift tickets to people over 65.

Other Influences

Average educational levels continue to rise in the United States and Canada, which appears to increase levels of recreational land use. More women are entering the work force, which increases personal income yet decreases overall levels of leisure time. For some years now, we have been becoming more health conscious, which increases the popularity of such activities as jogging, walking, and aerobic dancing. New types of recreational equipment are constantly evolving, sometimes, as in mountain biking, spurring the development of whole new forms of outdoor recreation. Finally, our greater consciousness of the disabled and the economically disadvantaged is influencing the establishment and development of new recreational areas.

Types of Recreational Resources

There are literally tens of thousands of recreational resources, ranging from the minuscule (a handball court) to the gigantic 5.3 million-hectare (13.2 million-acre) Wrangell-St. Elias National Park and Preserve in Alaska. Some are privately owned, others belong to the public. A comprehensive listing and discussion of these resources is beyond our scope. In this chapter, we focus on two categories of recreational resources: those found in urban areas and those administered by the federal governments. The majority of Americans and Canadians live in urban areas and have access to a variety of recreational facilities. As measured by number of visits, local sites are the most important source of recreational opportunities; it is estimated that more than half of all recreational use occurs at local sites (Figure 11.6). In the United States, the federal government is the most important provider of resource-based recreational activities—those that occur at sites of natural wonder or historical significance. Federal sites account for approximately 13% of leisure-time activity; in many states, they also occupy a significant amount of land.

Figure 11.4
The continued rise in recreational spending in the United States is illustrated by this recreational vehicle, a common sight on the nation's highways in the summer. This particular RV was photographed in Grand Teton National Park in Wyoming.

Figure 11.5
The exclusive use of public buses in parts of Yosemite National Park, California, was started in 1970. Future plans call for an essentially auto-free valley.

Figure 11.6
Golden Gate Park, San Francisco. A quiet walk through an uncrowded park is the most common type of outdoor recreation.

Recreational Resources

Urban Recreational Resources

People engage in most of their recreational activities at or close to home. Watching television, going to the movies, eating out, and other such activities are far and away the most common forms of recreation. Even outdoor recreation occurs mostly close to home—if not in backyards, at least on the sidewalks, in local parks, and in shopping centers, which increasingly have recreational functions.

Urban Parks

The natural landscape within a city plays an important part in urban recreation. One rarely sees pictures of Seattle without Mount Rainier in the background, San Francisco without its bays, Honolulu without beaches, or Denver without the Rockies. Although many cities lack such attractive natural features, some have made the most of what they do possess. For example, San Antonio, located on the hot South Texas plain, has made a small river running through the town the centerpiece of its tourism with River Walk.

Parks normally mean much more to the local people than to the out-of-town tourist, even if they are a Central (New York) (Figure 11.7) or a Golden Gate Park (San Francisco). Almost every large city has a famous park: for example, Lincoln Park in Chicago, Fairmount Park in Philadelphia, and Stanley Park in Vancouver. The main

Figure 11.7

Central Park in New York City, an island of green in the largest city in the United States. The park contains a variety of recreational resources, including a zoo, a conservatory, playgrounds, formal gardens, and wooded areas.

TRAVEL AND SIGHTSEEING

Often it is difficult to separate recreational travel from all other forms of travel. People go to professional meetings and spend much of their time sightseeing, which is one reason why Las Vegas and Hawaii are among the most important convention centers in the United States. It can also be difficult to assess the economic importance of travel and sightseeing to an area. One way is to divide the tourism into basic and service activities. As noted in Chapter 9, a *basic* industry brings in money from outside the region, while a *service* industry produces things or performs services for the resident population.

An amusement park is a basic activity when it attracts visitors from outside the area because it is bringing in money to support the people who live in the city. To the extent that it attracts local residents, who are spending money they earned in the area, the amusement park is a service (nonbasic) industry. In its ability to attract tourists and their dollars, Walt Disney World in Florida is a resource comparable to the oil fields of east Texas. The basic-nonbasic concept thus helps to distinguish tourist destinations of regional, national, or international importance from those of only local importance.

function of city parks is to offer green open space in an increasingly unnatural center city, although they are often the site of other forms of urban recreation, such as zoos, sporting fields, and museums.

Most city parks are artificially created, but a few large, natural parks do exist within city limits. One example is Mission Trails Regional Park in San Diego. Consisting mostly of mountains or steep-sided canyons, the park was set aside at the edge of town just as the city was starting to envelop it. The original plan for the park called for medium-density recreational development, such as off-road vehicle parks, campgrounds, and playgrounds, but as park money became scarcer and the enthusiasm for open space greater, only a trail and parking lot were constructed. At 2323 hectares (5764 acres), it is second in size in the western United States to South Mountain Park in Phoenix, which at 6475 hectares (16,000 acres) is called the world's largest municipal park.

Nature-Oriented Attractions

Zoos, aquariums, and botanical gardens are also primary tourist attractions in cities. There are zoos in almost every city of any size, and they are assuming new forms that should ensure even greater attractiveness in the future. The old zoo of pavement, cages, and pacing animals is fast becoming a relic of the past. The new zoo is creating a world of nature as fascinating in miniature as the larger world it is meant to reflect—the rain forest, the tropical savanna, the arctic tundra. To walk into the lush rain forest of several acres in Omaha, Nebraska, or to watch a Sumatran tiger reclining on the hillside in the San Diego Zoo is an event that should keep the zoo-going experience fresh (Figure 11.8). That zoos are increasingly functioning as breeding sites for endangered animals should also enhance their image.

Museums

Museums form an important part of a city's tourism. They bring tourists to the urban area for all kinds of special exhibits or ongoing attractions. The largest and best known assemblage of museums in the United States is that managed by the Smithsonian Institution in Washington, D.C., but most large cities contain several outstanding museums.

A special type of museum is the historical village, often recreated by rebuilding earlier structures and sometimes by moving them to the site. Plymouth Plantation and Sturbridge Village in Massachusetts; Williamsburg in

Figure 11.8
Tiger River in the San Diego Zoo, an exhibit bringing together plants and animals of the Southeast Asian tropical rain forest, tries to be as authentic an assemblage as possible and is illustrative of the new trend toward ecologically based zoos. A Sumatran tiger is shown.

Virginia; and the Ukrainian Cultural Heritage Village in Alberta are examples. Recreated urban neighborhoods of an earlier period are typified by Atlanta Underground and Seattle's Pioneer Square.

Amusement and Theme Parks

The old-style amusement park with roller coasters and ferris wheels, typified by pre–World War II Coney Island and once a feature of virtually every city in the country, was in danger of becoming obsolete in the 1950s under the twin impacts of television and increasing crime. Sharing this decline was the touring carnival, once an annual feature of small-town America. A different form of amusement park has come back strongly, however.

The pure amusement park with rides and games of chance still exists in parks like Magic Mountain north of Los Angeles, but **theme parks** like Walt Disney World in Orlando, Florida, now are setting the standard (Figure 11.9). In the case of the latter, a moderately important tourist area became almost overnight one of the world's

major tourist attractions. The "themes" in these parks are almost endless but seem to draw mostly on nostalgia (Main Street in Disneyland) and travel (Epcot Center in Walt Disney World). By enabling one to go anywhere in space and time, offering all the exciting rides of the amusement parks of the past, providing stage shows, restaurants and fast-food outlets, theme parks have ensured their popularity.

Spectator Sports

The ability to be a spectator at sporting events adds an enormous dimension to urban recreation. Although sports television has taken millions from the contests themselves, Texas viewed from the air on a Friday night still presents a line of headlights snaking toward the nearest high school football stadium, and college stadiums throughout the nation are packed for Saturday afternoon games. Adding to the popularity of spectator sports are the increasing numbers of professional teams and the stadiums, many of them indoors, being built in cities across the country.

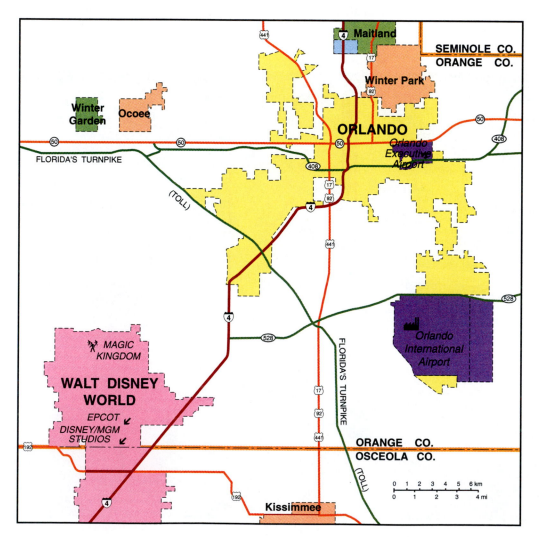

Figure 11.9

The Disney Corporation purchased 11,000 hectares (27,000 acres) near Orlando, Florida, to control the enormous induced investment in Orlando of Walt Disney World, which has become the world's most visited human-made attraction since its opening in 1971.

When a baseball team goes to the World Series or a football team to the Super Bowl, the host city receives millions of dollars of free publicity as well as tens of thousands of visitors. The stakes are even higher for the biggest of sporting events, the Olympics. Atlanta, Georgia, preparing for the 1996 Olympics, anticipated a total of more than 2 million spectators and visitors for the events, but actually got over 5 million, with 600,000 visitors jamming the city on peak days.

Performing Arts

The performing arts are an important recreational activity in many cities. Las Vegas seems to be the quintessential performing arts city, with its economy largely dependent on serving visitors seeking pleasure in one way or another. Gambling is the main attraction, but every casino and hotel has some sort of live entertainment. New York City, on a different scale, is the largest entertainment city. Broadway and off-Broadway theaters, the Metropolitan Opera, and Carnegie Hall are among the attractions that help bring millions of tourists to New York City each year.

Nightlife, restaurants, the cinema, and retail stores round out the "amusement" facilities in a city. Shopping is a significant recreational activity for many urban residents. Ghirardelli Square in San Francisco started the trend for amenity-based shopping, which seems to have reached its zenith, for now, in the West Edmonton Mall in Canada and the Mall of America in Bloomington, a suburb of Minneapolis. Rather than make amusement parks, shows, eating out, or even attending a sporting event an end in itself, these activities are being blended together in a pleasant central location. Lush vegetation, fountains, swimming pools with artificially produced waves, and even hourly sunrises and sunsets can be provided in such malls.

U.S. National Park System

Among the most important recreational resources in the United States are the **national parks** and other areas administered by the National Park Service (NPS). They contain some of the country's most spectacular scenery, preserved for the enjoyment of future generations.

History

The most copied American institution is not democracy, which started in Greece, or the automobile, first put together in France, but national parks, which at last count had been adopted by 125 nations. There is some argument about which was the first American national park, although most agree that it was Yellowstone in 1872. (Yosemite Valley was given to California to use as a park in 1868, but the state did such a poor job that the valley later was made part of Yosemite National Park, which was established in 1890.)

The founding of Yellowstone National Park was an unlikely event; the United States of the 19th century was the last

---

OUTDOOR RECREATIONAL ACTIVITIES

Americans and Canadians engage in a seemingly endless number of recreational activities, ranging from the simple and safe (walking for pleasure) to the potentially dangerous (mountaineering). Some require no equipment, others have high price tags. Some can take place anywhere at any time, others—such as river rafting—have geographical and/or seasonal constraints. Although some activities raise environmental concerns, the appreciation of the continent's recreational resources is probably greater than at any time in the past.

The classification presented below indicates the variety of ways we pursue pleasure.

Recreational travel
    Walking for pleasure, hiking
    Cycling
    Using off-road vehicles (jeeps, motorcycles, and all-
        terrain vehicles)
Water-oriented activities
    Swimming, surfing, wind surfing
    Parasailing
    Boating (canoeing, kayaking, sailing, motorboating)

Winter-oriented activities
    Sledding, tobogganing
    Downhill and cross-country skiing
    Snowmobiling
Camping
    Hunting
    Fishing
    Nature study and birding
Adventure travel
    Wilderness travel
    Mountaineering, including rock climbing
    Mountain biking
    River running
Air activities
    Hang gliding
    Skydiving
    Hot-air ballooning
    Bungee-cord jumping
    "Air surfing"
Sports
    Participatory
    Spectator

---

place one would expect such a landmark in environmental history to take place. The last half of the 19th century was perhaps the most rapacious era in the country's history: the prevailing attitude was that resources were inexhaustible. Americans had no form of environmental ethic to draw on, and there was no precedent for setting aside a vast tract of public land to form a scenic park.

The Yellowstone Plateau of 1872 was about as natural as it was possible to be in that period, despite its having been known and available to miners, trappers, and settlers for at least a half century. The only human imprints on the land were a network of trails, a cabin or two, and other marks of a nonpermanent population. The physical environment discourages human settlement. The area encompassing Yellowstone National Park is predominantly a volcanic plateau averaging 2440 meters (8000 ft) in elevation, which at this latitude signifies extreme cold much of the year. Attempts at agricultural settlement have been few and unsuccessful. The timber is three-fourths lodgepole pine, an inferior lumber tree. No minerals have been found that can be mined economically. All of this was important in the late 1800s, because to move from an economic policy where every place, no matter how inferior, had a price tag, to the total exclusion of economic use was an enormous step. Even today, a national park is rarely established if there is a major competing economic use for the area.

The first large exploring expedition to Yellowstone in 1870 was generally credited with generating the idea of a national park. Trappers, miners, soldiers, and various expeditions had passed through the region for half a century, but the information they brought back of wonders they encountered was neither accepted nor widely disseminated. This was not to happen with the 1870 Washburn expedition. The difference was

that it included the surveyor-general of the Montana territory, a man who was later to become governor of the new state, and N. P. Langford, who was to become the world's first park superintendent. The expedition members commanded a respect that gave credence to the claims they made about Yellowstone. Members of the expedition told their nation about it, and suddenly the country believed what it was told.

On the expedition's last night in what was to become the park, expedition members were discussing how they could profit from staking claims in the area. As the conversation progressed, they began to realize the effect this would have on the quality of the land. The expedition's route had touched on most of the scenic points of the region, and in the total wilderness of that day, the effect of seeing these wonders must have been overwhelming. The expedition members' feeling, simply stated, was that visitors of the future should be able to repeat that experience, which would be possible only if the area remained as it was—free of private claims, a pristine wilderness. And this could happen only by retaining the area in public ownership.

Once the idea of a national park was born, the public immediately became enthused. An expedition the following year contributed significantly to the birth of the park. The Hayden expedition of 1871 compiled an impressive record of the geology, flora, and fauna of the area. But more important, the expedition included Thomas Moran and William H. Jackson, who would expose the world to Yellowstone through painting and photography with a quality that scarcely has been equaled since. Reproductions of Moran's paintings of the Grand Canyon of the Yellowstone continue to be best-sellers in the park today, and Jackson's photographs are still considered among the finest ever taken of the Yellowstone region (Figure 11.10).

Figure 11.10
"The Grand Canyon of the Yellowstone," painted by Thomas Moran, helped persuade Congress to establish Yellowstone National Park in 1872. Moran's sketches and paintings, together with photographs taken by William H. Jackson, testified to a majestic new landscape, too valuable for private exploitation.

The goals of the bill establishing Yellowstone as the country's first national park, which was signed by President Ulysses S. Grant on March 1, 1872, were remarkably similar to those guiding the system today. The park was "dedicated and set apart as a public park or pleasuring ground for the benefit and enjoyment of the people; . . . regulations shall provide for the preservation from injury or spoliation of all timber, mineral deposits, natural curiosities or wonders within said park, and their retention in their natural condition." The legislation creating the National Park Service, which came almost half a century later in 1916, stated that "the purpose (of such parks) is to conserve the scenery and the natural and historic objects and the wildlife therein and to provide for the enjoyment of the same in such a manner and by such means as will leave them unimpaired for the enjoyment of future generations."

The early years of Yellowstone National Park, when very little money was spent on the park, were so bad that by 1886, the park was turned over to the U.S. Army, which controlled it until 1916, when the National Park Service was established. By then, the nation had 14 national parks, though not many people visited them. The first director of the NPS, Stephen Mather, decided that what the parks needed more than anything was congressional support and to get that, there had to be as many constituents as possible visiting the parks. His goals have been reached with a vengeance. The annual number of visitors to the national parks, less than half a million when he took office in 1916, doubled in three years, then doubled many more times to 275 million in 1997.

The original goal of the NPS—to save the most unique scenic areas of the country—has been met with 51 national parks and another 45 "natural areas." Although it is certainly possible to find spectacular natural areas outside the national park system, there are "samples" of each (mountains, canyons, deserts, and seashores) somewhere in the system. In recent years, the system has grown through the addition of historical areas, ecologically important areas, and samples of natural landscapes close to population centers.

Types of National Parks

The **National Park System** is a complex system of 367 parks containing 32.6 million hectares (80.6 million acres) in 49 of the 50 states (Delaware has none), U.S. territories, and the District of Columbia. Their names hint at their variety: national parks, national historical areas, national monuments, national battlefields, national seashores, national parkways, and more (Table 11.1). For the sake of simplicity, we can divide them into natural, historical, and national recreation areas.

Natural Areas

The parks started as beautiful, unique natural areas: the Yellowstones and the Yosemites. The national parks, monuments, and preserves that make up 93% of its area

are still the backbone of the system. At the core stand the 51 national parks, the crème de la crème of the system. Fairness would assign one park to a state, but in fact, Alaska and California have 14 of the 51, and the 26 states east of the Mississippi have only 7 among them (Figure 11.11).

National Monuments   The **national monuments,** 76 of them on almost 2 million hectares (5 million acres) of land, are a national park system in miniature, with natural and historical areas in equal numbers. What makes national monuments stand out, however, is not their variety but the way they are established: national monuments are the only parks in the system that can be designated by presidential proclamation rather than an act of Congress. The intent of such a procedure, established by the National Antiquities Act, was to allow the president to act quickly to save

Table 11.1 The National Park System		
Classification	**Number**	**Acreage[1]**
International historic site	1	35
National battlefield	11	13,143
National battlefield park	3	8728
National battlefield site	1	1
National historic site	72	19,821
National historical park	36	159,627
National lakeshore	4	228,745
National memorial	26	7951
National military park	9	37,882
National monument	76	4,787,744
National park	51	47,783,680
National parkway	4	170,547
National preserve	14	22,332,830
National recreation area	18	3,699,551
National reserve	2	22,407
National river[2]	6	362,152
National scenic trail	3	176,352
National seashore	10	592,532
National wild and scenic rivers and riverways[3]	9	212,682
Without designation[4]	11	46,809
Totals	367	80,663,217

Source: From The National Parks: Index 1993, *p. 13. National Park Service.*

[1]Acreages as of December 31, 1992.

[2]National Park System units only.

[3]National Park System units and components of the National Wild and Scenic Rivers System.

[4]Includes White House, National Mall, and other areas.

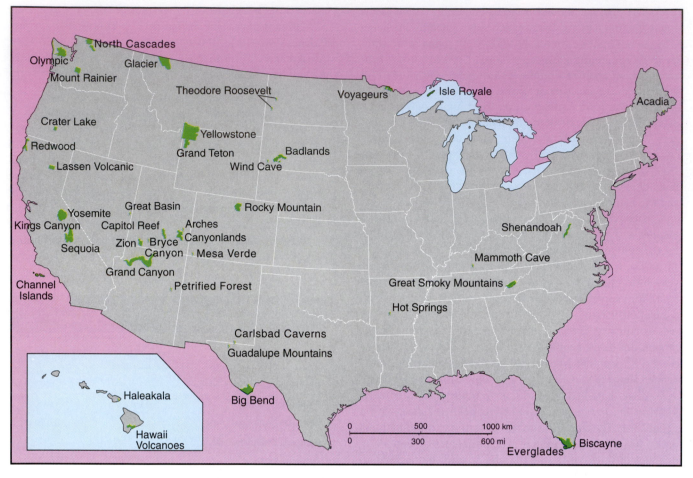

Figure 11.11
National parks of the United States. See Figure 11.12b for those in Alaska.

recently discovered archaeological sites on public lands before they could be plundered. It has also been a useful tool in the hands of environmentally oriented presidents to save any site of natural, historic, or scientific interest. Many of our best known parks, such as the Grand Canyon and Olympic National Parks, were first established as national monuments.

National Preserves A new park category first used extensively in Alaska, **national preserves** were established to allow noncompatible uses, such as sports and subsistence hunting by native people, to exist in parklands without jeopardizing traditional national park values. The largest national park in the United States, Wrangell-St. Elias National Park and Preserve in Alaska, is made up of a national park of 3,374,299 hectares (8,331,604 acres) and a national preserve of 1,966,972 hectares (4,856,721 acres) (Figure 11.12). All but four of the 14 preserves, which total almost 9 million hectares (over 22 million acres), are in Alaska.

Historical Areas
There are 172 historical parks in the NPS, making up almost half the number of park areas, yet only 0.3% of the area. Their importance to the nation is out of all proportion to their size, however. They include the most important battlefields of the Civil War, such as Shiloh, Antietam, Chickamauga, Vicksburg, and Gettysburg (Figure 11.13); parks from the nation's expansion, including forts, trading posts, and gold mining towns; and some 44 historical parks commemorating famous Americans.

National Recreation Areas
Recreation areas were originally added to the national parks system in the 1930s to enable National Park Service recreational expertise to be applied to reservoirs being formed behind huge federal dams, such as Hoover and Grand Coulee. Since then, they have included parks that serve new purposes, such as expanding and enriching our holdings in various ecosystems.

The Grand Canyon is one of the earth's most dramatic physical features, and on an average summer day, it seems as if everyone wants to see it at once. More than 4.5 million people visited the canyon in 1993, and there appears to be no limit to the number of future visitors. Unfortunately, there is only one readily accessible series of viewpoints, along the rim, and getting to know the canyon better requires a real commitment in time and effort. It can be explored by air, trail, and river.

To many, the best way to see the canyon is by plane. One certainly sees more of it this way, and it is quick, easy, and relatively inexpensive. Consequently, 821,000 people took air tours in 1997. The impact of the flights on the canyon has been enormous. At one time, it was hard to find a quiet moment to contemplate the canyon's grandeur, no matter how hard one worked to reach a remote spot, as small planes and helicopters blanketed the canyon with noise. In 1986, a midair collision killed 25 people and focused attention on the overflight controversy, which had been building for years. Now Federal Aviation Administration regulations restrict aircraft to a minimum height over the canyon (4425 meters; 14,500 ft) and to a certain zone, which ensures that almost all rim visitors and 90% of backcountry users are not bothered by overflights.

Walking into the canyon is much harder than it first appears. Because visitors begin by going down, once they get tired and start back, they have to double their efforts to retrace their steps. It is also hot in the canyon during the main summer visitor season, and with every meter descended, it gets hotter and drier. A true wilderness experience means an overnight trip in the canyon, which requires a wilderness permit, and the permits are hard to get. So most people take day hikes into the canyon, an often harrowing experience for the reasons mentioned. Yet hikers on the first 5 kilometers (3 mi) of the canyon's major trail, the Bright Angel, make it one of the most heavily used backcountry trails in the United States. Trying to preserve the natural environment while providing for the visitor's protection and comfort is difficult indeed.

Rafting the Colorado River through the Grand Canyon is widely considered as one of the country's great adventures. Yet it was almost a hundred years after Major John Wesley Powell made the first trip through the canyon before 1000 people had repeated the trip; the rapids on the river are numerous and difficult. The advent of better-quality equipment, more adventuresome travelers, and commercial rafting companies has led to more than 16,000 people a year now making the trip, a limit set by the National Park Service that is supposed to represent the *carrying capacity* of the canyon. This is the number of river runners that can tour the canyon without inflicting environmental damage. It presupposes a high level of low-impact use by the visitors. Everything, including human waste, must be carried from the canyon, and rafting companies must adhere to a rigid schedule of departures and takeouts.

A once-isolated beach where Deer Creek joins the Colorado River at the bottom of the Grand Canyon is now visited by some 16,000 people a year who traverse the canyon in rubber rafts, wooden boats, and kayaks.

Water is available at the Three Mile Shelter on the Bright Angel Trail, which leads from the south rim of the Grand Canyon to the Colorado River, a distance of 12.4 kilometers (7.7 mi).

Recreational Resources

Figure 11.12a

At 5489 meters (18,008 ft), Mount St. Elias, in Wrangell-St. Elias National Park, Alaska, is the second highest mountain in the United States.

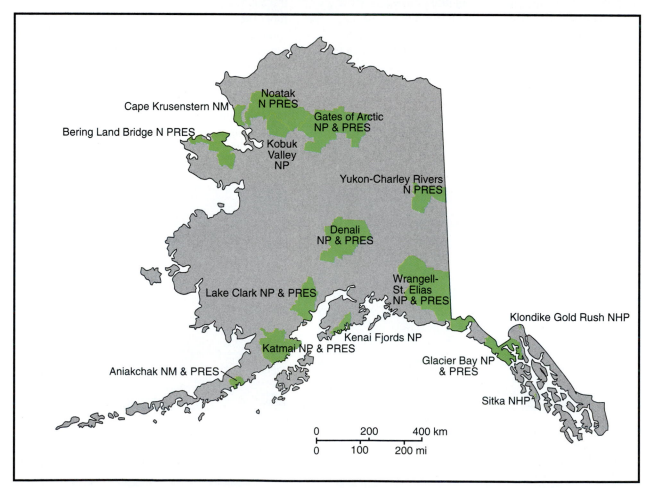

Figure 11.12b

The ten national preserves in Alaska occupy more than 86,000 square kilometers (33,200 sq mi). Seven of the preserves have adjoining national parks, counted as separate units of the National Park System. Other national preserves are located in Florida and Texas.

Figure 11.13
Civil War sites, such as Vicksburg National Military Park, Mississippi, shown here, are an important part of the system of National Historical Parks in the United States.

National Seashores These were established following the realization in the 1950s that if something was not done, a treasured part of the country would vanish into private ownership. The problem was especially critical on the Atlantic and Gulf Coasts, and it is there that all but one of the ten national seashores (Point Reyes in California) are located.

Rivers and Lakes These were protected for the same reason. During the big dam-building era following World War II, it looked as if a free-flowing stream was something future generations would see only in history books. The establishment of the National Wild and Scenic Rivers System in 1968 eventually obviated the need for such areas as Ozark National Scenic River, and establishing four national lakeshores around two of the Great Lakes fulfilled the most pressing need in that category.

Urban Parks The most unique and most heavily used units of the national park system are national recreation areas that lie within or just adjacent to large cities. They are also very expensive, so there are only four of them: Golden Gate (San Francisco), Gateway (New York City),

Santa Monica Hills (Los Angeles), and Cuyahoga Valley (Cleveland) (Figure 11.14). The basic purpose of such units is to "bring the parks to the people," because most national parks are located in lightly populated areas and there is still a sizable minority of Americans who have never visited a national park. Having large areas of reasonably scenic open land still available is the basic criterion for such parks.

Management of the National Parks

Recreation on 90% of the landscape of the United States takes place in areas that have been, or will be, altered by development or human use. The designated wilderness areas of the United States, discussed below, are an attempt to protect the natural landscapes by preventing *all* development and severely restricting use. The national parks are areas where an attempt has been made to do both: preserve the natural scene and make it available to the visitor. How that is to be achieved has been a source of conflict within the National Park Service for at least half a century, a conflict that reflects a broader divergence of views about how publicly owned land resources should be used.

Figure 11.14
The closing of the Presidio of San Francisco, a U.S. Army post, has led to a strong planning effort to ensure the best use of this 567-hectare (1400-acre) portion of the 29,638-hectare (73,180-acre) Golden Gate National Recreation Area.

Figure 11.15
The capacity of this park-type road between Tower Junction and the northeast entrance of Yellowstone National Park is low, permitting closer contact with the surrounding landscape than would a wider, more modern road.

The **preservationist ethic** calls for preserving such areas from commercial development, keeping them as ecological reserves, and using them only for nondestructive types of outdoor recreation. According to the **utilization ethic,** however, land resources are economic goods that should be developed to maximize profits. Some people argue that it is possible to strike a balance between preservation and utilization; the **multiple-use ethic** would permit the use of resources for a variety of purposes (for example, logging, recreation, and wildlife conservation) as long as such use did not damage or diminish the capacity of the land for self-renewal.

These philosophical differences pertain to how one views the purpose of national parks. Are they primarily repositories of biological and scenic diversity of unique national significance, recreational and tourist areas, or underused resources that could easily withstand more economic exploitation? The answer to that question bears directly on the goals and techniques of park management, relating to such issues as road and trail building, construction of hotels and other accommodations, and wildlife preservation. Currently, the NPS spends 90% of its budget servicing visitors rather than protecting resources.

All parks are reached by some form of mechanized transportation. In Gates of the Arctic National Park in Alaska, it may be an airplane, and in Isle Royale National Park in Lake Superior, it is by boat, but most national parks are reached by automobile. If preservationists have their way, the roads that have been built to serve the parks are all that ever will be constructed. Although there have been more or less continuous fights over the size of the roads, very few four-lane roads have been built or traffic lights installed in recent years (Figure 11.15). No roads have been built into wilderness areas of any of the national parks for over half a century. Future debates are likely to occur over whether or not public transportation, bus or rail, can replace the way people use some of our most crowded parks, like Yosemite, and whether some roads can actually be removed.

Figure 11.16
In the early decades of national park management, nature was frequently manipulated. In this case, trees were planted at the sagebrush-covered site of Mammoth Campground, Yellowstone National Park in Wyoming.

The building of accommodations peaked half a century ago. Although there has been considerable upgrading of accommodations in the national parks, the "pillow count" has remained about the same since before World War II. As the nation's transportation system has been upgraded, it has become easier to stay outside the parks, visit them, and return in one day. The chief question now is whether or not to reduce the number of accommodations to improve environmental quality.

At one point, the building of campgrounds was not considered an intrusion on the environment, and new ones were added according to demand. The result was a tremendous building program during the 1960s. The policy collapsed as three things occurred: (1) park managers realized that camping with the new metal-sided trailers and recreational vehicles is no more environmentally benevolent than hotels, (2) fees were instituted, and (3) private campgrounds proliferated in the areas surrounding the parks (Figure 11.16).

Increased attention is being given to improving the quality of buildings in the parks, from rangers' quarters to cafes. There are still recreational slums in the park service,

private and public, and the NPS is attempting to obtain funds to upgrade them. It is also trying to get a policy that will ensure that concessionaires' buildings will adhere to specified standards.

Trying to find the best policy for the preservation of wildlife is one of the most difficult problems the NPS has faced in the last few decades. At one point, an extreme anthropomorphic viewpoint was taken toward wildlife: "bad" animals like the wolf and cougar were exterminated, "good" animals like the buffalo were raised on ranches, and "interesting" animals like the grizzly bear were fed on floodlit platforms while visitors in bleachers watched. Trying to adopt a hands-off policy toward the animals and let nature take its course has proved a daunting task. One of the best examples of this, and the park faced with some of the most difficult management problems, is Yellowstone National Park.

Case Study: Yellowstone National Park
Yellowstone is the most active thermal area in the world. Due to the commercial use of the thermal areas of Iceland and New Zealand, Yellowstone also contains nearly all the

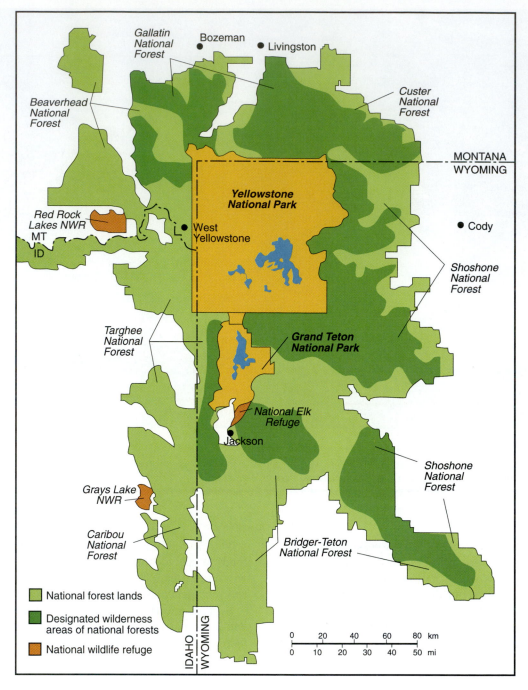

Figure 11.17

The Greater Yellowstone ecosystem is made up of Yellowstone and Grand Teton National Parks, national forests, wilderness areas, wildlife refuges, and private land, a total of 7.3 million hectares (18 million acres).

large geysers in the world. The Grand Canyon of the Yellowstone, with the falls of the Yellowstone River at one end, has a combination of scenic factors without parallel (see Figure 11.10). With the exception of Denali in Alaska, Yellowstone is probably the best place in America to see wildlife.

One of the major problems facing Yellowstone, and indeed all of the parks, is that park boundaries no longer ensure environmental protection from the area outside. Ani-

mals and pollutants move across borders, and ozone depletion and global warming pose long-term threats. Nowhere is this more evident than with the wildlife of Yellowstone. To ensure their survival, environmental attention must be given to what is called the **Greater Yellowstone ecosystem** (Figure 11.17), a 7.3 million hectare (18 million acre) area made up of Yellowstone and Grand Teton National Parks and the surrounding national forests, wildlife refuges,

Figure 11.18
The conflict over the reintroduction of the gray wolf to Yellowstone National Park has been intense, pitting local ranchers against wildlife advocates. Before their virtual extermination by government agents and bounty hunters over 60 years ago, hundreds of thousands of wolves inhabited the western United States.

and wilderness areas—one of the largest, relatively natural, temperate ecosystems on earth.

The most dramatic animal in the park, and the one most in danger, is the grizzly bear. At the top of the food chain, with no enemy but people, it is, ironically, the only animal that people have to fear in the park. The grizzly has been a part of the park from the beginning, less seen than its more docile cousin the black bear, but it was not endangered until 1970, when the NPS closed the open garbage dumps in Yellowstone, cutting off a food resource on which the grizzly had become dependent. With the loss of its habitual food supply, the grizzly went directly to the source, primarily campers. To protect campers, the NPS had to kill 23 grizzlies in the ten years following the dump closing. Whether this action alone made the grizzly an endangered species is doubtful, but the bear's future in the lower 48 states (there is a stable population in Alaska) is very much in doubt.

If humans and the grizzly are to coexist in Yellowstone, it will be necessary to separate them, and to do that, all human food sources must be secured. Every campground in the park stipulates that no food or article used in food preparation is to be left anywhere that bears can get to it, unless food is actually being prepared or eaten. This policy departs significantly from that of the 1960s, when black bears stood at the side of the roads in Yellowstone and begged for food. Officials hope that removing artificial food sources will do for the grizzly what it did for the black bear—make it wild again. Outside the park, however, only by keeping prime grizzly habitat natural in the face of potential competing ranching, logging, or mining demands will we ensure the bear's survival. That may happen in the Greater Yellowstone ecosystem as service industries and tourism, which put a premium on the natural landscape, increase relative to such basic industries as ranching.

The largest herd of elk in North America and one of the largest herds of buffalo anywhere also run up against the problem of boundaries in Yellowstone Park. The average elevation of the park is more than 2440 meters (8000 ft); it contains outstanding summer range but poor winter range for the animals. The normal migration patterns were north along the Yellowstone River to lower country in Montana and south down the Snake River into Wyoming. The migration routes were closed by both development and hunters at the park boundaries. In the past, animals that exceeded the carrying capacity of the winter range were removed; in the future, it is hoped that winter range outside the park can somehow be obtained.

The only animal no longer present in the Yellowstone ecosystem that was there historically is the gray wolf (Figure 11.18). Before about 1940 and an increased understanding of ecological balances, the National Park Service exterminated the gray wolf. The simple solution to refilling this vacant ecological niche would be to reintroduce the wolf, which is present in fair numbers a few hundred kilometers to the north. For years, that reintroduction was fought by ranching interests surrounding the park, although ranching produces far less income than tourism (which would benefit from the presence of the wolf) and the public in Wyoming, Montana, and Idaho surrounding the park has indicated it favors the reintroduction. In 1995, the U.S. Fish and Wildlife Service started a program that introduced 15 wolves per year for several years in Yellowstone and the same number in national forest lands in central Idaho.

As the oldest national park in the world, Yellowstone has always tended to be the leader in the application of new concepts in park management. The great Yellowstone fire of 1988, one of the largest wildfires in American history, may have changed forever the way we view forest fires. By regarding fires as harmful, we have allowed an enormous amount of unburned fuel to accumulate in the national parks and indeed throughout the forested areas of the country. So when the Yellowstone fires began, during one of the driest periods on record, nothing could stop them, despite the use of 10,000 firefighters and the expenditure of $120 million. The NPS had 16 years before set a "let burn" policy, but a "controlled burn" policy (a planned surface burn) may have to be instituted to make up for the years during which natural fires were extinguished. Yellowstone, incidentally, recovered beautifully from the fires, and even tourism has not suffered.

Other U.S. Federal Recreational Resource Lands

In addition to national parks, the federal government manages a variety of other types of publicly owned lands, six of which are described below. All allow some form of recreational use.

National Forests

The 153 **national forests** and the 18 national grasslands of the United States comprise more than 77 million hectares (191 million acres) and contain a majority of the country's wildlife, much of its timber, and many of the campgrounds, trails, wild rivers, and wilderness that Americans count on for outdoor recreation (Figure 11.19). A basic difference between the national parks and the national forests, however, has a major influence on the kind of recreational activities available in the national forests: national forests are open to commercial uses, such as logging, mining, grazing, and drilling for oil and gas, while national parks are not. In some areas, this may not be much of a problem. In areas of southern California, for example, where the slopes are steep, the timber is limited, and the population is large, the use of the land for watershed protection and recreation instead of for lumbering is seldom debated. Where there are large amounts of commercial timber, however, recreational pursuits, such as wilderness travel, nearly always take second place to timber production.

By law, the national forests are to be managed under the principle of *multiple use,* which is intended to balance the needs of recreation and wildlife with those of such developmental activities as logging and mining. Although no use is to be particularly favored over others, conservationists charge that the U.S. Forest Service has supported too much commercial logging and that the forests are being cut at an unprecedented rate. They are especially concerned that nearly half of the billions of board feet of timber taken from the national forests has come from irreplaceable "old growth" forests in Oregon and Washington—virgin forests containing trees that are among the tallest and oldest in the world. They claim, too, that the forest service allows destructive methods of harvesting (particularly clear cutting), that rates of reforestation are inadequate, and that the federal government loses an average of $500 million per year on timber sales because building and maintaining the logging roads costs far more than the timber companies pay for the wood (Figure 11.20).

The U.S. Forest Service, especially in the past, tended to lease out land for recreation. Summer homes, for example, never a feature of the national parks, are common in many national forests. The same is true of institutional camps (run by churches, schools, and other organizations), and private recreational ventures, such as ski areas. With the increasing demand being put on forest service areas by recreationists, however, leases are sometimes not renewed and new ones are not given. Other differences between the national forests and national parks include the permitting in the forests of hunting, mountain bikes on nonwilderness trails, and pets in most areas, along with generally less stringent environmental rules.

One example of recreational use of the national forests is the Laguna Mountain Recreation Area, a 3240 hectare (8000 acre) area within the 200,000 hectare (one-half million acre) Cleveland National Forest of southern California. It was established by presidential proclamation in 1926 to give the people of San Diego County a pleasant wooded portion of the peninsular range where recreation would have top priority. The forests of the area—oak, pine, fir, cedar—are in sharp contrast to the Mojave Desert visible down a 1220 meter (4000 ft) escarpment to the east. Developed and dispersed recreational activities are common; the recreation area contains four campgrounds, two picnic areas, private recreational cabins, institutional camps, and several kilometers of trails that are open for horseback riding, mountain biking, and hiking. A section of the Pacific Coast trail, which stretches from Mexico to Canada, runs through the area. Continuing controversies within the area include the potential use of an abandoned air force base, once proposed as a minimum security prison,

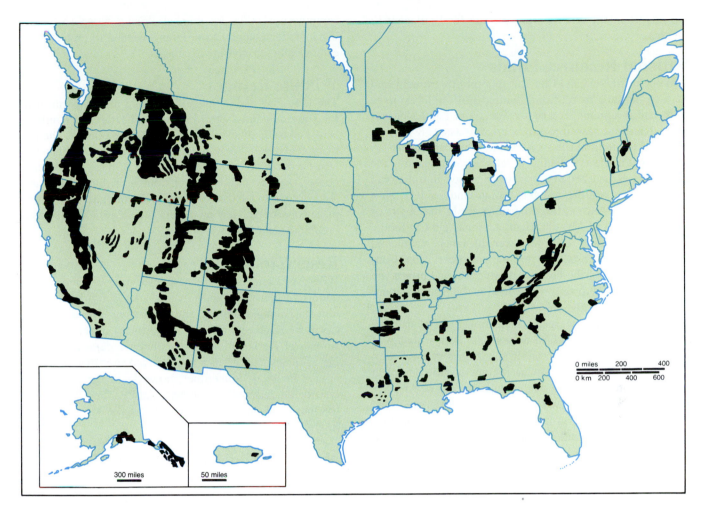

Figure **11.19**

National forests of the United States. Since their establishment in 1905, national forests have been devoted almost entirely to making grazing, mineral, and timber resources available to private industry. Conservation groups contend that the government loses money on the operation of the forests and hence subsidizes the destruction of resources it should protect.

Figure **11.20**

Clear cutting, one method of tree harvesting, removes all the trees from a given area at one time. The site is then left to regenerate naturally, or is replanted, often with fast-growing seedlings of a single species. In addition to reducing the recreational value of the area, excessive clear cutting destroys wildlife habitats, accelerates soil erosion and water pollution, and replaces a mixed forest with a wood plantation of no great genetic diversity.

Recreational Resources

and development threats from private land adjacent to the area.

National Resource Lands

A majority of land held by the federal government and open to outdoor recreation is called **national resource lands** and is administered by the Bureau of Land Management. National resource lands encompass 138 million hectares (341 million acres), a third of which are in Alaska. This is land that was essentially "left over" after land in the United States had become private or turned into national forests, parks, wildlife refuges, Indian lands, and defense lands. Besides Alaska, large portions are in the desert or semidesert areas of Nevada, Utah, Arizona, and California. The land was being used, however, by ranchers, miners, and recreationists, usually with minimal supervision. Consequently, much of the use was destructive, and efforts are being made today to restore the land and develop and enforce protective regulations.

The form and degree of recreational use of national resource lands have much to do with the lasting quality of these lands. Many of them are in the deserts of the Southwest, one of the fastest growing regions of the country. It was obvious that the expanding populations of Los Angeles, San Diego, and Phoenix would result in increased use of the lands surrounding the urban areas. What was not obvious was the explosive parallel growth in technology and popularity of off-road vehicles (ORVs)—especially jeeps, motorcycles, and various all-terrain vehicles (Figure 11.21). The landscape cover in these fragile environments began to disappear under the impact of increasingly numerous weekend visitors using ORVs for their recreation. Today, however, nature study, hiking, mountain biking, and other noncon-

sumptive uses of the desert are also rising in popularity, giving important support to programs needed to protect these ecologically sensitive environments.

Wildlife Refuges

The U.S. Fish and Wildlife Service in the Department of the Interior administers 35 million hectares (87 million acres) of land, almost 90% of which is in Alaska. The primary goal of wildlife refuges is to protect the wildlife while also encouraging the recreational use of the area. Wildlife observation and birding are obvious uses, but hunting and fishing are also allowed in many of the refuges. The "showplace concept" of viewing towers and identification signs is present in some places.

Reservoirs

Wherever dams are built and reservoirs created, recreationists usually follow, pursuing such activities as boating and fishing, camping, and picnicking. The U.S. Army Corps of Engineers and the Department of the Interior's Bureau of Reclamation are the major federal dam-building agencies and are usually involved in providing recreational facilities. Sometimes, however, if the reservoir is considered outstanding, the National Park Service is given the job, as in the Glen Canyon and Lake Mead National Recreation Areas.

National Wilderness Preservation System

The **National Wilderness Preservation System** (NWPS) is the country's first attempt at preserving totally natural landscapes. When the National Park Service was formed,

Figure 11.21
Motorcycles, dirt bikes, dune buggies, and other types of off-road vehicles can cause extensive damage to the sensitive ecosystems of deserts and seashores. They also reduce the aesthetic experience of those who come to enjoy nature.

JOHN MUIR AND THE HIGH SIERRA WILDERNESS

America's best-known naturalist was a man of Scottish birth named John Muir. From an early age, Muir showed a great love of nature, particularly when as a young man he moved to California and got his first glimpse of the Sierra Nevadas. Working to save portions of the Sierras as national parks and gaining lasting fame from his writing, he also saw that love of nature was not enough. As he watched natural beauty succumb to exploitation and development, he realized that people have a responsibility to protect that landscape. Muir became involved in numerous battles to save the Sierras from lumbering and overgrazing, and he helped found the Sierra Club in 1892 to both introduce wild country to its members and sell them the message of preservation.

Muir wrote of the Sierras: "And after ten years spent in the heart of it, rejoicing and wondering, bathing in its glorious floods of light, seeing the sunbursts of morning among the icy peaks, the noonday radiance on the trees and rocks and snow, the flush of alpenglow, and a thousand dashing waterfalls with their marvelous abundance of irised spray, it still seems to me above all others the Range of Light." (*Gentle Wilderness: The Sierra Nevada*, edited by David Brower, Sierra Club ed. [New York: Ballantine, 1968]).

Muir's High Sierras look much as they did a hundred years ago, thanks to his efforts and those of preservationists who followed. The most spectacular section of the Sierras is roadless from Lake Isabel near Bakersfield to the Tioga Road in Yosemite, a distance of 265 kilometers (165 mi). The High Sierra wilderness is made up of designated wilderness areas in Yosemite, Kings Canyon, and Sequoia National Parks, seven national forest wilderness areas, and roadless areas of the U.S. Forest Service and Bureau of Land Management. It contains 1,134,000 hectares (2.8 million acres) of great forests, including the giant sequoia; the deepest gorge in North America (Kings Canyon); the tallest mountain in the conterminous United States, Mount Whitney; lakes, streams, waterfalls, meadows, and wildlife. It is American wilderness at its finest. Most of the area is linked by the John Muir Trail, a 350-kilometer (220-mi) trail from the summit of Mount Whitney to Yosemite Valley.

The beginning of the 350-kilometer (220-mi) John Muir Trail, along Vernal and Nevada Falls in Yosemite National Park, California, is spectacular. The end is even more spectacular: the summit of Mount Whitney, the highest peak in the conterminous United States.

Recreational Resources

the major goal was to preserve nature from commercial exploitation, such as logging, mining, agriculture, and water-resource development; recreational development was not considered a problem. The visitation explosion that hit the parks after World War II showed this to be a naive assumption, and protecting park wilderness from tourist development was one of the purposes of a "wilderness bill."

The main impetus for wilderness preservation came, however, from areas in the national forests. In the late 1920s, Aldo Leopold, working for the U.S. Forest Service, persuaded it to establish a Gila Wilderness Area in a national forest on the border of Arizona and New Mexico. This was later expanded to a national system of roadless areas throughout the United States, a system improved by Bob Marshall, the head of recreational services in the forest service during the 1930s. Unfortunately, the system was only as strong as the chief of the forest service wanted it to be, and after 1945, the increasing demand for lumber led to the building of roads into some wilderness areas. The battle to establish a legally protected, national system of wilderness areas on federal land began, dominating conservation attention between the late 1950s and 1964, when President Lyndon B. Johnson signed the Wilderness Act.

That legislation defined **wilderness** as "an area of undeveloped land which is affected primarily by the forces of nature, where man is a visitor who does not remain; it contains ecological, geological, or other features of scientific or historic value; it possesses outstanding opportunities for solitude or a primitive and unconfined type of recreation; and it is an area large enough that continued use will not change its unspoiled, natural conditions." The NWPS is a complex system within which certain rules apply. Essentially, there is to be no development: no roads, buildings, motors, or vehicles. As in the national parks, there is to be no commercial use: no logging, mining, grazing, or dam construction. It differs from the national parks system in that no specific government agency administers the system; each agency is responsible for the areas it controls.

Initially, the NWPS consisted of wilderness areas that had been established in the national forests. Procedures were adopted for mapping, studying, and holding hearings on potential additions to the system from areas in the national parks, wildlife refuges, and public domain. The process was to be completed by 1984, but as late as 1999, some states' entire portions of the NWPS still were not established. As might be expected, each proposed wilderness area addition pits those who want more wilderness against those who want less. There are currently 36 million hectares (88 million acres) in the NWPS, almost two-thirds of them in Alaska (Figure 11.22).

The value of the wilderness areas includes everything from high-quality recreation, wildlife preservation, and watershed protection to slowing global warming and saving biodiversity, yet most people are just happy that there are some areas in the country that will remain natural as long as the country exists. The major attraction of wilderness areas is not that one can see landscape features not accessible by car but that one can view landscapes that are free of roads, power lines, logged hillsides, buildings, and all the trappings of civilization, which makes the most ordinary wilderness landscape special to many people.

National Wild and Scenic Rivers System

In 1968, when the bill establishing the **National Wild and Scenic Rivers System** was signed into law, it looked as if our children would have to consult a book or travel abroad to see what a free-flowing river looked like. The United States was in an era of dam building, and a look at a list of all the authorized dams in the country presaged a day when undammed rivers would be very rare indeed (Figure 11.23). The new bill attempted first of all to prevent building dams on streams included in the system and classified streams as wild, scenic, and recreational, protecting them from development. After a rather slow start, the speed of authorization has escalated in recent years, with nearly 16,000 kilometers (9936 mi) of river now in the system.

The Future

Two trends in recreation in the United States are on a collision course: the development of recreational facilities to meet the increasing demand and the fight to preserve the natural landscape, which is threatened by resource exploitation and development, including that of recreational facilities. The future will see more Disneylands, picnic tables, bike trails, campgrounds, and other developments designed to make the recreational areas and facilities in our cities more usable. The days are long gone when demands for recreation could be met simply by building a new road into a previously unused area.

San Francisco illustrates a few things that can be done. In the 1950s, a freeway was being built that would have sliced the downtown into several islands; massive high-rises were being erected along the waterfront area, which eventually would have created a private wall between the city and the bay; and finally, shallow parts of the bay itself were being filled in to create new land. Local environmental activists stopped the freeway, no more high-rises were built adjacent to the bayfront, and filling the bay was prohibited. Instead, the first major rapid transit system built in this country in half a century, the Bay Area Rapid Transit, was created. An already good bus and trolley system was revitalized, directly benefiting the recreational use of the area and indirectly preserving the quality of the downtown area. The halt on bayfront high-rise construction has allowed the development of an enormous tourist complex along the bay that

Figure 11.22

The Absaroka-Beartooth Wilderness Area, north of Yellowstone National Park. These undeveloped areas, left in their natural state, provide a refuge for wildlife, a laboratory for ecological research, and an opportunity for solitude and primitive forms of recreation.

includes the famous landmarksof Ghirardelli Square, the Cannery, and Pier 39 (Figure 11.24). Preventing the filling of the bay has preserved the major feature of the San Francisco landscape.

As the United States enters the 21st century, we must realize that our resource base is not the same as at the start of the 20th century. Our mineral resources have been severely depleted, our first-growth forest is essentially gone, and no good agricultural land is unused. Our scenic resources, though—the Grand Canyons, Rainiers, and Yosemites—are all still there, thanks to our ancestors' foresight in establishing the national parks over a century ago. Vast tracts of wilderness remain, especially in Alaska, but also in every western state and in many of the rest. These will form an increasingly valuable element in the balance of trade in this country, as the world comes to see the many scenic wonders in our public parks as well as such features as Disneyland, the Golden Gate Bridge, and the skyline of New York City. They will also add immeasurably to the quality of life of our citizens, which in the long run is what it is all about.

Recreation in Canada

Canada, the second-largest country in the world, has a correspondingly large number of recreational resources. With a population density a tenth that of the United States, moreover, it is rich in wilderness, natural beauty, and spaciousness. According to the United Nations list of national parks and protected areas, Canada is behind only Australia, the United States, and Greenland in acreage of protected areas: national parks, wildlife preserves, and wilderness. Canada also has a large number of world-class urban tourist destinations in such cities as Montréal, Ottawa, Québec City, Toronto, and Vancouver.

Atlantic Provinces and Québec

The easternmost provinces of Canada contain a number of famous islands and peninsulas, such as Nova Scotia, Prince Edward, and Newfoundland, and some of the most impressive coastal scenery in the country. The connection with the sea is strong, with fishing villages, excellent sport fishing, and uncrowded beaches. The French, English, and Scottish backgrounds of many of the people give a European flavor to the landscape of farms and small villages.

Québec, a province larger than Alaska that includes much of the Labrador Peninsula, is best known for its dominance by French-speaking people. A trip to Québec City is like visiting a typical French city. The city has a beautiful location not far from the mouth of the St. Lawrence River and derives part of its fame from the

Recreational Resources

Figure 11.23

The Kern River in California, included in the U.S. Wild and Scenic Rivers System, is undeveloped from this point, a few miles above Lake Isabella, to the river's source in Sequoia National Park.

spectacular Château Frontenac, a 19th-century hotel that overlooks the city (Figure 11.25).

Montréal, the second-largest French-speaking city in the world, was home to a world's fair in 1967 and the summer Olympics in 1976, events that introduced millions of visitors to one of North America's most interesting cities. Sites from both events, Expo 67 and Parc Olympique, are important tourist attractions today.

Ontario and the Prairie Provinces

Ontario contains the most prosperous section of Canada and a thriving tourist trade. The primary areas for recreation and tourism are the Great Lakes, Toronto, Niagara Falls, and Ottawa. Ontario adjoins all the Great Lakes but Lake Michigan and has close contact with major population centers in the United States. Toronto, the largest city in Canada, has an impressive skyline and downtown and the world's tallest structure, the 553-meter (1815-ft) CN Tower. Niagara Falls, near Toronto, is one of the continent's most visited natural features. The falls are surrounded by tourist attractions on both Canadian and Ameri-

can sides, such as *Maid of the Mist,* a boat that takes tourists close to the base of the falls.

Ottawa, the Canadian capital, besides containing the parliament building and the National Gallery of Art, is a livable city with bicycle paths, parks, and traffic-free streets in the downtown. The Rideau Canal and changing of the guard in front of the houses of Parliament are other major attractions.

The open expanses of Canadian prairie and wheat country play host to a modest number of tourists. There are also thousands of lakes in Saskatchewan and Manitoba with associated wildlife and uncrowded water-based recreation. Winnipeg, isolated from other major cities, is the most important tourist center in the two provinces. In southern Saskatchewan, a Grasslands National Park is being established to preserve a representative sample of the Canadian Prairies, once the richest wildlife area in the country.

Alberta and British Columbia

For sheer scenic beauty, few areas in the world can exceed that encompassed in the five Rocky Mountain national

Figure 11.24

The conversion of the Ghirardelli Chocolate Factory (the tower to the right) into a commercial and tourist complex may have kept a wall of apartment buildings, like that to the left, from being built around San Francisco Bay.

Figure 11.25

The Château Frontenac, built in 1892 for the Canadian Pacific Railroad and located above Québec City, is a hotel, landmark, and tourist attraction. The older portion of the city, shown below the Château, reflects the architecture of French cities of the 18th century.

Recreational Resources

parks—Banff, Jasper, Yoho, Kootenay, and Waterton Lakes—along the Alberta–British Columbia border. Banff, once a railroad siding of the Canadian Pacific Railroad located near some hot springs, has the distinction of being the first Canadian national park, established in 1885, only 13 years after Yellowstone. The incomparable Lake Louise, Moraine Lake in the Valley of Ten Peaks, the Columbia Icefield, and the abundant wildlife are among the major attractions (Figure 11.26).

Calgary, in the plains to the east of the Canadian Rockies, hosted the winter Olympics in 1988 and is fa-mous for the annual Calgary Stampede, one of the world's great rodeos and western exhibitions. Edmonton is known for, among other attractions, an enormous enclosed shopping and leisure center with such features as a 2-hectare (5-acre) water park. Vancouver, besides its urban attractions, contains the surprisingly natural 405-hectare (1000-acre) Stanley Park close to the city center. Victoria, the British Columbia capital, is a beautiful city with gardens, museums, and the province's parliament building.

Figure 11.26
A major attraction of Banff National Park in Alberta, Canada, is glacier-fed Lake Louise, situated at 1731 meters (5680 ft) in the Canadian Rockies.

Northwest Territories, Nunavut, and Yukon

Northern Canada comes as close to being uninhabited as any large area on the earth's surface outside of Antarctica. The Northwest Territory and Nunavut, with an area larger than India, have a combined population of only 68,000, not much more than the capacity of the Toronto Blue Jays' Skydome. There is plenty of spectacular country, but the difficulty of transportation and lack of tourist infrastructure have limited their recreational potentials. The expected growth of adventure travel in the years ahead will help as fishers, river rafters, climbers, and nascent explorers tap this rich recreational resource. Kluane National Park in the Yukon Territory is accessible from the Alaska Highway and contains Canada's highest peak, 6050-meter (19,856-ft) Mount Logan.

Canadian National Parks and Protected Areas

The objective of the Canadian national parks is "to protect for all time, representative areas of Canadian significance in a system of national parks, and to encourage public understanding, appreciation, and enjoyment of this national heritage so as to leave it unimpaired for future generations." Although this is similar to the U.S. national park mandate, visitors to Canadian national parks, such as Banff with an entire town within its borders, may receive the impression that Canadian parks are not as well protected as those in the United States. This is largely a matter of many of the parks being isolated, with no nearby areas outside their boundaries where park facilities can be located; newer parks have received the same protection as those in the United States. In addition, Canada has plans for a significant expansion of its park system, which, when completed, would protect the biophysical diversity of Canada by locating national parks in each of 39 "National Park Natural Regions" based on physiography and vegetation. There are currently 34 national parks in only 21 of the areas, but study areas have been established or new parks proposed for the rest (Figure 11.27).

The 34 national parks total slightly more than 200,000 square kilometers (77,200 sq mi). They are part of a larger system of protected areas, including about 1200 provincial parks totaling 300,000 square kilometers

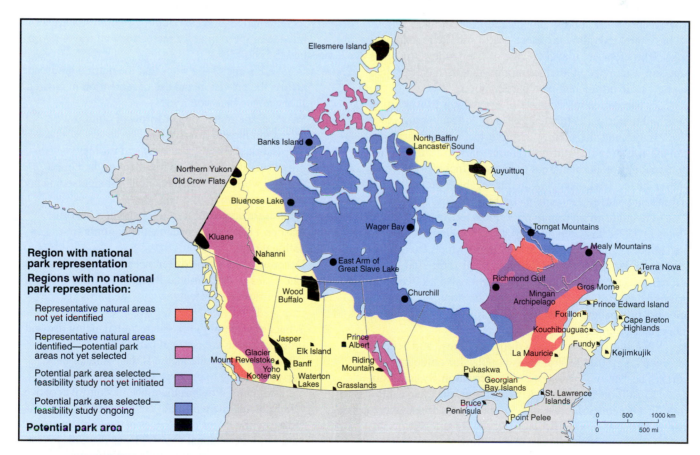

Figure 11.27
National park and natural regions in Canada.

Recreational Resources

(115,800 sq mi) and a number of national wildlife areas, bird sanctuaries, and nature reserves. Several Canadian protected areas have been recognized by the United Nations as biosphere reserves and world heritage sites. They include Wood Buffalo National Park, a 45,000-square kilometer (17,370 sq mi) park on the border of Alberta and the Northwest Territories that contains the world's largest herd of buffalo (6000) and is the summer home of the last of the whooping cranes.

Summary

Recreation is important because of the large amount of time we spend participating in sports and fitness activities, traveling for pleasure, and just taking it easy. Recreation is also important because of its financial aspects, from the manufacture of sporting equipment to the operation of recreational areas. Government agencies such as the National Park Service and the U.S. Forest Service spend much of their time and money managing federal land used for recreation. Privately owned land in the United States, two-thirds of the total, receives a majority of the country's recreational use.

Recreational activities in America have increased as the population has grown and become more urbanized and as work has become less physically demanding. The work week has become shorter and vacation time longer. Mobility has increased, allowing more use of remote recreational areas. The increase in average age of the country's population and a number of socioeconomic factors have resulted in changes in our recreational participation.

Urban recreation consists of everything from visiting zoos and theme parks to attending an opera and going shopping. Increasing the quality of the urban environment enhances the amount and enjoyment of recreation there. Recreation in nonurban areas commonly includes activities with a strong relationship to the physical geography of the area: skiing, surfing, mountain climbing, and river running.

The idea of national parks, started in the United States in 1872, has been adopted by 125 nations. There are 367 parks in the National Park System, which can be described under the headings of natural, historic, and recreational areas. Management of the national parks consists primarily of trying to balance the preservation of a reasonably pristine natural landscape with an annual visitation of more than a quarter billion people.

The national forests, wildlife refuges, and federal reservoirs are other federal areas popular for outdoor recreation. The National Wilderness Preservation System contains 36 million hectares that will be kept in their natural state in perpetuity for various forms of wilderness recreation. The National Wild and Scenic Rivers System attempts to ensure the existence of free-flowing streams.

Canadian recreational resources are concentrated along the Canadian–U.S. border and in the Rocky Mountains. There is an extensive Canadian national parks system and an even larger system of provincial parks. Two important foci of tourism in the country are the Banff-Jasper series of five national parks in the Canadian Rockies and the cities of Montréal, Ottawa, Québec, Toronto, and Vancouver. The Yukon, Northwest Territories, and Nunavut contain vast areas of spectacular landscapes, very few people, and a high recreational potential.

Key Words

Greater Yellowstone ecosystem	National Wild and Scenic Rivers System
multiple-use ethic	National Wilderness Preservation System
national forest	
national monument	preservationist ethic
national park	recreation
National Park System	theme park
national preserve	utilization ethic
national resource lands	wilderness

Gaining Insights

1. What are some of the economic implications of recreation? How do recreation and tourism affect your community?

2. What are the major influences on the demand for recreational resources? Would you opt for a higher income or more time for recreation and travel if you had the choice?

3. What are some of the ways that the natural landscape affects the provision and use of recreation in an area? In the area where you live?

4. What are the most popular forms of outdoor recreation in your area? Why?

5. The National Park Service has real problems preserving the natural environment of the Grand Canyon while making it available for sightseeing and outdoor recreation. Explain.

6. What were some of the geographic and historic reasons Yellowstone became the nation's first national park?

7. Describe some of the differences between the several categories in the National Park System. How do these differences affect the way the parks are managed?

8. What are some of the problems in balancing use and preservation in the national parks?

9. What are some of the methods the National Park Service is using to preserve the grizzly bear in Yellowstone? Why has there been a controversy over reintroducing the wolf into Yellowstone?

10. What are the major differences between the national parks and the national forests? Does the U.S. Forest Service have two incompatible assignments: protecting the forests while making their resources available for commercial use?

11. Where would you expect to find the major wilderness areas in the United States? What values do such areas have?
12. Why is tourism concentrated in the southern, southeastern, and western parts of Canada? Briefly describe Canada's national recreational resources.
13. Why do the Canadian national parks tend to be more "developed" than American parks, and why might this change over time?

Selected References

Canadian Environmental Advisory Council. *A Protected Areas Vision for Canada.* Ottawa: Minister of Supply and Services Canada, 1991.

Carothers, Steven W. and Bryan T. Brown. *The Colorado River Through Grand Canyon.* Tucson: University of Arizona Press, 1990.

Chubb, Michael and Holly R. Chubb. *One Third of Our Time?* New York: John Wiley & Sons, 1981.

Cordell, H. Ken, John C. Bergstrom, Lawrence A. Hartman, and Donald B. K. English. *An Analysis of Outdoor Recreation and Wilderness Situation in the United States: 1989–2040.* Fort Collins, Colo.: U.S. Department of Agriculture, 1990.

Environment Canada Parks Service. *National Parks System Plan.* Ottawa: Minister of Supply and Services Canada, 1991.

Everhart, William C. *The National Park Service.* Boulder, Colo.: Westview Press, 1983.

Foreman, Dave, and Howie Wolke. *The Big Outside.* Tucson: Ned Ludd Books, 1989.

Ise, John. *Our National Park Policy, A Critical History.* Baltimore: The Johns Hopkins University Press, 1961.

Junkin, Elizabeth D. *Lands of Brighter Destiny: The Public Lands of the American West.* Golden, Colo.: Fulcrum, 1986.

Keiter, Robert B. and Mark S. Boyce, eds. *The Greater Yellowstone Ecosystem.* New Haven and London: Yale University Press, 1991.

Knudson, Douglas M. *Outdoor Recreation.* New York: Macmillan, 1984.

National Park Service. *The National Parks: Index, 1993.* Washington, D.C.: Government Printing Office, 1993.

O'Brien, Bob R. *Our National Parks and the Search for Sustainability.* Austin: University of Texas Press, 1999.

Palmer, Tim. *Endangered Rivers and the Conservation Movement.* Berkeley and Los Angeles: University of California Press, 1986.

Rafferty, Milton D. *A Geography of World Tourism.* Englewood Cliffs, N.J.: Prentice-Hall, 1993.

Robinson, J. Lewis. *Concepts and Themes in the Regional Geography of Canada.* Vancouver: Talonbooks, 1989.

Rosenow, John E. and Gerreld L. Pulsipher. *Tourism: The Good, the Bad and the Ugly.* Lincoln, Neb.: Media Productions and Marketing, 1979.

Runte, Alfred. *Yosemite, the Embattled Wilderness.* Lincoln, Neb.: University of Nebraska Press, 1990.

Smith, Stephen L. J. *Recreation Geography.* London and New York: Longman, 1983.

Wright, R. Gerald. *Wildlife Research and Management in the National Parks.* Urbana and Chicago: University of Illinois Press, 1992.

Zaslowsky, Dyan and T. H. Watkins, the Wilderness Society. *These American Lands.* Covelo, Calif.: Island Press, 1994.

The object to be buried was large and heavy, weighing more than 1000 tons. It had taken 200 workers over three years to ready it for its funeral voyage, which was made on a 320-tire trailer welded to a barge called the *Paul Bunyan.* Setting off from the outskirts of Pittsburgh, Pennsylvania, the barge traveled down the Ohio and Mississippi Rivers, across the Gulf of Mexico, through the Panama Canal, northward along the Pacific Coast, and up the Columbia River to the outskirts of Richland, Washington—a 13,000 kilometer (8100 mi) trip. When the *Paul Bunyan* finally docked, workers with blowtorches scrambled aboard to unweld the transport trailer from the barge so it could make the last leg of the journey, a mere 48 kilometers (30 mi). The five enormous trucks required to haul the heavy trailer made a surreal funeral procession as they inched their way along dusty roads. Finally, they reached the gravesite, a large trench labeled simply "Burial Garden," and the object was placed inside (Figure 12.1). At a cost of $98 million, the country's first major nuclear funeral was over.

What was buried at the Hanford Military Reservation that spring day in 1989 was the spent core of the country's first commercial nuclear reactor. Built in 1957, the reactor at Shippingport, Pennsylvania, was intended to usher in an age of nuclear-power generation for peaceful purposes. Energy use, it was predicted, would be too cheap to monitor. Instead, it has become clear that nuclear power has created a host of problems for which we lack ready solutions. The reactor vessel buried at Hanford will continue to emit radiation for at least 100,000 years. No permanent storage site exists for the thousands of spent fuel rods of that reactor or the dozen others about to be retired, all of which are far more radioactive than when they were inserted in their reactors. And the Hanford Reservation is an environmental nightmare; its millions of cubic feet of nuclear waste have released millions of curies of radioactivity into the air, water, and soil.

As the preceding chapters have made clear, Americans and Canadians have used their bountiful natural resources to create wealthy, technologically advanced societies. At the same time, the use of those resources has had severe environmental impacts. Mining slag heaps, agricultural fertilizers, industrial waste water, and hundreds of other by-products of society have affected and continue to affect the quality of the water, air, and soil on which our existence depends. In this chapter, we examine some of the ways people have altered the natural physical environment.

Figure **12.1**

The spent reactor vessel from the Shippingport nuclear power plant in Pennsylvania lies alongside other radioactive waste in a burial trench, north of Richland, Washington. The trench will be filled in after more waste has been added.

Impact on Water

The supply of water on earth is constant, and it is over 4 billion years old. The system by which it continuously circulates through the biosphere is called the **hydrologic cycle** (see Figure 12.2). In that cycle, water may change form and composition, but under natural environmental circumstances, it is purified in the recycling process and is again made available with appropriate properties to the ecosystems of the earth. *Evaporation* and *transpiration* (the emission of water vapor from plants) are the mechanisms by which water is redistributed. Water vapor collects in clouds, condenses, and then falls again to earth. There, it is reevaporated and retranspired, only to fall once more as precipitation.

People's dependence on water has long led to efforts to control its supply. Such manipulation has altered both the quantity and the quality of the water supply.

Availability of Water

Most of the coterminous United States west of the 98th meridian is arid or semiarid, receiving less than 52 centimeters (20 in.) of precipitation a year. The availability of water is a concern in every western state, but California exemplifies the situation best, primarily because of two factors: (1) it has grown rapidly in recent decades to become the most populous of all states, home to roughly one in nine Americans, an expansion that has placed unprecedented demands on water supplies; and (2) a persistent drought plagued the state from 1986 to 1993.

California's Water Supply

Even in a normal precipitation year, Los Angeles receives only 37 centimeters (15 in.) of rainfall; San Diego, 26 centimeters (10 in.); and Bakersfield, 16 centimeters (6 in.). Contrast these precipitation averages with those in such cities as New York (106 cm, 42 in.) and Chicago (81 cm, 32 in.). Nearly all of California's rain falls in the winter, from November to April. It is not the rain that falls on the cities that provides most of the water needed to sustain the state, however. That comes from the runoff of melting snow in the High Sierras and the Rocky Mountains.

The cities, agriculture, and industry can exist only by moving water from those distant sources to where it is needed. More than 1400 dams tame such rivers as the Colorado, Owens, and Feather. Pipelines and concrete canals called aqueducts link reservoirs behind these dams with towns and cities. They snake their way around, over, and through the mountains, carrying water as much as 960 kilometers (600 mi) from the northern and eastern parts of the state to the more heavily populated south and west (Figure 12.3).

Surprisingly, it is not the cities that consume most of the water. Urban residents use less than 10% of developed

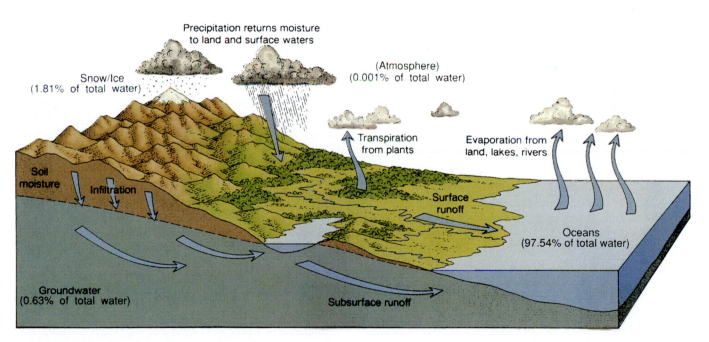

Figure 12.2

The hydrologic cycle. The sun provides energy for the evaporation of fresh and ocean water. The water is held as vapor until the air becomes supersaturated. Atmospheric moisture is returned to the earth's surface as solid or liquid precipitation to complete the cycle. Because precipitation is not uniformly distributed, moisture is not necessarily returned to areas in the same quantity as it has evaporated from them. The continents receive more water than they lose. The excess returns to the seas as surface water or groundwater. A global water balance, however, is always maintained.

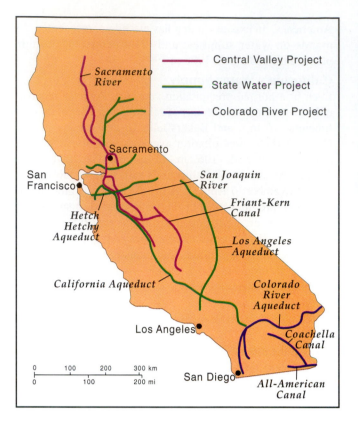

Figure 12.3

The network of dams, reservoirs, and aqueducts shown here has helped make California the richest agricultural state in the United States. Three huge canal systems carry the runoff of melting snow to the state's farms and cities: the State Water Project, the Colorado River Project, and the Central Valley Project (CVP). The CVP, run by the federal Bureau of Reclamation, is the single largest supplier of water, providing about 25% of all agricultural water in the state.

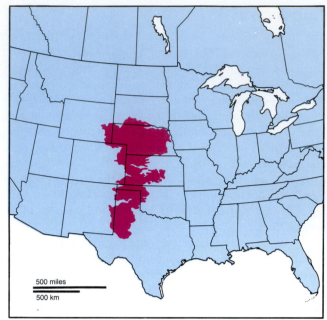

Figure 12.4

The Ogallala aquifer, the largest underground water supply in the United States. Water from the aquifer is used on 20% of U.S. irrigated land. The level of the water table is falling because irrigation systems are depleting the water faster than precipitation can replenish it.

water supplies, while farmers account for 80% of the total. Almost nothing of value could grow in California without irrigation. It has been estimated that it takes more than 4500 gallons of water to produce the typical food one person consumes in a day. The fruits and vegetables raised in the 800-kilometer (500-mi) long Central Valley account for a good deal of that water, but the four biggest uses of water are to irrigate pasture and alfalfa (for livestock), cotton, and rice.

Those four crops require nearly two-thirds of all water used in the state for agriculture and about one-half of the water consumed for all purposes. Each year, more than 4 million acre-feet of water are used to grow a single crop, alfalfa—more water than the state's 34 million residents use for all their household needs. (One acre-foot is the amount of water that it would take to cover an acre of land with one foot of water, or 325,851 gallons.)

Water in California is allocated through a combination of ancestral rights, federal subsidies, state and municipal laws, and other obstacles to its more efficient use. Some irrigation districts pay as little as $3 an acre-foot, while urban residential users pay as much as $600 for an equivalent amount.

The common perception is that the demand for water now exceeds available supplies, a belief fostered by the water rationing and conservation measures imposed in many California cities. One summer, for example, residents of Marin County were limited to 50 gallons of water per day per person, Santa Barbarans were forbidden to water their lawns, and the construction of new housing developments in the community of Lake Elsinore came to a halt.

The drought forced Californians to think deeply about their need for dependable supplies of water. Some argue that if water were allocated in a more efficient way, the already developed supplies could meet the demands for decades to come. If California farmers used just 10% less water, for example, that would double the total available to all residential consumers. The state has begun to create a market in water rights, allowing farmers to sell water to new users who value it more. Some cities have provided monetary incentives for water conservation and reclamation. More recently, the U.S. Congress passed legislation that will allow water trading on the Central Valley Project.

The High Plains

Another area where farming is placing extreme demands on the water supply is in the High Plains region, which stretches from South Dakota to Texas. Agriculture there depends on drawing irrigation water from a vast underground formation called the **Ogallala aquifer** (Figure 12.4). The largest underground water supply in the United States, the Ogallala aquifer spreads beneath 8 million hectares

(a)

(b)

Figure 12.5
The Kissimmee River in Florida before (a) and after (b) the U.S. Army Corps of Engineers turned it into a straight canal. The habitat once supported thousands of fish, waterfowl, and such wading birds as the wood stork, snowy egret, and great blue heron.

(20 million acres) in eight states. It supports nearly half the country's cattle industry, a fourth of its cotton crop, and a great deal of its corn and wheat.

The water bed is less than 60 meters (200 ft) thick in part of the region, but the 170,000 wells that now puncture the aquifer cause the water table to fall from 15 centimeters (6 in.) to 1 meter (3.3 ft) each year in some areas—a rate far greater than that at which the aquifer can be replenished by nature. Hydrologists expect that as much as 40% of the acreage irrigated by water from the aquifer will be lost in the next 15–20 years, causing economic hardship in the region.

Modification of Streams

To regulate the water supply for agriculture and urban settlements, to prevent flooding, or to generate power, North Americans have manipulated rivers by constructing dams, canals, and reservoirs. Although they generally have achieved their purposes, these structures can have unintended environmental consequences. These include reduction in the sediment load downstream, followed by a reduction in the amount of nutrients available for crops and fish and an increase in the salinity of the soil.

Channelization, another method of modifying river flow, is the construction of embankments and dikes and the straightening, widening, and/or deepening of channels to control floodwaters or to improve navigation. Like dams, these systems can have unforeseen consequences. By reducing the natural storage of floodwaters, they can aggravate flood peaks downstream. They also disrupt natural life in the former floodplain.

Until 1960, for example, the Kissimmee River in Florida twisted and turned as it traveled through a floodplain between Lakes Kissimmee and Okeechobee (Figure 12.5). The habitat supported hundreds of species of birds, reptiles, mammals, and fish. At the request of ranchers and farmers who had moved onto the wetlands and were disturbed by the tendency of the river to flood, the U.S. Army Corps of Engineers dammed and dredged the stream, turning 166 kilometers (103 mi) of meandering river into a dirt-lined canal only 90 kilometers (56 mi) long.

After the completion of the canal in 1971, the wetlands flanking the river disappeared, oxygen levels in the old channel plummeted, and alien plant species moved in. Fish populations declined drastically, and 90% of the waterfowl, including several endangered species, disappeared. In 1992, Congress approved a $280 million project to restore a 36-kilometer (22-mi) portion of the river. The canal will be filled with dirt to allow the plain to flood, and in time, the river will again form curves and oxbows on its way south.

Channelization and dam construction are deliberate attempts to modify river regimes, but other types of human action also affect river flow. Urbanization, for example, has significant hydrologic impacts, including a lowering of the water table and increased flood runoff. Likewise, the removal of forest cover increases runoff, promotes flash floods, lowers the water table, and hastens erosion. Nevertheless, the primary adverse human impact on water is felt in water quality. People draw water from lakes, rivers, or underground deposits to use for drinking, bathing, agriculture, industry, and many other purposes. Although the

The United States built its largest dams in the 1930s and '40s, when the Tennessee, Colorado, Columbia, and other rivers were dammed to prevent flooding and to provide safe navigation, hydroelectric power, and a dependable water supply. It was not a time when environmentalists held much sway, and the negative consequences of dam construction went largely unrecognized. The benefits, on the other hand, were evident. Folksinger Woody Guthrie's tribute to the damming of the Columbia River ran, in part:

> *And on up the river is Grand Coulee Dam,*
> *The mightiest thing ever built by a man,*
> *To run the great factories and water the land,*
> *It's roll on, Columbia, roll on!**

The best sites for water development in the United States have already been exploited, but this is not the case in Canada. Under construction in northern Québec is one of the world's biggest hydropower undertakings: the James Bay project. Sponsored by Hydro-Québec, the provincial utility company, the project began in 1971. The first phase of the project, known as James Bay I, entailed the diversion of three rivers to feed La Grande River and the construction of nine dams and five major reservoirs. The dams have flooded thousands of square kilometers of land—an area larger than Connecticut. All this was accomplished without any environmental impact assessments before construction began.

If Hydro-Québec has its way, James Bay I will be just the beginning. Under the next phase of the scheme (James Bay II, or the Great Whale project), four more rivers will be diverted and 21 more dams and additional reservoirs built to feed three large generating stations. Even farther into the future are plans to divert the Nottaway and Rupert Rivers into the Broadback, although growing opposition to the entire James Bay project makes completion of that phase uncertain.

Critics of the project include the region's indigenous population, the Cree and Inuit. The latter would be affected by James Bay II, but much of the Cree's ancestral burial and hunting grounds has already been submerged. Environmental experts charge that further flooding would not only threaten the Cree's source of livelihood but also endanger many species of wildlife that depend on the area's forest and wetlands. These include caribou, black bears, freshwater seals, and waterfowl. Mercury from submerged vegetation has already leached into reservoirs, and the level of mercury in the region's freshwater fish has increased sixfold since completion of the first phase. The fish are a dietary staple of the Cree, many of whom now have mercury levels that exceed the safety standards set by the World Health Organization. Finally, project critics note that Québec does not need the power that James Bay II would generate and that, in an effort to develop a market for the surplus power created by James Bay I, the utility encouraged the siting of water-polluting smelting plants along the St. Lawrence River valley.

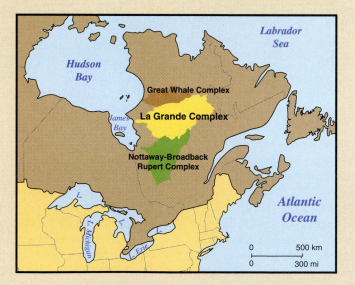

*ROLL ON COLUMBIA. Words by Woody Guthrie. Music based on "Goodnight Irene" by Huddie Ledbetter & John A. Lomax, TRO–© Copyright 1936. (Renewed), 1957 (Renewed) and 1963 (Renewed) Ludlow Music, Inc., New York, NY. Used by Permission.

water that is withdrawn returns to the water cycle, it is not always returned in the same condition as it was at the time of withdrawal. Like other segments of the ecosystem, water is subject to serious pollution problems.

Water Quality

As a general definition, **environmental pollution** by humans means the introduction into the biosphere of wastes that, because of their volume or their composition or both, cannot be readily disposed of by natural recycling processes. In the case of water, pollution exists when water composition has been so modified by the presence of one or more substances that either it cannot be used for a specific purpose or it is less suitable for that use than it was in its natural state. Pollution is brought about by the discharge into water of substances that cause unfavorable changes in its chemical or physical nature or in the quantity and quality of the organisms living in the water. *Pollution* is a relative term. Water that is not suitable for drinking may be completely satisfactory for cleaning streets. Water that is too polluted for fish may provide an acceptable environment for certain water plants.

Human activity is not the only cause of water pollution. Leaves that fall from trees and decay, animal wastes, oil seepages, and other natural phenomena may affect water quality. There are natural processes, however, to take care of such pollution. Organisms in water are able to degrade, assimilate, and disperse such substances in the amounts in which they naturally occur. Only in rare instances do natural pollutants overwhelm the cleansing abilities of the recipient waters. What is happening now is that the quantities of wastes discharged by humans often exceed the ability of a given body of water to purify itself. In addition, people are introducing pollutants, such as metals or inorganic substances, that cannot be broken down at all by natural mechanisms or that take a very long time to break down.

The four major contributors to water pollution are agriculture, industry, mining, and municipalities and residences. Table 12.1 shows the kinds of water pollutants associated with each source. It is helpful to distinguish between "point" and "nonpoint" sources of pollution. As the name implies, *point sources* enter the environment at specific sites, such as a sewage treatment facility or an industrial discharge pipe. *Nonpoint sources* are more diffuse and therefore more difficult to control; examples include runoff from agricultural fields and road salts.

Agricultural Sources of Water Pollution

In the United States and Canada, agriculture is the leading nonpoint source of water pollutants. Agricultural runoff carries three main types of contaminants: fertilizers, biocides, and animal wastes.

Fertilizers   Agriculture is a chief contributor of *excess nutrients* to water bodies. Pollution occurs when nitrates and phosphates that have been used in fertilizers and are present in animal manure drain into streams and rivers, eventually

---

Table 12.1
Sources and Types of Major Water Pollutants

Source	Type of Pollutant
Agriculture	Fertilizers (principally, nitrogen and phosphorous), biocides, animal wastes, sediments
Industry	Synthetic organic chemicals, including polychlorinated biphenyls (PCBs) and dioxin; toxic inorganic chemicals, including heavy metals; radioactive materials; heat discharge
Mining	Acids, chlorides, heavy metals
Municipalities and residences	Nutrients, organic materials, heavy metals, toxic chemicals, chlorides, sewage

---

accumulating in ponds, lakes, and estuaries. The nutrients hasten the process of **eutrophication,** or the enrichment of waters by nutrients. Eutrophication occurs naturally when nutrients in the surrounding area are washed into the water, but when the sources of enrichment are artificial, as is true of commercial fertilizers, the body of water may become overloaded with nutrients. The end result may be an oxygen deficiency in the water.

Figure 12.6 illustrates one form that overfertilization of a water body can take. Algae and other plants are stimulated to grow abundantly. When they die, the level of dissolved oxygen in the water decreases, primarily because of the bacteria acting on the dead and decomposing vegetation. Fish and plants that cannot tolerate the poorly oxygenated water are eliminated. In addition to being potentially lethal for fish, eutrophication affects the suitability of water for drinking and bathing, because excess nutrients have been shown to pose a health hazard to humans.

Scientists have estimated that as many as one-third of the medium- and large-size lakes in the United States are affected by accelerated eutrophication. Symptoms of a eutrophic lake are prolific weed growth, large masses of algae, fish kills, rapid accumulation of sediments on the lake bottom, and water that has a foul taste and odor.

The overfertilization of marine (salt) waters helps produce **red tides,** population explosions of microscopic algae. Depending on the makeup of the organism, these algal blooms also can be brown, yellow, or green. They occur most often in shallow, nearshore waters, coastal bays or estuaries and frequently are associated with fish kills. If the algae are thick enough, they block the sunlight, forcing out other small organisms and killing shellfish beds. As the algae die, their decay begins to consume much of the water's dissolved oxygen, and other marine life suffocates. In addition, some planktons produce toxins that can poison

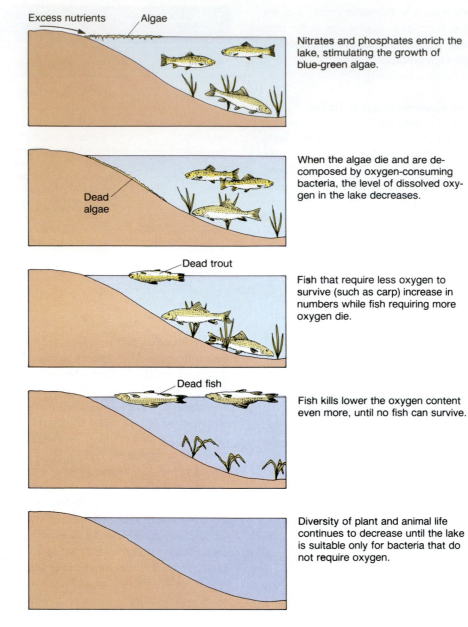

Excess nutrients   Algae

Nitrates and phosphates enrich the lake, stimulating the growth of blue-green algae.

Dead algae

When the algae die and are decomposed by oxygen-consuming bacteria, the level of dissolved oxygen in the lake decreases.

Dead trout

Fish that require less oxygen to survive (such as carp) increase in numbers while fish requiring more oxygen die.

Dead fish

Fish kills lower the oxygen content even more, until no fish can survive.

Diversity of plant and animal life continues to decrease until the lake is suitable only for bacteria that do not require oxygen.

Figure 12.6

Eutrophication is hastened by artificial sources of nutrients. Although accelerated eutrophication is primarily a result of agricultural activities, additional sources of nitrates in both surface and underground water supplies are urban drainage, industrial wastewater, and septic tanks.

other marine life and cause respiratory, neurological, or gastrointestinal illness in people. Red tides have occurred along all American coasts but in recent years have been most severe along the Atlantic, forcing beach closings at sites from Florida to Maine (Figure 12.7).

Biocides   The herbicides and pesticides used in agriculture are another source of chemical pollution of water bodies. Runoff from farms where such *biocides* have been applied contaminates both ground and surface waters. One of the problems connected with the use of biocides is that the

long-term effects of such usage are not always immediately known. DDT, for example, was used for many years before people discovered its effect on birds, fish, and water plant life. Another problem is that thousands of these products, containing more than 600 active ingredients, are now in wide use, yet very few have been reviewed for safety by the Environmental Protection Agency (EPA).

Biocide contamination of groundwater exists in at least 34 states. An EPA survey found biocides in about 10% of all community water systems, with 1% containing potentially unsafe concentrations. The situation is often

Figure 12.7
Red tides off the coast of central California. These sudden blooms of microscopic algae in the area around the boat, are becoming more common as populations around harbors increase. Shellfish poisoning following red tides has forced the closure of many mussel beds.

more serious in farm areas where residents depend on wells for their drinking water. Surveys in Minnesota and Iowa, for example, indicate that 30–60% of private wells may be tainted by runoff from farm herbicides and pesticides.

Animal Wastes A final agricultural source of chemical pollution is animal wastes, especially in areas where animals are raised intensively. This is a problem particularly in feedlots, where animals are crowded together at maximum densities to be fattened before slaughter. Large feedlots, such as the one pictured in Figure 12.8, produce as much waste as a large city. It is estimated that animal wastes in the United States total about 1.5 billion tons per year, with feedlots generating about half the total. If not treated properly, the manure pollutes both soil and water with infectious agents and excess nutrients.

Industrial Sources of Water Pollution
As Table 12.1 indicates, agriculture is only one of the human activities that contribute to water pollution. In the United States, about half the water used daily is used by industry. Many industries discharge organic and inorganic wastes into bodies of water. These may be acids, highly toxic minerals (such as mercury or arsenic), or, in the case of petroleum refineries, toxic organic chemicals. The nuclear power industry has caused some water pollution when

radioactive material has seeped from tanks in which wastes have been buried, either at sea or underground. Such pollution can have a variety of effects. Organisms not adapted to living in contaminated water may die; the water may become unsuitable for domestic use or irrigation; or the wastes may reenter the food chain, with deleterious effects on humans.

Among the pollutants that have been discharged into the water supply in the United States are **polychlorinated biphenyls (PCBs),** a family of related chemicals used as lubricants in pipelines and in a wide variety of electrical devices, paints, and plastics. During the manufacturing process, companies have dumped PCBs into lakes and rivers, from which they have entered the food chain. Several states have banned commercial fishing in water where fish have higher levels of PCBs than are considered safe. While not all of the effects of PCBs on human health are known, they have been linked to birth defects, damage to the immune system, liver disease, and cancer. Although the EPA banned the direct discharge of PCBs into U.S. waters in 1977, immense quantities remain in water bodies.

PCB-contaminated sediments are found on the bottoms of harbors and channels in all of the Great Lakes. The International Joint Commission, a U.S.-Canadian agency that oversees management of the Great Lakes, has identified

Human Impact on the Environment

Figure 12.8
The Montfort Beef Company feedlot near Greeley, Colorado. The sanitary disposal of organic wastes generated by such concentrations of animals is a problem only recently addressed by environmental protection agencies.

more than 40 highly toxic areas (Figure 12.9). They include Saginaw Bay in Michigan, Waukegan and Sheboygan Harbors in Wisconsin, the St. Louis River and Bay in Minnesota, the Grand Calumet River in Indiana, the Ashtabula and Cuyahoga Rivers in Ohio, the Buffalo River in New York, and Hamilton Harbor, the St. Clair River, and Jackfish Bay in Ontario.

The petroleum industry is a significant contributor to the chemical pollution of water. Although massive oil spills like that from the tanker *Exxon Valdez* in March 1989 command public attention, smaller spills routinely dump millions of gallons of oil into U.S. waters each year. More than half the oil normally comes from oil tankers and barges, usually because of ruptures in accidents. Much of the rest comes from refineries, the discharge of tank flushings and ballast from tanker holds, and seepage from offshore drilling platforms. The Gulf of Mexico, the site of extensive offshore drilling, is among the most seriously polluted major bodies of water in the world.

Acid precipitation (often called acid rain), a by-product of emissions from factories, power plants, and automobiles, has affected the water quality and ecology of thousands of lakes and streams. Because the precipitation is caused by pollutants in the air, it is discussed later in this chapter.

Many industrial processes, as well as electric power production, require the use of water as a coolant (Figure 12.10). **Thermal pollution** occurs when water that has been heated is returned to the environment and has adverse effects on the plants and animals in the water body. Many plants and fish cannot survive changes of even a few degrees in water temperature. They either die or migrate. The species that depend on them for food must also either die or migrate. Thus, the food chain has been disrupted. In addition, the higher the temperature of the water, the less oxygen it contains, which means that only lower-order plants and animals can survive.

Mining
Mining for coal, copper, gold, and other substances contributes to contamination of the water supply through the wastes it generates. Rainwater reacts with the wastes, and dissolved minerals seep into nearby water bodies. The exact chemical changes produced depend on the composition of

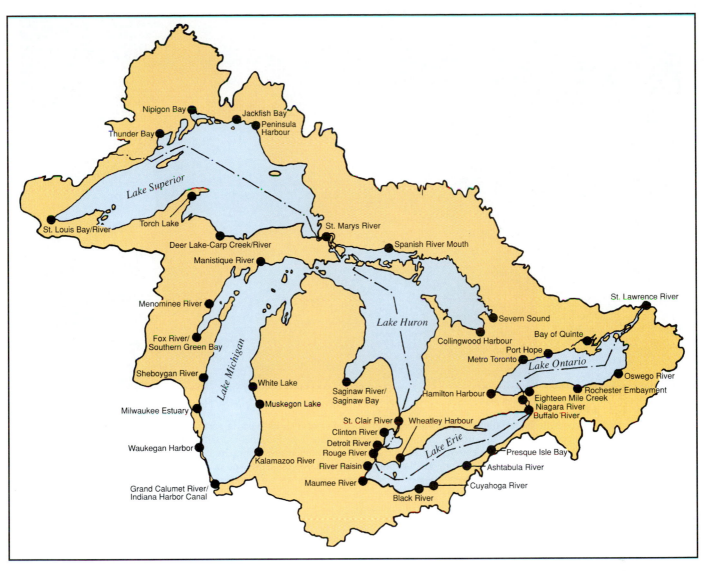

Figure 12.9

Toxic sites in the Great Lakes region. Contaminants contributed by industries include PCBs, mercury, heavy metals, and dioxin, while agriculture is responsible for phosphates and pesticides such as dieldrin and chlordane. Among the effects of the pollution are degradation of fish and wildlife populations and restrictions on the consumption of drinking water and fish.

the slag heaps and the reaction of the minerals with sediments or river water.

Strip mining for coal in states such as Kentucky, Tennessee, and West Virginia involves digging a trench, excavating the coal, digging another trench, depositing the soil and waste rock in the first trench, and so on. Mixed in with the discarded soil and rock are minerals and heavy metals, such as sulfur, manganese, aluminum, and iron. Rain or snow combines with these elements to form an acidic brew that turns streams the color of dried blood and destroys fish and plant life downstream.

Mining wastes are also a problem in the West, where they cover an area two-thirds the size of Connecticut, or some 800,000 hectares (2 million acres). The wastes are the

result of mining and processing gold, silver, lead, copper, and other hard-rock minerals in states such as Utah, Nevada, and New Mexico. The pulverized rock is laced with such toxic materials as lead, arsenic, and cyanide. The acidic solutions that form when moisture mixes with these materials have polluted an estimated 16,000 kilometers (10,000 mi) of streams, killing fish and tainting underground supplies of drinking water.

In addition to altering the quality of the water, the contaminants have secondary effects on plant and animal life. Each year, for example, thousands of animals and migratory birds die in Arizona, Nevada, and California after drinking cyanide-laced waters at gold mines. The toxic water is poured over mounds of crushed rock to leach the

Human Impact on the Environment

Figure 12.10

Cooling towers at Three Mile Island nuclear plant near Harrisburg, Pennsylvania. Electric power generating plants account for about three-fourths of all water used for cooling purposes in the United States. Other industries that use and release large amounts of cooling water are metal smelters, paper mills, petroleum refineries, and chemical manufacturing plants. Heated wastewaters are often significantly warmer than the waters into which they are discharged, disrupting the growth, reproduction, and migration of fish populations.

gold and then settles into ponds and lakes that attract wildlife.

Municipalities and Residences

A host of water pollutants derives from the activities associated with urbanization. Water runoff from towns and streets contains contaminants from garbage, animal droppings, litter, vehicle drippings, leaking fuel-oil storage tanks, and the like. Pollutants include nutrients and organic materials from landscaped areas, heavy metals, and toxic chemicals. The use of detergents has increased the phosphorus content of rivers, and salt (used for deicing roads) increases the chloride content of runoff. Because the sources of pollution are so varied, the water supply in any single area is often affected by diverse contaminants. This diversity complicates the problem of controlling water quality.

Sewage can also be a major water pollutant, depending on how well it is treated before being discharged. Raw, untreated human waste contains viruses responsible for dysentery, polio, hepatitis, spinal meningitis, and other diseases. Although sewage causes less water pollution in de-

veloped countries than in those where sewage treatment is not practiced or is not thorough, the United States and Canada are not without such sources of water pollution. Victoria, British Columbia, for example, still discharges its raw waste into the sea. Each day, nearly 20 million gallons of untreated sewage are pumped into the Strait of Juan de Fuca.

Only half of the U.S. population lives in communities with sewage-treatment plants that meet the minimum goals set by the federal Clean Water Act. When the aged sewer system of Dade County, Florida, ruptures, as it does periodically, millions of gallons of raw sewage pour into the Miami River, which empties into Biscayne Bay in downtown Miami.

In many communities, special problems arise after heavy rains, when storm water filled with animal wastes, street debris, and lawn chemicals floods the sewers. As treatment plants become overloaded, both the runoff and raw sewage are diverted into rivers, bays, and the ocean. New York City alone has more than 500 storm water outlets that overflow in heavy rains, pouring some 65 billion gallons of untreated sewage (about 10% of the city's total sewage) into the Hudson River and Long Island Sound each year. But 1200 other cities in the East, Midwest, and Northwest also have sewage networks that are combined with the overflow systems for storm water.

Controlling Water Pollution

As long as there are people on earth, there will be pollution. Thus, the problem is one not of eliminating pollution but of controlling it. In recent years, concern over increased levels of pollution has brought about major improvements in water quality.

The U.S. government in 1972 took the lead in regulating water pollution with the enactment of the Clean Water Act. Its objective was "to restore and maintain the chemical, physical, and biological integrity of the nation's waters." Congress established uniform nationwide controls for each category of major polluting industry and directed the government to pay most of the cost of new sewage-treatment plants. Since 1972, industries have spent billions of dollars to comply with the Clean Water Act by reducing organic waste discharges.

The gains have been impressive. Many rivers and lakes that were ecologically dead or dying are now thriving. Once dumping grounds for all kinds of human and industrial waste, the Hudson, Potomac, Cuyahoga, and Trinity Rivers are cleaner, more inviting, and more productive than before, and they now support fishing, swimming, and recreational boating. Similarly, Seattle's Lake Washington and the Great Lakes are healthier than they were two decades ago. Authorities have announced ambitious plans to clean up the waters of Chesapeake Bay, the country's largest estuary, and to undo much of the damage that has been inflicted on Florida's Everglades by improving the water quality of the Kissimmee River and Lake Okeechobee.

Such gains should not mislead us. While some of the most severe problems have been attacked, the EPA estimates that serious pollution still plagues at least 10% of the river, stream, coastal water, lake, and estuary mileage in the United States. The solution to water pollution lies in the effective treatment of municipal and industrial wastes; the regulation of chemical runoff from agriculture and mining; and the development of less polluting technologies. Although pollution-control projects are expensive, the long-term costs of pollution are even higher.

Impact on Air

The troposphere, the thin layer of air just above the earth's surface, contains all the air that we breathe. Every day, thousands of tons of pollutants are discharged into the air by cars and incinerators, factories and airplanes. Air is polluted when it contains substances in sufficient concentrations to have a harmful effect on living things.

Air Pollutants

Truly clean air probably has never existed. Just as there are natural sources of water pollution, so are there sub-

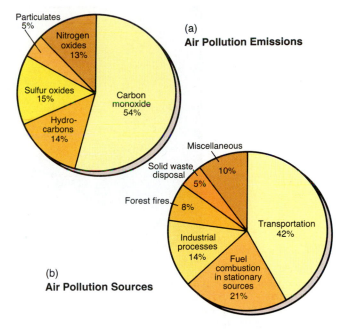

Figure 12.11

Emissions of the principal air pollutants in the United States (a). Carbon monoxide, one of the most lethal gases known, comes primarily from vehicle exhausts. **Sources of air pollution (b).** If pollution from vehicles were eliminated, emissions would be reduced by 42%.

Source: National Air Pollution Control Administration; U.S. Department of Health and Human Services.

stances that pollute the air without the aid of humans. Ash from volcanic eruptions, marsh gases, smoke from forest fires, and windblown dust are natural sources of air pollution.

Normally, these pollutants are of low volume and widely dispersed throughout the atmosphere. On occasion, a major volcanic eruption may produce so much dust that the atmosphere is temporarily altered. The 1991 eruption of Mount Pinatubo in the Philippines sent ash 30 kilometers (19 mi) into the stratosphere; within weeks, a dense atmospheric haze blanketed the equatorial and tropical zones of the earth. In general, however, the natural sources of air pollution do not have a significant, long-term effect on air, which, like water, is able to cleanse itself.

Far more important than naturally occurring pollutants are the substances that people discharge into the air. These pollutants result primarily from burning fossil fuels (coal, gas, and oil) and other materials. Fossil fuels are burned in power plants and many industries, in home furnaces, and in cars, trucks, buses, and airplanes. Scientists estimate that about three-quarters of all air pollutants come from burning fossil fuels. The remaining pollutants largely result from industrial processes other than fuel-burning, forest fires, the incineration of solid wastes, and the evaporation of solvents. Figure 12.11 depicts the major sources of the five primary air pollutants.

Human Impact on the Environment

Figure 12.12

An example of the effects of temperature inversion. A brown cloud hovers over Denver when temperature inversions keep air pollutants from dispersing. Denver frequently has the worst carbon monoxide level in the country from mid-November to mid-January, when cold winter air over the city is trapped by warm, still air at higher altitudes. The city has recently embarked on a campaign to reduce air pollution, implementing a wood-burning ban, a mandatory oxygenated fuel program, and voluntary restrictions on driving on days when pollution is high.

Factors Affecting Air Pollution

Many factors affect the type and the degree of air pollution found at a given place. Those over which people have relatively little control are climate, weather, wind patterns, and topography. These determine whether pollutants will be blown away or are likely to accumulate in the area where they are produced. Thus, a city on a plain is less likely to experience a buildup than is a city in a valley.

Wind Currents

Winds can carry air pollutants long distances. In North America, most of the prevailing winds are westerlies. As air masses move across the continent from west to east, they pick up, transport, and deposit contaminants. This means that the pollutants generated in one place can affect areas hundreds of kilometers away.

The worst effects of the air pollution that originates in New York City are felt in Connecticut and parts of Massachusetts. The chemical reaction that produces smog takes a few hours, and by that time, air currents have carried the pollutants away from New York. In a similar fashion, New York receives pollutants produced elsewhere. Much of the acid rain that affects New England and eastern Canada originates in the coal-fired power plants along the lower Great Lakes and in the Ohio Valley that use extremely high smokestacks to disperse sulfurous emissions.

One of the biggest power plants in the United States, the Navajo Generating Station near Page, Arizona, burns more than 20,000 tons of coal a day. The combustion releases more than 10 tons of sulfur dioxide every hour. Under normal weather conditions, the emissions are blown toward the Grand Canyon and are largely responsible for the haze that blankets the canyon for much of the year.

Inversions

Unusual weather can alter the normal patterns of pollutant dispersal. A *temperature inversion* magnifies the effects of air pollution. Under normal circumstances, air temperature decreases away from the earth's surface. A stationary layer of warm, dry air over a region, however, will prevent the normal rising and cooling of air from below. During an inversion, the air becomes stagnant. Pollutants accumulate in the lowest layer instead of being blown away, so that the air becomes more and more contaminated.

Normally, inversions last for only a few hours, although certain areas experience them much of the time. Temperature inversions occur often in Los Angeles in the fall and Denver in the winter (Figure 12.12). If an inversion lingers long enough, say, several days, it can contribute to the accumulation of air pollutants to levels that seriously affect human health.

Urbanization and Industrialization

Other factors that affect the type and degree of air pollution at a given place are the levels of urbanization and industrialization. Population and traffic densities, the type and density of industries, and home-heating practices all help to determine the kinds of substances discharged into the air at a single point. In general, the more urbanized and industrialized a place is, the more responsible it is for pollution. The United States may contribute as much as one-third of the world's air pollution, a figure roughly equivalent to the proportion of the world's fossil fuel and mineral resources it consumes.

Figure 12.13

A smoggy day in Los Angeles (1978). Rush-hour traffic produces large amounts of hydrocarbons and nitrogen oxide. The latter reacts with oxygen to form nitrogen dioxide, which is what gives smog its characteristic reddish-brown tint. The 1990 Clean Air Act stipulates that passenger cars must emit 60% less nitrogen oxide and 40% fewer hydrocarbons by the year 2003.

The sources of pollution are so many and varied that we cannot begin to discuss them all in this chapter. Instead, we will focus on three types of air pollution and their associated effects.

Photochemical Smog

More than one-third of the U.S. population lives in metropolitan areas that have **photochemical smog** one or more times during the year.

Smog is a hazy mixture of gases and solid particles. While it may contain as many as 100 different chemical compounds, three ingredients are crucial to its formation: nitrogen oxides, hydrocarbons, and sunlight. Smog is created when oxides of nitrogen react with the oxygen present in the air to form nitrogen dioxide. In the presence of sunlight, nitrogen dioxide reacts with hydrocarbons to form new compounds, such as **ozone.** Ozone is *the* major component of smog.

Warm, dry weather and poor air circulation promote ozone formation. As long as sunlight continues to bake the mixture of hydrocarbons and nitrous oxides, smog continues to form. In general, therefore, more ozone is produced during the summer months than during the rest of the year.

Although nitrogen oxides and hydrocarbons are generated by a variety of sources, motor vehicles are the chief source, being responsible for about 40% of smog (Figure 12.13). Engine exhaust contains both hydrocarbons and nitrogen oxides. In addition, hydrocarbons come from

Human Impact on the Environment

gasoline vapor that escapes when cars are refueled. Other sources of smog are dry cleaners (the solvents emit fumes), furniture makers and paint shops (hydrocarbon solvents, paints, and varnishes), and bakeries (sunlight changes yeast by-products into ozone). These establishments account for another 40% of smog formation. Power plants and industrial facilities typically contribute 15% of the ingredients found in smog.

Although all of America's largest cities have too much smog, four regions that are characterized by high ozone levels are California, the Texas Gulf Coast, the Great Lakes area, and much of the Northeast. A number of factors determine whether an area will have smog. The size and density of the population, the amount of traffic, and the number of industries in the area are crucial since they all help determine the kinds and amounts of substances discharged into the air.

The climate and topography of California are particularly conducive to ozone pollution. Many of its valleys are at least partially encircled by mountains that help hold air pollutants in the basins. When temperature inversions occur, the pollutants are effectively trapped, unable to escape to the stratosphere. At one time ozone levels in Los Angeles exceeded the limit established by the federal Clean Air Act (0.12 parts of ozone per million parts of air) nearly half the days in the year and sometimes reached three times the acceptable level. Other cities adjacent to mountains and subject to temperature inversions and the accumulation of smog are Salt Lake City, Phoenix, and Denver. These and other metropolitan areas that experience ozone pollution are shown in Figure 12.14.

People who have experienced a really smoggy day know that smog makes their eyes sting and their throat feel raw, but they may not be aware of its more harmful effects. Ozone reduces lung capacity, scars lung tissue, and lowers the lungs' resistance to infection. Ozone is especially hazardous to children because of their smaller breathing passages, to the elderly, and to those who already suffer from asthma, emphysema, or other respiratory diseases.

Smog is also a significant cause of damage to crops and trees. Because ozone damages plant cell membranes, crops grown under smoggy conditions mature more slowly and yield less than normal. The EPA estimates that ozone pollution reduces crop yields by as much as $3 billion per year.

Acid Rain

Although acid *precipitation* is a more precise description, **acid rain** is the term generally used for pollutants, chiefly oxides of sulfur and nitrogen, that are created by burning fossil fuels and that change chemically as they are transported

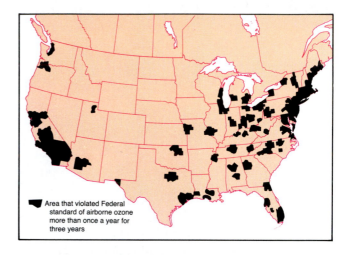

Area that violated Federal standard of airborne ozone more than once a year for three years

Figure 12.14

Urban areas where ozone levels violated the U.S. federal standard more than once a year for three years. By far, the worst ozone pollution in the United States occurs in the large Los Angeles basin, which often exceeds the federal ozone standard nearly half the year. Los Angeles, Anaheim, and Riverside, California, are the only major U.S. cities in the *extreme* urban smog category. Among the cities in the *severe* category are Baltimore, Philadelphia, New York, Muskegon, Gary, Chicago, Milwaukee, and San Diego.

Source: Environmental Protection Agency.

INVERSIONS AND PUBLIC HEALTH: THE DONORA VALLEY TRAGEDY

In late October 1948, a heavy fog settled over Donora, Pennsylvania. Stagnant, moisture-filled air was trapped in the Donora Valley by surrounding hills and by a temperature inversion that held the cooler air against the ground under a lid of lighter, warmer upper air. Smoke and fumes from the town's zinc and steel plants filled the air, gradually building in concentration. The fog turned to dense smog as sulfur dioxide emitted by the zinc works was converted to deadly sulfur trioxide by contact with the air.

Both old and young, with and without past histories of respiratory problems, reported difficulty in breathing and unbearable chest pains. Before the rains washed the air clean nearly a week after the smog buildup, 20 people were dead and hundreds hospitalized. Almost half of the town's 12,300 residents became ill. A tragic union of natural weather processes and human activity converted a normally harmless, water-saturated inversion to deadly poison.

through the atmosphere and fall back to earth as rain, snow, fog, or dust. When sulfur dioxide is absorbed into water vapor in the atmosphere, it becomes sulfuric acid. Sulfur dioxide contributes about two-thirds of the acid in the rain. About one-third comes from nitrogen oxides, transformed into nitric acid in the atmosphere. Once the pollutants are airborne, winds can carry them hundreds of kilometers, depositing them far from their source.

When washed out of the air by rain, snow, or fog, the acids change the *pH factor* (the measure of acidity/alkalinity on a scale of 0 to 14) of soil and water, setting off a chain of chemical and biological reactions (Figure 12.15). The average pH of normal rainfall is 5.6, slightly acidic, but acid rainfalls with a pH of 2.4—approximately the acidity of vinegar and lemon juice—have been recorded. It is important to note that the pH scale is logarithmic, so that 4.0 is ten times more acidic than 5.0, and *100* times more acidic

than 6.0. Acid rain also coats the ground with particles of aluminum and toxic heavy metals, such as cadmium and lead.

Acid rain is a worldwide problem. In North America, it is most serious in the northeastern United States and southeastern Canada, the result of geology and location (Figure 12.16). These areas have large amounts of igneous bedrock, which is not as able as sedimentary rock to buffer the effects of acid deposition. In addition, they are downwind from the major source region for the two main components of acid precipitation. Ten states in the central and upper Midwest (Missouri, Illinois, Indiana, Tennessee, Kentucky, Michigan, Ohio, Pennsylvania, New York, and West Virginia) produce more than half the sulfur dioxide and one-third of the total nitrogen oxides emissions. Approximately half of the acid rain that falls on Canada originates in the United States, which has become a sore point in

Figure 12.15
The formation and effects of acid rain. The acids in the precipitation damage soil, vegetation, and water.

Human Impact on the Environment

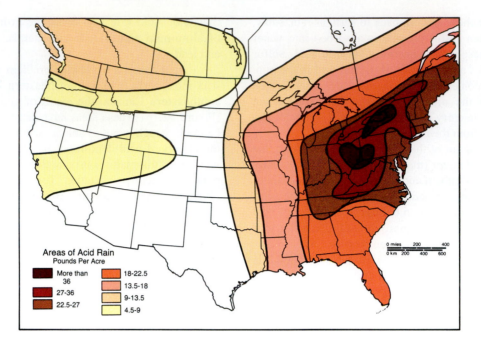

Figure 12.16

Where acid rain falls. In general, the areas that receive the most acid rain in the eastern United States and Canada are those that are least able to tolerate it. Their surface waters tend to be acidic rather than alkaline and are unable to neutralize the acids deposited by rain or snow.

Source: Data from Canadian Government.

U.S.–Canadian relations because the effects of the precipitation on water bodies and forests are deadly.

The *aquatic effects* of acid rain are clear-cut. The acidity of a lake or stream need not increase much before it begins to interfere with the early reproductive stages of fish. Also, the food chain is disrupted as acidification kills the plants and insects upon which fish feed. As the water becomes more acidic, there is a progressive loss of many kinds of organisms. Lakes with a pH of 4.5 or lower are nearly sterile. Acid rain has already damaged at least 11,000 lakes in the United States and 14,000 in Canada, rendering them almost fishless. Tens and perhaps hundreds of thousands more are in peril.

As Figure 12.17 indicates, acid deposition also has *terrestrial effects,* harming soils and vegetation. It leaches toxic constituents, such as aluminum salts, from the soil and kills microorganisms in the soil that break down organic matter and recycle nutrients through the ecosystem. Significant forest damage has occurred in northeastern United States, particularly at higher elevations, where for days at a time trees are bathed by clouds saturated with pollutants. Spruce and fir, often the dominant species above 2000 meters (6500 ft), are in severe decline in much of the Appalachian Mountains.

The *material effects* of atmospheric acid are evident in damage to buildings and monuments. The acid etches and corrodes marble, limestone, and metals, such as iron and bronze. The Lincoln and Jefferson Memorials in Wash-

ington, D.C., are just two of the thousands of structures that are slowly being dissolved by acid fumes in the air.

Controlling Air Pollution

In the last 40 years, significant progress has been made in cleaning up America's air. A series of Clean Air Acts (1963, 1965, 1970, 1977) and amendments identified major pollutants and established national air quality standards. After many years of debate, Congress in 1990 passed a Clean Air Act that represented the most sweeping legislation to date. It set forth goals to protect public health and the environment by reducing the amount of air pollutants that can be released, and it established a timetable for reaching those goals. Major provisions called for:

- reducing urban smog by 15% by 1996 and 3% each year thereafter until federal air quality standards are met;
- stipulating that passenger cars must emit 60% less nitrogen oxide and 40% less hydrocarbons by the year 2003;
- using cleaner burning fuels in the most polluted cities; and
- requiring utilities to reduce the release of sulfur dioxide and nitrogen oxides.

Reaching these goals will require reducing the type and volume of air pollutants from both stationary and non-stationary (mainly vehicle) sources. A number of strategies

Figure 12.17

Dead trees on Mount Mitchell, North Carolina. A combination of acid rain and ozone pollution has caused extensive damage to forests along the crest of the Appalachians from Maine to Georgia. The pollution weakens trees to the point where they cannot survive such natural stresses as temperature extremes, high winds, drought, or insects. In addition, the high levels of lead and other heavy metals in the soil make it difficult for the forests to regenerate.

can be employed to clean up stationary sources. Technological options include switching to cleaner-burning fuels (for example, natural gas and low-sulfur coal); coal washing, which removes much of the sulfur in coal before it is burned; and removing pollutants from effluent gases in the smokestack by using scrubbers, precipitators, and filters. Another possibility is to replace fossil fuels with alternative energy sources, such as hydroelectric power, nuclear energy, and wind or solar power. Although such alternative sources are "cleaner" than the fossil fuels, the two most extensively used—hydro and nuclear—have their own adverse environmental impacts. A more promising approach appears to be conserving energy. Increasing energy efficiency reduces emission levels without reducing economic productivity.

As we have seen, vehicles of all types are major air polluters, particularly in congested urban areas. The goals of the Clean Air Act of 1990 could be met through several different strategies. These include conforming to tighter tailpipe emission standards by using catalytic converters to treat exhaust gases, retiring older automobiles, phasing out leaded gas, and implementing rigorous vehicle inspection programs. Alternative energy sources for powering vehicles hold promise; they include fuels that burn cleaner than gasoline, such as methanol, ethanol, and natural gas, and hydrogen; and electric batteries. Finally, travel can be made

more efficient if a community is committed to rewarding those who carpool or use alternative means of transport, such as bicycles or mass transit.

Anyone who compares air quality in, for example, Los Angeles in the year 2000 with that of the 1960s will see that these legislative efforts have led to significant improvements. These advances have not been easy in the face of increased population in a consumer society, but they are unmistakeable. With public backing, the future should see a continuation of such positive trends.

Impact on Soils

By design or by accident, people have brought about many changes in the physical, chemical, and biochemical nature of the soil and altered its structure, fertility, and drainage characteristics. The exact nature of the changes in any region depends on past practices as well as on the original nature of the land. Two problems that are a matter of current concern are soil erosion and salinization.

Soil Erosion

Over most of the continent's surface, the thin layer of topsoil upon which life depends is not very deep, usually less than 30 centimeters (1 ft). Below it, the lithosphere is a complex mixture of rock particles, inorganic mineral matter,

Human Impact on the Environment

organic material, living organisms, air, and water. Under natural conditions, soil is constantly being formed by the physical and chemical decomposition of rock material and by the decay of organic matter. It is simultaneously being eroded, for **soil erosion**—the removal of soil particles, usually by wind or running water—is as natural a process as soil formation. Under natural conditions, however, vegetation holds most soil in place, and what little is moved by water or wind is replaced by decomposition of the rock below the surface. In other words, the rate of soil formation will equal or exceed the natural rate of soil erosion, so that soil depth and fertility tend to increase with time. Soil formation is a slow process; it can take a century or more to replace 2.5 centimeters (1 in.) of topsoil.

When land is cleared and planted to crops, or when the vegetative cover is broken by overgrazing or other disturbances, the process of erosion accelerates. When its rate exceeds that of soil formation, the topsoil becomes thinner and eventually disappears, leaving behind only sterile subsoil or barren rock. At that point, the renewable soil resource has been converted into a nonrenewable and dissipated asset.

In recent years, soil erosion in the United States has been at an all-time high. As Figure 12.18 indicates, wind and water are blowing and washing soil off pasturelands in the Great Plains, ranches in Texas, and farms in the Southeast. The country's croplands lose almost 2 billion tons of soil per year to erosion, an average annual loss of more than 9 metric tons per hectare (4 tons per acre). In some areas, the average is several times that. Of the roughly 167 million hectares (413 million acres) of land that are intensively cropped in the United States, nearly 40% are losing topsoil faster than it can be replaced naturally.

Some figures reveal the size and implications of accelerated erosion.

- Every year, more than 325 million tons of topsoil wash into the Mississippi River and are transported to the Gulf of Mexico. This is equivalent to the removal of a layer of topsoil approximately

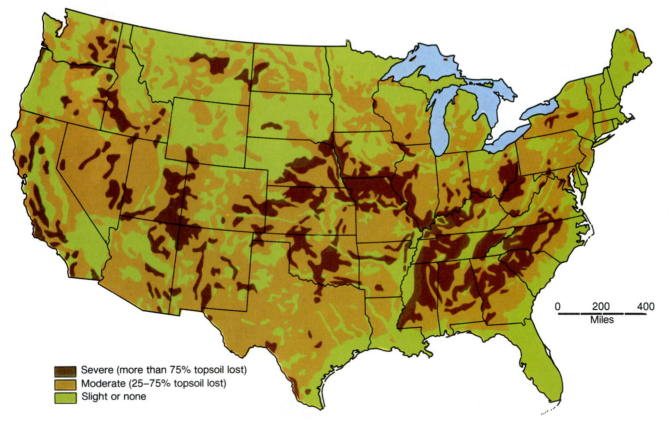

Severe (more than 75% topsoil lost)
Moderate (25–75% topsoil lost)
Slight or none

0   200   400
Miles

Figure 12.18
Soil erosion in the United States. Although many activities contribute to erosion, agriculture and deforestation are particularly significant. In recent years, erosion has been most severe in the rolling-hill regions of western Mississippi, western Tennessee, and Missouri. Each of these areas loses about 10 tons of topsoil per acre of cropland annually. However, soil resource stress and depletion affect all parts of the country.

Source: Based on data from 1934 Reconnaissance Erosion Survey of the United States and other soil conservation surveys by the Soil Conservation Service—U.S. Soil Conservation Service.

1 millimeter thick from the entire central region of the United States.

- For every bushel of corn produced from Iowa farmland, two bushels of soil are lost.
- It is estimated that farmers in the hilly Palouse area of eastern Washington have lost more than 500 pounds of winter wheat per acre because severe erosion has impoverished their soils.
- North America's "breadbasket" is the area of rich prairie soils that extends from the Midwest through the Great Plains into Canada. In a century of farming on these lands, we have lost, on average, half the topsoil and soil carbon. Although the soils remain fertile, increasing amounts of chemical fertilizers and water are needed to maintain yields.

Like most processes, soil erosion has secondary effects. As the soil quality and quantity decline, croplands become less productive and yields drop. Streams and reservoirs experience accelerated siltation. If the topsoil is heavily laden with agricultural chemicals, erosion-borne

WIND EROSION AND THE DUST BOWL

Although water and wind are both important movers of soil, wind erosion is not usually as evident as water erosion. It does not create a Grand Canyon or carve deep gullies in a Kentucky hillside. Under certain conditions, however, wind can equal water as an erosive force, moving immense amounts of soil. The crucial ingredients are a dry climate, relatively flat land, exposed soil, and strong winds. In the early 1930s, these factors came together in the portion of the Great Plains where five states (New Mexico, Texas, Oklahoma, Colorado, and Kansas) meet. They created what came to be known as the Dust Bowl.

The dust storms of the 1930s had their origin years earlier, in the soaring wheat prices of World War I. As prices dropped after the war, farmers tried to recoup their losses by breaking more sod to plant more crops.

When the drought came—and droughts in that area are cyclical—crops died and the fertile soil turned to dust. Without roots to hold it in place, it blew away. Scouring winds lifted hundreds of millions of tons of topsoil off the land. A single storm in April 1934 is reported to have whipped up a dust cloud more than 2000 kilometers (1200 mi) long and transported it northeastward to leave a film of dust on store shelves in New York City and on ships at sea.

For much of the 1930s, the skies in the Dust Bowl were saturated with dust clouds. Chickens roosted at noon, streetlights were needed even at midday, and trains collided in the black blizzards that rolled across the Plains. Scores of people died from respiratory infections. In his novel about the Dust Bowl, *The Grapes of Wrath*, John Steinbeck wrote, "The dawn came, but no day."

Human Impact on the Environment

silt pollutes water supplies, increasing the operating costs of water-treatment facilities. The danger of floods increases as bottomlands fill with silt, and the cost of maintaining navigation channels grows (Figure 12.19).

Soil Salinization

Accelerated erosion is a primary cause of soil deterioration, but in arid and semiarid climates, salt accumulation can be a contributing factor. **Salinization** is the concentration of salts in the topsoil as a result of the evaporation of surface water. It occurs in poorly drained soils in dry climates, where evaporation exceeds precipitation and dehydrates the land. As water evaporates, some of the salts are left behind to form a white crust on the surface of the soil (Figure 12.20).

Like erosion, salinization is a natural process that has been accelerated by human activities. Poorly drained irrigation systems are the primary culprit, because irrigation water tends to move slowly and thus to evaporate rapidly. Mild or moderate salinity makes soil less productive and lowers crop yields; extreme salinity can ultimately render the land unsuitable for agriculture.

Salinization affects portions of both Canada and the United States. Approximately 1.6 million hectares (4 million acres) of cultivated soils in Saskatchewan and Alberta are classified as overly saline. There are also areas of serious salinization in the U.S. Southwest, particularly in the Colorado River drainage basin, and in the San Joaquin Valley of California. Ironically, the irrigation water that transformed that arid, 430-kilometer-long (270-mi) valley into one of the country's most productive farming regions now threatens to make much of it worthless again. Irrigation water picks up salt from the soil, and a hard layer of clay

concentrates the polluted water near the surface, reducing the soil's fertility. Some farmers have siphoned off the water, depositing it in ditches and open fields and making them unfit for cultivation.

Excessive groundwater withdrawals in coastal areas can also foster salinization. Saltwater intrusion occurs when seawater infiltrates an aquifer (Figure 12.21). It is a significant problem in several southeastern and Gulf coastal states. Coastal areas of Texas, for example, are consuming fresh groundwater at such a high rate that salt water from the Gulf of Mexico is replacing fresh water in the Texas Gulf aquifer.

Maintaining Soil Productivity

Economic conditions have contributed to the accelerated rates of soil erosion seen in recent years. To maintain or increase their cultivable land and yields and to maximize their short-term profits, many farmers have neglected conservation practices. They have plowed under marginal lands, torn down windbreaks and fencerows, and abandoned crop rotation patterns in favor of continuous monoculture cropping. Some have converted land from cattle grazing to the production of row crops, such as corn and soybeans, which leave soil exposed for much of the growing season and increase soil loss. This is especially true if heavy applications of herbicides create weed-free fields.

Conservation techniques have not been completely forgotten, of course. They are persistently advocated by farm organizations and soil conservation groups and practiced by many farmers. Techniques to reduce erosion by holding the soils in place are well-known. They include erecting windbreaks and constructing water diversion channels, contour plowing, terracing, no-till farming (allowing

Figure 12.19
Soil erosion in Iowa.

Figure 12.20
Salinization in Death Valley, California, has left a white precipitate on the soil surface. Depending on the degree of salinity, soils can support salt-tolerant (halophytic) vegetation or none at all.

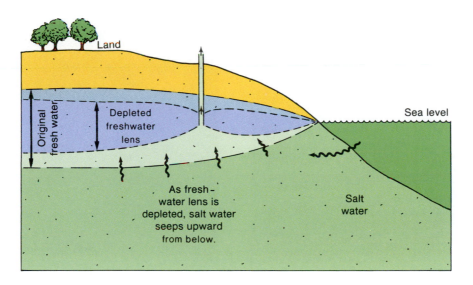

Figure 12.21
Saltwater intrusion occurs in coastal areas when fresh groundwater is withdrawn faster than it is replenished. As the freshwater supply decreases, it is replaced in the aquifer by salt water, and wells may begin pumping it instead.

Human Impact on the Environment

crop residue, such as cut corn stalks, to remain on the soil surface throughout the winter), strip-cropping, and crop rotation (Figure 12.22). In addition, government subsidies can encourage farmers to idle marginal, highly erodible land. Only by employing such practices can the country maintain the long-term productivity of soil, the resource base upon which all depend.

Solid-Waste Disposal

One of the most vexing problems technologically advanced countries such as the United States and Canada face is one of the most mundane: what to do with the solid wastes produced by residential, commercial, and industrial processes. Although these account for much less tonnage than the wastes produced by mining or agriculture, they

(a)

(b)

(c)

(d)

Figure 12.22
Methods of controlling soil erosion include intercropping (a), contour plowing (b), strip cropping (c), and no-till farming (d).

are everywhere, a problem with which each individual and municipality must deal.

Municipal Waste

The wastes that communities must somehow dispose of include newspapers and beer cans, toothpaste tubes and old television sets, broken refrigerators and rusted cars. Solid-waste disposal is a greater problem in the United States than in any other country, for Americans throw away more trash per person than do any other people. As Figure 12.23 indicates, the amount of household waste doubled between 1960 and 1990. In 1996 it stood at some 210 million tons per year, or about 740 kilograms (1600 pounds) per person. Solid-waste disposal costs are now the second-largest expenditure of most local governments.

The volume of trash is the result of three factors: affluence, packaging, and open space. Craving convenience, Americans rely on disposable goods that they throw away after very limited use. Thus, although readily available substitutes are more economical, we annually throw out some 16 billion baby diapers, 2 billion razors, 1.6 billion pens, and a million tons of paper towels and napkins. In addition, nearly all consumer goods are encased in some sort of wrapping, whether it be paper, cardboard, plastic, or foam. One-third of the yearly volume of trash consists of these packaging materials. Finally, the United States has traditionally had ample room in which to dump unwanted materials. Countries that ran short of such space decades ago have made greater progress in reducing the volume of waste.

The sheer volume of the trash is one of the two main problems that American communities are facing in disposing of their wastes. A second is the toxic nature of much of it. Although ordinary household trash does not meet the governmental definition of **hazardous waste**—discarded materials that may pose a substantial threat to human health or to the environment when improperly stored, transported, or disposed of—much of it is hazardous nonetheless. Products containing toxic chemicals include paint thinners and removers, furniture polishes, bleaches, oven and drain cleaners, used motor oil, and garden weed killers and pesticides.

Methods of Waste Disposal

American communities employ three methods of disposing of solid waste: landfills, incineration, and recycling (Figure 12.24). Each has its own impact on the environment.

Landfills Approximately 55% of U.S. municipal solid waste is deposited in sanitary *landfills,* where each day's waste is compacted and covered by a layer of soil (Figure 12.25). *Sanitary* is a deceptive word here. Until recently, there were no federal standards to which local landfills had to adhere, and while some communities and states regulated the environmental impact of dumps, many did not.

Even if no commercial or industrial waste has been dumped at them, most landfills eventually produce *leachate* liquids that contaminate the groundwater. The EPA estimates that only 15% of the landfills have liners to prevent contaminants from seeping into groundwater supplies. Indeed, more than two-thirds of the dumps lack any type of system to monitor groundwater quality. Leachate forms when precipitation entering the landfill interacts with the decomposing materials. Thus, heavy metals are leached from batteries and old electrical parts, while vinyl chlorides come from the plastic in household products. Typical leachate contains more than 40 organic chemicals, many of them poisonous.

New York City's largest landfill, the Fresh Kills on Staten Island, illustrates the problem. Each day, some 27,000 tons of waste are trucked to the site. Over 40 years

Figure **12.23**

Municipal solid waste generated each year in the United States, in millions of tons. The figures do not include waste from agriculture, mining, or industry. Until recently, U.S. and Canadian communities did not have to worry too much about the trash they produced. There was plenty of open space in which to dump it, and there was not as much trash even a generation ago as there is today. Furthermore, most of it was biodegradable and not nearly as hazardous as it is now.

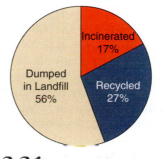

Figure **12.24**

Methods of solid-waste disposal in the United States.

Human Impact on the Environment

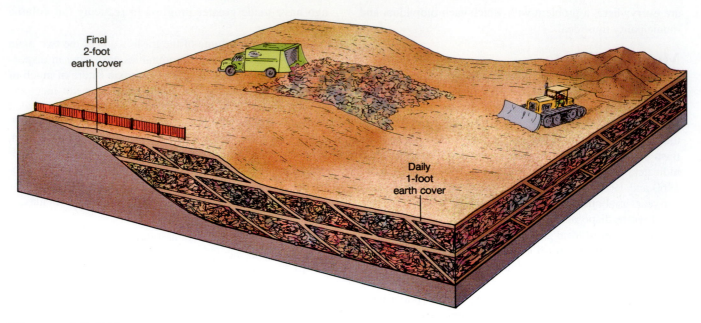

Figure 12.25

A sanitary landfill. Wastes are sealed between successive layers of clean earth each day. Although far more desirable than open dumps, sanitary landfills pose environmental problems of their own, including groundwater contamination and seepage of methane and hydrogen sulfide, gaseous products of decomposition.

old, the dump was not constructed to hold its contents securely and does not comply with current state landfill regulations. It is located in an ecologically sensitive wetland area and is adjacent to residential communities. Every day, thousands of gallons of contaminated leachate seep into the groundwater beneath the landfill. The decomposition of organic material generates 140,000 cubic meters (5 million cubic feet) of methane gas annually. The rotting garbage produces odors that include the fishy smells of amines, the goatlike aroma of organic acids, and the sickeningly sweet scents of aldehydes. When Fresh Kills is full to capacity about the year 2005, it will cover 160 hectares (400 acres) and stand 170 meters (505 ft) high. Taller than the Statue of Liberty, taller even than the largest of the great pyramids of Egypt, Fresh Kills will be a veritable monument to trash.

Municipal landfill capacity is shrinking dramatically. The number of landfills fell from 18,000 in the late 1970s to 6000 in 1990 and 3600 in 1995, increasing pressure on those that remain open. Many of those remaining will be closed within the next ten years, either because they are full or because their design and operation pose a threat to the environment. Dwindling capacity is particularly serious in the densely populated East Coast states, where roughly half of the cities no longer have local landfill sites. At considerable expense, they export their trash to neighboring communities and even other states. New Jersey, for example, sends more than half of its solid waste out of state. Although the problem is most serious in the East, it is not limited to that area.

The problem is not one of space. There are vast amounts of open space in the country. But as people become more aware of how landfills contaminate water, soil, and air, public opposition makes it increasingly difficult to site new landfills. Everyone wants them, but NIMBY—"not in my backyard." More careful attention must now be paid to both siting and construction. Landfills are being built away from floodplains, aquifer recharge zones, and sites with highly permeable rock formations. To avoid costly cleanups later on, modern, well-designed landfills must contain impermeable bottom layers, monitoring systems to collect methane gas, and leachate collection systems to protect the groundwater. All these factors have increased dramatically the cost of constructing and operating landfills.

Incineration The quickest way to reduce the volume of trash is to burn it, a practice that was common at open dumps until it was halted by the Clean Air Act of 1970. Concern over air pollution also forced the closure of old, inefficient *incinerators* (facilities designed to burn waste), providing an impetus to design a new generation of incinerators. More than 110 municipal incinerators, mostly in the Northeast, burn approximately 17% of the national total of trash.

Connecticut burns about half of its household trash, and Massachusetts, New Jersey, and New York may soon do likewise. A Dade County, Florida, facility burns 25% of the waste from 26 municipalities, including Miami. Most

Figure **12.26**

A waste-to-energy incinerator at Peekskill, New York. The heat derived from burning refuse produces steam that can be used to generate electricity or heat buildings. Although incineration significantly reduces the volume of trash, the residual ash is an environmental hazard.

municipal incinerators are of the waste-to-energy type, which use extra-high (980° C, 1800° F) temperatures to reduce trash to ash and simultaneously generate electricity or steam that is then sold to pay operating costs (Figure 12.26).

A decade ago, incinerators were hailed as the ideal solution to overflowing landfills, but it has become apparent that they pose environmental problems of their own by generating toxic pollutants in both air emissions and ash. Air emissions from incinerator stacks have been found to contain an alphabet soup of highly toxic elements, ranging from arsenic to zinc, and including, among others, cadmium, dioxins, lead, and mercury, as well as significant amounts of such gases as carbon monoxide, sulfur dioxide, and nitrogen oxides. Emissions can be kept to acceptably low limits by installing electrostatic precipitators, filters, and scrubbers to capture pollutants before they are released into the outside air, although the devices add significantly to the cost of the plant.

The concentration of toxins in the ash residue of burning creates a greater problem. Incinerators typically re-

duce trash by only 75%. One-fourth remains as ash, which must then be buried in a landfill. Ideally, the ash would be treated as hazardous waste, to lessen the danger of contaminating groundwater supplies.

Managing Municipal Waste

The problems associated with landfills and incinerators have spurred interest in two alternative waste-management strategies: source reduction and recycling. By *source reduction,* we mean producing less waste in the first place so as to shrink the volume of the waste stream. Manufacturers can reduce the amount of paper, plastic, glass, and metal they use to package food and consumer products. Over the last 40 years, for example, the weight of plastic soft-drink bottles and of aluminum beverage cans has been reduced by 20–30%. Some products, such as detergents and beverages, can be produced in concentrated form and packaged in smaller containers.

Recycling, the reprocessing of discarded materials into new products, reduces the amount of waste needing

Human Impact on the Environment

disposal by making it available for reuse. Studies indicate that at least 85% of household solid waste is recyclable and need never go to a landfill or incinerator. Most paper products (40% of the waste stream) are recyclable, so is virtually all yard waste (20% of municipal solid waste). Leaves, grass clippings, tree trimmings, even discarded Christmas trees can be composted or shredded into wood chips and used as mulch. Thousands of businesses and at least 600 communities have initiated recycling programs for a variety of materials. Most programs collect paper, aluminum cans, and glass. Although recycling of plastics is more difficult, plastic milk and soda bottles are easily recycled. Recycling has made great gains since 1990, the percentage of municipal waste recycled increasing from 10% to 27% in 2000.

Recycling has a beneficial impact on the environment. It saves natural resources by making it possible to cut fewer trees, burn less oil, and mine less ore. Because it takes less energy to make things out of recycled material than out of virgin materials, recycling saves energy. It reduces the pollution of air, water, and soil that stems from the manufacture of new materials and from other methods of waste disposal, and it saves increasingly scarce landfill space for materials that cannot be recycled—including hazardous wastes.

Radioactive Waste

The EPA has classified more than 400 substances as hazardous, that is, posing a threat to human health or the environment. Industrial wastes have grown steadily more toxic. Currently, about 10% of industrial waste materials are considered hazardous. None presents a more serious concern than radioactive waste; its disposal is a major environmental problem.

Sources of Radioactive Waste

Every facility that either uses or produces radioactive materials generates *low-level waste*. Nuclear power plants pro-

---

RISKY BUSINESS

Just weeks before the country's first permanent repository for radioactive waste was to open, the Department of Energy (DOE) indefinitely postponed operations at the facility. One billion dollars and ten years' worth of effort had gone into building the *Waste Isolation Pilot Plant (WIIP)* 650 meters (2000 ft) beneath the desert sands of southern New Mexico, 42 kilometers (26 mi) east of the town of Carlsbad.

Mined from salt beds, the plant is a giant underground cavern with 1.6-kilometer-long (1 mi) corridors, 56 vast storage rooms and deep ventilation shafts. It was designed to store chiefly *low-level* military waste from the production of nuclear weapons. The machinery, clothing, and other items are contaminated with plutonium and will remain dangerous and radioactive for hundreds of thousands of years. The facility was meant to stay dry forever, but in 1987, scientists noted that moisture was seeping into the underground chamber, raising the possibility that moisture would mix with salt and form brine that would corrode the steel drums containing the waste. In addition, safety investigators from the DOE questioned whether the facility was designed adequately to withstand an earthquake. A ruptured joint on a pipe in the fire-control system raised questions about the general quality of the construction. In the face of all these doubts about the adequacy of the plant, DOE officials acknowledged in 1989 that they were not yet ready to open the New Mexico repository.

At the same time it was encountering difficulties with the WIPP, the DOE was running into problems planning another repository, the Yucca Mountain facility. Yucca Mountain is situated in southern Nevada, 160 kilometers (100 mi) northwest of Las Vegas and 32 kilometers (20 mi) from the California border. In 1987, Congress had designated Yucca Mountain as the country's first permanent repository for *high-level* radioactive waste.

Intended to safely store waste from commercial as well as military reactors for 10,000 years, the project was to have opened in 2003. Most of the waste would be in the form of radioactive fuel pellets sealed in metal rods; these would be encased in extremely strong glass and placed in steel canisters entombed in chambers 300 meters (1000 ft) below the Nevada desert. The amount of radiation involved would be far greater than that at the WIPP in New Mexico.

Many doubt that the Yucca Mountain facility will ever be completed and licensed. Three areas of concern have emerged. The area is vulnerable to volcanic and earthquake activity, which could cause groundwater to well up suddenly and flood the repository. Rainwater percolating down through the mountain could penetrate the vaults holding the waste and corrode the canisters. Finally, the Yucca Mountain site lies between the Nevada Test Site, which the DOE used as a nuclear bomb testing range, and the Nellis Air Force Base Bombing and Gunnery Range. Questions have been raised about the wisdom of locating the waste repository just a few kilometers from areas subject to underground explosions or aerial bombardment.

---

duce about half the total low-level waste in the form of used resins, filter sludges, lubricating oils, and detergent wastes. Industries that manufacture radiopharmaceuticals, smoke alarms, radium watch dials, and other consumer goods produce low-level waste consisting of machinery parts, plastics, and organic solvents. Military facilities and research institutions, including universities and hospitals, also produce radioactive waste materials. Altogether, these facilities produce some half-million cubic meters (650,000 cubic yards) of low-level radioactive waste each year. While some low-level waste will lose its contamination within months, other waste will remain radioactive for centuries.

High-level waste is nuclear waste with a relatively high level of radioactivity. It consists primarily of spent power-reactor fuel assemblies and waste generated from the manufacture of nuclear weapons. Because the waste can remain radioactive for 10,000 years and more, its disposal is especially troublesome. The volume of high-level waste is not only great but increasing rapidly. Approximately one-third of a reactor's rods need to be disposed of every year.

By 1996, nearly 30,000 metric tons of spent-fuel assemblies were being stored in the containment pools of commercial nuclear power reactors, awaiting more permanent disposition. Many more tons are added annually. "Spent fuel" is a misleading term: the assemblies are removed from commercial reactors not because their radiation is spent but because they have become too radioactive for further use. The assemblies will remain radioactively "hot" for thousands of years.

Disposal Methods

Unfortunately, no satisfactory method for disposing of any radioactive waste has been devised. Between 1945 and 1970, some 90,000 barrels of low-level waste were dumped in the ocean along both the Pacific and Atlantic Coasts. Many of these barrels have either rusted through or been crushed by water pressure, releasing their toxic contents. Even without such physical damage, the life expectancy of the barrels must be presumed to be far shorter than the half-life of their radioactive contents.

Cardboard boxes containing wastes contaminated with plutonium (which remains dangerously radioactive for 240,000 years) have been rototilled into the soil on the assumption that the earth would dilute and absorb the radioactivity. The armed forces have been criticized by the EPA for the ways they have disposed of radioactive and other wastes at nearly 2000 military installations around the United States. From the Aberdeen Proving Grounds north of Baltimore to Pearl Harbor, Hawaii, military bases may be among the most dangerously polluted places in the country. For decades, the armed forces have drained solvents, PCBs, and other toxic chemicals into waterways, dumped both liquid and solid wastes into unlined landfills or open-air storage pits, and generated huge amounts of fuel spills.

Because low-level waste is generated by so many sources, its disposal is particularly difficult to control. Evidence indicates that much of it has been placed in landfills, often the local municipal dump, where the waste chemicals may leach through the soil and into the groundwater. By EPA estimates, at least 25,000 legal and illegal dumps contain hazardous waste. As many as 2000 are deemed potential ecological disasters.

Much low-level radioactive waste is now being placed in tanks and buried in the earth at 13 sites operated by the U.S. Department of Energy and three sites run by private firms (Figure 12.27). Millions of cubic feet of high-level military waste are temporarily stored in underground tanks at four sites: Hanford, Washington; Savannah River, South Carolina; Idaho Falls, Idaho; and West Valley, New York (Figure 12.28). Government plans to construct permanent repositories for radioactive wastes at Carlsbad, New Mexico, and in the Yucca Mountains of Nevada have encountered serious difficulties and have been at least temporarily postponed.

Solid waste will never cease to be a problem, but its impact on the environment can be lessened by reducing the volume that is generated, eliminating or reducing the production of toxic residues, halting irresponsible dumping, and finding ways to reuse the resources that waste contains. Until then, current methods of waste disposal will continue to pollute soil, air, and water.

Summary

People are part of the natural environment; our lives literally depend on the water, air, soil, and other resources the biosphere contains. In our efforts to enhance our quality of life, to build a strong society and a flourishing economy, we have significantly altered the intricately interconnected systems of the biosphere: the hydrosphere, the troposphere, and the earth's crust. All human activities affect the environment. An action that impinges on any part of the web of nature inevitably triggers chain reactions, the ultimate consequences of which appear never to be fully anticipated.

Efforts to control the supply of water have altered both the quantity and quality of water. Agriculture and urbanization are placing extreme demands on available water in California and the High Plains. Pollutants associated with agriculture, industry, and other activities have degraded the quality of water supplies in many areas, although regulatory efforts have brought about major improvements in some areas in recent years.

Combustion of fossil fuels has contributed to serious problems of air pollution. Too many Americans breathe polluted air, and acid rain has already damaged too many lakes, streams, and forests. Poor agricultural practices have accelerated soil erosion and salinization.

Finally, all common methods of disposing of the solid wastes we produce release contaminants into the surrounding environment. The disposal of hazardous wastes in particular—such as the spent core of the reactor at Shippingport—threatens serious environmental deterioration.

Temporary storage of
▲ high-level military waste

Low-level military and
● commercial waste

Permanent storage of military
▫ waste (under construction)

Permanent storage of high-level
■ commercial waste (under construction)

Figure 12.27

Sites where low-level radioactive wastes are stored in the United States. Low-level waste includes contaminated fluids, resins, tools, machine parts, building materials, and clothing. Some of it will remain radioactive for centuries.

Key Words

acid rain	photochemical smog
channelization	polychlorinated biphenyls (PCBs)
environmental pollution	recycling
eutrophication	red tides
hazardous waste	salinization
hydrologic cycle	soil erosion
Ogallala aquifer	thermal pollution
ozone	

Gaining Insights

1. Sketch a diagram of or briefly describe the *hydrologic cycle.* How do people have an impact on that cycle?

2. If the supply of water on earth is constant, why is the availability of water a problem in parts of the United States?

3. Describe the chief sources of water pollution and some of their effects. In what parts of the United States is water pollution a problem?

4. What steps can be taken to control water pollution?

5. What factors affect the type and degree of air pollution found at a place?

6. Describe the relationship of *ozone* to *photochemical smog.* What cities are particularly characterized by smog, and what are its effects?

7. What are the causes and effects of *acid rain,* and where in the United States is it a problem?

Figure 12.28

Storage tanks under construction in Hanford, Washington. Built to contain high-level radioactive wastes, the tanks are shown before they were encased in concrete and buried underground. The Hanford Military Reservation stores more defense waste than any other site in the country. Releases of radioactive and toxic wastes at Hanford have contaminated the Columbia River and area groundwater supplies and soil.

8. What are the goals of the U.S. Clean Air Act of 1990?
9. How have people diminished the amount and productivity of soil? Why has soil erosion in the United States accelerated in recent decades?
10. What are ways of maintaining soil productivity?
11. What methods do communities use to dispose of solid waste? What ecological problems does that disposal present?
12. Distinguish between *hazardous waste* and *radioactive waste*. How is the latter disposed of?

Selected References

Bailes, Kendall E., ed. *Environmental History: Critical Issues in Comparative Perspective.* Lanham, Md.: University Press of America, 1985.

Cunningham, William P., and Barbara Woodworth Saigo. *Environmental Science: A Global Concern,* 4th ed. Dubuque, Iowa: WCB/McGraw-Hill 1996.

Cutter, Susan and William H. Renwick. *Exploitation, Conservation, Preservation,* 3rd ed. New York: John Wiley & Sons, 1998.

Ehrlich, Anne H. and Paul R. Ehrlich. *Earth.* London: Methuen, 1987.

Enger, Eldon D. and Bradley F. Smith. *Environmental Science: A Study of Interrelationships,* 6th ed. Dubuque, Iowa: WCB/McGraw-Hill, 1997.

Firor, John. *The Changing Atmosphere.* New Haven: Yale University Press, 1992.

Goldman, Benjamin A., et al. *Hazardous Waste Management: Reducing the Risk.* Council on Economic Priorities. Washington, D.C.: Island Press, 1986.

Goudie, Andrew. *The Nature of the Environment,* 5th ed. Cambridge, Mass.: Blackwell, 1999.

Mason, Robert J. and Mark T. Mattson. *Atlas of U.S. Environmental Issues.* New York: Macmillan, 1990.

Miller, G. Tyler, Jr. *Living in the Environment,* 11th ed. Pacific Grove, CA. Calif.: Brooks/Cole, 1999.

National Geographic Magazine, special edition: *Water,* November 1993.

Neal, Homer and J. R. Schubel. *Solid Waste Management and the Environment.* Englewood Cliffs, N.J.: Prentice-Hall, 1987.

Postel, Sandra. *Air Pollution, Acid Rain, and the Future of Forests.* Worldwatch Paper 58. Washington, D.C.: Worldwatch Institute, 1984.

Pye, Veronica I., Ruth Patrick, and John Quarles. *Groundwater Contamination in the United States.* Philadelphia: University of Pennsylvania Press, 1983.

ReVelle, Charles and Penelope ReVelle. *The Environment,* 3rd ed. Boston: Jones and Bartlett, 1988.

Wellburn, Alan. *Air Pollution and Acid Rain.* New York: John Wiley & Sons, 1988.

World Resources Institute. *The Information Please Environmental Almanac.* Boston: Houghton Mifflin, annual.

Young, John E. *Discarding the Throwaway Society.* Worldwatch Paper 101. Washington, D.C.: Worldwatch Institute, 1991.

CHAPTER

13

Culture Regions of the United States

One of the authors of this text likes to tell the following story:

Back in the early 1960s, every spring break my friends and I would pile into someone's old jalopy and leave the snowy Ohio State University campus to head for the sun and sand of Fort Lauderdale. While the attraction of Florida at that time of year was due primarily to its physical geography, we would be passing through some very interesting culture regions as well. In those days, there were few freeways, and each winding road from central Ohio to southern Florida brought vistas of different types of houses, farms, and towns, not to mention foods and dialects. Along the way, we played a game that we called "how do you know you've come to a new region?" There were many clues beyond the more obvious physical ones, such as the transition from plains to mountains, and most of these involved observations of varying "cultures."

Not far south of Columbus, Ohio, everything started to change. The houses became more traditional and primitive, and country and gospel music dominated the airways, making it frustratingly difficult to find our normal daily infusion of rock and roll. As we crossed the Ohio River into West Virginia, coal towns and lumber trucks increased in number, and when we stopped for gas and Twinkies, the attendant spoke with a not-quite-southern nasal twang. We were in a region that my geography professors called Appalachia. Since we usually left Ohio in the late afternoon (after the last final exam), we were ready for breakfast somewhere in the Carolinas. A strange substance appeared on our plates—tasteless and basically inoffensive but hardly delightful. They called it grits and served it automatically. We had crossed the "grits line" into the South. We also noticed that the gas stations and restaurants had three types of restrooms, labeled "men," "women," and "colored." Confederate and American flags appeared everywhere, while huge banners urging us to 'meet our maker before it was too late' were occasionally draped across the country roads.

As we drove on through the coastal areas of South Carolina and Georgia, more changes appeared. The architecture looked vaguely tropical and exotic, especially when it was covered with flowers and vines. Many of the people talked so slowly we could barely stand to wait for an answer when we asked directions. We drove through Charleston and Savannah and, although none of us had yet had an urban geography course, we couldn't help noticing that they were not at all like Cleveland.

Just as we were getting used to being in this exotic place called the South, things began to change again. Somewhere around Daytona Beach, we noticed that a lot of people sounded more like New Jersey than Georgia, and grits were a thing of the past. As we entered southern Florida, we stopped at an art deco Jewish deli and tried to befriend a waitress from Boston. Florida, it seemed, was a tropical Northeast.

After a week in the sun, we would reluctantly head north to the excitement of a new academic quarter. To avoid boredom and to increase our knowledge of America, we usually chose a different route home. Often, we would pass through Atlanta, which even at that time was an anomaly. It didn't seem to fit our regional transect. In fact, we observed, it seemed to be very much like Columbus. We even heard some rock and roll.

Twenty-five years later, I drove through many of the same places. While southern accents could still be heard, they were softer and grammatically decipherable. The radio announcers spoke the American equivalent of BBC English (uniform from coast to coast) and played the same hits heard in California. The freeways were lined with chain motels and restaurants, and the small towns often had fern bars and art galleries. The cities had skyscrapers and malls, and long gone were the "colored" restrooms. Religious banners had been replaced by "Bibleland" theme parks more reminiscent of Disney than the backwoods signs I had once encountered. Along Atlanta's beltway, corporate "Edge Cities" seemed to announce the arrival of a new universal urban form, while Charleston seemed full of northerners seeking to preserve and enhance a quaint past à la Beacon Hill or Georgetown. I could not help but wonder if the idea of culture regions still made sense in this rapidly changing landscape.

On the Nature of Culture and Culture Regions

It could be argued that even during the peak of the variations discussed above, the United States has never really had culture regions. At least since the time of European colonization, America has been characterized by fluid cultural patterns that are always in a state of flux. Cultural variations are as much illustrative of a slice of time as a slice of place. Boston may have started out as a Puritan experiment, but it soon became a center of Yankee entrepreneurs, Irish politicians, Italian restaurateurs, and, most recently, high-tech yuppies. Today, Boston may be more like San Francisco than Bar Harbor. In America, places and regions change as people and ideas pass through them. Culture regions are necessarily transitory. The remainder of this chapter is devoted to a discussion of those regions in the United States; Canada is treated in the following chapter.

The contrast between the United States and much of the rest of the world is now being made even more evident by events in such areas as Eastern Europe and the former Soviet Union. In Bosnia and Serbia, long-standing and deeply felt attachments to culture and place have seemingly resisted all attempts to create socialist utopias. Similarly, the Scots and the Welsh (not to mention the Irish) have, at least to some degree, managed to survive 800 years of anglicization. By comparison, American culture regions are but lightly felt flashes in the pan. On the other hand, many of those regions are actively and purposefully being recreated and redefined to encourage tourism and regional promotion. Since the completion of its River Walk and Mexican Market district, for example, San Antonio has arguably become more exotic and "regional" than it was 50 years ago. Similarly, Seattle and Portland have sought to create a regional architecture and ambiance only in the last 30 or so years. During the same period, Tucson began to deemphasize the grass lawn in favor of desert landscaping and "Southwestern" architecture. Both lists—evidence of regional homogeneity and evidence of regional heterogeneity—are endless, and the boundaries change continually.

Two problems arise when we attempt to define and discuss culture regions in the United States in the same way that we might for more traditional parts of the world. First, most of the components of American culture are relatively unimportant. While we may define culture regions in Europe or Asia on the basis of language, religion, food-ways, agriculture, house form, and the like, such variables are relatively invisible in America. Catholic areas exist in Wisconsin, Louisiana, and Vermont, but for very different reasons and no one much cares about boundaries. The situation is very different from, say, Bosnia or Ireland. Second, many of the important boundaries involve micro rural-urban or modern-traditional dichotomies rather than large regions. That is, Hispanic East Los Angeles may be more like West San Antonio than either is like the Anglo neighborhoods nearby. In a way, this is nothing new. Five hundred years ago, Florence may have had as much in common with Brugge as it had with rural Italy. The difference is that today, the vast majority of people (76%) rather than a tiny, elite minority live in urban areas, and so the "culture" of the vast interstitial areas loses importance. Maps of large culture regions must therefore be taken with a grain of salt. Having said that, however, culture regions still can be worth some analysis and discussion, if only to organize our thoughts on cultural identity and change.

If we are to define culture regions in the United States using something other than such traditional components of culture as language and religion, what variables do we use? We suggest a combination of current and historical variables so that regional "sense of place" is based on not only current population and landscape characteristics but historical differences as well. Heritage provides certain cultural baggage that serves to "reproduce" particular behaviors and attitudes that might be factored into some definition of culture. Life in Charleston is, to some degree, still "Charlestonian," even though the city's population, economic condition, and functions have changed greatly since its early days as genteel capital of the Deep South. Similarly, a New York of Pakistani cabdrivers is still New York.

All of this is necessarily subjective. What follows is a breakdown of some possible culture regions in the United States. This breakdown relies on some current "hard data," but it also depends on interpretations of historical sense of place. To a degree, such regions are based upon imagery and representation (as in art and film) rather than on numerical data. In the United States, our consensus regions are based as much on the depiction of places in movies as on actual characteristics. The South is "Gone with the Wind" and the West is "Wagon Train." Such regions exist because we need them to exist. We want to go "out west" and to believe that such a place is really there. In spite of current trends toward homogeneity, we will always be creating regions in our minds.

Current Culture Regions for the United States

Most regional geographies of the United States utilize traditional regional breakdowns usually following state lines, that is, New England, Midwest, Great Plains, and so on. Such regions make some sense, and using state boundaries makes data compilation easier. On the other hand, if "culture" is our main concern, state boundaries will not do. South Boston and northern Vermont are too different to coexist in a culture region called New England. But how many regions do we need to show the essential current and historical variations across the great expanse that is the United States? As anyone who has watched the New York Marathon on television knows, there are at least 200 culture regions in that city alone (with another several hundred in the metropolitan area), each containing more people than the state of Wyoming. On the other hand, some have suggested that we really have only two or three very gross culture regions in the United States today, such as the Manufacturing Belt, the Deep South, and the West. As Figure 13.1 indicates, we choose a middle path with 13 regions, but keep in mind, it could just as easily be 8 or 27.

We begin in the Northeast. In 1961, Jean Gottmann coined the term *Megalopolis* for the northeastern seaboard, and it remains a meaningful name for the region (see Chapter 9 for another discussion of the term).

Megalopolis: Headquarters of America

While many countries have a primate city, that is, one city that effectively dominates the entire nation in population, economic importance, educational opportunities, health facilities, and a variety of central-place functions, the United States has a primate region (Figure 13.2). While London, Paris, Vienna, and Mexico City act as primate centers for their respective countries, in the United States, this role is distributed, albeit unevenly, through the region called Megalopolis.

Megalopolis extends roughly from southern New Hampshire to northern Virginia and contains dozens of metropolitan areas. The Census Bureau has recently grouped most of these urban areas into consolidated metropolitan areas, such as Boston-Lawrence-Salem and New York-northern New Jersey-Long Island, so that the entire region now contains only 15 metropolitan statistical areas.

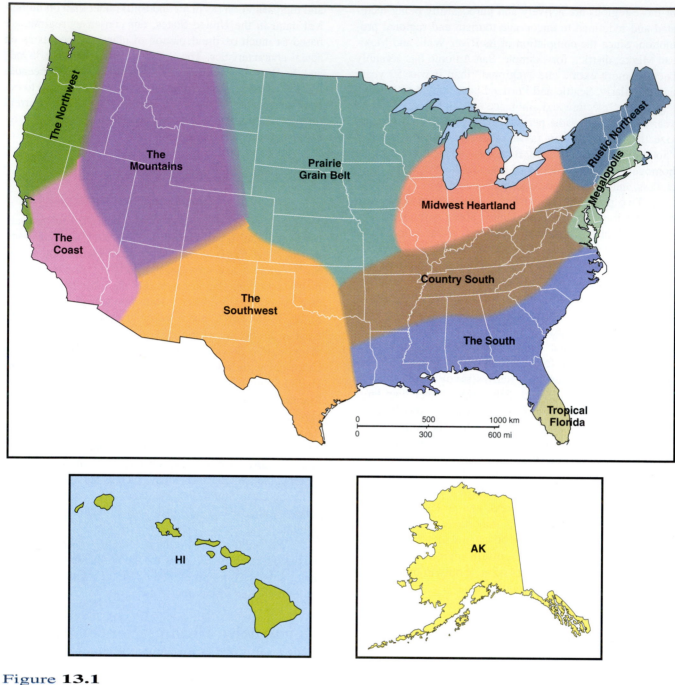

Figure 13.1
Culture regions of the United States.

To a degree, the whole idea of metropolitan areas focusing on traditional cities and suburbs is becoming meaningless in this region as new centers emerge between the coalescing older cores. This is the land of Edge Cities—suburban, multipurpose nodes that are bigger and more important than most of the older cities. Tyson's Corners (a Virginia suburb of Washington, D.C.) and Stamford, Connecticut, for example, have more office space than all but a handful of the region's downtowns. Other Edge Cities, such as Ballston (another Virginia suburb of Washington, D.C.), have more

high-density housing and better subway connections than all but a few urban cores, such as Manhattan and central Boston. Still, the essential characteristic of Megalopolis is that it is virtually all urban and suburban, whatever new forms these terms may encompass. Nearly every county from New Hampshire to northern Virginia is part of some metropolitan area, and most of these urban areas have grown together functionally, if not physically. Freeways and good rail transportation make Boston and Providence, as well as Baltimore and Washington, D.C., increasingly

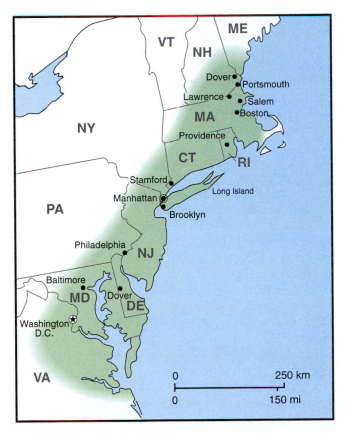

Figure 13.2
Megalopolis.

parts of the country, Megalopolis still contains the highest proportion of major decision makers. To a very real degree, this is the essence of the region's "culture." While there are still remnants of past architecture, dialects, foods, and religion, these characteristics have been overwhelmed by symbols of power and modernity. This is primarily a region of people who believe that they are "in charge." New York bankers, Washington insiders, Ivy League educators, and Philadelphia's Mainline families all have a role in America that is palpably different from the leaders of other regions. Not only do they make the important decisions today, but they have always made the important decisions—they are supposed to make important decisions. It is an old tradition. Of course, only a small percentage of the Megalopolitan population actually participates directly in this leadership (although nearly everyone still views the rest of the nation as sort of a big outback), but the same is true of most regional cultural characteristics. Relatively few people in Tennessee are actually country musicians, and only a tiny percentage of New Englanders have ever lived in a saltbox house or collected maple sap for syrup. Still, such things give a region its character.

The historical leadership role of Megalopolis can be seen in nearly every aspect of American life, from art and architecture to science and industry. Skyscrapers, apartment buildings, sprawling housing tracts, subways, expressways, department stores, universities, city parks, and pizza all diffused out of Megalopolis over the years. It remains to be seen if pretenders to the throne such as Los Angeles will greatly affect this role in the future.

Megalopolis has other important cultural characteristics as well. It is now and always has been the gateway to America—the "front door" through which the vast majority of immigrants have entered the United States. While immigration from Asia and Latin America has recently diminished the dominance of Megalopolis in this regard, its long-standing tradition of ethnic diversity is still being constantly revitalized by new arrivals. Spanish is spoken in many of the former mill towns of Massachusetts, and Ethiopian restaurants thrive in Washington's Adams-Morgan district. Parts of Brooklyn have been named "Little Odessa" because of the flood of Russian immigrants. America began in Megalopolis, and many Americans begin there, too. The newly refurbished facilities on Ellis Island commemorate this reality.

Not all ethnic diversity was due to voluntary foreign immigration. The cities of Megalopolis were the first in America to have large black communities. Boston had a significant black population at the time of the American Revolution, and Washington, D.C., gained a huge influx of refugees from Virginia during the Civil War. Indeed, the first ghetto-making techniques, such as blockbusting and racial steering, were perfected in Harlem near the turn of the century, and that community soon became the cultural center of black America. The landscape of immigration is not as dazzling as the landscape of power, but it cohabits

parts of the same functional area. Shared airports and sports teams help tie together once-distant cities.

Megalopolis contains about 50 million people, or close to one-fifth of the total population of the United States. While this is a much smaller percentage of the total than was the case 100 or 50 years ago, it is not true that this Northeastern urban core is a declining region. All of the largest metropolitan areas grew between 1990 and 1996. Only four of the smaller ones (Providence, Hartford, New London and Springfield) actually lost population (about 1% to 2% each). Although it has large pockets of poverty and obsolescence, Megalopolis epitomizes the trend toward a "transactional" civilization based on information, services, and high technology. Midtown Manhattan remains, by far, the single leading concentration of office space in the nation (and probably the world), while Boston's tally of 27 universities is not matched anywhere else in the country. More than one-half of the 150-meter (500-ft) skyscrapers in the United States are in Megalopolis.

Megalopolis is where decisions are now and always have been made. The White House, the Pentagon, Fifth Avenue, Madison Avenue, Broadway, Harvard, and Yale, to mention only a few, are still the control centers of America. In spite of the rise of important regional centers in other

Culture Regions of the United States

the region of Megalopolis. The tenements of New York and the alley houses of Washington were invented to house the impoverished masses, and the tiny, brick row houses of Baltimore and Philadelphia served the same purpose at lower densities. The Italian food markets of South Philadelphia and the teeming streets of New York's Chinatown serve to keep alive ethnic traditions, and so assimilation sometimes seems slower in Megalopolis, in part due to the sheer numbers of people involved.

A final and important cultural characteristic of Megalopolis and one that relates very strongly to the two aspects of tradition (power and ethnic identity) discussed above is its intense interest in history and historic preservation. Megalopolis is thus a landscape of contrasts—a region of new and constantly changing symbols of power and ethnicity but also a land of careful heritage monitoring and preservation. Beacon Hill in Boston had the first urban preservation guidelines in the nation (1920s), and Williamsburg, Virginia, was the first totally reconstructed historic place (1920s). Throughout Megalopolis, and often in the shadows of nearby skyscrapers, there are preserved neighborhoods, such as Society Hill in Philadelphia, South Baltimore,

Greenwich Village, Georgetown and Capitol Hill, North Boston and the Boston Common, and hundreds more in big cities and small towns alike. Because Boston is perceived as the cradle of the American Revolution, patriotic duty sometimes acts as a break on unmitigated growth and change. Ethnic identity is being celebrated too, as the creation of a Black Heritage Trail in Boston demonstrates. Much that is America has Megalopolitan roots.

Capital   Every region should have a capital or epitome place that focuses and enhances its essential characteristics. In Megalopolis, such a place would be Midtown Manhattan, perhaps near the corner of Fifth Avenue and 50th Street in the midst of Rockefeller Center (Figure 13.3). A complex of 18 office towers built from the 1930s on, Rockefeller Center, because of its size and fame, has served to anchor and stabilize the office core of New York for over half a century. Even the name "Rockefeller" epitomizes the power and prestige that drive Megalopolis. The skating rink and other amenities attract all kinds of people, and one can view the passing diversity of New York while sitting on a nearby bench. Midtown Manhattan is nothing if not lively and

Figure 13.3
Rockefeller Center, New York City. The skating rink becomes an outdoor cafe in summer.

diverse. Finally, Rockefeller Center illustrates good and historic design. Its older buildings are art deco classics and have inspired many preservation efforts aimed at heightening awareness of this genre. It is part of the heritage of the region. Spending a few hours at Rockefeller Center is experiencing the essence of Megalopolis.

Rustic Northeast: Recreation Area for Megalopolis

The Rustic Northeast is a state of mind; if it did not exist, we would probably have to create it. It is now, and to some degree always has been, a recreation area for Megalopolis (Figure 13.4). It is "the wood lot," "the dairy shed," and "the lungs of the city," an ever-retreating refuge from the ever-expanding sprawl of the Edge Cities to the south. The Rustic Northeast thus includes not only the portions of the traditional New England states not already identified as part of Megalopolis, but also parts of upstate New York and northeastern Pennsylvania and New Jersey. It is the romantic Northeast that has so far escaped the clutches of megalopolitan sprawl, and it extends from the beaches of Cape Cod and the coast of Maine to the wooded uplands of central Pennsylvania. It plays the role of the small, homogeneous, quaint, picturesque, and out-of-the-way to Megalopolis's role as the central, powerful, diverse, and blatantly historical.

The Rustic Northeast is a place to get away from it all. A variety of mostly seasonal retreats, vacation homes, and factory outlets make it seem self-consciously unimportant and peripheral to the carryings-on of the modern world. The only major metropolitan area (Albany-Schenectady-Troy, with a population of 879,000) is really a collection of small cities, each with its own identity; the Syracuse metropolitan area, on the western edge of the region, adds another 750,000 people. But even these centers seem somewhat remote from the "real" Rustic Northeast of fishing villages and ski resorts.

The Rustic Northeast is a sort of cultural attic where the activities and behaviors of the past are stored. It is the coast of Maine where once-important fishing and boat-building towns occupy a seemingly endless array of isolated coves and inlets. Stately homes now turned bed-and-breakfast inns overlook graying wooden docks and sheds covered in fishing nets and adorned with signs like "Yankee Whaler Lobster House." Towns with names like Kennebunkport and Bar Harbor are known as much for the important personages who have vacationed there as for any current functions. Indeed, most of these towns are "summer places," very active only when the summer people are there. The same is true of many other coastal areas of the Rustic Northeast, such as Cape Cod and Nantucket. Many of the cafes and galleries are open only during "the season," as the owners and managers follow the tourists to Florida and southern California during the colder months. In these settings, the summer people are seeking not only beauty and relaxation but images of a quieter, more predictable past.

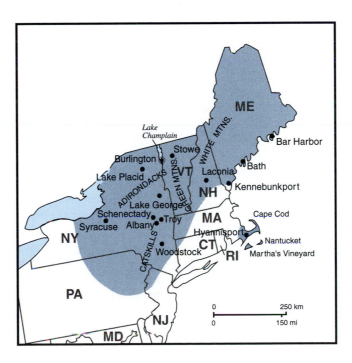

Figure 13.4
The Rustic Northeast.

Culture Regions of the United States

Farther inland, the recreation theme goes uphill to such places as Pinkham's Notch, Mount Washington, and Mount Baxter (the latter now lends its name to a winter parka distributed by Land's End). As the coastal areas shut down for winter, life picks up in such places as Stowe, Vermont, and Lake Placid, New York (deep in the heart of the Adirondack Forest Preserve), which have become famous as ski resorts and winter sports centers. Lake George, the Finger Lakes, and the "romantic" Hudson River Gorge round out the rustic imagery of this Megalopolitan periphery. Much of the Rustic Northeast acts as a retreat for a certain clientele within Megalopolis. For more than a half century, for example, the Catskills represented a sort of off-off Broadway for Jewish comedians and songsters from New York City who could sharpen their acts on New York audiences who were in a vacation mode. The universities of the Rustic Northeast, while not quite Harvard or Yale, provided a kind of woodsy Ivy League where the children of Megalopolis could study well away from the distractions of urban life. Such colleges and universities as Dartmouth, Cornell, Bennington, Bucknell, and Lehigh dot the hills and dales throughout the region. Similarly, sailing clubs on Martha's Vineyard or Hyannisport served to reunite the elites of Megalopolis in a vacation setting, while afficionados of horse racing could gather at Saratoga. Even the generation of peace and love celebrated its muddy "be-in" near the small town of Woodstock, New York.

The Rustic Northeast is a place for the purposeful reproduction of culture—especially, desirable traits associated with Americana. The Rustic Northeast is "Fall Color Country," a land of near-perfect countryside landscapes ideal for a variety of rustic activities, from picking apples to watching the production of maple syrup. It is a humanized landscape of picturesque villages with white American Baroque churches nestled in well-manicured town commons. At least that is the postcard imagery, and an increasing number of places are trying to live up to the image as nonconforming (business) buildings, (industrial) activities, and (immigrant labor) populations become obsolete or move on, leaving the way open for historical enhancement. Here, culture can be enthusiastically reproduced by people who move to the region precisely and even primarily because they want to be part of this culture. They want to live in traditional houses, collect blackberries, and wear tweed jackets.

Capital   It is much harder to pick an epitome spot for the Rustic Northeast than for Megalopolis because the former is characterized by the quality of being peripheral rather than by centrality. Nevertheless, there are many contenders: small, quaint, out-of-the-way places with traditional images. It is tempting to choose small places like Bath, Maine, or Laconia, New Hampshire, but perhaps a compromise in the direction of representational diversity is provided by Burlington, Vermont (Figure 13.5). Nestled between the Green Mountains and Lake Champlain, it offers a

sense of rural refuge while at the same time serving as the home of the University of Vermont (founded in 1791). As a gateway to Montréal, it also reminds us of the role of French Canadians in the cultural history of the region.

The South: Changing Values

Somewhere south of the Chesapeake Bay, Megalopolis gives way to the South, a region consisting of an area that has been called such things as the south coastal plain, the Deep South, the Bible Belt, and the Sun Belt. It stretches from central Virginia to southeastern Texas, with its northern border roughly following the fall line through Alabama (Figure 13.6). While its physical geography is relatively simple, it is a difficult region to describe in cultural terms in part because of its consistent inconsistencies. It is a region of good manners and politeness but also a land of high rates of violence; a region that prides itself on unchanging traditional values but is perhaps the most rapidly changing of all American regions; a region of relative cultural uniformity and simplicity yet culturally the most different area of the nation; a region that tried to secede from the United States and still displays Confederate flags yet is arguably the most patriotic section of the country, whose officers dominate the armed forces; it is a land of primitive sharecropper shacks and dazzling new golf resorts. It is the heart of the "New South."

Compared to the nation as a whole, the "culture" of the South is as simple as black and white. With the exception of southern Louisiana, the South has been dominated by cultural traits from Britain and West Africa. The vast waves of immigration that brought rapid change and diversity farther north largely avoided the South due to its plentiful supply of cheap labor and low rate of industrialization. The people of the South have names like Miller and Scott and are predominantly Protestant. While much of this is changing, especially in "border towns" like Atlanta and Houston, it remains a valid generalization. The South is the only section of the country that traditionally has had large areas with predominantly black populations. From South Carolina to the Mississippi Delta, many counties are now or once were largely black. The music, dialect, foods, architecture, dress, and mannerisms of the South have been shaped to some degree by Africa, but the chaos of the slavery period has made the systematic study of the origin and character of African traits difficult. The search for this missing heritage is important to many blacks.

The other major geographic source of cultural traits in the South was Britain, especially Celtic Britain. It has been argued that the antiauthority trait often associated with southerners arrived on this side of the Atlantic with Scots-Irish, who were never enamored of English authority and welcomed any opportunity to flaunt their resistance to it. On the other hand, most of the region's leadership over the centuries (such as the Virginia presidents) was English and thus put an emphasis on proper, orderly behavior. Therein

Figure 13.5
Burlington, Vermont, is situated on the east shore of Lake Champlain.

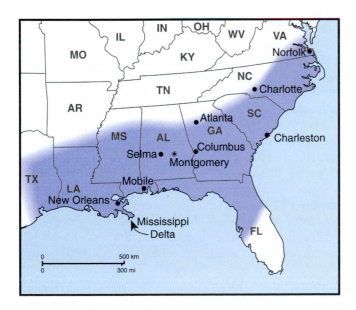

Figure 13.6
The South.

lay the inconsistency. Only in the South did these two traits hybridize in relative isolation over the years. This blending shows up in many ways. No other region adheres so strongly to the teachings of the Bible, yet no other region (with the possible exception of neighboring Country South) has generated so many new, highly individualistic religious sects. No other region claims to dislike so strongly both excessive federal laws and anyone who disobeys them. In no other region have underpaid and underrepresented workers been so uninterested in (and even hostile toward) labor unions.

All of this goes beyond the simple explanation of political conservatism. It is the hybridization of African and British culture as well as the historic tension between them that gives the South both its charm and its problems. The current problems of many blacks are related to the disruptions of slavery, when groups of loved ones were broken up and individuals sold separately. On the other hand, compared to the often gruff, dour, or rude demeanors associated with some northern populations, the tradition of southern hospitality is not entirely mythical, and gracious living is

Culture Regions of the United States

still a valued goal. The South is also a region of exuberant self-expression and, as such, has contributed mightily to the popular culture of the United States. A variety of oral traditions, including jazz, zydeco, blues, tap dancing, and storytelling, originated in the South. Unlike the rich traditions of the Northeast, which largely came from someplace else, the contributions of the South are home-grown.

The unique culture traits of the South are largely rural. With the exception of New Orleans, there have never been any big cities in the heartland of the South. Indeed, some have argued that "southern city" is a contradiction in terms. Even Charleston and Savannah, which ranked high on the list of American urban places by 1800, were never really dynamic centers of economic growth and change on the level of Boston or New York. More often, they provided a seasonal refuge for the rich from life on the plantation. They were designed more for genteel living than industrial efficiency. Today, there are large cities, such as Charlotte and Atlanta, on the edge of the region, but most southerners prefer the image of slow-paced, gracious living in small towns. With increasing prosperity and diminishing discrimination, many blacks and whites who left the region decades ago for high-paying jobs in the North are returning.

In recent years, the South has been renamed the Sun Belt. Indeed, while many writers have suggested that this name extends across the country from Florida to California, it is in the South where the name itself is most common. Florida, Texas, Arizona, and California have their own "sunny" nicknames and have not needed a new regional identity. In the South, on the other hand, Sun Belt provides a positive antidote to more disparaging regional nicknames. It implies amenity and prosperity, where many others implied backward isolation. Today, the name Sun Belt appears in phone books and business directories all across the South.

The per capita income figures for American regions are converging as the poorer areas, such as the South, rise to join the American average. The new-found prosperity of the South is a result of many factors. Military bases have played an important role in many local economies from Norfolk, Virginia, through Columbus, Georgia, to Texas. More recently, a variety of resort and retirement communities have sprung up throughout the region to take advantage of mild winters and the low cost of living. More important, however, the South has developed an infrastructure of highways, hospitals, universities, office parks, and airports necessary for continued economic growth. The South is still dominated by relatively poor, small towns, but it is on its way up. As blacks begin to play a more important role in the political and economic life of the region, and as a more diverse group of outsiders is attracted to its amenities, the culture of the South is bound to change even more in the near future than it has in the recent past.

Capital   As with the Rustic Northeast, there are many possibilities for the capital of this region. Some of the gracious old ports, such as Charleston and Mobile, have seen the entire history of the region and wear their pasts well, but today, they are a bit too sanitized and international to epitomize the South. In addition, they have always had aspirations to urbanity that do not sit well with the South's rural tradition. Montgomery, Alabama, is perhaps the best choice. While it is a sizable city, it avoids looking or acting like one (Figure 13.7). It values its role as the Confederate capital while sitting astride the Black Belt, with its black soil, high concentration of blacks, and route of the march on Selma. It thus epitomizes both continuity and change. When Montgomery loses all aspects of its deep southern traditions, the region of the South will be unrecognizable as a different kind of place.

Tropical Florida

It makes sense to discuss Tropical Florida at this point in the chapter because it is on the East Coast and we are generally going from east to west (Figure 13.8). Chronologically, however, we are jumping way ahead. Tropical Florida is among the newest and most artificial of the American regions. It is a very different kind of place than the ones we have just discussed or those that follow. It is appropriate that Walt Disney World is located here because, to a degree, the entire region is a sort of tourist center. As recently as 1900, Tropical Florida was largely undeveloped swampland. The Everglades constituted a wilderness, and many of the beaches were covered with dense mangrove thickets. Miami had fewer than 1700 people, and Miami Beach was not yet founded. The boom came after World War I, as entrepreneurs like Henry Flagler and architects like Addison Mizner began to create a new kind of American landscape. Flagler pushed the railroad all the way to Key West, making southern Florida a resort destination for the elite.

The creation of new communities, such as Palm Beach and Boca Raton, involved inventing a new architectural style that would make the area seem exotic as well as warm. Architects were recruited who had studied abroad, especially in Latin America and the Mediterranean countries, to develop a look that would differentiate Tropical Florida from the rest of America. That look began to take shape around 1920 as a sort of "Spanish-Moorish-Italian-Hawaiian" design combining Mediterranean texture with tropical openness. The exotic architecture of Tropical Florida continued to evolve as colorful art deco, Boomerang Modern, and thatched-roofed coffee shops appeared. The monumental art deco hotels that were constructed along Miami Beach in the 1950s brought fame and, for some, fortune. Between the buildings were alligator farms, snake shows, seashell stores, and other bits of exotica all designed to lure tourists.

Tropical Florida is a region designed by outsiders to attract visitors. The nature of these visitors, however, has changed over the decades. Originally, it was the northeastern

Figure **13.7**

Montgomery, Alabama: the state capitol building. Jefferson Davis took the oath as President of the Confederate States of America here in 1861.

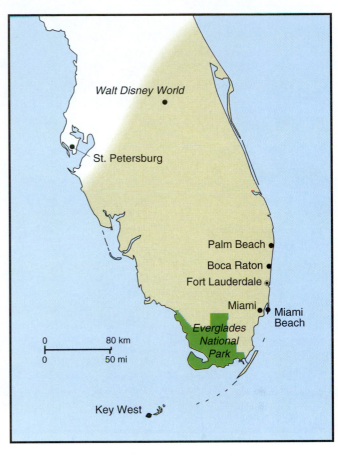

Figure **13.8**
Tropical Florida.

elite who were supposed to frequent the private clubs and tasteful golf courses of Tropical Florida. Later, both the east and west coasts of south Florida became retirement magnets for people seeking refuge from harsh northern winters. The increasing volume of people supported by Social Security and other retirement incomes gave cities from St. Petersburg to Fort Lauderdale a new image. Sometimes, entire communities moved—as when, for example, neighborhoods of Jews from New York were reassembled in Miami Beach along with delis and other cultural support systems. Elsewhere, trailer parks house entire bowling teams from Ohio or former 4-H clubs from Iowa. In spite of serious environmental problems, Tropical Florida continues to attract migrants. Between 1990 and 1996, five of the 30 fastest-growing metropolitan areas in the country were in Tropical Florida.

Not everyone has come from northern climes. Most recently Tropical Florida has been the destination of many Latin American migrants. There are many areas in the United States that are now Latin in the sense of having large Spanish-speaking populations or recent immigrants from Latin America. From Spanish Harlem in New York to East Los Angeles, there is an increasing Hispanic presence. In Tropical Florida, however, there is a difference. Not only are the percentages higher there than in almost every other region except for parts of the Southwest, but the Hispanic population is in control to a much greater degree. The early waves of migration from Cuba were made up of educated and well-off families that quickly started businesses and became political leaders throughout southern Florida. More recently, Latin American elites from a variety of countries have purchased second homes and made investments in the economy of Tropical Florida.

Most important, Tropical Florida has become the capital of the Caribbean. The Miami airport is a hub for travel for everyplace from Cartagena to Trinidad. People from throughout the region come to Tropical Florida to shop, invest, party, and visit relatives. Trade fosters capital accumulation in Spanish Tropical Florida at unprecedented

Culture Regions of the United States

Figure 13.9
Fort Lauderdale, Florida.

levels. Banks have proliferated down Miami's Brickell Boulevard. Tropical Florida seems destined to become not only a more important population and financial center within the United States, but also a vital link between it and the lands to the south.

Capital   While there are many contenders for the region's capital, a likely choice is Fort Lauderdale. Even with a metropolitan population of more than 1 million, it refuses to look like a major city (Figure 13. 9). Nestled around a large yacht basin and strung out along the beach, the high-rise condominiums and hotels do not conform to standard ideas of city structure. Density follows amenity rather than centrality. It is a major urban center with few visible means of support other than the imported capital of retirees and tourists. It thus epitomizes Tropical Florida.

Midwest Heartland

Various aspects of early American culture came together west of the Appalachians in the eastern portion of the Midwest (Figure 13.10). In Ohio, for example, the northeast was settled by New Englanders, while the southeast was settled by southerners. This contrast in origin shows up (or did show up) in everything from architecture and dialect to foodways and music. Between the two, downtown Columbus was laid out on the model of Philadelphia, complete with a Broad Street. Gradually, this region became the most average of all regions as the regional idiosyncracies of the East and the South gave way to a sort of Norman Rockwell America in speech patterns, the mixture of religions, the mixture of natives and immigrants, architecture, and agriculture. There is nothing "wild West" or "Deep South" here. Even today, such cities as Columbus remain important test markets for new products and ideas because they represent "average" America.

In recent decades, the Midwest Heartland has tended to lag behind the national average in both population and economic growth. Often the region is called the Rust Belt to signify a certain degree of industrial obsolescence, with such places as Pittsburgh and Cleveland heading the list of large, declining metropolitan areas. Perhaps part of the problem is that the region is average. As increasingly footloose people and industries have sought exotic locations in the mountains, deserts, and coastal margins of the country, "back home in Indiana" has had a hard time competing.

Another factor, however, also relates to key aspects of the region's sense of place and uniqueness: The Midwest Heartland is the land of overspecialized cities. When the

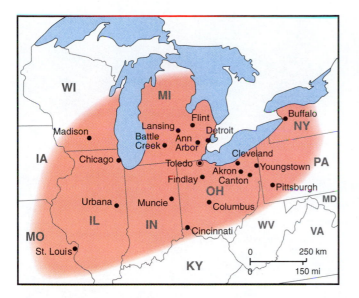

Figure 13.10
The Midwest Heartland.

tire industry was booming, Akron boomed, but today, there are no longer any producers of tires in the city, and no other employers have leaped into the breach. The sprawling factories sit empty. Since the Midwest Heartland has more specialized industrial cities than any other region and since many of the industries associated with those cities are declining, the region as a whole has not prospered.

The Midwest Heartland has more large metropolitan areas that lost population between 1990 and 1996 than any other region. Pittsburgh, Toledo, Youngstown, Dayton, Buffalo, and many smaller places all lost people during this period. Even with massive out-migration, the prospects are not necessarily bright for those who have remained. Unfortunately, many of the specialized cities are filled with now-redundant specialized people. Metropolitan areas such as Buffalo, Cleveland, Detroit, and St. Louis have large neighborhoods of southern blacks and whites and second- or third-generation European immigrants who only know making cars or steel. In most cases, the shift to services or modern industrial technologies has been slow and the hemorrhaging of jobs severe. These declining cities mask the existence of many strong and stable local economies, but their images loom large in the current view of the Midwest Heartland.

In a sense, the relatively specialized nature of places in the Midwest Heartland is what gives the region its identity. While eastern cities such as Boston and New York have tended to be diverse historically, acting as ports, industrial centers, financial centers, university towns, and sometimes state capitals, the cities of the Midwest Heartland tended to push one predominant identity. Of course, there were early attempts at specialization in the East, such as the mill towns of Massachusetts, but they were the exception rather than the rule. It was in the Midwest Heart-

land that the idea of special-function cities was perfected. Akron made tires, Canton made roller bearings, Toledo made glass, and Findlay was the home of Marathon Oil, while Muncie made transmissions and Battle Creek specialized in cereals. Mining towns and steel towns also remained specialized, and even universities were placed in their own cities from Ann Arbor to Urbana. State capitals were usually located well away from the major industrial centers and ports, in such places as Lansing and Madison. While relatively diverse central-place cities such as Chicago and Cincinnati have held their own over the decades, many of the more specialized places have a built-in inflexibility that makes it difficult to roll with the punches in a modern world economy. There are just too many cities and not enough functions. The situation is made worse by declining farm populations and a diminishing need for agricultural service centers. It remains to be seen whether the problem of over-specialization will be repeated in other regions. What will happen to the growing number of resort cities in Tropical Florida and the Southwest when the diminishing ozone layer finally makes sunshine truly dangerous?

The long-term prospects for the Midwest Heartland are far from bleak. Rich in land, water, and other resources and with plenty of picturesque and habitable space, the Midwest Heartland stands ready to benefit if and when overcrowding, pollution, environmental catastrophes, and high living costs finally overtake such regions as Megalopolis, Tropical Florida, and the Coast. Being average may very well become a new trend.

Capital　The capital of the Midwest Heartland should be a specialized city of some kind. It could be a classic small town centered on a courthouse square, a university town, a state capital, a river or lake port, or a declining industrial center. There are many possibilities. A good choice might be Toledo, Ohio, a once-booming lake port and industrial city that is now searching for a new role in the world economy (Figure 13.11). It does not have the serious problems of Detroit or the optimism of Columbus. It is like the Midwest Heartland as a whole; it is in transition.

Country South

As settlers moved westward from the culture hearths along the East Coast beginning in the late 1700s, Country South was largely bypassed (Figure 13.12). Daniel Boone and Davy Crockett notwithstanding, the heart of Country South remained a land devoid of major population centers and important economic foci until very late in the 19th century, when the exploitation of such resources as timber and coal led to the development of railroads and company towns. Even then, it was only the border cities, such as Atlanta, Birmingham, and Louisville, that boomed. Eastern Kentucky and the Ozarks remained well away from the beaten path. Even Memphis, the largest city in the region, was nearly closed down during the 1870s after a severe outbreak of yellow fever.

Figure 13.11
Toledo, Ohio. A ship is being loaded with grain on the Maumee River.

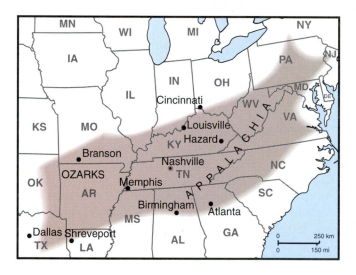

Figure 13.12
The Country South.

During the 1930s when federal Works Progress Administration programs brought a variety of road-building and other infrastructure improvement projects to Country South, government officials occasionally encountered nearly unbelievable levels of isolation, especially in parts of eastern Kentucky and Tennessee. Residents were described as "contemporary ancestors" because their dialects, music, house types, clothing, and religious practices were more in tune with the early 19th century than the 20th. Harry Caudill's *Night Comes to the Cumberlands* provides an excellent analysis of these forgotten ways of life. Of course, not all of Country South as we have defined it fits these descriptions, but there are plenty of sample places from Cades Cove, North Carolina, and Hazard, Kentucky, to parts of the Mississippi Delta and Arkansas Ozarks to give the stereotypes some validity.

By the 1950s and 1960s, the culture of Country South had become a national caricature. Snuffy Smith and Li'l Abner portrayed the inhabitants in the newspaper comic strips and on Broadway, while talented musicians donned hayseed attitudes and outfits for appearances on the Grand Ole Opry radio (and later television) show. The hit songs "Sixteen Tons" and "John Henry" spoke of life in the company coal towns, while "The Beverly Hillbillies" arrived in Los Angeles. Appalachia became nationally recognized as an officially defined and somewhat quaint poverty area. High birth rates and massive out-migration served to make Country Southerners a recognized immigrant group in many cities of the North and West.

Most of the traditional musicians who gave rise to the country music industry were from this region. White musicians such as the Carter Family and Bill Munroe gathered in Nashville and created everything from bluegrass to country and western music. Meanwhile, in Memphis, black musicians gathered to create first the blues, then rhythm and blues. Both cities later played a major role in the formative years of rock and roll with stars like Elvis Presley and Jerry Lee Lewis from the Memphis area and the Everly Brothers from Nashville. Border cities, such as Atlanta, Shreveport, Dallas, and Cincinnati, also played important roles in the creation of a music industry based on traditional styles of "pickin' and singin'." Music was the aspect of traditional culture most easily commercialized, and so it serves as the most recognizable element of the folk culture of the region. Still, there were many others.

In recent decades, the image of relic culture has proved to be quite profitable. Nashville's Grand Ole Opry is a nationally recognized theme park, and individual stars such as Dolly Parton have opened competing entertainment centers. Elvis Presley's home, Graceland, attracts pilgrims from around the world, and Eureka Springs in the Ozarks specializes in old-fashioned passion plays and religious music. Recently, Branson, Missouri, has emerged as a major center of country entertainment. The list is a long one. Many of the poor, "backward" people have become very wealthy indeed.

The new relative prosperity of Country South is not due to the revival of folk culture alone. Most of the region is now considered to be part of the Sun Belt. Actually, the term Sun Belt is used more widely in Country South than in such states as Florida, California, and Texas that have always been associated with the sun. Sun Belt is a new description for a new region. Much of the growth in Country South/Sun Belt is due to overflow from the high-cost regions to the north. As industries have sought to relocate away from high taxes, high land and labor costs, and strict environmental controls, many have taken close looks at the small- and medium-sized towns and cities of Country South. Domestic and foreign auto plants have located in Kentucky and Tennessee, where labor and the community are perceived to be more flexible than in the big cities to the north. A significant return migration of Country Southerners from such places as Detroit and Chicago is also fueling growth.

As in the South, there is a certain amount of catch-up growth in this region as universities, major office buildings, big hospitals, airports, shopping malls, and so on, are built. While Country South is still not a wealthy region and many social problems persist, the gap is closing. Will the culture remain identifiably unique as affluence grows?

Capital   As the central place for the institutionalization of the region's native folk culture as well as a busy and prosperous metropolitan area of just over a million people, the capital almost has to be Nashville (Figure 13.13). More

specifically, perhaps the capital could be just behind the stage at the Grand Ole Opry.

Prairie Grain Belt

Hopes were high for the Prairie Grain Belt during the early decades of settlement in the late 19th century. This was the region that would feed the world (Figure 13.14). The vast prairies with their rich, dark soil and apparently adequate levels of moisture could ensure that the United States would forever be an exporter of basic foodstuffs once railroads were built. All of this did eventually come to pass, but it has not always been easy. No sooner was the region settled than the blizzards of the 1880s wreaked havoc and misery on the unsuspecting population. The cold winds of the Northern Plains have never been conquered, and so, in a sense, much of the Prairie Grain Belt remains a frontier. Population has long been stable or declining north of a Minneapolis-Denver line. Indeed, North Dakota has fewer people today than it had in 1920. No other state can make that claim.

Periodic droughts also took their toll, especially the long and severe period of the 1930s Dust Bowl. Attempts to remedy environmental problems have led to further declines in the agricultural population coupled with modest increases in the larger towns and cities since there are probably more irrigation engineers and soil consultants today than there are farmers. Farms are large and heavily capitalized because they must rely on mechanized irrigation, planting, and harvesting.

Small towns are dying. Unlike the situation east of the Mississippi, where many small towns are holding their own as they serve a variety of purposes from exurban retreats to settings for quaint seasonal festivals, the towns of the Plains are usually too far from employment centers to attract exurbanites and too architecturally nondescript to appeal to visitors. Factor in the harsh winters and scorching summers, and there is little demand for housing. Suitcase farmers who live in Minneapolis or Omaha come in periodically to supervise the farms. There may still be too many people in the region as aquifer depletion and soil erosion continue to take a toll. To a very real degree, the region is dependent on federal subsidies and the artificially high crop prices they bring. We are afraid to abandon the breadbasket even though we do not really need the bread. Beyond a handful of major cities, there is some doubt as to whether a regional culture still exists, or at least whether it is still being reproduced. The young are leaving. On the other hand, at least until recently, a relatively uniform and persistent rural culture has long characterized the Prairie Grain Belt.

Since the population of many rural counties has been declining for the better part of a century and no new groups have moved in, the original German-Scandinavian-Czech culture traits have tended to ossify in quiet isolation. Politically, these culture traits have often translated into a sort of

Culture Regions of the United States

Figure 13.13
Nashville, Tennessee. Country music at Opryland.

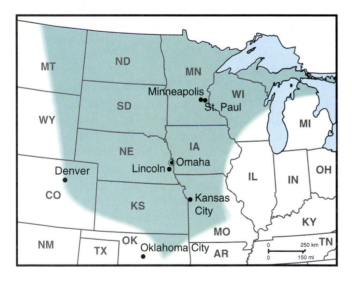

Figure 13.14
The Prairie Grain Belt.

progressive pragmatism, largely devoid of the emotional appeals associated with politics in much of the rest of the country. Ethnicity here is usually down-to-earth and un-self-conscious. While German neighborhoods and Little

Italys exist throughout urban America, they must be constantly redefined in the context of the changing city. In the Prairie Grain Belt, there has been no such need. The "Norwegian bachelor farmers" made famous on public radio's "Prairie Home Companion" resist the commercialization and commodification that often occur in big-city ethnic neighborhoods. Still, in the face of continued out-migration and the influence of mass media, things may be changing. There may be no more Lawrence Welks in the making.

There are, however, some nicely preserved cities. Population growth in the Prairie Grain Belt has been concentrated in a few large metropolitan areas, such as Minneapolis-St. Paul, Kansas City, and Omaha. Minneapolis-St. Paul (by far the largest urban area, with a metropolitan population of nearly 2.8 million) had a growth between 1990 and 1996 of a respectable 9%. These are the essential Prairie Grain Belt cities, rising to fame and fortune processing the animal and vegetable resources of the prairies. Even here, however, these are no longer really growth industries, and the mills of Minneapolis are rapidly losing ground to higher-tech activities. In spite of the Twin Cities' title as the world's coldest major metropolitan area, their water-related amenities and reputation for good planning have ensured their continued growth. Denver, on the margins of the

Figure 13.15
Omaha, Nebraska.

region, had a higher rate of growth, but it is culturally more integral to an adjacent region than to the Prairie Grain Belt.

Capital   Even though Minneapolis and Kansas City have higher profiles and growth rates, it is perhaps Omaha that best reflects the Prairie Grain Belt identity (Figure 13.15). The stockyards and silos of the Nebraska metropolis give it a character that epitomizes the Plains. Even the football flagship of the state, at the University of Nebraska in nearby Lincoln, is known as the Cornhuskers. Agricultural roots run deep here, and farm reports dominate the airwaves.

The Mountains

As we enter the Mountains, we leave behind the eastern regions of the country where rural settlement was more or less continuous and uniform (Figure 13.16). In the Mountains, settlement is discontinuous and spotty, with agriculture, in most areas, playing a distant second fiddle to mining, mountain recreation, or grazing. As one commentator has pointed out, there is no American West but rather a variety of different American Wests. The mining towns of Montana and the Mormon settlements of southern Utah have little in common culturally except for a certain re-

moteness and attitude of conservative self-reliance. Historically, most of these "wests" were far more connected to outside regions than to other areas in the Mountains.

An exception now, and in the past, is the Mormon culture region, which dominates the center of the Mountains. Stretching from northern Arizona through Utah and eastern Nevada to southern Idaho, the Mormon region exhibits a cultural unity that stands in stark contrast to the rough-and-ready mining towns and lumber camps that dominate the periphery of the region. The Salt Lake City metropolitan area, although smaller than the more peripheral Denver–Boulder area, has more than 1 million people and is becoming an important transit hub. As mining and irrigated agriculture have given way to ski resorts and office complexes, the economic power of the Mormon church has grown.

While population tends to be dispersing in many eastern regions as big cities lose out relative to smaller towns, population is still imploding in the Mountains. A large proportion of the people live within 100 kilometers (60 mi) of either Denver or Salt Lake City. The energy booms and busts associated with Wyoming and, to a lesser degree, other parts of the region have made living beyond the urban cores a tricky business. Under such circumstances, can the

Culture Regions of the United States

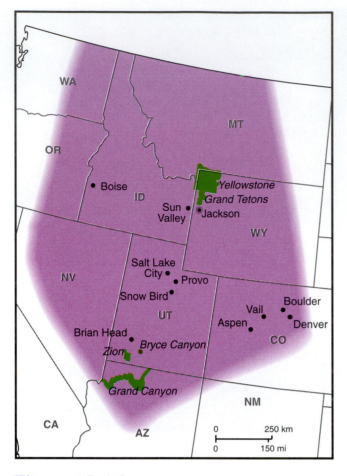

Figure 13.16
The Mountains.

self-reliant mountain man-yeoman rancher-lumberjack culture of the Mountains survive?

Perhaps the only indigenous culture group with the ability to expand vigorously is the Mormons. Since 15 of the top 20 counties with the highest birth rates in the nation are in the Mormon culture region (some competing only with East Africa for highest in the world), the Mountains seem destined to become increasingly Mormon. As for other aspects of its "wild West" past, the successful reproduction of local culture traits may lie in the commercialization of heritage.

The Mountains, as the name suggests, has always been a region associated with recreation. In recent decades, however, it has become a virtual playground for the residents of the booming urban centers all around it and, increasingly, for people from all parts of the world. Yellowstone, Grand Teton, Zion, Bryce Canyon, Grand Canyon, and a variety of other national and state parks are attracting record numbers of people from far and wide, allowing foreign tourists to enjoy wide-open spaces not often available in other parts of the developed world. Similarly, Aspen, Vail, Snow Bird, Brian Head, Sun Valley, and hundreds of

other downhill and cross-country ski resorts have boomed in recent years. For the Japanese, even ranching has become a kind of exotic recreation. The name Montana conjures up wild West images in Japan, where it has become fashionable to buy an American ranch and to become a cowboy. In addition to the thrill, profits can be made by exporting western products and exotica to Japan. American movie stars and other celebrities are also buying land and opening businesses in the Mountains as private jets make the "back to the land" movement more attractive. The Mountains, with its ghost towns, ranches, national park lodges, and Mormon villages, could become a vast and sprawling theme park for the world.

Not all of the Mountains is picturesque. Open-pit mining and lumbering have left scars in many locales, and bleak reservations for American Indians dot the region. While looming large in the imagery of the region, these actual landscapes are harder to make into commodities and sell. There is also a darker side to the region's reputation as a last vestige of pure American values—self-reliance, low crime rates, big families, and remote small towns. There is a small, but noticeable trend for the Mountains to become a sort of refuge for those seeking to escape the perceived chaos and confusion of an urban, multiethnic America. Such places as Boise and Provo are seen as escape paths for those who are no longer able to cope with Los Angeles. The Mountains may be the last Turnerian frontier in mainland America—a place for people to visit for as long as possible, the mythical, old-fashioned America of forests and farms.

Capital The place where all the trends come together in the Mountains is Jackson, Wyoming. It is a cowboy town adjacent to the Grand Tetons and other scenic wonders of the Mountains region (Figure 13.17). It is remote and quintessentially western, yet it is in danger of being overrun by developers seeking to cash in on its cultural heritage and natural beauty by building seemingly endless numbers of seasonal homes and ski resorts. It appears to be a restaurant row in the wilderness. It is the capital of our emerging theme park.

The Southwest: Hispanic Heritage and Booming Regional Capitals

The Southwest is one of the few regions of the United States where major settlements and even large cities existed before the coming of Europeans (Figure 13.18). Chaco Canyon in New Mexico and the pueblos of northeastern Arizona are reminders that not all American Indians lived in minimal and temporary shelter. In addition, the region varies from the norm in that initial European settlement came from the South rather than the East. Spanish explorers linked the Upper Rio Grande Valley with Mexico at an early point in time, and such cities as Santa Fe were more similar to Mexico City than to Boston. While this cultural

Figure 13.17
The landscape features that give the Mountains region its name are reflected in Jackson Lake, Grand Teton National Park. The lake is about 48 kilometers (30 mi) north of the town of Jackson, Wyoming.

tradition has been greatly romanticized in recent decades, there is still a great deal of heritage in the Santa Fe–Albuquerque–Taos area that is a unique hybridization of American Indian and Spanish influences. The same can be said of Tucson, located in the part of southern Arizona that remained part of Mexico until the Gadsden Purchase in 1853.

Those who recall the story of the Alamo know that Texas was also part of the Hispanic culture realm, as well as the nation of Mexico. While Texas has had a much greater influx of outsiders over the years than has New Mexico, many traditions remain. San Antonio is considered to be the capital of the Mexican–American Southwest in that it is a producer of many Mexican-oriented products, such as salsa, and because it is the focal point of major migration routes between the border and northern metropo-

lises such as Chicago. Its landscape, food, and music reflect this role. A major attraction for both locals and tourists is the Mexican Market complex adjacent to the downtown River Walk and convention hotels. Roughly half of the metropolitan population shares an Hispanic heritage to some degree.

Yet another major center of Hispanic culture in the Southwest lies along the U.S.–Mexico border from Brownsville to El Paso. Throughout much of this zone, there are relatively few cultural differences between one side of the border and the other. The influences of both nations come together here and permeate everything. With rapid inmigration, however, the influence of Mexico is growing faster and will continue to increase with the North American Free Trade Agreement. The El Paso–Juarez area is the

Culture Regions of the United States

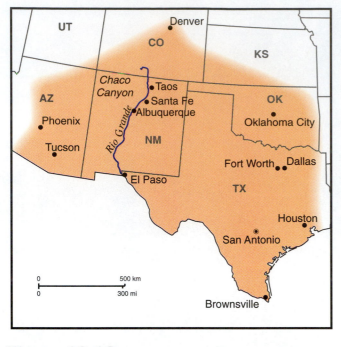

Figure 13.18
The Southwest.

second largest (after San Diego-Tijuana) paired-city complex on the border, with most of the others lying to the east.

Most of the rapid population and economic growth in the Southwest has occurred in the large metropolitan areas that encircle the region. Dallas-Fort Worth, Houston, Oklahoma City, Denver, and Phoenix all lie on the edge of the Southwest and share some aspects of its culture. While Hispanic culture is relatively less important in these cities, it is very important in absolute terms since large ethnic barrios dot the cityscapes. Much of the regional dress associated with the cowboy look of Texas originated in the Spanish-Mexican tradition. The boots, hats, chaps, spurs, bandanas, and, of course, guitars that have come to symbolize the cowboy were not part of the Anglo migration into Texas from the East. There is some doubt about the yodel, however. It may have been introduced by German-Swiss settlers.

Although Mexico has given the region much of its ambiance, there are a goodly number of non-Hispanic culture traits as well. While American Indians adopted at least some Hispanic traits in most of the region, such was not the case in Oklahoma. Oklahoma was the destination for the "trail of tears" that led many tribes from the Southeast on a forced march to new reservations. Parts of Oklahoma are thus North Carolina once removed. The names and faces may seem southwestern, but the story is different.

Over the past 60 years, white and black Americans from eastern regions have poured into the Southwest to participate in the energy booms. The Southwest began its seri-

ous trek toward urbanization and industrialization with the discovery of vast reserves of oil during the 1930s. While considerable numbers of rough-and-ready workers headed for the oil patches during that period, the boom was not immediate. The Dust Bowl was in progress, as was the Great Depression, and more people were leaving the northern parts of the Southwest than were coming in. There was a great culture shift over the decades of the 1930s and 1940s, as much of the population of the Southwest went to California only to be replaced by mass migrations into the region from the South and Country South. Country music became western music when it hit Fort Worth.

World War II brought a massive demand for petroleum products as well as the construction of a number of military bases from Texas to Arizona. Test pilots liked the clear, blue skies. By 1950, regional growth had reached phenomenal proportions. Census enumerators could not believe the figures for Houston. Between 1900 and 1950, the city grew from 45,000 to nearly 600,000, while Dallas made a similar jump, but only to 435,000. It was time to develop a new regional image. It was time to have skyscrapers and suburbs. Much of the culture of the Southwest is culture on the move. The drive-in restaurant was invented in Dallas during the 1930s, and mobility became an important theme in the legends of the land. Cowboys, oilmen, truckers, and railroaders, all on the move: the region was settled by the unsettled. Progress and the pursuit of happiness often meant moving around. In this sense, the Southwest, and especially Texas, epitomizes American culture as a whole.

In keeping with the tradition of mobility and change, many of the cities of the Southwest have pioneered new forms of urban growth. Houston has never seen the necessity for zoning and relies instead on deed restrictions and other "gentlemen's agreements" for a modicum of order. Both Houston and Dallas have specialized in the creation of Edge Cities, with major nodes of activity spread more or less evenly throughout the metropolitan areas. It is no accident that the tallest building in America located outside a traditional downtown is in the Galleria District of Houston. Similarly, Phoenix has been designed as a city of realms or "villages" rather than a central city with suburbs. None of this is unique, of course, but the trends are far advanced in the Southwest. Walled and gated communities, age-specific and childless retirement zones, and vast trailer parks also characterize many parts of the Southwest. Conditions, covenants, and restrictions have shaped many a neighborhood in the absence of more traditional, elected governments.

Capital Since it represents a symbolic midway point between the boomtown images of the bigger Texas cities and the exotic, Spanish culture of the highland cities farther west, San Antonio would seem to be the ideal capital or epitome city (Figure 13.19). It also relies heavily on the military and has recently pursued the commodification of its heritage for tourism and the convention business.

Figure 13.19
San Antonio, Texas. Latino ambiance at the street markets.

The Northwest

In the minds of many of its residents, from roughly Big Sur, south of San Francisco, to the Canadian border, the Northwest is an ecological utopia (Figure 13.20). Obviously, the region is not really an ecological utopia, but it may be as close as we can get to one within the confines of the United States. The region has always had at least a degree of environmental degradation, from the scars on the landscape resulting from large-scale lumber operations to the pollution of air and water around pulp mills. In recent years, the rampant growth of metropolitan areas from San Francisco to Seattle has led to the usual concerns about traffic congestion and overdevelopment. Yet the positive image of the Northwest remains largely intact in the minds of thousands of annual refugees from other parts of the country.

Perhaps the country needs this environmentally oriented region, even though it is at least partly exaggerated if not mythical. Just as we once needed a frontier to reinforce our image of a pristine, unsullied place where one could get away from it all and start over, so today we need to believe there is a place not only relatively free of grimy factories, slums, and pollution but also where people value the environment and greed is mitigated by an appreciation of nature. The Northwest is seen as a land of tall trees and wet wool. It is a place where people live in redwood cabins and collect blackberries, but it is not the wild West. It is a place where civilized people can reside in civilized cities and live wholesome lives. It is a place to start wineries. At least, perhaps, this is what the country needs to believe. It is a place that is not New York, Texas, or southern California.

While the Northwest image is definitely exaggerated, it has some basis in fact. As we have seen, all regions from the Rustic Northeast and Megalopolis to Tropical Florida and the Southwest attract migrants interested in participating in a life-style. Twentieth-century America has been a place where migration is not just for economic betterment or survival, as it might be in, say, Somalia, but a search for a life-style. Eugene, Oregon, tends to attract people who are different from those heading to New York or Houston. San

Culture Regions of the United States

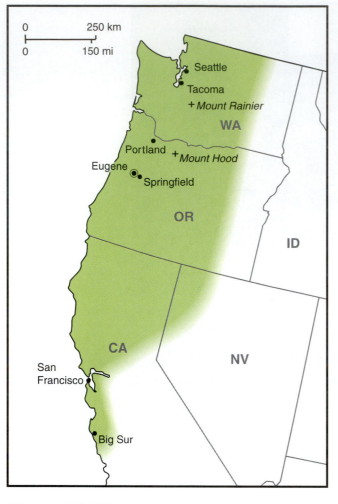

Figure 13.20
The Northwest.

Francisco tends to attract people who are not particularly enamored of Los Angeles. Life-styles are created by people gathering together in search of a life-style. Selective migration means a lot since it helps to reproduce the culture of places.

From the beginning, the Northwest attracted settlers from the Northeast, especially from New England, who were, in a sort of noblesse oblige way, concerned with protecting the natural beauty of their new-found home. The Sierra Club was founded in San Francisco, and it was there that such photographers as Ansel Adams began to produce stunning depictions of America's natural wonders. Artists and writers intrigued by the Northwest splendors soon gathered from Big Sur in the south to the San Juan Islands in the north, and the region became associated with the sound of whales breaching. Of course, there are many other stories associated with the making of the region, such as the radical labor movements of Seattle in the early decades of the century, but it is the ecological one that has staying power. It is the story that the rest of the country wants to hear.

The Northwest is also one of the few regions of the nation that has long been associated with social tolerance. The beatniks and hippies were drawn to San Francisco for a variety of reasons, but it was the tolerant tradition of the city that encouraged a lot of cultural experimentation. Seattle and Portland have become well-known as cities that have led the way in the sensitive revitalization of their skid row districts. Both have revitalized the downtown fringe while developing ample facilities for low-income and homeless populations. There have been relatively few attempts at "poor removal," and life-style diversity is respected. While there have been signs, of late, that the tolerant image is cracking, it has proved to be remarkably durable. While other parts of the country, such as southern California, have become much more intolerant as new migrants have moved in and tried to close the door behind them, mellowness has remained largely intact in the Northwest.

In much of the region, cultural and economic homogeneity have facilitated this equanimity. Compared to, say, Detroit, Houston, and New York, Portland and Seattle are very much white, Protestant, middle-class cities. Perhaps there is more room for a few eccentric environmentalists and hippies when there are fewer ghettos and barrios to worry about. Even the diverse city of San Francisco has few real urban problems on the scale of Cleveland or Newark. Why not let people play flutes and sing odes to the trees?

While the Northwest is not growing as rapidly as the Sun Belt, many of its parts have boomed in the recent past. The Seattle-Tacoma area grew by nearly 12% between 1990 and 1996 and now, with more than 3.4 million people, constitutes one of the major metropolitan areas in the nation. It is now bigger than Cleveland, and, like that city, its downtown bristles with mighty skyscrapers and sports arenas, and suburban malls dot the landscape. Boeing, a genuine industrial giant, seems even more important with the decline of many eastern industrial cities and corporations. While Seattle still sits in the shadow of Mount Rainier and Portland in the shadow of Mount Hood, the northwoods ambiance of these cities may give way as hordes of southern Californians continue to arrive in BMWs stocked with white wine and brie. The Redwood Curtain may prove to be quite permeable.

There was a time when the Northwest spurned excessive growth and preferred to live on its traditional activities and resources, but hard times have changed all that. Trees grow a lot faster in the southeastern states, and they can usually be harvested more easily there. The lumber industry has been in decline for some time in spite of the demand for plywood in Japan. The region has sought to develop new employment by imitating success stories elsewhere. The area south of Portland is thought to be either another Silicon Valley or Research Triangle, while downtown San Francisco and Seattle have become financial centers for the boom in world trade with the Pacific Rim.

Figure 13.21
Eugene, Oregon.

Capital In spite of the growth, old attitudes are hard to change. One of the best places to examine the increasingly complex assortment of attitudes and values concerning the possible conflict between the Northwest tradition and the need for growth and change is Eugene, Oregon (Figure 13.21). The Eugene–Springfield metropolitan area is home to both the University of Oregon and a number of lumber-related industries, and both are having problems in the face of a California-inspired property tax revolt. The Northwest values are deeply embedded, but jobs are needed, and change has got to come.

The Coast: Southern California Extended

People in the eastern part of the nation rarely go to a coast. They may go to the shore or to the cape, sound, bay, or beach, but not to the coast. "The Coast" has come to stand for a vague region associated with Hollywood, surfing, television studios, the recording industry, white wine, bean sprouts, volleyball, and other aspects of life in southern California (Figure 13.22). This is what people in New York or Washington, D.C., mean when they say that they have been transferred to the Coast.

Perhaps the term *coast* emerged because the actual destinations of early migrants to southern California, espe-

cially those in the entertainment world, were vague and ill-defined. Few people headed for Los Angeles; rather, they went to Hollywood, Malibu, Beverly Hills, Santa Monica, Westwood, the "Valley," Pasadena, or, if they knew an actual work address, to a studio in Culver City. People heading for New York went to New York. People heading west just aimed for the Coast and hoped for the best.

If anything, current destinations are even more vague and scattered. In the common vernacular, the Coast now extends from the Silicon Valley of San Jose to the biomedical laboratories of La Jolla. The Coast is also expanding inland to include such "overflow" cities as Fresno and Bakersfield, where land costs and taxes are lower. In addition, resorts such as Palm Springs and recreation zones such as Yosemite and Las Vegas are joining the Coast as freeways and air shuttles make more places accessible.

The Coast is now a widely scattered network of interconnected corporate centers, elite residential enclaves, industrial zones, resorts, and barrios. Never before has areal specialization occurred on such a scale. The Coast is an unevenly settled region containing more than 25 million people in a handful of sprawling communities. Between the settlements are vast tracts of empty desert and high mountain ranges. Much of its fame, as well as its attractiveness for the movie and television industries, is related to its

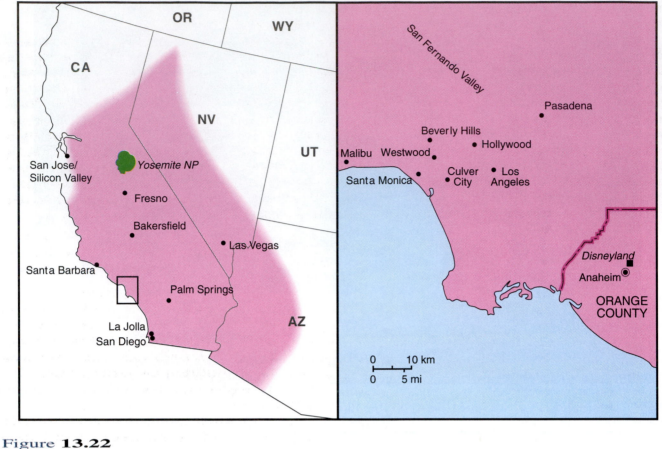

Figure 13.22
The Coast.

physical diversity. Sunny beaches lie within sight of snow-capped peaks. Here, one can ski in the morning and golf in the afternoon.

The core of the Coast is a southern California "megalopolis II" stretching from Santa Barbara in the north to San Diego (and perhaps Tijuana) in the south. This giant urban complex contains nearly 20 million people and is the "second city" of the nation in competition only with the original Megalopolis. The Coast has become the dominant area for media with not only its long-standing Hollywood movie industry, but also the television and music industries. Only serious programs, such as the national news and "Late Night with David Letterman," come from New York. Although political power is still concentrated in Washington, D.C., California is now by far the biggest political plum in any national election, and so a great deal of attention is focused there. The Coast is also an economic powerhouse in terms of industry and corporate headquarters. Indeed, at one time the Coast seemed to be destined to capture the vast majority of the "industries of the future," with everything from computers and airplanes to hot tubs and wine being produced there. This vision has declined somewhat since 1990, but it is still a very dynamic region.

The many attractions of the Coast—jobs, entertainment, beaches, sunshine, mountains, innovative life-styles, excellent educational institutions, off-beat religions, cutting-edge technologies, and a variety of other opportunities—have traditionally led to high rates of migration to the region, but up until about 1970, most of the migrants came from other parts of the United States. While intranational migration to the Coast has slowed, the region has become the destination of choice for immigrants from around the world.

The once-teeming cities of Megalopolis now seem homogeneous compared to the cultural diversity of the Coast. Some school districts have student populations from countries as diverse as Iraq, Somalia, Afghanistan, Russia, El Salvador, Cambodia, Vietnam, and Samoa, often speaking a total of 70 or more languages. There is much talk of the Coast becoming "majority minority" in the near future, but it is already impossible to determine what *majority* and *minority* mean. There are so many foreign whites, middle-class blacks, Hispanics of all colors and social classes, and such a variety of languages, skin tones, and cultures that old categories make little sense. An upper-class suburban street may house a Japanese computer specialist, a Mexican importer, a black basketball player, a television preacher from Oklahoma, a rock star from New York, a business person from Hong Kong, and a French-Canadian retiree. Who is a minority?

The cultural diversity of the Coast involves more than the immigration of people from a variety of backgrounds. The place itself foments and encourages the evolution of new fads, fashions, foods, follies, and philosophies. Architecturally, the Coast influenced America throughout the 20th century as such innovations as California bungalows, Las Vegas-style commercial signage, patios and wooden decks, and signature architecture (such as doughnut shops shaped like giant doughnuts) spread throughout the nation. The entertainment industry has long been a major player in the creation of new images of the ideal way of life, but the process has speeded up today. MTV, "infotainment" news programs, films and music by such talents as Spike Lee and Ice T, cable channels featuring X-rated comedians, fashions derived from street gangs and sports activities, and a variety of styles and colors from foreign lands all combine to make the cauldron of the Coast a place where ever-changing creative innovations are most likely to influence national and international popular culture. Unlike Megalopolis, with its Smithsonian Institution and Metropolitan Museum, which guard and protect the culture of the past, the Coast churns out the culture of the moment.

Of course, the Coast has never been exactly culturally "normal." Over the past 100 years, real estate booms, the Dust Bowl, the elusive promise of Hollywood stardom, scientific research contracts, organized crime connections, surfing competitions, planned golf and retirement communities, and a rather porous border with Mexico have all helped to create a population that was very different from, say, Iowa's. Over the past 30 years, however, the diversity has intensified. This diversity could become a tremendous advantage for the region as America endeavors to compete in the new global economy. Culturally and economically, the Coast is representative of the world economy, with everything from Mexican sweatshops deep in the Los Angeles garment district to Asian entrepreneurs maneuvering to take advantage of new technological breakthroughs. In the short run, however, there are many problems to be solved.

There is a danger that the modern, antiurban form typically found in the Coast is not as conducive to the assimilation of immigrants as the traditional cities of Megalopolis and the Midwest Heartland were a century ago. Immigrants could learn the ropes in Lower Manhattan and North Boston just by being there, since the economic activity of the city was going on all around them. Though their living conditions were bad, at least they could watch America in action. This is no longer so easily accomplished in most American cities, and it is very difficult to do in a place like Los Angeles. The jobs have not just moved a few miles out to the suburbs, they have often gone vast distances, even over mountain ranges, to locations entirely inaccessible to low-income workers. Housing prices some-

times start at $500,000 in these remote Edge Cities, and public transit is virtually nonexistent.

The new culture groups are left to fend for themselves, and sometimes they do. In vast areas of the Coast, small Asian and Hispanic businesses have sprung up along the old commercial strips abandoned by the mega-developers, creating exotic landscapes. More often, however, the zones left behind are in deep trouble. Neighborhoods of poverty are sometimes larger than entire cities in other parts of the world. Ghettos and barrios can cover more land area than such cities as Boston or Cincinnati. There is a danger that smoldering, permanent polyglot, depressing "riot zones" are emerging on an unprecedented scale. Like London in 1850 or New York in 1900, the Coast is where the new society is being invented. As always, both the best and worst qualities of the experiment become exaggerated, but in the Coast, there is an entirely new scale.

Capital An ideal capital for a region as diverse as the Coast should be diverse, but since there is so much areal specialization, that becomes a problem. One possibility is Anaheim, since it has at least been very diverse over time, experiencing nearly every type of attraction from agricultural settlement to idyllic suburb to immigrant magnet (Figure 13.23). It is embedded in the sprawling Los Angeles–Orange County conurbation, yet it retains significant identity. Perhaps most important, it is the home of Disneyland, one of America's major culture machines.

Hawaii: Polynesian Culture, American Politics, and Japanese Money

The United States extends well into the Pacific Ocean and includes a variety of island groups. Only the Hawaiian Islands are officially part of the country, however, and so the culture region of Hawaii focuses there (Figure 13.24). Since there is no need to create a new regional boundary and since Hawaii is such an isolated and unique part of America, the discussion can be quite short.

Hawaii is an increasingly important link between Asia and the "European" world of Washington, D.C. Its language is from Malaysia; much of its work force came originally from China, Japan, or the Philippines; its capital investment now comes from a variety of Asian as well as American sources; and tourists arrive from every part of the world. Its agricultural crops and decorative landscapes have come from all over the Pacific Rim and, indeed, the world. With more than 1.2 million people in a small and confined habitable area, population densities are now, or soon will be, midway between the American norm and that of East Asia. High-rise hotels, apartments, and condominiums already line the beaches of Oahu, and traffic congestion exceeds that of Singapore. East Asian food is served along with burgers and fries at fast-food chains, and the music and dance traditions are remotely connected to those of Java and Bali, although the songs are most often sung in English.

Figure 13.23
Anaheim, California, gateway to Disneyland.

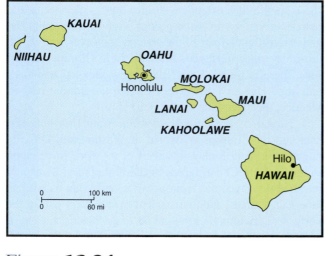

Figure 13.24
Hawaii.

After decades of Americanization, there is a movement afoot to enhance the role of "native" culture in the region, but as in the Coast, it is increasingly difficult to tell what *native* is. Hawaiian traditions have been mixed with Asian and European practices for so long that purity could

be elusive. In a region dependent on tourism, however, culture commodification is usually profitable, and so, authentic or not, it is likely that "things Hawaiian" will become increasingly visible in the future.

Carrying capacity is an issue in Hawaii. No other region is so dependent on long-distance commerce for basic needs. Most of the land is in parks and commercial monoculture crops, and so nearly every food must be imported over thousands of kilometers. No other island group in the middle of the ocean has more than a million people, so it will be interesting to see what infrastructure will be needed to support continued growth.

Capital   Honolulu is the primate city and transportation center for Hawaii, and nearly every important event that happens in the region occurs there (Figure 13.25). Perhaps the ideal place for the capital would be right in front of the East-West Center at the University of Hawaii. It is the main think tank for scholars interested in the Pacific Rim and the cultures there.

Alaska: Petroleum and the Last Frontier

Alaska is the largest and emptiest state in the union (Figure 13.26). During the 1980s, Alaska surpassed Wyoming

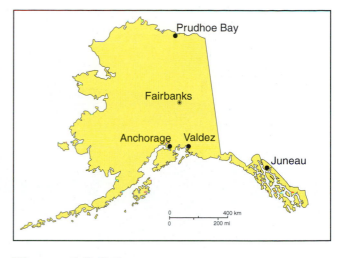

Figure 13.26
Alaska.

Figure 13.25
Honolulu, Hawaii. Lei occupations.

to move out of last place in population, but it remains to be seen how much of that growth is permanent. The oil pipeline has been completed, and many of the jobs associated with its construction have left. Fairbanks's Nordstrom department store, the one place to buy stylish clothes in the far north, closed when the boom ended.

Alaska is mostly empty of people, and it seems very likely to remain that way. Over half the population of the region resides in the Anchorage area, and the dominance of that city may become more pronounced in the future as there are few options. The capital, Juneau, is located in a fjord-like landscape with little flat land and no place to grow. Northern and central portions of the region are just too cold and inhospitable for most people. With diminished needs for over-the-pole defense systems and military operations, remote settlements may dwindle.

Nevertheless, the region looms large in the minds of many Americans. It is the genuine last frontier, a place to make a new start far from the crowd. It is a place where people take great pride in their ability to survive the elements, a feat made more difficult than it might be by the sprawling, chaotic nature of the towns. The compact architecture of Switzerland or Finland is nowhere to be found. Getting a car started and the tires unfrozen when it is –45° C (–50° F) to drive to a store for a loaf of bread is part of the challenge. It is not a place for trendiness and laid-back living.

As the television program "Northern Exposure" once suggested, the region has more than its share of characters and individuals. It can be a land of bleak absurdity, such as one lone shack of a store selling nothing but ice in the middle of a wintry, wind-blown plain or urbanites from the lower 48 trying to train dog teams for the Iditarod race.

Alaska is a land of contrasts in that it is a battle-ground for individualistic, antigovernment frontierspeople seeking to develop the region's resources and environmentalists who see it as the last bastion of unspoiled nature in America. It is a land of elk and oil spills. The negative impact of human activity is exceptionally visible in Alaska because the permafrost makes burying waste impossible in many areas. The vast distances make transporting used materials away unlikely as well, so anything that ever came to the far north remains there to rust in the open air. It is a good argument for keeping population to a minimum.

Of course, the role of the region could change drastically with the new world order. The frontier mentality of Alaska was political as well as physical because the former Soviet Union lurked just beyond. Even hardy frontierspeople could go no farther than Nome. In the future, there could be heated, all-weather freeways to Vladivostock or high-speed tourist trains connecting Denali National Park

373

Figure 13.27
Fairbanks, Alaska, contains many quonset huts like these in Barrow.

to Lake Baikal. These developments are not on the near horizon, however.

Capital Locating a capital presents an especially difficult problem in Alaska since the region is so large and fragmented and because the state is pondering a new site between Anchorage and Fairbanks. The new capital would be a planned city, perhaps a sort of "Tundrilia" carved out of the wilderness. In the meantime, Fairbanks makes an ideal capital (Figure 13.27). It is a small city on the edge of civilization where both the wind and the oil come out of the north.

Summary

In a country as new and mobile as the United States, it is difficult to speak about culture regions in the traditional sense. We do not have hard-and-fast consensus regions based on language, religion, political systems, and other important dimensions of ways of life. Rather, we have ever-changing regions that rise and fall with the construction of new highways, cities, resorts, and communication networks.

In spite of the strong tendencies toward cultural homogenization and place obliteration, however, regional identities persist. In some cases, real local traditions simply refuse to die, while in other cases, long-abandoned traditions are revived to capture tourist dollars or satisfy some political agenda. Whatever the reason, the different regions of the United States continue to have their own personalities and senses of place. Finding them is basic to the joy of travel. We have identified and discussed some possible culture regions in this chapter, but the list is meant to be suggestive rather than exhaustive. Creating regions helps us to think more clearly about the relationship between culture and place.

Gaining Insights

1. What is a culture region? Why do geographers create regions to focus their thinking? What regions can you create in your neighborhood, city, or area?
2. What is culture? Why is America different from many other parts of the world when it comes to equating culture and region?
3. Different localities often have a unique sense of place; that is, they look and feel like no other places. What kinds of things do you notice that give places character? Are these things easy to spot, or do you have to look for them?
4. What do you think the United States will be like in 50 years? Will the culture regions be more or less obvious than they are today?
5. What are the distinguishing characteristics of Megalopolis? What is its relationship to the Rustic Northeast?

6. What are the "consistent inconsistencies" of the South to which the text refers? How may one account for them?

7. Why can Columbus, Ohio, and other cities of the Midwest Heartland be said to represent "average" America? Do you believe there is such a thing?

8. Describe some of the elements of the folk culture of the Country South. To what factors can we attribute the new prosperity of the region?

9. What does it mean to say there is a variety of different American wests?

10. What are the origins of the culture traits that define the Southwest? Do you think the region is becoming more culturally diverse or more culturally homogeneous?

11. What is the dominant image of the Northwest? Do you believe the image is justified?

Selected References

Annals of the Association of American Geographers, special issue on regions of North America, 62 (June 1972).

Carney, George, ed. *The Sounds of People and Places: Readings in the Geography of American Folk and Popular Music.* Lanham, Md.: University Press of America, 1987.

Caudill, Harry. *Night Comes to the Cumberlands.* Boston: Little, Brown, 1963.

Conzen, Michael, ed. *The Making of the American Landscape.* Boston: Unwin Hyman, 1990.

Garreau, Joel. *The Nine Nations of North America.* Boston: Houghton Mifflin, 1981.

Heat-Moon, William Least. *Blue Highways: A Journey into America.* Boston: Little, Brown, 1982.

Kerouac, Jack. *On The Road.* New York: Viking Press, 1957.

Lee, Harper. *To Kill a Mockingbird.* Philadelphia: Lippincott, 1960.

Meinig, D. W. *Southwest: Three Peoples in Geographical Change.* New York: Oxford University Press, 1971.

————, ed. *The Interpretation of Ordinary Landscapes.* New York: Oxford University Press, 1979.

Prisig, Robert. *Zen and the Art of Motorcycle Maintenance.* New York: Morrow, 1974.

Relph, E. C. *Place and Placelessness.* London: Pion, 1976.

Rooney, John, Wilbur Zelinsky, and Dean Lowder, eds. *This Remarkable Continent: An Atlas of United States and Canadian Society and Culture.* College Station: Texas A&M University Press, 1982.

Steinbeck, John. *Travels with Charley: In Search of America.* New York: Viking Press, 1962.

Thurber, James. *The Thurber Carnival.* New York: Harper and Brothers, 1945.

Zelinsky, Wilbur. *The Cultural Geography of the United States.* Englewood Cliffs, N.J.: Prentice-Hall, 1973.

Zukin, Sharon. *Landscapes of Power: From Detroit to Disneyland.* Berkeley: University of California Press, 1991.

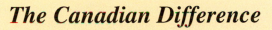

CHAPTER

14

The Canadian Difference

Throughout most of the 1980s, a printer in Montréal violated Québec's language laws by displaying his store window signs in English. These laws were introduced in 1977 for the express purpose of safeguarding the French language in Québec. Official business could be conducted only in French; all migrants (including those from English-speaking provinces in Canada) were compelled to educate their children in French; and signs—ranging from traffic signs to retail signs such as the printer's—could contain only French (Figure 14.1).

The issue of French-only signs has become a rallying point for both language groups. Many English speakers feel that the laws are unfair, a kind of "reverse discrimination" as they are asked to pay for the injustices of their forebears. French speakers believe that these laws ensure the survival of the French language in a country where they are outnumbered by three to one. The other provinces in Canada wonder why they must adhere to a double standard: they must scrupulously conduct business in English and French, while Québec is allowed to officially shut the English language out. And officials in the federal government worry that

Québec's action frays the fabric of Canada's carefully crafted bilingual identity.

In 1988, the Canadian Supreme Court ruled that the law was unfair, arguing that the Canadian Charter of Rights and Freedoms protects the interests of English and French speakers throughout the country. However, this same charter also has a loophole allowing Québec to override certain supreme court decisions when it feels they threaten its culture. After the supreme court struck down the law, Québec's premier overrode the court's decision and once again imposed the French-only legislation.

The Canadian writer Margaret Atwood describes how each culture can be said to revolve around a single unifying symbol—a core ideal that provides a rallying point for all members of a society. The symbol for the United States is the Frontier, the promise that old habits, life-styles, and landscapes can be discarded in favor of something newer and better just over the hill or around the corner. For the United Kingdom, the symbol is the Island, fostering safety and insularity from the corrupting influences of the European Continent. For Canada, the central symbol is Survival. Early

Figure 14.1
Québec's language laws are strict, and many English language signs violate a 1977 provincial law stipulating that public signs and commercial advertising must be solely in French. In 1988 and again in 1992, this law was altered allowing languages other than French on commercial signs as long as French predominates. Some zealous residents even take the law into their own hands, obscuring or spray-painting slogans over objectionable words.

colonists to this harsh and forbidding land were forced to survive marginal soils, endless winters, and hostile natives. Today, survival is defined in cultural and political terms: can Canada survive as a distinct society alongside the behemoth to the south, and can it survive as an integrated society in the face of fissures linguistic, ethnic, and regional?

The issue of survival was revisited in October 1992, when Canadians voted on a series of amendments to the constitution. The constitution had been around only since 1982 and had not been fully approved since Québec had yet to sign it. This package of amendments was supposed to change that, but it failed resoundingly, and Canada was left in limbo once again.

That the constitution has yet to be fully ratified testifies to the tenuous hold Canada has on its own identity. The vastness of the land and the small size of the population have magnified strong regional and ethnic identities. More so than perhaps any other country, Canada is a land shaped by geography. Geography has given its inhabitants a standard of living among the highest in the world. Yet geography has also served as an agent of cultural, political, and social fragmentation—separating the Canadian people from each other by stretches of unbridgeable space.

Canadian Settlement Patterns

Canada is a sparsely settled country. Each square kilometer of land contains less than three Canadians (6 per sq mi), compared with a population density of 27 per square kilometer (68 per sq mi) for the United States and 45 per square kilometer (115 per sq mi) for Mexico. Yet this overall density conceals a great deal of variation. Nine out of every ten Canadians live in the far south, within 160 kilometers (100 mi) of the U.S. border; each individual living in the Yukon, Nunavut, or Northwest Territories has several square kilometers just to him or herself.

These population patterns result mainly from the physical and climatological environment. Few agricultural products can be grown in the great Canadian Shield, where the temperature dips well below the threshold of human comfort. So most Canadians hug the south, with its more moderate climate and fertile soils.

Canada's settlement pattern is discontinuous. The country is made up of distinct population clusters separated by large stretches where settlements are sparse (Figure 14.2). The main population center is defined by an axis that runs from Québec City, down the St. Lawrence River

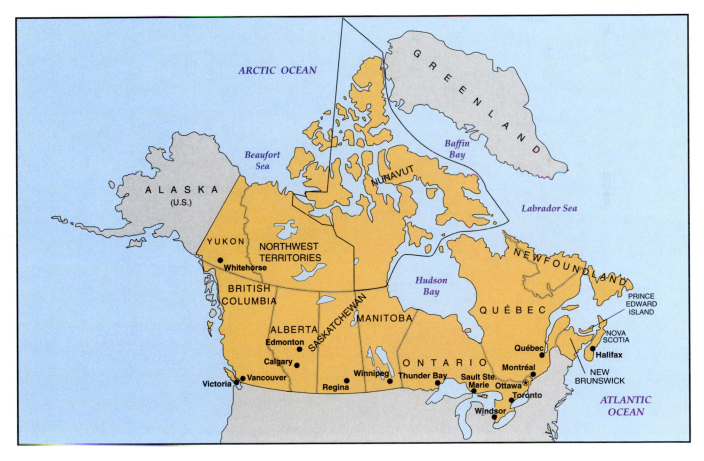

Figure 14.2

Canada contains the second largest landmass in the world, yet has a population smaller than that of California. Most people cluster along the southern border with the United States, with the largest concentration living in a wedge between Windsor, Ontario, and Québec City, Québec (see Figure 14.20)

to Montréal, and then into the southern Ontario "boot" bounded by the St. Lawrence, Lakes Ontario and Erie on the south, and Lake Huron and the Ottawa River on the north. This has been described as Canada's "Main Street" and contains more than half its total population. But it is also cut off from the rest of Canada by a physiography and climate not conducive to settlement. The Canadian Shield ensures a sparse population in the nearly 1000 kilometers (600 mi) between southern Ontario and the grain lands of the Prairie Provinces, which are broken off from British Columbia by the Canadian Rockies. A combination of water, climate, landforms, and an international boundary isolates the population concentration in the Atlantic Provinces from central Canada.

Pre-European Settlements

Canada was first encountered by tribes migrating over the land bridge that once connected Asia and North America. This event occurred anywhere from 45,000 to 20,000 years ago. By the time Europeans sighted the land that would become Canada, the aboriginal population numbered more than 250,000 and was organized into 50 tribes, each with a distinct language and culture (Figure 14.3).

The natural environment influenced the economy, social organization, and culture of the native peoples. The Arctic regions were inhabited by members of the Inuit (Eskimo) culture. The Inuit organized into tribal groups, containing 500–1000 members, which in turn were divided into several bands, composed of two to five families. Dwellings alternated between snow houses or igloos in the winter and skin tents—constructed of stretched seal or caribou skins—in the summer. Agriculture was out of the question in this forbidding climate. Inuits hunted seals, walruses, and whales and navigated the icy rivers and lakes in kayaks. Farther south, subarctic First Nations (Indian) tribes likewise focused on hunting and fishing. Bands of 25–30 people would move through the boreal forest, seeking thinly distributed game. Snowshoes, toboggans, and sleds were essential for moving cargo over long distances in the winter, while canoes provided transportation in the summer.

The northwestern coast—in present-day British Columbia—enjoyed a far more temperate climate. Salmon was especially plentiful and was eaten fresh or dried year-round. Villages were located near the sea or along the many streams, and tribal groups followed the migrations of sea

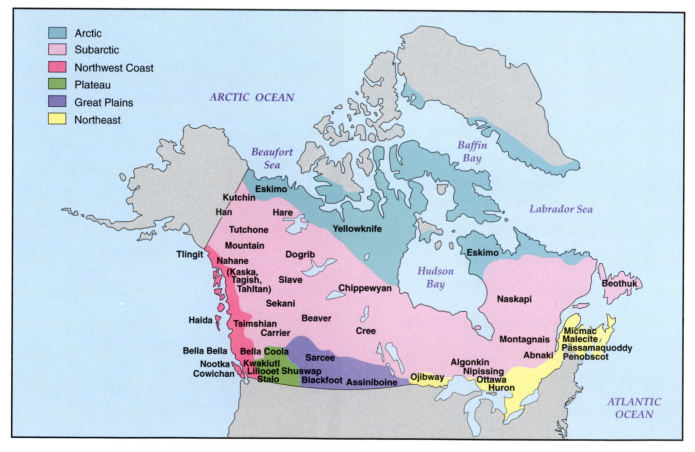

Figure 14.3

The area of what is now Canada enjoyed tremendous cultural diversity before Europeans appeared on the scene. These are just some of the more prominent native tribes that occupied various environmental niches.

mammals and waterfowl. Dugout canoes provided transportation, and a variety of harpooning, spearing, and netting techniques were used to capture fish and sea mammals. Giant totem poles (Figure 14.4) were probably the most distinctive feature of the Northwest Coast culture. Natives of the Columbian Plateau also fished for salmon in the major rivers and hunted a variety of game: deer, caribou, bear, mountain goats, and smaller animals.

Buffalo hunting was the predominant activity among men in the Canadian Great Plains. Women gathered edible roots and berries and processed the game. Fish were less available in the drier climate. The Algonkin tribes of the Northeast Woodlands cultivated corn, beans, and squash. This type of economic activity facilitated greater population densities and allowed for towns of as many as 2500 people. Villages contained several long-houses, within which dwelt several related families.

Of course, all aspects of Canadian First Nations economy, society, and culture would change dramatically after the encounter with Europeans. Arguably, the changes were perhaps less negative for Canadian than for American Indians. Europeans settled the land more slowly here, and many tried to get along with the neighboring tribes since a successful fur trade depended on Indian knowledge and assistance. But the difference between the cultures was so extreme and their goals so different that any equilibrium could only be maintained for short periods of time. In the long run, each of the native cultures suffered from contact with the more technologically advanced and aggressive cultures of Europe.

European Settlement Nodes

Canada was probably the first part of the New World "discovered" by Europeans. In the late 10th and early 11th centuries, Norsemen hopscotched across the northern Atlantic, colonizing Iceland, then Greenland, and finally anchoring at l'Anse aux Meadows on the northern tip of Newfoundland (Figure 14.5). The Norse did not seek precious metals and spices but rather promising sites to raise cattle and establish villages. The venture lasted only two years, but the Norse probably traveled to Newfoundland for several centuries in search of fish and timber.

The main wave of European exploration after Columbus largely bypassed Canada. The country had little to offer in the way of gold and silver, the climate was much harsher than that of continental Europe, and glaciers had scraped the land clean of topsoil. Canada's first extensive explorer—the French seaman Jacques Cartier—summed up his impressions of the Labrador and Newfoundland coast: "I think that this is the land which God gave as his portion to Cain."

Despite such drawbacks, Europeans returned to Canada for three main reasons. The first was access to the Far East. Columbus had dreamed of a quick passage to Asia, and when John Cabot bumped into Cape Breton Island or Newfoundland in 1497, he thought he had encountered northern China. Though Cabot was proved wrong, many believed that access to China was simply a matter of cutting through or getting around the North American landmass. In their pursuit of this Northwest Passage, Englishmen like Martin Frobisher in 1576 and John Davis in 1585 expanded European knowledge of the northernmost part of North America. Their travels helped chart the frigid

Figure 14.4
Giant totem poles are now featured in cities like Vancouver, British Columbia, where once they served to welcome visitors to a Haida village or homestead.

The Canadian Difference

Figure 14.5
L'Anse aux Meadows, on the northern tip of Newfoundland, was occupied by Norse between A.D. 990 and 1050. It was never really intended as a permanent settlement, but probably functioned as a staging point from which expeditions sought out more promising sites.

coasts of Labrador, Baffin Island, and Hudson Bay. They failed to accomplish their goal, however, and the search for a water passage was finally laid to rest by 1631.

The second reason why Europeans returned to Canada was fish, or more precisely, cod. The waters around Nova Scotia and Newfoundland are relatively shallow, since they rest on the continental shelf. Here, the cold Labrador current mixes with the subtropical Gulf Stream to create one of the world's greatest fishing banks, rich in haddock, redfish, mackerel, herring, and cod. John Cabot noticed this bounty on his first voyage, and it did not take long before fisherman of all nationalities—English, Breton, Spanish, and Portuguese—rushed to take advantage. Cod became a cheap source of protein for European consumers, and no one company or country had exclusive rights to the Newfoundland fishery. A free-for-all ensued as fishing fleets competed to see who could harvest the most cod. At first, most fishermen did not even bother to land in Canada,

traversing the Atlantic from far-flung ports in Western Europe. Those who did establish outposts on land did not penetrate very deeply into the Canadian interior, a fact reflected in the coastal orientation of Newfoundland settlement to this day.

The third and perhaps most enduring reason for exploring Canada was fur. The vast boreal forests contained abundant wildlife, and beaver fur became a particularly valued commodity in Europe, especially as it could be fashioned into beaver hats. In the headlong pursuit of fur, Europeans depended on the native tribes who would catch the animals, prepare the furs, and then exchange the pelts for various trinkets. Both French and English were involved in this trade, and fur quickly became the most lucrative commodity in all of Canada. French traders followed the path of Samuel Champlain down the broad path of the St. Lawrence River, up the Ottawa River, and into the interior Great Lakes (Figure 14.6). Later, settlements

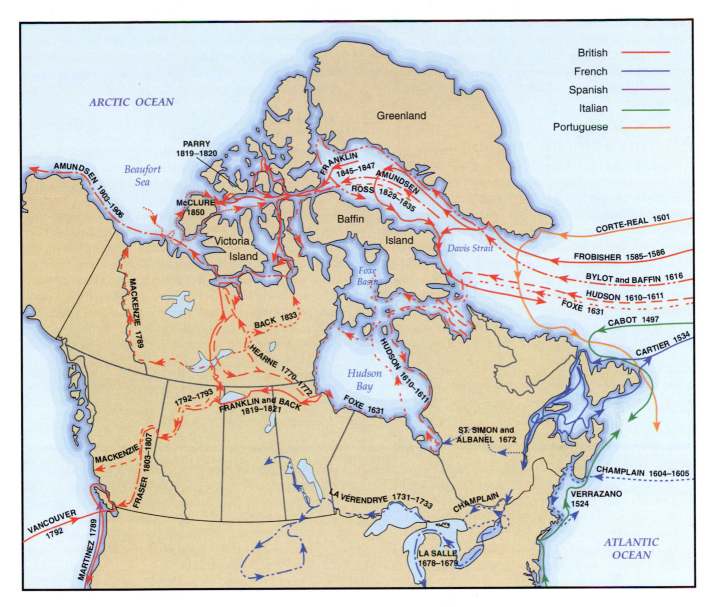

Figure 14.6
The European exploration of Canada was a joint affair of many different men representing several nationalities. The Italian John Cabot was the first to observe the eastern coastline, which would later be penetrated by the Frenchman Jacques Cartier. British sailors took the lead in charting out the Hudson Bay region, while Spaniards initially looked over the Pacific coastline.

would be established along this axis: Québec where the St. Lawrence narrowed, Montréal on a large riverine island about 240 kilometers (150 mi) upstream, and Fort Frontenac at the outlet of Lake Ontario. The fur lines were stretched as intrepid traders pushed farther and farther into the continental wilderness.

English traders operated under the auspices of the Hudson's Bay Company, formed in 1670 after English sailors seeking the Northwest Passage had instead uncovered a rich source of fur along the banks of the Hudson Bay. Forts were duly established along the river outlets to this formidable and forbidding water body, each serving as a funnel for furs coming in from the drainage basin. Unlike

the French voyageurs, these traders made sure they never strayed far from the water.

By 1750, the main settlement nodes were along the peninsulas and islands of Acadia, a few fishing villages on the Atlantic side of Newfoundland, a string of settlements hugging the St. Lawrence River between Montréal and Québec, and a series of trading forts on Hudson Bay. As was true of the United States, the direction of colonization was from east to west. But there were key differences between the two countries. Canada had much less usable land than the United States, and this restricted the paths of expansion. And unlike the United States, with its hopeful contingents of religious dissenters, utopians, and

The Canadian Difference

those seeking a better life in the New World, Canadian colonization always served as a handmaiden to the real activity that defined both French and English colonies: the exploitation of natural resources for export to the home country. It was a pattern that would define Canada from that point on.

External Influences on Canada

A major issue for many Canadians is that of self-determination: the ability to fully control their country's destiny. Such a goal may seem surprising in light of the fact that Canada is a sovereign country and has been so since 1867, when the British North America Act proclaimed Canada's independence from the United Kingdom. But external powers have always played a crucial role in shaping Canada: first as a part of New France, then as a British colony, and most recently as what some see as an economic adjunct of the United States.

It is impossible to understand Canada without understanding its relationships with these three external entities.

The pressing issues that still bedevil it—the political and geographical divisions between French and English speakers, the ambiguity of national identity, the concern over Canada's economic and cultural dependence—are a legacy of these historical ties.

New France

By 1700, France owned land from the Atlantic Ocean to the mouth of the Mississippi River but had few people to show for it. Early settlement of Canada's Atlantic Coast and the St. Lawrence River valley was haphazard, propelled mainly by the demands of the fur trade and the need for protection from the hostile Iroquois. Most of the early colonists came to Canada to seek a quick fortune in this enterprise. They called themselves **coureurs de bois,** independent traders who roamed the forests from Lake Erie to Lake Winnipeg and down the Mississippi River to the Ohio River in search of fur and markets. To survive both economically and physically, the coureurs de bois immersed themselves in Indian culture and lore. This type of activity did not lend itself to stability or deference to authority. A report sent to France

TWO FRENCH FUR TRADERS

In the 18th century, many Frenchmen sought their fortune in the fur trade. Colonial residence in Canada may have seemed dreary, but it was an important opportunity at a time when opportunities were few and far between. François Havy and Jean Lefebvre were the Québec City-based **factors** (or agents) for Dugard and Company, a major import and export concern based in Rouen. The Atlantic trade conducted by Dugard and Company formed a link in an elongated chain of intercontinental exchange that extended from Indian trappers and hunters in the Great Lakes frontier to the frontiers of Parisian fashion. Between 1732 and 1747, Havy and Lefebvre coordinated between one-tenth and one-fifth of the colony's import trade. They could not have formally immigrated to Canada—permanent settlement was open only to Catholics, and Havy and Lefebvre were Protestant Huguenots—but they ended up spending most of their working lives in the colony.

Like other metropolitan factors, Havy and Lefebvre supplied French products to the colony and bought items for export overseas. Since the ships arrived between July and October, factors had to anticipate colonial needs through the following winter and spring. Canadian residents needed a huge assortment of goods, from blankets and plowshares to wine and spirits, all sold at high markups. In return, factors like Havy and Lefebvre

purchased products for export to France. Fur was by far the most important commodity, and the two agents helped finance trading expeditions with necessary credit, to be repaid once the fur cargo reached Montréal. They also determined the price for pelts and assessed their quality, privileges that suggested that French factors held the upper hand over colonials.

The colonial trade was a high-risk endeavor for all involved. A European war and a number of captured ships disrupted the Atlantic trade and dulled the luster of the fur trade. By 1747, Dugard and Company terminated its Canadian enterprises and its association with Havy and Lefebvre. But these two agents continued to try their hand at various Canadian investments—real estate, mortgages, seal oil—before finally giving up. Havy, the older of the two men, left for La Rochelle in 1756 and died brokenhearted in 1766. Lefebvre survived the British siege of Québec City in 1759 but died during a return to France in 1760. While never Canadian, these two Frenchmen had a major impact on the Canadian economy and in turn were influenced by this hybrid society, an implanted European civilization in an American wilderness. "I will always love your country and its people" were the last words Havy wrote to a Canadian correspondent. More than 200 years later, Canada continues to be influenced by its French connection.

in 1689 complained that "the people of this country, neither docile nor easy to govern, are very difficult to constrain," and it is not surprising that colonists, having left a rigid feudal structure in the Old World, welcomed the freedoms and opportunities of the New.

While the French government enjoyed the revenues brought on by the fur trade, it disapproved of this social fluidity. It also worried that the small population of the colony could be overwhelmed by the burgeoning British colonies to the south. By 1663, only about 2500 people lived along the St. Lawrence and fewer still in Acadia. King Louis XIV demanded a new colonial arrangement and sent Jean Talon to Canada as a royal **intendant.** The intendant was one of the positions established to administer France's overseas colonies. Nominally, the intendant was second in rank to the colonial governor, but because he controlled the civil government and reported directly to the king, he often carried preeminent authority within the colony. Before taking office, Talon was invested with a specific set of tasks: increase the population, improve agriculture, and transform a rough society into an overseas extension of France itself.

To accomplish this transformation, a type of feudal organization was introduced. This organization—termed the **seigneurial system**—was first established under the previous authority of the Company of New France. Tracts of land, or seigneuries, were granted to persons (seigneurs) on the condition that they cede most of the land to settlers in the new colony. The settlers were considered **censitaires,** or tenants, and were expected to pay a yearly rent and to treat the seigneurs as landed nobility. In return, a seigneur was supposed to provide his tenants with a viable agricultural infrastructure—especially a gristmill to grind flour. The company had been unable to fill its seigneuries, but the new intendant aggressively pursued colonization on these lands. New seigneuries were established, and seigneurs were put under increased pressure to fill their lands with colonists. Settlers were forbidden to sell land unless it had been cleared and contained one or more buildings.

The colonization goals were only partly met. Within the next 100 years, only about 9000 people immigrated to Canada—about one-third as indentured servants, one-third as released soldiers, and the remainder as released prisoners and women shipped over to become wives. Exceptionally high fertility rates were the primary cause of the population rising to about 65,000 by 1760.

The geography of the seigneurial system was different from most other settlement patterns introduced in the New World. Seigneuries were set up to maximize the number of holdings with frontage on the St. Lawrence River. Each grant abutted a portion of the river and then extended deep into the hinterland. As Figure 14.7 indicates, the spatial structure within each seigneurie followed the same format. Each settler was given a roture: a strip of property that was narrow and deep, usually containing between three and 15 acres, about ten times longer than wide. The seigneur occupied a larger than average strip in the middle that he could exploit, although most chose to live not on their land but in the more hospitable confines of Québec or Montréal.

At first, settlement stayed close to the St. Lawrence River. Later, it would push as far east as the Gaspé peninsula and spread to a few tributary rivers, such as the Chaudière, the Richelieu, and the Ottawa. Once the riverfront lots were taken, the pattern of settlement could also be pushed inland as new strips were carved out along a row behind the original farmsteads and parallel to the river. But for the most part, the vast holdings of New France were deceiving—the extent of colonization stretched for 320 kilometers (200 mi) along two strips that were only 1.6 kilometers (1 mi) deep. A traveler cruising up the St. Lawrence between Québec and Montréal would have been able to see the entire colony (see Figure 14.8). Seine nets in the river gave way to water meadows; then came fields, the house and barn, pasture, orchards, and finally a large expanse of woodland.

Each string of farms was known as a **rang.** A house would appear every 100 or 200 meters, forming a chain of settlement broken by such physical features as a stream or inadequate soil but almost never by a seigneurial boundary. Each rang had its own school, chapel, and mutual aid services. In essence, these rangs formed one continuous village with the houses and farms punctuated by a few service centers.

The preference for rang settlement was found mostly among the settlers themselves. A river location provided for better than average soils, fishing opportunities, and transportation access. It also enabled many of the farmers to dabble in the fur trade, since the Canadian wilderness loomed just behind their farms. Talon and future intendants were less pleased by the arrangement and attempted to encourage compact villages that were thought to be more secure and more easily administered. But the rang form of settlement persisted for as long as riverfront land was available.

The French controlled the bulk of what is now Canada from the late 1600s until 1763, at which time they lost their claim on North America (they were more concerned with maintaining a toehold in the Caribbean and preserving some fishing rights in the North Atlantic). But despite its relatively short tenure, the French regime left a tenacious legacy on the portion of Canada most heavily settled (which was renamed Québec). The descendants of seigneurs and habitants forged a New World identity and came to call themselves **Canadiens.** They continued to retain the French language and Catholic religion of their ancestors, despite British domination. And they shaped a distinctive landscape, reflected in the pattern of fields and settlement along the St. Lawrence and other rivers in Québec today.

British North America

From the time of Cabot's first sighting, England controlled some territory in Canada. Lands around the Hudson Bay,

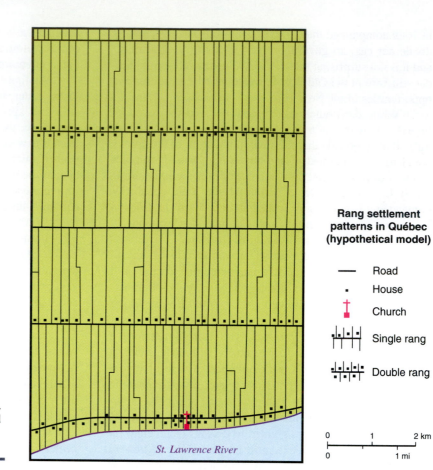

Rang settlement patterns in Québec (hypothetical model)

— Road

▪ House

✝ Church

Single rang

Double rang

0   1   2 km

0   1 mi

St. Lawrence River

Figure 14.7

New France was divided into seigneuries, each
further divided into a series of long lots or rotures.
The lots were laid out first along the shoreline and
then along roads roughly parallel
to the river. See also Figure 3.4.

Figure 14.8

The St. Lawrence riverfront near Québec City, as it appeared in 1787. A variety of activities from fishing to lumbering occurred within
each long and narrow lot. The shape of settlement made the relations between settlers and their nearest neighbors especially important.

T. Davies, *A View of the Château-Richer, Cape Torment, and Lower End of the Isle of Orleans near Québec.*

Newfoundland, and, after 1713, Nova Scotia were a part of the emerging British Empire, though of far less consequence than such colonies as Virginia and Massachusetts Bay farther south. The French territories joined these scattered domains in 1760, and from that point on, all Canada technically belonged to the United Kingdom. As a culture region, Canada was far more complex. Indigenous peoples still held sway over much of the interior, and the bulk of the European settlers were French. The Canadiens had reason to eye British dominion with suspicion. Only six years earlier, a British governor of Nova Scotia (which then included New Brunswick) banished all of the French-speaking Acadians (see "The Expulsion of the Acadians").

The British treated the Canadiens with a far lighter touch than the Acadians. Any aspirations they may have had to assimilate the French-speaking population were frustrated by the relative strength of the French presence. The Québec Act of 1774 made it official that French Catholics could keep their language and culture provided they kept clear of British economic designs and did not interfere with the political governance of Canada. In fact, the British buttressed the official Roman Catholic church as a means of safeguarding their dominance. Church catechism neatly dovetailed with British designs. It was a deeply conservative movement rooted in agriculture, deference to priestly authority, and a suspicion of all modernizing influences. And with one notable exception—the Rebellion of 1837–38—it left English Canadian interests largely unchallenged until the mid-20th century.

The British were not content to simply administer a French colony. They realized that while Canada lacked spectacular agricultural potential, it was still rich in land and marine resources. These resources were thought best secured through a stable, prosperous, loyal population. From its very inception as a British colony, Canada contrasted this loyalty with the alleged perfidy of the United States. Moreover, officials viewed Canada as culturally English—despite its large French population—and strove to maintain this identification even after Canadian independence in 1867. Even the first French Canadian prime minister, Wilfrid Laurier, stated that "the future of Canada is to be English. I do not share the dreams or illusions of the small number of my countrymen of French origin along the banks of the Saint Lawrence."

THE EXPULSION OF THE ACADIANS

This is the forest primeval; but where are the hearts that beneath it
Leaped like the roe, when he hears in the woodland the voice of the huntsman?
Where is the thatch-roofed village, the home of Acadian farmers,—
Men whose lives glided on like rivers that water the woodlands,
Darkened by shadows of earth, but reflecting an image of heaven?
Waste are those pleasant farms, and the farmers forever departed!
Scattered like dust and leaves, when the mighty blasts of October
Seize them, and whirl them aloft, and sprinkle them far o'er the ocean
Naught but tradition remains of the beautiful village of Grand-Pré.

That verse comes from Henry Wadsworth Longfellow's *Evangeline,* written in 1847, almost a century after the British governor expelled the French-speaking Acadians from their homes along the Minas Basin and the Annapolis Valley off the Bay of Fundy in Nova Scotia.

Acadia, as this region was known to the French, was first settled by Europeans in 1606, when the colony of Port Royal was established. Over the next 105 years, it became a political football, alternating between French and British control a total of seven times. For the most part, Acadia was largely peripheral to both French and English concerns. The inhabitants were free to enjoy the relatively mild climate and high fertility of the soils, achieving a level of prosperity unknown to most French peasants. The Acadians also benefited from plentiful land; the population was never more than 10,000.

The good times lasted until about 1750, at which time the British began to press the French inhabitants to declare their loyalty to the United Kingdom. Most Acadians were reluctant to shed their neutrality in the larger European conflict between France and England, and when the British served an ultimatum demanding compliance, most Acadians believed that the threats of expulsion would not be carried out. But they gambled and lost. In August 1755, nearly all were forced to flee to other English colonies or land still controlled by France. A sizable segment migrated down to the Isle of Orleans in the mouth of the Mississippi River. Their descendants form the Cajun population that lives in Louisiana today.

The Canadian Difference

The French prospered demographically in British Canada. They maintained their status as the single largest nationality and kept pace with the overall population growth of English Canada. Immigration from France played no role in this trend. Few European French chose to move to Canada, and any migration activity that existed was out of Québec into New England. As was the case under the French regime, Canadien expansion was a product of exceptionally high birth rates. Some demographers estimate that the average French-Canadian woman had nearly eight children, far more than the average English Canadian.

Canada remained a British colony until 1867. With the strong encouragement of the United Kingdom, which had earlier granted Canada substantial autonomy, Ontario, Québec, New Brunswick, and Nova Scotia combined to form the Dominion of Canada. They were soon joined by Prince Edward Island and the western territories (Newfoundland did not join until 1949).

The birth of Canada was very different from that of the United States in another significant way. It was marked by an absence of intensity and an emphasis on compromise. While Canada won self-government, it still remained very much a part of the British orbit. For 50 years following **Confederation,** Canada's economy was devoted to providing raw materials to Britain in exchange for finished products. Canada was also expected to provide military troops whenever Britain got into war, even against an enemy like the South African Boers, as well as to contribute to the British navy. Canadians had no independent diplomatic corps until the late 1920s; did not secure a separate status as citizens, distinct from British subjects, until 1947; possessed no national flag until 1965; and did not gain full control of their constitution until 1982.

British dependence provoked a great deal of ambivalence. By the turn of the 20th century, many Canadians sensed that their destiny lay apart from Britain and its affairs. The growth of western Canada in particular made many feel that hidebound ties to the home country impeded national progress. Still other Canadians reveled in their cloak of "Englishness" so flaunted in cities like Victoria, British Columbia. For them, Canada was distinctly counter-revolutionary; this section of North America would proudly fly a Union Jack.

A Sometimes Friendly Neighbor

Today, neither the United Kingdom nor France exerts as much influence on Canada as does the United States. The two countries are neighbors, sharing an 8900-kilometer (5525-mi) border (including the boundary between Canada and Alaska); they are certainly allies, partners in the North American Treaty Organization and both tucked safely under the North American Air Defense (NORAD) shield. Yet there is much more to the relationship, something almost familial. The United States attends to its wider affairs, seemingly secure in the knowledge that Canada will tag along. Canada, on the other hand, worries incessantly about being overshadowed by its rambunctious big brother and has sought a definition distinct from the looming U.S. presence.

Today, both countries pride themselves on friendly relations, but this has not always been the case. Before independence, Canada felt itself under constant pressure from the United States. In the early 1800s, expansion-minded "warhawks" in the U.S. Congress argued that, with a thirteen-to-one population advantage, an American takeover would be quick and painless. During the War of 1812, American militias ravaged the towns and villages of Ontario, and later the developing notion of *manifest destiny* worried Canadians, especially as the United States flexed its military muscle against Mexico. After the American Civil War, Canada found itself alongside the largest and best equipped army in the world. Confederation emerged as a response, since Britain figured that it would be difficult for the United States to justify invading a newly independent country.

The military menace did indeed dissipate after Confederation, only to be replaced by an economic and cultural challenge. At this time, Canada was a weak and disjointed society with little to connect a collection of old colonies and frontier settlements. As a first step in forging a coherent nation, a railroad was proposed, paid for with public funds, that would join Canada's east, center, and west. This was considered an important step in building Canada's underdeveloped economy, strengthening east-west links within Canada, and stanching the flow of migrants to U.S. mill towns. And so in 1885, the last iron spike of the Canadian Pacific Railway was driven, and together with the already completed Intercolonial Railway (connecting the Maritimes with Québec), Canada finally possessed a transcontinental railroad, sixteen and a half years after the United States. Not long before, a **National Policy** had been introduced to protect infant Canadian industries with high tariffs on imported goods and to patch together a single, independent national economy. This protectionist policy continued into the 20th century but was not enough to offset the powerful tugs of the U.S. economy. In recent decades, American interests have controlled up to a third of some Canadian industries, and the Canada-U.S. Free Trade Agreement, introduced in 1989 to set up a tariff-free zone, as well as NAFTA, augur for an even closer relationship between the two economies.

The United States increased its pull on the Canadian economy through the late 19th and early 20th centuries. Canada may have achieved unity as a state, but economically, it seemed more like a series of appendages, each inserted as separate points in the larger American economy. Canadian exchanges were more likely to flow north-south across the international border than east-west across provincial lines: New Brunswick and Nova Scotia to New

England, Québec to New York, Ontario to Michigan and Ohio, Manitoba and Saskatchewan to the American Plains. As the American population grew and the American economy expanded, industrialists searched for sources of raw materials—wood pulp for metropolitan newspapers, lumber to build midwestern cities, metals (nickel, iron, and zinc) for the steel industry. And the influence of American capital expanded as well. By 1922, the United States overtook the United Kingdom as Canada's chief foreign investor and, at about the same time, became its largest trading partner. As of 1998, Canada sold 84% of its exports to the United States. What is more, exports to the United States constituted about one-fifth of Canada's gross national product, making them far more important to the Canadian economy than is the case with most industrialized countries (Figure 14.9).

What worried and still worries many Canadians is not the purely economic aspects of the relationship but what this portends for Canadian culture. *Culture,* of course, is a loosely defined term encompassing a whole panoply of attributes that help define a people. As it is, English-speaking Canadians already share many of these attributes with the United States: both societies are composed of immigrants who have assimilated into a largely Anglo-American culture. Similarities abound in language, dialect, religion, and ancestry. Cultural differences must be found in the smaller details. More than any single criterion, Canadians fear they will be vulnerable to "Americanization," a process defined by one 1908 writer as "the gradual but steady development of a non-British view of things; a situation in which public opinion here regarding the heart of the Empire and Imperial policy is formed along the lines of United States opinion, and, therefore of an alien viewpoint."

Culturally, the influence of the United States has been nothing short of phenomenal. A trip to Ontario can seem no different from a trip to Michigan. The landscape shares many of the same features, the same commercial chains line the roads and highways, and billboards advertise the same products (see Figure 14.10). So many Canadians live within

Figure 14.9

About 15 million Canadians and more than $250 billion in exports cross into the United States each year, an activity that occurs by air, sea, and across myriad border crossings. That shown here, at Niagara Falls, is one of the largest.

The Canadian Difference

150 kilometers (95 mi) of the border that America's electronic media are readily accessible. Even if they were not, Canadians receive American programming through Canadian television—only 4% of TV dramas are Canadian, although there is plenty of domestic sports, news, and public affairs programing. And Canadian radio plays mainly American popular songs, although some of these pop stars are Canadian. Print media are also dominated by the United States; Canadian newsstands prominently display such American magazines as *Time, Newsweek, The Atlantic Monthly,* and *Cosmopolitan.* Canadian magazines, with a relatively tiny domestic market (and virtually no interest across the border), have had trouble competing with the enormous circulation of American publishers.

This has engendered a certain defensiveness among Canadians, who understandably feel that they know far more about the United States than Americans know about them. The demographic, economic, and cultural power of the United States ensures that this imbalance will continue, and many Canadians resist the erosion of their own set of values. The 1988 Canadian election pivoted on the free trade agreement, and many opponents feared that free trade would mean "massive dumping of American products and protection of the American cultural domination of [the Canadian] cultural scene." Implementation of the pact and the inclusion of Mexico did little to assuage this anxiety. Canadians continue to cast a wary eye on the cultural and economic influences of their sometimes friendly neighbor.

Canadian Cultural Diversity

While Canada is a land of extraordinary cultural variety, this diversity is tempered by the desire of many immigrants, First Nations peoples, and even some French Canadians to assimilate into the dominant English-language culture. One might argue that English Canada has developed a "civic nationalism" based not on ancestry or heritage or even religion, but on common institutions and a sense of political unity. This still leaves a number of Canadians who are unassimilated into the English-language culture in one way or another. Most significant are the French-speaking Canadians who officially have equal cultural status with English Canadians and predominate in Québec—Canada's largest province by area and second largest by population. But Canada also includes a polyglot mixture of people who possess no attachment to the English and French categories. Canada is a multicultural state, and various peoples have found it possible to preserve important attributes of their identity. First Nations peoples are found throughout the country but are clustered in the north. Many descendants of

Figure 14.10
The influence of the United States on Canadian social, cultural, and economic life is ubiquitous. American products, such as magazines and movies, are sometimes easier to obtain than their Canadian counterparts.

overseas immigrants—in cosmopolitan cities like Toronto and isolated Saskatchewan farmsteads—speak their original language and carry on many of the old traditions. In recent years, they have been joined by newcomers from places as different as Hong Kong and Haiti, each adding a distinct flavor to Canada's cultural medley.

The Linguistic Divide

People can be categorized in many different ways, and Canadians are no exception. Canadians are grouped by race, region, occupation, religion, and nationality, among other things. But historically, Canadian society has pivoted on the division between those who speak English, termed **Anglophones,** and those who speak French, termed **Francophones.** (Canadians who use a language other than English or French are called **Allophones,** or Other speakers.) These broad linguistic categories subsume a great deal of variety—each includes many people who have little in common with each other—but they also reflect the dominant political realities that have defined Canada since the 1763 Treaty of Paris.

On the macro scale, French and English Canadians exist together. As Canada's two "founding peoples," they share responsibility for settling, developing, and forging the Canadian state. Canadian Confederation was patched together by the concerted efforts of both groups. Since then, Anglophones and Francophones have served in the parliament and the federal bureaucracy. In fact, three of Canada's most noted and longest serving prime ministers, Wilfrid Laurier, Louis St. Laurent, and Pierre Trudeau, have been French-Canadian.

On nearly every other scale, French and English Canadians exist apart. The differences are deeper than language alone. By and large, they subscribe to different religions, they come from separate ethnic stock, and they claim a separate set of values. For a long time, there were even distinct differences in such items as occupation, birth rates, and death rates between the two peoples. These auxiliary indicators, largely related to economic welfare, have equalized as French-Canadian society has become more industrial and modern.

The populations of French- and English-speaking Canadians have changed a great deal over time. As was mentioned before, few French speakers came over to Canada after 1763, and so immigration was primarily Anglophone in nature. By 1834, the size of the British population surpassed the French population, and from that point on, French speakers were a numerical minority in Canada. Today, they are roughly one-quarter of the population.

Figure 14.11 displays the growth in population of Canada's British (including English, Irish, and Scottish), French, and other ethnic stock. Until recently, data on language were unavailable, so they must be inferred from data on ethnic origins, but these still give a good approximation since language shifts were infrequent. By 1851, there were almost twice as many British as French ethnics in Canada. Over the next 130 years, the absolute size of each group expanded, but the proportion of British ethnics has slowly

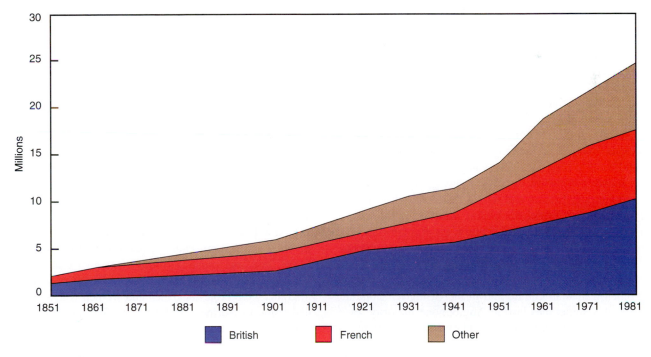

Figure 14.11

Evolving ethnic composition of Canada, 1851–1981. Note the increasing significance of so-called "ethnic" Canadians with roots outside of France or the British Isles.

declined, from 66.7% in 1861 to 43.6% in 1981. Higher birth rates allowed the French-Canadian population to keep pace, and their proportion remained steady at about 30% throughout much of this period, although it has fallen in recent decades to about 27% as fertility levels have declined. The largest increase has been registered by non-French, non-British ethnics, who surpassed the French population in 1981. We will discuss this population in more detail ahead. Their significance in the linguistic context comes from their propensity to learn English and shun French, even where French speakers are in the majority. As a result, immigrants, their children, and their grandchildren have helped bolster the Anglophone population in Canada and have allowed it to continue its numerical dominance. So while Canada has become ethnically less British in the last 130 years, it has not become less English-speaking.

Geography of Language

Language in Canada is of more than sociological or demographic interest. It also has a clear geographic dimension. The descendants of the original French colonists remained close to their original home on the St. Lawrence. As land ran out, expansion proceeded by the slow and steady settlement of adjoining areas, much of which were of poorer quality. French Canadians first penetrated the wilderness in strips parallel to the great river. When this was exhausted, they exploited the shoreline along the smaller tributaries and the arable reaches of the Gaspé peninsula, finally moving east into the forests and shores of northern and eastern New Brunswick, west along both sides of the Ottawa River, and south into northern New England.

Beyond this hearth were scattered small pockets of French-Canadian settlement. Descendants of the original Acadians, some of whom had resisted expulsion in 1755 and some of whom straggled back after political restrictions eased, persisted in Nova Scotia, New Brunswick, and Prince Edward Island. In the 19th century, a number of French Canadians settled what would become the provinces of Manitoba and Saskatchewan. The first of these were Canadiens involved in the fur trade, an occupation that does not lend itself to high densities. Many intermarried with the native Indian population, creating a new group of **Métis** (a people of mixed French and native descent), who carried many elements of French Catholic culture on to the Canadian Prairies. Later in the 20th century, the Roman Catholic church, seeking to expand its mission, established agricultural colonies and urban parishes in the west.

The Catholic church played a vital role in perpetuating French-Canadian culture and in increasing its scope in the face of a sometimes hostile Anglophone majority. Between 1840 and 1950, French Canada was one of the most religious societies on earth. Within and without Québec, the chief geographical unit was the parish, led by a chief curé who looked after all of the civil and religious needs of the parishioners. The supremacy of the clerics was acknowledged by all the parishioners; for most of them, the church was the only link with the outside world. Each developed parish contained a cluster of religious and educational buildings. Economic institutions were also established, such as banks and agricultural cooperatives (Figure 14.12). Within Québec, the parishes served a homogenizing function, keeping all of the self-contained communities within the same cultural framework. Outside Québec, the parishes served to maintain islands of French Catholicism, especially since all of the western provinces turned unabashedly anti-French around the turn of the century, abolishing public French-language education and conducting all official business in English. But the Church provided priests and the requisite institutions, making these "little Québecs" viable for some time and allowing the small populations to survive.

Early settlement goals sought to strengthen Canada's British heritage, and most early immigrants came either from the United States or the British Isles. Many of the American immigrants were refugees of the American Revolution, *Loyalists* who had sided with the Tory faction and moved north during or after the war. They settled in the Atlantic colonies of Nova Scotia and New Brunswick, in the Eastern townships (an area between the St. Lawrence Valley and northern New England where British authorities surveyed rectangular townships and lots), and in what became southern Ontario. The availability and relative cheapness of land just across the international border soon enticed some Americans not motivated by political considerations. At this time, the U.S.–Ontario boundary was at the frontier of western expansion and was quite porous, so these "Late Loyalists" moved freely between 1790 and 1812, when renewed war between the United States and the United Kingdom sealed the border to any future American settlement. Canadian immigrants from overseas came primarily from England, Wales, Scotland, and Ireland. Most immigrants were neither wealthy nor impecunious (after all, they needed some funds to make the journey), but they moved to Canada because they were worried about their future at home. For this reason, Canada attracted many more immigrants from the less developed Celtic fringe than it did from southern England or the industrializing Midlands.

These immigration patterns defined Canada by 1871, as seen in Table 14.1. Irish constituted a plurality in Ontario and New Brunswick; Scots were the largest group in Nova Scotia. "Irish" is misleading since it includes two very different (and antagonistic) groups of people: Catholics from Southern Ireland and Protestants from Ulster. Most Irish immigrants were Protestant, men and women whose ancestors had come to Northern Ireland from Scotland. Along with Scots, they imparted to English Canada a strong Scottish flavor. Together, these migrants pushed the frontier of Canadian settlement, although because of the physiographical barriers, colonization leapfrogged over harsh territory rather than unfolding continuously.

Figure 14.12

Although the majority of Québécois live in cities like Montréal and Québec, their hearts belong in the countryside, in places like St. Fidele Charlevoix. The spires of the church dominate these villages just as the Roman Catholic church once dominated provincial culture and society.

Table 14.1
Ethnicity and Religion of Canadians, 1871

	Canada	Ontario	Québec	New Brunswick	Nova Scotia
Total Population	3,485,761	1,620,851	1,191,576	285,594	387,800
Ethnicity		Percent	Percent	Percent	Percent
French	1,083,000	5%	78%	16%	8%
Irish	846,000	35	10	0.5	16
English	706,000	27	6	29	29
Scots	550,000	20	4	14	34
German	203,000	10	0.7	2	8
Dutch	30,000	1	—	2	1
Indian	23,000	0.8	0.6	0.5	0.4
Black	22,000	0.8	—	0.6	2
Religion	Canada	Ontario Percent	Québec Percent	New Brunswick Percent	Nova Scotia Percent
Roman Catholic	1,492,000	17%	86%	34%	26%
Methodist	562,000	29	3	9	10
Presbyterian	545,000	22	4	14	27
Episcopalian	494,000	20	5	16	14
Baptist	239,000	5	0.7	25	19

From J. L. Granatstein et al., Nation: Canada Since Confederation, *3d edition. Copyright © 1990 McGraw-Hill Ryerson, Ltd., Toronto. Taken from 1871, 1881, and 1891 Census of Canada. Reprinted by permission.*

The Canadian Difference

Most new settlers chose to locate first in southern Ontario, others populated arable regions of the Maritimes, and some found land in the Québec Eastern Townships, in the Ottawa River valley, and even along the Gaspé. When these agricultural lands began to fill up, about the time of Confederation, settlement proceeded farther west. These settlers were whole families looking for fertile land on which to establish farms and communities, and they soon overwhelmed the French-speaking Métis, dooming any hopes of extending French Canada into the western realms. With the opening of the transcontinental railroad and an easier path across the Rocky Mountains, British Columbia also attracted a number of settlers seeking livelihoods in gold, timber, salmon fishing, and farming. Very few of these settlers were French. Immigrants from other places, predominantly Europe, began to increase in the late 19th century. Many of them also moved west into the Prairies, contributing to the Anglophone predominance in western Canada.

Settlement was not an exclusively rural phenomenon, especially as Canada began to industrialize in the mid-19th century. New cities were established, and already existing cities grew bigger, enticing native and immigrant alike. Up until very recently, Montréal was the most important of these. Before the British conquest, Montréal was a significant **entrepôt** (or distribution and trading center) for shipping, warehousing, and all the functions associated with the import and export of goods. The British fostered Montréal's growth and placed it in the center of the Canadian economy, establishing commercial partnerships (some exiled from the American colonies), factories, banks, and insurance companies. The Lachine Canal (which allowed oceangoing ships to travel straight through to the Great Lakes) and the frenzy of railroad construction secured Montréal's position as a transportation hub. English and Scots quickly took the reins of this burgeoning economy, many building enormous mansions on the slopes of Mont Royal and Westmount. Irish immigrants swelled the ranks of the laboring classes. Together, these gave Montréal an Anglophone majority by the mid-19th century and a legacy of political dominance that persisted well into the 20th century.

These historical processes have created the linguistic distribution we see today. On the most general scale, French speakers predominate in Québec, and English speakers form a majority in the rest of Canada. Higher resolutions show this pattern to be more complicated. As Figure 14.13 illustrates, substantial linguistic minorities exist throughout Canada, and they too have played a role in regional growth and development. These include Anglophones in Québec and Francophones outside of Québec, as well as peoples who speak in neither tongue.

English speakers in Québec are found in three major regions. Despite a recent exodus, the Montréal metropolitan area includes nearly half a million Anglophones, and it attracts a substantial number of immigrants as well, many of whom are Allophones. The original settlement of the East-ern Townships imbues this region with a strong English character. The landscape shows square farmsteads rather than long strips, many of the place names are English, and a number of English-language institutions, including a university and a newspaper, still exist. The Anglophone presence diminished as French expansion claimed much of the land, but the region still contains a little more than 10% English-speaking population. The Ottawa River valley includes a mixture of rural Anglophone settlements and more modern communities made up of English-speaking professionals who commute to Ottawa but live in Québec because of the availability of affordable housing there.

Most of the Francophone population outside of Québec lives in the bilingual regions of Ontario and New Brunswick, reflecting French expansion from the St. Lawrence hearth. The area around Ottawa contains more than 20% Francophones, some rural villagers (Prescott and Russell counties contain a majority French population), and some federal civil servants. Because it is the federal capital of an officially bilingual nation, Ottawa is the most aggressively bilingual part of Canada, a fact evident in signs, institutions, and the ability of many residents to converse easily in French and English. Up the river in northeast Ontario lives a larger proportion of French speakers, although they are more geographically dispersed and enjoy less institutional security. And large French populations reside also in some of Ontario's biggest cities, but the greater distance from Québec makes them less culturally secure.

French-language enclaves are found in each of the Maritime Provinces. Francophone settlements in Nova Scotia (along Cape Breton Island and the southwest coast near Yarmouth) and Prince Edward Island (in the westernmost section of the island near the town of Summerside) are visible but are not large. But along the coast and in the interior of northern and eastern New Brunswick, French speakers predominate. In fact, this is the only area outside Québec to boast a majority Francophone population. Descendants of the original Acadian population are joined here by descendants of colonists who arrived from Québec. A rural lifestyle, relative remoteness, and a high birth rate enabled this Francophone population to sustain itself for many decades. It has also given the French residents of New Brunswick a substantial measure of political power that has resulted in the only truly bilingual province in Canada.

Each of the western provinces contains French speakers, but—with a few exceptions—their prospects for long-term cultural survival appear bleak. There have been some positive developments. Provincial governments have grown more tolerant, and the federal government now struggles to preserve French-speaking minorities. In addition, Alberta and British Columbia have enjoyed years of strong economic growth that have enticed French- as well as English-speaking migrants. The tide of modernization has washed over these minuscule French enclaves, however, resulting in the erosion of village institutions, the migration of rural residents to the cities, and continued assimilation into the English language.

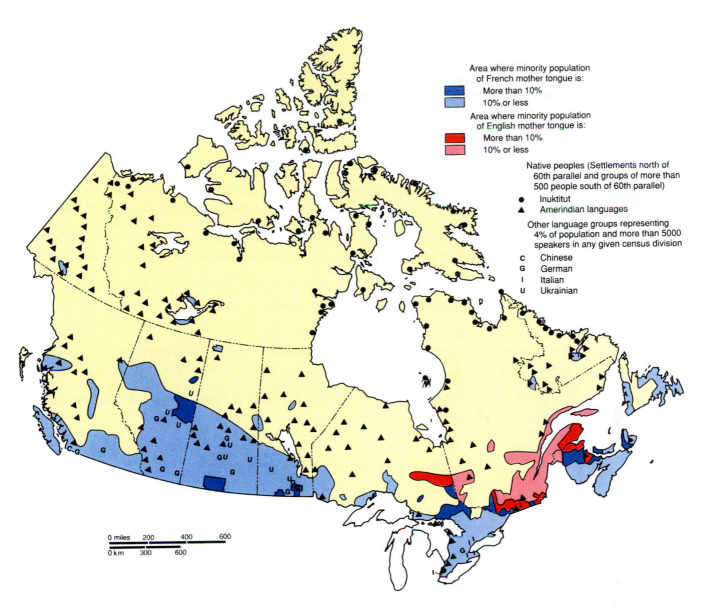

Figure 14.13
English and French speakers divide along provincial lines, with most Francophones clustered in Québec. But the picture is complicated by the presence of what are officially called *linguistic minorities,* French speakers outside of Québec, English speakers within Québec, and speakers of more than 100 other languages sprinkled throughout the cities and countryside.

Politics of Language

The geographical concentration of Francophones in Québec combined with the historical competition between French Canada and English Canada has had a combustible impact on Canadian politics. Language has always been the Canadian dilemma. It led to Canada's only civil rebellion in 1837–38, when French-Canadian **patriotes**—mainly first-generation urban professionals resentful of English dominance—attempted but failed to seize control of Québec. French Québec was then an overwhelmingly rural and Catholic society, and the patriotes lacked crucial support from the French-Canadian peasants and the Roman Catholic church, both groups having been co-opted

by the British. After that time, linguistic accommodation was cultivated by the mutual cooperation of British and French leaders and facilitated by the introspection of the French-Canadian people. But linguistic harmony has never been achieved.

Until 1960, a widespread movement for economic and social rights within the wider society remained elusive. Much of French Canada was firmly planted in a rural, stagnant economy; in the banks, offices, factories, stores, and even on city streets, English reigned supreme. Any large workshop in a Québec town would have an English-speaking manager and a number of French laborers. In any smart shop in downtown Montréal, all of the signs would

The Canadian Difference

be in English and the clerks would address customers in that language. French speakers simply had to adapt, even though they were the majority population, and woe to those who were not conversant in Canada's majority tongue. English was the language of commerce and industry, and native English speakers had an enormous advantage over their French-speaking compatriots. In 1961, the average Francophone Montréaler earned only 67 cents for every Anglophone dollar.

The situation began to change slowly over the early 20th century and then swiftly through the 1960s and '70s. An expanding economy lured many from the farms to the cities in search of industrial jobs. As they came into contact with English businesspeople and managers, urban Francophones became more aware of their economic inferiority. Urban life also increased the number of choices available, undercutting the church's conception of the proper French-Canadian life. In time, a new French-Canadian middle class emerged to serve a more powerful and active Québec provincial government that claimed to represent the economic interests of its French-speaking constituency. The **Quiet Revolution,** in which an entire society turned its back on the old ways and embraced a more modern outlook, had begun.

Changes in French Canada were so rapid that it seemed that everywhere one looked, the traditional props were being toppled. French Québec changed from one of the most religious societies to one of the most secular; its birth rate plummeted from one of the highest in the industrialized world to one of the lowest; it metamorphosed from what had been described as a feudal society in 1944 to an aggressively cosmopolitan one by 1970.

For many, French Canada's aspirations seemed increasingly out of step with its membership in an Anglophone-dominated Canada. The notion that Québec should separate—dormant for nearly 130 years—began to pick up support. The new French-Canadian nationalism sought more provincial authority, with or without the consent of the federal government, and some argued that only sovereignty would suffice. In 1976, this view gained enormous credibility with the provincial election of the separatist **Parti québécois** and its firebrand leader René Levesque. Although the Parti was unsuccessful in fulfilling its primary mandate—Québec voted to stay in the Confederation in 1980—renewed calls for independence in the wake of the defeated **Meech Lake Accord** in 1990 and the referendum on the constitutional amendments in 1992 testify to the continuing influence separatism has on Canadian society. The Parti québécois has forever changed the politics of language in Canada.

Canadian Immigrant Influences

French Canadians are not Canada's only minority. Indeed, a tourist may be struck by just how fresh is the influence of peoples from Europe, the Near East, South Asia, and East Asia. Such cities as Toronto, Montréal, Vancouver, and Winnipeg sport their "Little Italys," "Chinatowns," and enclaves of almost every nationality—Greek, Haitian, Pakistani, to name just a few. Rural landscapes in Canada are painted in ethnic hues that are still quite vibrant: the Prairie Provinces in particular offer up a pastiche of Eastern, Northern, and Central European communities.

This is surprising since most Canadians do not care for immigration; historically, they have preferred that their country retain its character—defined as British or French, depending on the source. A survey in the 1990s indicated that most Canadians felt there was too much immigration, and other studies have shown less tolerance for immigration in Canada than is found in the United States. Immigration has flowed from necessity, not desire, as Canada has sought to increase its numbers. And lately this increase has extended far beyond the types of people who founded Canada.

Until Confederation, Canada was clearly divided into French and British stock, which together comprised 90% of the population. In the 19th century, few immigrants came from outside the British Isles, and in each decade from 1851 to 1901, more people left Canada than arrived. There were two main reasons for this. Most immigrants (and many Canadians) viewed the United States—with its warmer climate, more bustling cities, and seemingly more open society—as the more attractive destination. And English Canadians were determined to preserve their British heritage. **Anglo-conformity** was the rule, as immigrants were expected to readily assimilate into the mainstream culture and were recruited in terms of their supposed ability to fit in. This principle held up through the mid-20th century, well summarized by Prime Minister Mackenzie King's statement in 1947 that "the people of Canada do not wish to make a fundamental alteration in the character of their population through mass migration."

But a fundamental change was in the offing. As indicated in Figure 14.14, immigration levels increased over the course of the 20th century, and the British share of immigration decreased, from more than 50% in the late 19th century to less than 10% by the end of the 20th. The immigrants arriving in the 1990s were primarily from outside Europe.

Table 14.2 lists some of the major immigrant groups that have arrived in Canada. The first wave of new Canadians came from Scandinavia and the Slavic lands of Central and Eastern Europe. Agricultural workers were favored to people Canada's vast interior and to shield the countryside from various "threats," especially a "yellow peril" of Chinese and Japanese. Eleven percent of British Columbia's population was Asian in 1900—laborers who had helped build railroads, mine coal, and fell trees for disgracefully low wages—but the government considered them "unfit for full citizenship," "obnoxious to a free community," and "a continual menace to health." A prohibitive head tax was

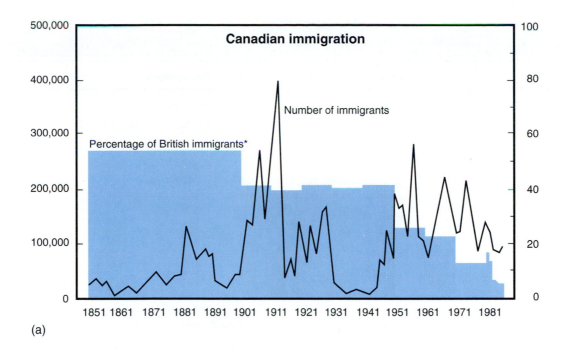

Canadian immigration

Percentage of British immigrants*

Number of immigrants

(a)

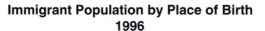

**Immigrant Population by Place of Birth
1996**

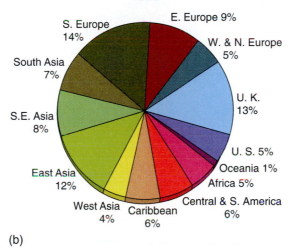

E. Europe 9%

W. & N. Europe 5%

U. K. 13%

U. S. 5%

Oceania 1%

Africa 5%

Central & S. America 6%

Caribbean 6%

West Asia 4%

East Asia 12%

S.E. Asia 8%

South Asia 7%

S. Europe 14%

(b)

Figure 14.14

Canadian immigration has fluctuated with economic conditions (the depressions of the 1890s and 1930s were immigration troughs) and with political restrictions imposed by Canada and the United States (which was often the immigrants' first choice) (a). The proportion of British immigrants continues to decline, with most newcomers hailing from Asia or Latin America (b).

(a) *Source:* Graph derived from data in Edward N. Herberg, *Ethnic Groups in Canada: Adaptations and Transitions,* table 3.3. Nelson Canada, Scarborough, Ontario, 1989.
(b) *Source:* Pie chart derived from data in Census of Canada.

placed on Chinese immigration in 1904, and other laws restricting Asian immigration were put into place shortly thereafter, not to be eased until the late 1960s.

To encourage European immigration, the government established a Dominion Lands Act, granting 65 hectares (160 acres) to each settler in exchange for $10 and the willingness to work the rich prairie soil. Canadian officials were most eager to attract English yeoman farmers, considering them to be best suited to Canadian culture. But British farmers were singularly uninterested in the task of

Table 14.2
Some Notable Immigrant Groups

Group	Dates	Major Occupations
Native peoples	pre-1600	Self-contained societies
French	1609–1755	Fishing, farming, fur trading, supporting occupations
Loyalists from the United States	1776–mid-1780s	Farming
English and Scots	mid-1600s on	Farming, skilled crafts
Germans and Scandinavians	1830–50, c. 1900, 1950s	Farming, mining, city jobs
Irish	1840s	Farming, logging, construction
African Americans	1850s–70s	Farming
Mennonites and Hutterites	1870s–80s	Farming
Chinese	1855, 1880s	Gold panning, railway work, mining
Jews	1890–1914	Factory work, skilled trades, small business
Japanese	1890–1914	Logging, service jobs, mining, fishing
East Indians	1890–1914 1970s	Logging, service jobs, mining Skilled trades, professions
Ukrainians	1890–1914 1940s, 1950s	Farming Variety of jobs
Italians	1890–1914 1950s, 1960s	Railway work, construction, small business Construction, skilled trades
Poles	1945–50	Skilled trades, factory jobs, mining
Portuguese	1950s–70s	Factory work, construction, service jobs, farming
Greeks	1955–75	Factory work, small business, skilled trades
Hungarians	1956–57	Professions
West Indians	1950s 1967–	Domestic work, nursing Factory work, skilled trades, professional, service jobs
Latin Americans	1970s–	Professions, factory and service jobs
Vietnamese	late 1970s–early 1980s	Variety of occupations

From E. Herberg, Ethnic Groups in Canada: Adaptations and Transitions. *Copyright © 1989 Nelson Canada, Scarborough, Ontario. Reprinted by permission.*

cultivating the hard, cold lands, so most of the new arrivals were Slavs from Eastern Europe, where the climate and the topography were much like that of continental Canada. As a group, Slavs fell behind British, Americans, and other Northwest Europeans in the preference pecking order, but they were willing to resettle, and they became some of the most enthusiastic and loyal citizens of their adopted country.

Ukrainians made up a large proportion of a stream that included Scandinavians, Mennonites, Germans, and Hutterites (Figure 14.15); more urban-oriented migrants at this time included Poles, Jews, and Italians. Settlers assembled entire villages that in many ways mirrored the places they had left—from the *buda,* or log hut, that served as the first shelter, to more established structures with plastered walls, thatched roofs, and dirt floors regularly tamped down with cow dung and water (a curiously odor-free method of keeping floors clean and shiny). The legacy of Ukrainian settlement patterns is revealed in the present-day landscape.

Onion-dome churches, representing both the Eastern Catholic and Orthodox Christian faiths, dot the treeless landscape, and Ukrainian is still a common language in western Canada (Figure 14.16).

Between 1945 and 1957, the complexion of immigration changed somewhat as 1.7 million Europeans left their homes to arrive in what they hoped would be a land of second chances. Canada was also largely a land of second choice, since many immigrants who would ordinarily have entered the United States were blocked by that country's restrictive policies. Political persecution was a major factor driving people from Eastern Europe and the Soviet Union. Economic opportunities beckoned residents of the war-ravaged countries of Italy, Greece, Germany, the Netherlands, and Great Britain. These new migrants had primarily urban destinations, and they tended to favor the more industrial provinces of Ontario and Québec. Italian newcomers in particular left their mark on large cities; today, they

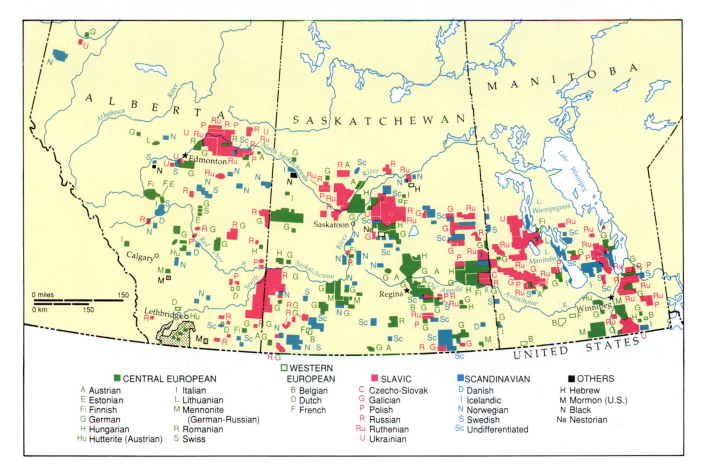

CENTRAL EUROPEAN

A Austrian	I Italian
E Estonian	L Lithuanian
Fi Finnish	M Mennonite
G German	(German-Russian)
H Hungarian	R Romanian
Hu Hutterite (Austrian)	S Swiss

WESTERN EUROPEAN

B Belgian
D Dutch
F French

SLAVIC

C Czecho-Slovak
G Galician
P Polish
R Russian
Ru Ruthenian
U Ukrainian

SCANDINAVIAN

D Danish
I Icelandic
N Norwegian
S Swedish
Sc Undifferentiated

OTHERS

H Hebrew
M Mormon (U.S.)
N Black
Ne Nestorian

Figure 14.15

The Canadian prairies are seasoned with the flavors of various European nationalities, many of whom still live in self-contained communities and continue to speak the ancestral language.

are the second largest ethnic group in Toronto (after the British) and possess nearly all of the requisites of a self-contained community: grocery stores, newspapers, television shows, churches, social clubs, and health and welfare services. In Montréal, 150,000 Italians have emerged as a true "third force" behind French and English residents.

In the last three decades, Canada has experienced a fundamental shift in immigration, away from a narrowly European bias toward global breadth. After 1967, Immigration Canada changed the preference quotas. Asians, until recently excluded from entering Canada, now make up the largest share of arrivals. Between 1991 and 1996, more than three out of ten newcomers arrived from such places as Taiwan, Hong Kong, India, Sri Lanka, the Philippines, and Vietnam. Jamaicans and other Caribbeans, Chileans and Guyanans, and an increasing number of Africans are adding more garnishes to the Canadian salad bowl.

The results of this influx are evident nearly everywhere. Toronto now sends out property tax notices in six languages, Sikh Canadian Mounties are allowed to wear a turban instead of the regulation trooper hat, and many of the skyscrapers in Victoria and Vancouver have been built with dollars from Hong Kong emigrés (Figure 14.17).

Especially noteworthy has been the increased racial diversity within Canada. Of course, Canadian society has been multiracial since French fur trappers collaborated with the Algonkin. Later, a number of black refugees from the American slave states would seek sanctuary in Nova Scotia and Ontario, and Chinese and Japanese migrants arrived on the Pacific Coast. But together, these groups were never more than 3% of the total population, and their small size allowed Canada to avoid facing issues of race. In recent years, the number of **visible minorities** (nonwhites) has risen markedly, and by 1996 they comprised 11% of the overall population. Most racial minorities now suffer disadvantages in employment and income, and their increase will force Canada to come to grips with issues of racial prejudice and discrimination.

Partly as a consequence of greater ethnic and racial complexity, Canada has begun to revise how it regards itself. Anglo-conformity, the prevailing ideology from Confederation until World War II, is giving way to a policy of

The Canadian Difference

Figure 14.16

Onion-dome churches are a common sight in the Canadian Prairies, and may represent either Orthodox Christian or Eastern Catholic faiths. Shown here is the Ukrainian Orthodox Church of St. Elias in Spirit River, Alberta.

multiculturalism. The recognition of Canada's multicultural character—not really articulated until 1967—lagged well behind the fact that more and more non-French, non-British, and non-Europeans were entering Canada. This differs from the American model of assimilationism, which focuses on the individual and individual rights. Canadian policy recognizes the uniqueness of each ethnic group and seeks to protect its culture. And while official multiculturalism includes the stated goal of eventually integrating all Canadians into either the French or English spheres, it also accords them what would seem to an American to be an extraordinary menu of rights and privileges, including funding for various ethnic organizations and allowing some students to attend part-time schools that teach in their native language.

Immigration has been geographically uneven, as is indicated by Table 14.3. The Maritime Provinces contain few residents born outside Canada, while about a quarter of the populations of Ontario and British Columbia are of foreign origin. In fact, more than half of all immigrants end up in Ontario, which has emerged as the crucible of Canada's new plural identity. The tendency of immigrants to go to certain provinces has a great deal to do with where they view the greatest opportunity, and so they favor large cities and regions with dynamic economies.

Public attitudes vary over the increasingly diverse nature of Canadian society. Some ethnic groups think that the government does not go far enough and that it should affirm their rights as a constitutionally protected population. Many Anglophones worry that an expanded number of recognized ethnic groups will create even more problems than Canada already has. The sharpest criticisms, however, have come from French Canadians. To them, multiculturalism is more than an irritant, it jeopardizes their very existence. According to one angry observer, it was as if the "Immigration Minister had introduced the Trojan Horse to French Canada" and used immigration to destroy it from within.

Statistics show why immigration worries French Canadians so much. Only about 15% of all foreigners move to Québec, and those go mainly to Montréal. Most of these migrants are Allophones (or Other speakers), and few decide to acculturate into the French language—they are more than twice as likely to switch to English, and more retain their native language in Québec than in any other province (Figure 14.18). This is particularly galling for Francophones, who feel that many newcomers are simply refusing

(a)

(b)

Figure 14.17
More and more, western Canada is identifying itself with the Pacific Rim, confirmed by the presence of Chinese immigrants, as with this Victoria school (a), and Chinese capital, as in this Vancouver bank (b).

	% Born Outside Canada	% Total National Immigrant Population
Canada	*17.4%*	—%
Newfoundland	1.5	0.2
Prince Edward Island	3.3	0.1
Nova Scotia	4.6	0.8
New Brunswick	3.3	0.5
Québec	9.3	13.4
Ontario	25.3	54.8
Manitoba	12.2	2.7
Saskatchewan	5.3	1.1
Alberta	15.0	8.1
British Columbia	24.2	18.2
Yukon	10.4	0.1
Northwest Territories	4.8	0.1

Table 14.3
Percentage of Population Born Outside Canada by Province, 1996

Source: Data from 1996 Census of Canada.

to learn the language of the majority, which in Québec is French. But more immigrants are concerned with learning what the majority of people on the continent speak, and that is English.

Québec's government has attempted to resolve this dilemma by pushing for more French-speaking immigrants and also by enacting language legislation that requires all immigrants to send their children to French-language schools. Needless to say, this new education policy has met with some resistance by immigrant parents who worry that their children are not getting the skills they need. At the same time, the movement of different nationalities into the French Catholic school system has forced a change in the nature of Francophone identity. For decades, most French Canadians could trace their ancestry to when Canada was part of New France. Now, if Québec is to be successful in broadening the meaning of Québécois identity, it must be capable of embracing a variety of new arrivals.

Canada's Regional Variety

As we have seen, Canada is marked by discontinuities in its settlement pattern, a legacy of powerful influences, and an extraordinary cultural diversity. With so many things pulling the country apart, creating an integrated, cohesive society becomes quite difficult. **Regionalism** is a potent phenomenon in Canada, and it could be argued that many Canadians feel more allegiance to their local unit than to the country as a whole. Certainly this is true for residents of Québec—many more say they are "Québécois" than claim

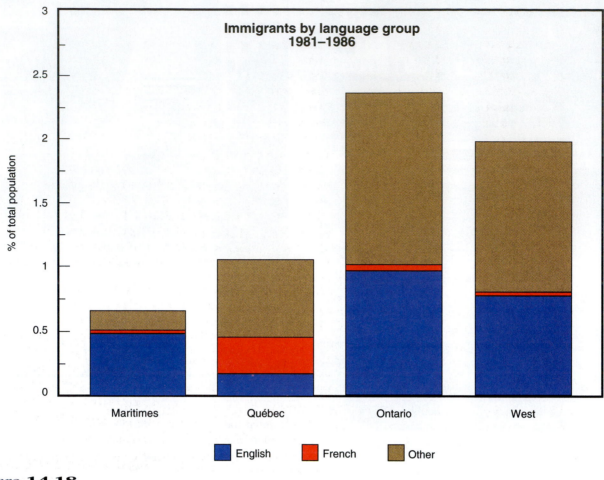

**Immigrants by language group
1981–1986**

% of total population

Maritimes   Québec   Ontario   West

■ English   ■ French   ■ Other

Figure 14.18

Few immigrants to Canada come speaking French, and that worries Québec. Québec's government has acted assertively to reverse this trend by gaining control over who gets into the province and on making sure newcomers learn to operate in French. There was some success: during the 1980s, the French-speaking countries of Haiti, Vietnam, and France were major sources of Canadian immigration.

Source: Data from *Statistics Canada.*

to be "Canadian"—but it also applies to less culturally distinct areas. Canada is a bit like the proverbial elephant. Just as each blind man defines the creature in separate ways, so do residents from Newfoundland to the Yukon Territory offer distinct and perhaps irreconcilable conceptions of Canadian nationhood.

Much of this has to do with the nature of the Canadian economy, which tends to emphasize exports and imports rather than integration within the domestic economy. Much also has to do with vast social and cultural divisions between different regions, which exacerbate existing feelings of difference. And much has to do with the fact that regionalism has been enshrined in the Canadian political system; decisions are made by regional coalitions rather than as a function of majority will.

A Resource Economy

Over the years, Canada has developed as a source of raw materials, first as a colony of France, then of Great Britain,

and then for the global market. Fish and furs, lumber and wheat were exported from the country with very little processing. These have been joined by resources necessary for modern industry, such as zinc, oil, and hydroelectric power. This dependence on resources continues to this day. Exports account for 35% of Canada's gross domestic product (GDP), and close to half of all exports are nonmanufactured or unrefined products. The export of raw goods, combined with the fact that Canada imports mostly finished products (valued at 34% of GDP) has led some to term it "the world's richest underdeveloped country." As Table 14.4 makes clear, Canada is also one of the biggest **branch plant economies.** U.S. and other foreign companies, from IBM to Mitsubishi, have set up Canadian subsidiaries. And while Canada has some multinational corporations of its own—Seagrams, Gulf Canada, and a few others—the Canadian economy is more controlled than controlling.

The dependence on export and branch plant production tends to shortchange domestic economic integration

CANADA'S NEW TERRITORY, NUNAVUT

On April 1, 1999, the political map of Canada changed with the creation of the Territory of Nunavut. This new entity was formerly a part of the Northwest Territories, which remains a political unit with a considerably smaller land area. Nunavut's capital is Iqaluit, once Frobisher Bay, located on Baffin Island at approximately 63 degrees north latitude. The population of Nunavut, which means "our land," is only about 20,000, of which 85% is Inuit (formerly called Eskimo).

Nunavut's creation was first proposed in 1976 as part of a comprehensive settlement in Inuit land claims, and in 1980 an Inuit organization formally called for its establishment. Twelve years later the Tungavik Federation of Nunavut signed a land claims agreement, a portion of which also committed Canada to create Nunavut, subject to certain conditions, including a vote on boundaries. Shortly voters approved the proposed borders and an accord for the creation of the new territory on April 1, 1999. These plans were approved by Canada's Parliament in 1993 and received the Royal Assent.

Under the 1993 Inuit Land Claims Agreement, the population received the right to self-government and self-determination. The new government incorporates Inuit values, which are promoted through the Department of Culture, Language, Elders and Youth; the latter helps all government departments develop and implement policies that reflect Inuit values. The working language for the new government is Inuktitut, but other languages used are Inuinnaqtun, English, and French.

Among the problems faced by Nunavut are high costs for goods and services, high unemployment, low educational levels, and a young workforce. The government has formed an employment plan, which includes training programs, to reach a long-term goal of 85% Inuit employment in Nunavut. Reflecting Inuit culture, the Department of Sustainable Development promotes resource development, such as exploration for minerals, while working to protect and preserve the environment.

The creation of this new territory within Canada, with a government based on indigenous values, is a remarkable change that will be watched closely by aboriginal groups in many parts of North America.

---

and heighten regional separateness. Historically, each region developed independently from the others. Economic fortunes varied with the resources at hand and the demands of the outside world. In the earliest years, Québec traders exported beaver furs to the European nobility, while Newfoundlanders supplied the peasants with cod. The Maritimes achieved prosperity in the 1850s and 1860s by selling wood to Britain. By the late 19th century, Manitoba and Saskatchewan had become one of the world's breadbaskets. Alberta's economy was transformed by the burgeoning demand for petroleum in the mid-20th century. And more recently, British Columbia has grown wealthy from its extensive supplies of salmon, pulpwood, zinc, and copper (see Figure 14.19). Most of these resources are unconnected to the rest of the economy; 90% of British Columbia's exports make their way to foreign countries.

Beyond the different kinds of staples produced, Canada's regions are divided by other criteria. Some provinces are fiscally comfortable, while others are chronically poor. Some provinces garner the lion's share of Canadian industry, while others have little manufacturing and rely mainly on products of the farm, forest, and sea. These inequities show up in such indicators as unemployment and income, as indicated in Table 14.5. The federal government's attempt to flatten these disparities through such policies as

transfer payments from Ottawa to the poorer provinces have not erased considerable differences between provinces.

One way to understand Canada's economic inequalities is through the concept of **heartland** and **hinterland** (terms analogous to *core* and *periphery*). This perspective describes a situation in which the bulk of wealth, capital, and opportunity are located within a specific region or set of regions. These heartland regions have an enormous influence over the economic fate of various, outlying regions (hinterlands), which are developed to service the demands of the heartland. These concepts have greater salience in Canada since resource dependence has made the entire country into an international economic hinterland.

In the United States, the heartland region was once located in the Northeast and extended westward to Chicago and St. Louis. By 1960, part of this American heartland had grown so populous and wealthy that it formed a megalopolis. In Canada, the heartland can be defined in any number of ways, but most definitions include the more populated areas of Ontario and Québec. The geographer Maurice Yeates has described the Windsor, Ontario-to-Québec City axis as Canada's "Main Street" (see Figure 14.20). As of 1971, this 1000-kilometer (621-mi) ribbon constituted only a minuscule proportion of Canada's land area, but it contained more than 60% of the population. Financial

Table 14.4
Canada's Top Exporters, 1985

Company	Exports ($ millions)	Exports (% of sales)	Ownership
Canadian Wheat Board (a)	3610	88%	Federal government
Canadian Pacific (m)	2682	18	21% foreign
Alcan Aluminum (m)	1710	22	54% foreign
MacMillian Bloedel (f)	1326	57	Noranda 49%
Alberta & Southern Gas (e)	1300	83	100% foreign
Shell Canada (e)	1244	21	72% foreign
Mitsui & Co. (Canada) (e)	1028	57	100% foreign
Inco (m)	1010	50	65% foreign
Mitsubishi Canada (e)	980	77	100% foreign
Imperial Oil (e)	966	11	75% foreign
IBM Canada	922	29	100% foreign
PetroCanada (e)	903	17	Federal government
Mobil Oil (e)	898	48	100% foreign
Noranda (m,f)	860	25	Brascan 43%
Nova Corporation (e)	852	26	5% foreign
TransCanada Pipeline (e)	827	18	Bell 48%
B. C. Resources (e,f)	818	78	Canadian

From Wallace Clement, "Debates and Directions: A Political Economy of Resources" in The New Canadian Political Economy, edited by Wallace Clement and Glen Williams. Copyright © 1989 McGill-Queen's University Press, Montreal. Reprinted by permission.
Code: a=agriculture, e=energy, f=forestry, m=minerals
Note: Data excludes exports from Canadian subsidiaries of U.S. auto companies.

Table 14.5
Selected Economic Indicators by Province, 1996

	Unemployment Rate	Average Family Income	% Labor Force in Resources
Newfoundland	17.9	$42,993	6.8%
Prince Edward Island	13.9	$47,125	13.7%
Nova Scotia	10.7	$46,110	6.3%
New Brunswick	12.1	$45,010	6.0%
Québec	10.4	$48,261	3.0%
Ontario	7.2	$59,830	3.0%
Manitoba	5.7	$50,236	7.6%
Saskatchewan	5.9	$49,483	16.5%
Alberta	5.7	$56,916	7.7%
British Columbia	8.9	$56,527	4.6%

Source: Census of Canada, 1996.

power was concentrated in Montréal and Toronto (also Canada's two largest cities); political power was vested in Ottawa (Canada's capital); and industrial power was centered in such cities as Windsor (with its automobile plants) and Hamilton (with its steel industries). Unlike the United States, which has experienced a shifting of wealth and population to the south and west over the past 20 years, Canada's heartland continues to dominate the domestic economy. As 1996 data show, Ontario and Québec control 78% of manufacturing and about two-thirds of all economic output.

Figure 14.19
Goods being loaded in Vancouver Harbor. The yellow sulphur comes from Alberta, the potash from Saskatchewan, and the coal from British Columbia.

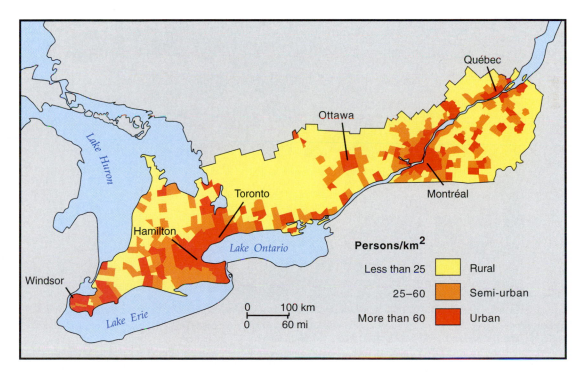

Figure 14.20
Canada's "Main Street," the most densely populated, economically productive, and politically powerful part of Canada.

The Canadian Difference

Maritimes. Poor and isolated, the people in the Maritime Provinces have always felt separated from the Canadian mainstream. For many, Boston and New York seem more accessible than Montréal and Toronto, and the majority of the English-speaking population is cut off from the rest of its co-linguists by French Québec. Given their geographical situation, some—including the premier of Nova Scotia—have suggested that, should Québec secede, the Maritimes would have little choice but to join up with the United States. Geography, a staple economy, and a long independent history have also created a distinct culture, but economic dependence on the rest of Canada makes independence unlikely. This is most true in Newfoundland, where islanders have lived for generations with few newcomers and have created a culture and dialect separate from mainlanders.

Québec. The presence of a majority French-speaking population makes Québec act more like a nation. Emblems of identity—from the fleur-de-lis flag to small parish towns dominated by a Roman Catholic church—abound here; it is fitting that Québec's motto is "Je me souviens" (I remember). The provincial government has taken on many of the trappings of a sovereign state, including separate Québec consulates, a Napoleonic legal code, and huge state enterprises such as Hydro-Québec. Prospects for Québec independence have been enhanced recently with the emergence of Québécois entrepreneurs, comfortable in conducting business in French and now employing 61% of the provincial work force. Whatever happens, this province will remain one of the unique and fascinating corners of North America.

Prairies and the West. While Ontario and Québec are tethered to their respective cultural legacies, western Canada has always been the most boldly "American" region. The spectacular growth of the Prairie Provinces at the turn of the century created a life-style at odds with the old ways. Heavy and varied immigration helped nurture the notion that this was a unique society in the making; a sense of economic grievance over federal policies has sustained this feeling. Albertans in particular have been emboldened by their growing oil fortune and have argued for their rights to exploit this resource as they see fit. British Columbians have followed California in seeing their destiny dawn along the Pacific Rim. Ontarians may be leery of United States influence, but western Canadians identify at times more closely with the western United States than they do with the "Eastern establishment" headquartered in Ottawa, Toronto, and Montréal.

The Canadian North. The Canadian North encompasses more than just the Yukon, Nunavut and Northwest Territories; it also includes the northern reaches of provinces from Labrador to British Columbia. Defined by the Canadian Shield and the arctic tundra, this land has become Canada's new resource frontier. The North has a vast amount of territory occupied by only a tiny population. Most of the people are native, primarily of Cree, Dene, and Inuit descent; non-native whites tend to be far more transient. Culturally distinct and geographically fragmented, the peoples of the North have historically had little leverage over decisions affecting them. But they have become more assertive lately. In the last few years, Crees in northern Québec stalled the flooding of their homeland by a massive water project, and the Inuit succeeded in gaining autonomy for a vast portion of northern Canada, Nunavut.

The Canadian Difference

The heartland is located precisely along the arc where Canada's first major settlements were established: the St. Lawrence valley and the towns and farmsteads on the northern shores of Lakes Ontario and Erie. These two regions, culturally divided but economically integrated, enjoyed an environment suited to agriculture and were poised to participate in overseas and intercontinental trade. The wealth accumulated through these initial advantages laid the groundwork for later industrial expansion. Political factors also played a role. In the late 19th century, the Canadian government buttressed the power of the center with a protectionist National Policy. This increased the power of Montréal financiers and accelerated manufacturing in Ontario, but it also hurt the Canadian fisherman, lumberjack, and farmer—located in the Maritimes, the Prairies, the Pacific coast, and the North. These hinterlands were confined to growing or extracting a small number of staples, while in turn they were forced to buy finished goods from factories in the heartland. The tariff wall was meant to incubate Canada's developing industries from international competition. But as it turned out, it was less effective in fostering industry than in promoting inequality within Canada itself.

Since that time, Canadian policy has changed, but the economic distinctions between regions remain. The key differences have to do with what proportion of the population is involved in resource extraction and what types of resources they are involved with. Ontario and Québec have a much lower proportion of their population dependent on primary industry than do the other provinces. The nature of the resources varies and has a bearing on provincial wealth. Alberta's high income reflects the lucrative nature of the oil industry; Newfoundland's low income can be attributed to a fishing economy.

The nature of the Canadian economy is evident in migration. Historically, the poor provinces in the Maritimes have lost population to regions with more economic opportunity. Ontario has traditionally captured a huge share of migrants, and between 1981 and 1986, more than one-fourth of all English-speaking migrants moved to the southern boot of Ontario. The Montréal area is quite attractive to French speakers, but linguistic considerations dull its appeal for Anglophones who have abandoned Québec in the last few years. Alberta and British Columbia are also popular destinations. Alberta enjoyed a tremendous influx in the late 1970s, due to opportunities in the oil industry, but the oil glut that followed made it considerably less attractive. British Columbia, which also has a resource-based but more diversified economy, has continued to attract more migrants than it sends forth, but the numbers have declined a bit lately. The most resource-dependent economy exists in the Yukon, Nunavut, and Northwest Territories, and this translates into significant volatility. For any given five-year period, in-migrants and out-migrants each equal between one-quarter and one-half of the population of these northern territories.

Federalism and Political Accommodation

With all its diversity, Canada is a peaceful country. While the United States was founded in a fiery War of Independence and forged in the crucible of the Civil War, the struggles over Canada's future have been waged in parliamentary deliberations and in electoral politics. The last internecine strife—the Rebellion of 1837–38—occurred long ago and was not very bloody at that. Since that time, accommodation has ruled Canadian politics. Of course, accommodation is an essential strategy of any government that speaks for the people, but with its various linguistic, ethnic, and regional groups, Canada has raised accommodation to a high art form.

Confederation was a good example of this, since it hinged on the cooperation of Canada West, which became the province of Ontario, and Canada East, which became the province of Québec. These two provinces represented English Protestant and French Catholic Canada, respectively, (while Anglophones dominated Québec's economy, political power was ceded to the Francophone majority) and followed a period where the two groups had collaborated in responsible government. Canada's other provinces joined Confederation on terms determined by the two dominant territories.

Under the **British North America Act,** a parliamentary system was set up, with a significant degree of authority granted to each of the new provinces, including jurisdiction over most law. The act and certain supplemental conventions balanced relations between the two linguistic populations in the following ways:

- It made French and English the official languages in the federal Parliament in Ottawa and in Québec's provincial parliament, although not elsewhere.
- It allowed Québec to establish a French type of property and civil law (Napoleonic code) different from the English-type laws set up in the other provinces.
- It provided guarantees for denominational, or religiously affiliated, schools in Ontario and Québec. The effect of this was to establish a French Catholic and English Protestant school system in Québec, but not in Ontario, where most Catholics were of Irish descent. Similar guarantees in new western Canadian provinces were soon stripped away.
- By convention, French Canadians were expected to control about 25% of the seats in the cabinet; general efforts would be made to include them throughout government in numbers proportional to their population.

Although Confederation was a collaboration between Canada's two *charter groups,* power was not shared equally; the Anglophone majority assumed a dominant role that was enhanced with the addition of English-speaking provinces, such as New Brunswick, Nova Scotia, and Prince Edward Island. But Confederation did display a spirit of compromise toward the linguistic and religious needs of the French-Canadian population and a recognition that the "French Fact" was in Canada to stay. The creation of a federal organization was important, because under this system, provinces were granted substantial autonomy. As a majority in Québec, French Canadians could exercise this authority to serve their cultural interests. Francophones living elsewhere were largely left to the mercy of English-speaking majorities, and most of their rights were successively stripped away.

Since Confederation, the question of federalism and what it means has been debated extensively. Most of the discussion has taken place within the context of the French-English divide, but other issues, such as regional autonomy and the place of ethnic minorities, have been affected as well. As established, Confederation probably bestowed no more power to the Canadian provinces than was granted to American states by the U.S. Constitution. Historically, however, Canadian provinces have accumulated remarkable control over revenues and have been more likely to challenge federal authority. The current question boils down to whether Canada should strengthen its central government or grant more power to the provinces. This is a fundamental issue that reverberates through many democracies, but the tentative nature of Canadian nationalism raises the stakes.

Those favoring more central power have traditionally been interested in strengthening the nation as a whole and fostering a keener sense of Canadian identity. People like former Prime Minister Pierre Trudeau have looked toward a country that resembles a vast mosaic of linguistic and ethnic groups, but where provincial initiatives are muted in favor of national policies. The federal government in Ottawa would play a bigger role in setting policy and would function as an impartial referee among the competing claims of various groups while seeking to bolster those groups least capable of helping themselves. The code phrase *multiculturalism* (pointing to all of the different ethnic and racial groups) *within a bilingual framework* (acknowledging the primacy of the French and English languages) summarizes this view of the state, a view most clearly represented by the 1982 Constitution Act.

This act was not signed by Québec, however, because of the fear that the linguistic and cultural rights of Québec's French majority would be needlessly curtailed. Québec's leaders contend that whatever its stated intentions, English Canada has always been a melting pot into which French minorities have been compelled to assimilate. It becomes the best interest of French Canada to construct its own melting pot, a territorially defined "nation" within which minority language rights are severely circumscribed. This view was more fully accommodated by the Meech Lake Accord of 1987, which granted Québec status as a "distinct society" and gave it an unprecedented measure of control over provincial matters, along with a veto over future constitutional amendments.

Meech Lake failed because it attempted to refashion the old collaboration between English Ontario and French Québec. But Canada has changed in the 135 years since Confederation. There is more ethnic diversity and a growing clamor for group rights. More provinces chafe at allegedly overweening dominance by the federal government and resent what is viewed as solicitousness toward Québécois demands. Many Other-speaking peoples and hinterland provinces would also like to be recognized as distinct societies so as to enjoy the same degree of autonomy offered Québec. Meech Lake was defeated in June 1990 by an alliance of the disaffected, scuttled by the Newfoundland Parliament and by a Cree legislator in Manitoba. And the most recent constitutional reform proposal—the so-called **Charlottetown Agreement,** which promised to accomplish such seemingly irreconcilable objectives as expanding minority rights, buttressing provincial power, and upholding the special status of Québec—proved even more unpopular than Meech Lake and was routed in an October 1992 referendum (Figure 14.21).

Fissures between linguistic groups and separate regional allegiances have created a much more complicated political party dynamic in Canada than in the United States. Most striking is the division between *national* and *regional* parties. National parties field candidates for election to the House of Commons in Ottawa (Figure 14.22). (Members of the Senate are appointed until age 75 on the prime minister's recommendation and have far fewer powers than members of the House of Commons.) Because Canada has a parliamentary system of government, the leader of the party that garners the majority of seats in the House of Commons becomes prime minister and usually chooses his or her cabinet from current members. The leading national parties in Canada are the Liberals (who have held control for most of Canada's history and are considered the centrist party), the Progressive Conservatives, and the New Democratic Party (a socialist party that usually comes in third in the polling but has regions of strength). The Reform Party, which is based exclusively in the western provinces, has emerged as a voice for populism and against federal concessions to Québec.

The provincial governments largely replicate the national parliamentary system except there is no senate. The leader of the winning party becomes premier. National parties compete in provincial elections, but they must share the hustings with more regionally based parties. Québec has the separatist Parti québecois, the Union Nationale (representing an older form of French Canadian nationalism that dominated the legislature through the mid-20th century), and a Liberal Party that charts a different course from the federal Liberals. At one point, Manitoba,

The Canadian Difference

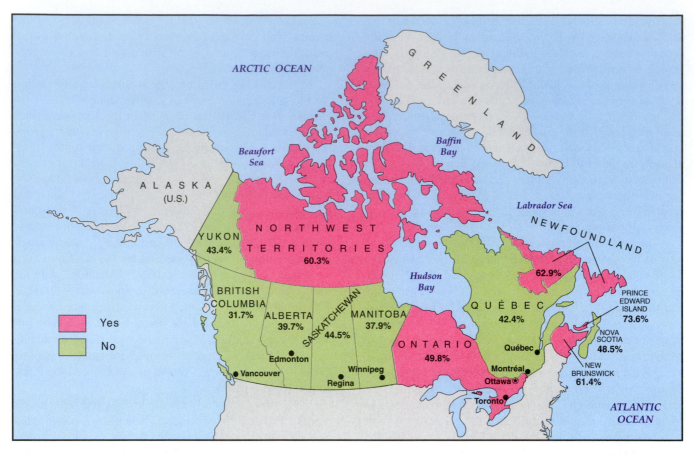

Figure 14.21

The defeat of the Charlottetown Agreement in October 1992 was a blow to Prime Minister Brian Mulroney and others who had hoped to find a lasting solution to Canada's constitutional crisis. To succeed, the measure had to pass in each province and territory; it prevailed in only five.

Saskatchewan, Alberta, and British Columbia contained a politically potent Social Credit Party, which continues to rule British Columbia. The New Democratic Party has seen recent strength in Ontario. The plethora of party choices may have something to do with the electorate's high vote turnout (about 75% go to the polls) and its volatility, with the sudden advent and eclipse of parties like the Union Nationale, Social Credit, and the Parti québécois. In these several ways, the Canadian system varies considerably from the United States, which suffers from voter apathy and has been dominated by the same two parties for as long as Canada has been a country.

Summary

We began this chapter by questioning the nature of Canadian identity. The immense size of the country and the diversity of the people have made this endeavor much more difficult for Canada than it is for most societies, but it would be wrong to suggest that there is nothing that distinguishes Canadians from people in the rest of the world in general

and from the United States in particular. A few brief and tentative hallmarks might be put up for discussion.

Bilingualism: The "French Fact" is probably the most obvious point of distinction, and it is this fact that distinguishes French and English Canada alike. Canada was founded by two very different populations that have continued to forge separate societies. But bilingualism may be a point of pride as well. Former Prime Minister Trudeau has argued that Canada could become a beacon for all of the other multilingual, multicultural states if it is able to resolve its differences democratically. Because of the size and geographical concentration of the Francophone population, Canada has been forced to take extraordinary steps to reconcile its majority with its largest minority. This experience has honed the Canadian attitude of compromise and has given it an international reputation as a fair and impartial player in world affairs.

Multiculturalism: French and English are not the only Canadians. Canada is a nation of immigrants, like the United States, with a large number of native peoples. These groups have come to demand greater recognition and a

Figure 14.22
Parliament Hill, Ottawa. The parliament buildings are an enduring and impressive symbol of Canadian governmental authority.

greater share of power. Both of Canada's charter groups have come to accept this viewpoint grudgingly: Anglophones as a result of declining British migration, and Francophones with the realization that they needed to broaden the nature of French-Canadian identity if they hoped to compete with English Canada.

An orderly society: It has been said that the United States is the country of the revolution, Canada of the counterrevolution. In essence, Canadians descend from people who decided to uphold the existing governmental order, and this has engendered a national personality much more receptive to authority and has allowed Canadians to achieve such things as nationalized health care, gun control, and metropolitan consolidation that have continually eluded American policymakers. This makes Canada particularly attractive to many people, and it is a feature that Canadians worry about losing as they become more closely integrated in a North American economy.

Geographic immensity: The sheer size of Canada, coupled with its small population, has produced one of the most lightly populated societies in the world, and this has created a special awareness. Most of the people live in the far southern tier of the country, but even here, distances are so vast as to have created several distinct regional identities. Northward lies an enormous, mostly uninhabited fund of land. Canada is much too orderly for this region to be a "wild North," just as the Royal Canadian Mounted Police ensured that there was no Canadian "wild West," but the North gives Canada much more resource flexibility than other countries and ensures that it will continue to enjoy many options in future years. The recent creation of the Inuit territory of Nunavut, for example, allowed Canada to satisfy a long-standing demand by one native group. State-run corporations such as Hydro-Québec have grown rich from the vast supplies of water and land resources, increasingly in demand from an overpopulated world. The presence of so much unexploited territory puts Canada in a special class, and slow population growth will keep open much of this endowment.

These are just a few of the special attributes that help define "Canadianness." Canada being Canada, these characteristics vary a bit by region, ethnicity, and language group. More significant, they do not add up to a strong sense of nationalism in the same way that the American bundle of myths, traits, symbols, and heroes crystallizes into a powerful group identity. But this may not be all to the bad. In a world where ethnic differences have deepened into hostile divides, where tribalism threatens to destroy political order, and where pluralism has become a feature most societies must reckon with, the Canadian dilemma may be a forerunner to a common experience. And its successful resolution, if that occurs, could prove a model for countries everywhere.

Key Words

Allophone	hinterland
Anglo-conformity	intendant
Anglophone	Meech Lake Accord
branch plant economy	Métis
British North America Act	multiculturalism
Canadien	National Policy
censitaire	Parti québécois
Charlottetown Agreement	patriotes
Confederation	Quiet Revolution
coureurs de bois	rang
entrepôt	regionalism
factors	seigneurial system
Francophone	visible minorities
heartland	

Gaining Insights

1. In which areas are Canada's people concentrated? What factors have shaped this population distribution? Why would these settlement patterns affect Canadian relations with the United States?

2. What were the main paths of European exploration into Canada, and how did the French and the British diverge in their continental penetration? For what three reasons were Europeans attracted to Canada?

3. How did the seigneurial system define the French settlement of the St. Lawrence River valley? What were the consequences of this system for the shape of farmsteads and villages?

4. Describe the present-day geography of language groups in Canada. What are some reasons why French speakers are found primarily in Québec? Where are some places where French speakers are found outside of Québec? Where do English speakers in Québec tend to concentrate?

5. What have been some of the changes in the types of people who have immigrated to Canada over the last 120 years? How has this forced a change from the original ideology of Anglo-conformity to the current notion of multiculturalism?

6. How has French-Canadian identity evolved over the last three centuries? What role did the Quiet Revolution and the Parti québécois play in fueling current demands for Québec's independence?

7. Where is Canada's heartland located, and how does it dominate the Canadian economy? What are the various Canadian hinterlands, and what economic function do these fulfill?

8. How does the Canadian federalism differ from the governmental structure of the United States? How did the British North America Act of 1867 balance relations between English and French speakers and between the central government and the provinces? What recent attempts have been made to alter the constitution? Why have these been unsuccessful?

9. What special attributes help define "Canadianness"? Why might "survival" surface as the central symbol of Canadian society?

Selected References

Arnopoulos, S., and D. Clift. *The English Fact in Quebec.* Montréal: McGill-Queen's University Press, 1984.

Breton, R., J. Reitz, and V. Valentine. *Cultural Boundaries and the Cohesion of Canada.* Montréal: Institute for Research on Public Policy, 1980.

Caldwell, G., and E. Waddell, eds. *The English of Quebec: From Majority to Minority Status.* Québec: Institut Québécois de Recherceh sur la Culture, 1982.

Clement, W., and G. Williams, eds. *The New Canadian Political Economy.* Kingston, Ontario: McGill-Queen's University Press, 1989.

Clift, D. *The Secret Kingdom.* Toronto: McClelland and Stewart, 1989.

Dreidger, L., ed. *Ethnic Canada: Identities and Inequalities.* Toronto: Copp Clark Pitman, 1987.

Granatstein, J. L., I. M. Abella, T. W. Acheson, D. Bercuson, R. C. Brown, and H. B. Neatby. *Nation: Canada since Confederation,* 3rd ed. Toronto: McGraw-Hill Ryerson, 1990.

Harris, R. C., and J. Warkentin. *Canada before Confederation: A Study in Historical Geography.* Ottawa: Carleton University Press, 1991.

Hiller, H. *Canadian Society: A Macro Analysis.* Scarborough, Ontario: Prentice-Hall of Canada, 1991.

Innis, H. A. *Essays in Canadian Economic History.* Toronto: University of Toronto Press, 1956.

"In Search of Canada," *Daedalus* (Fall 1988).

Joy, R. *Languages in Conflict.* Toronto: McClelland and Stewart, 1972.

Levine, M. *The Reconquest of Montréal: Language Policy and Change in a Bilingual City.* Philadelphia: Temple University Press, 1990.

Li, P., ed. *Race and Ethnic Relations in Canada.* Toronto: Oxford University Press, 1990.

Lipset, S. *Continental Divide: The Values and Institutions of the United States and Canada.* New York: Routledge, 1990.

McCann, L. D., ed. *Heartland and Hinterland: A Geography of Canada.* Scarborough, Ontario: Prentice-Hall of Canada, 1982.

McRoberts, K. *Quebec: Social Change and Political Crisis.* Toronto: McClelland and Stewart, 1988.

Verney, Douglas V. *Three Civilizations, Two Cultures, One State: Canada's Political Traditions.* Durham, N.C.: Duke University Press, 1986.

Yeates, M. *Main Street: Windsor to Quebec City.* Toronto: Macmillan of Canada, 1975.

GLOSSARY

A

acid rain   Precipitation that is unusually acidic; created when oxides of sulfur and nitrogen change chemically as they dissolve in water vapor in the atmosphere and return to earth as acidic rain, snow, fog, or dry particles.

agriculture   Cultivating the soil, producing crops, and raising livestock; farming.

air mass   A large body of air with little horizontal variation in temperature, pressure, and humidity.

Allophone   A Canadian who speaks a language other than French or English.

alluvium   The sediments carried by a stream and deposited in a flood plain or a delta.

alpine glacier   A glacier formed in the mountains, in contrast to continental glaciers, which are formed in plains regions.

amenities   Those things that are considered attractive about an area, usually such physical characteristics as beaches, lakes, mountains, and favorable climates.

Amtrak   The semipublic corporation that has provided most of the intercity rail passenger service in the United States since 1971.

Anglo-conformity   The expectation that immigrants to Canada would assimilate into the mainstream culture.

Anglophone   A Canadian who speaks English.

annexation   The act of adding territory to a city; incorporating adjoining areas into a city.

aquifer   Underground porous and permeable rock that is capable of holding groundwater, especially rock that supplies economically significant quantities of water to wells and springs.

areal specialization   The production of particular goods or services in a particular location, ideally where that production is most efficiently carried out.

B

baby boom   The dramatic rise in the birth rate in the United States and Canada following World War II and lasting well into the 1960s.

basic sector   Those products of an urban unit that are exported outside the city itself, earning income for the community.

blizzard   A heavy snowstorm accompanied by high winds.

blue-collar in suburbs and small cities   Workers living outside the central city and employed in jobs involving manual labor, such as mechanics, plumbers, or nurses.

boreal forest   The vast needleleaf (conifer) forests of subarctic North America and Eurasia.

bracero   A Mexican laborer admitted legally into the United States for a short time to perform seasonal, usually agricultural, labor.

branch plant economy   An economy in which many of the industrial firms are under foreign ownership.

British North America Act   The act passed by the British Parliament in 1867 that created the Canadian federation from the colonies of Nova Scotia, New Brunswick, Québec, and Ontario.

broadleaf trees   One of two major categories of trees (the other being needleleaf), in which the leaves are wide, flat, and planar in form. In temperate climates, broadleaf trees are usually deciduous; in the tropics, they are evergreen.

C

Canadian Shield   A large area of ancient rock in eastern Canada, with two small extensions into the U.S., most of which has been weathered and eroded down nearly to a plain. Considered to be the landform around which the rest of North America developed.

Canadien   An old term for a Canadian of French ancestry.

cash grain   Grain sold from the Corn Belt farm as its commercial product without its use as livestock feed on the farm.

censitaire   A tenant under the seigneurial system, so called because he was obligated to remit a *cens* or payment to the landowner.

central business district (CBD)   The center or "downtown" of an urban unit, where retail stores, offices, and cultural activities are concentrated and where land values are high.

central city   That part of the urban area contained within the boundaries of the main city around which suburbs have developed.

centrifugal forces   In political geography, those forces that tend to tear a country apart, such as the presence of more than one language or of a dissatisfied minority group.

centripetal forces   In political geography, those forces that tend to unite a country, such as a common sense of history or uniformity of language.

channelization   The modification of a stream channel; specifically, the straightening of meanders or dredging of the stream channel to deepen it.

chaparral   A natural vegetation consisting of dense evergreen shrubs, found near the coast in the Mediterranean climate of California.

Charlottetown Agreement   A Canadian constitutional reform proposal that would have upheld the special status of Québec; failed in 1992.

chemical weathering   The decomposition of earth materials due to chemical reactions that include oxidation, hydration, and carbonation.

Chinese Exclusion Act   An 1882 act of Congress that excluded Chinese laborers from the United States for a period of ten years; extended in 1890 and 1902.

circumferential freeway   A limited access road that circles, or partly circles, a metropolitan area, usually near the edge of the built-up area.

city   A multifunctional nucleated settlement with a central business district and both residential and nonresidential land uses.

climate   The long-term average weather conditions in a place or region.

Coastal Ranges   The mountain ranges along the Pacific Coast of North America, extending from southern Alaska to southern California.

combine   A machine that harvests, threshes, and cleans grain while moving over the field.

community of interest   A territorially defined group of people with common economic, social, political, or cultural interests.

concentric zone model   The idea that there is a series of circular belts of land use around the central business district, each belt containing distinct functions.

Confederation   The 1867 federal union of the Canadian provinces of Nova Scotia, New Brunswick, Québec, and Ontario.

conifers   Trees that bear seed cones; they are often needleleafed and almost always evergreen.

continental glaciation In the Northern Hemisphere, the southward movement from the polar regions of vast sheets of ice covering much of northern North America and Eurasia, which occurred at times when the earth was in a colder period than at present.

continentality In physical geography, the characteristic of being located far from the moderating effects of large bodies of water, resulting in hot summers and cold winters.

conurbation A large metropolitan complex formed by the coalescence of two or more urban areas.

convection precipitation Rain or some other form of precipitation produced when heated, moisture-laden air rises and then cools below the temperature at which condensation occurs.

cordillera A mountain chain or system.

core The nucleus of a region or country, the main center of its industry, commerce, population, political, and intellectual life; in urban geography, that part of the central business district characterized by intensive land development.

cotton gin A machine that separates the seeds, hulls, and any foreign matter from the cotton.

coureurs de bois Independent French traders who exchanged furs with the Indians; *see* voyageurs.

crude birth rate The ratio of the number of live births in a given year to the total population in that year, usually expressed in terms of a base of 1000 people.

crude death rate The ratio of the number of deaths in a given year to the total population in that year, usually expressed in terms of a base of 1000 people.

culture hearth A nuclear area within which an advanced and distinctive set of culture traits develops and from which there is diffusion of distinctive technologies and ways of life.

Cumberland Gap A pass through the Appalachian Mountains near the junction of Virginia, Kentucky, and Tennessee, used by many settlers moving west after the American Revolution.

cyclonic precipitation Rain or some other form of precipitation produced when the moist air of one air mass is forced to rise over the edge of another air mass.

D

deciduous trees Those trees that lose their leaves each year in the fall.

density rim A ring of relatively high population density that typically is located a mile or so beyond the nonresidential central business district in American and Canadian cities.

deregulation The elimination of government regulation over an industry. The intent is that market forces, rather than government regulation, will determine what happens in that industry in the future.

direct democracy In political geography, where eligible voters can personally cast their ballots on issues of public interest (in contrast to representative democracy).

Driftless Area A region in southwestern Wisconsin that was missed by all the glaciers; the landforms in this area strongly reflect lack of glaciation.

E

earth plates Major divisions of the lithosphere that slide or drift slowly over the partially molten, plastic lower levels of the earth.

economic base The mix of manufacturing and service activities performed by the labor force of a city to satisfy demands outside the city and earn income to support the urban population.

economics of agglomeration Savings that accrue to economic activities because needed goods and services are located within easy reach.

effective votes In political geography, the number of votes cast that are needed to win an election.

elderly suburban Senior citizens living in moderate-income communities on the outskirts of cities.

elite Those with the greatest social status, wealth, and influence in society.

entrepôt A place into which goods are imported and stored and from which they are traded or distributed.

environmental pollution *See* pollution.

Erie Canal A waterway 584 kilometers (363 mi) long through central New York state, from Albany to Buffalo; completed in 1825.

eutrophication The increase of nutrients in a body of water. The nutrients stimulate the growth of algae, whose decomposition decreases the dissolved oxygen content of the water.

excess votes Those votes cast which are above the minimum number needed to win an election.

extractive industry Primary activities involving the mining and quarrying of nonrenewable metallic and nonmetallic mineral resources.

F

factor inputs The different kinds of resources (raw materials, fabricated parts, labor, land, and capital) used to make a product.

factors Agents of French import-export companies who often spent a long time in colonial Canada.

farm business-related town A town that provides the nearby farm community with wholesale and retail supplies and services.

fault A break or fracture in rock produced by stress or the movements of lithospheric plates.

federal States Countries where the responsibilities of government are divided formally between central authorities in the national capital and lower levels of government, such as states and provinces.

flash floods Rapid flooding that results from intense precipitation.

flexibly specialized production system One that is designed to efficiently produce small quantities of a product and to easily convert to producing a different good.

Fordist production system A system characterized by large manufacturing plants with high capital investments dedicated to making a few standardized products using parts brought together in an assembly-line process.

fossil fuels Any of the fuels derived from decayed organic material converted by earth processes; especially, coal, petroleum, and natural gas.

Francophone A Canadian who speaks French.

friction of distance A measurement indicating the effect of distance upon the extent of interaction between two points. Generally, the greater the distance, the less the interaction or exchange or the greater the cost of achieving the exchange.

G

gatekeeping An immigration policy concerned primarily with keeping those that are not desired out of a country.

gathering industry Primary activities involving the harvesting of renewable natural resources of land or water; commercial gathering usually implies forestry and fishing industries.

Gentlemen's Agreement A 1907 understanding between the United States and Japan that the latter would tightly regulate emigration to the United States.

gentrification The process by which middle- and high-income groups refurbish and rehabilitate housing in deteriorated inner-city areas, thereby displacing low-income populations.

geographic information system (GIS) A method of storing and manipulating geographic information in a computer; the three major components of such systems are the digital map data, the hardware used to handle those data, and the associated software.

geography of uneven development The tendency for firms to move their industrial activities from well-developed to less-developed regions.

gerrymandering The deliberate manipulation of political district boundaries to achieve a particular outcome in elections.

glacial drift A term referring to any type of glacial deposit.

Greater Yellowstone ecosystem One of the largest temperate ecosystems on earth, composed of Yellowstone and Grand Teton National Parks and the surrounding national forests, wildlife refuges, wilderness areas, and private land.

H

hail Precipitation in the form of concentric layers of ice.

hazardous waste Discarded solid, liquid, or gaseous material that may pose a substantial threat to human health or the environment when it is improperly disposed of, stored, or transported.

heartland A central region; one of economic, strategic, or cultural importance.

heavy rail Urban rail passenger systems that have grade-separated rights of way; for example, subway and elevated systems.

high-tech manufacturing A term usually referring to the production of high value per pound items, particularly in the computer-electronics-communications-aerospace industries.

hinterland Outlying regions developed to furnish raw materials or agricultural products to the heartland; the market area or region served by a town or city.

horizontally integrated production A system of production in which specialized products manufactured at many places are brought together and assembled to make a marketable product.

hybrid seed The product of parent plants of two different varieties imparting consistent superior characteristics to the seed.

hydrologic cycle The system by which water continuously circulates through the biosphere.

I

ice cap A situation, or climatic region, where the land is permanently covered by an ice sheet.

infant mortality rate A refinement of the death rate to specify the number of deaths of infants under age 1 year per 1000 live births.

information-technology industry One that manufactures semiconductors, computers, computer software, communications equipment, or electronic machines.

integrated steel mill A plant that starts with raw materials and produces a full range of steel products in a continuous manufacturing process.

intendant A French colonial administrator who controlled the civil government and reported directly to the king.

intermodal traffic Freight that uses more than one form of transportation between its point of origin and its destination.

J

just-in-time delivery A manufacturing system in which inputs reach an assembly plant shortly before they are needed.

L

land bridge services Where maritime containers are taken across North America by rail from a West Coast port to an East Coast destination, or vice versa. Originally, the term referred to containers moving across the continent on a trip between Asia and Europe.

least cost theory of location (*Syn:* Weberian analysis) The view that the optimum location of a manufacturing establishment is at the place where the transportation costs of assembling inputs and marketing products plus all other costs are least.

leeward The side of a mountain or island opposite that of the prevailing direction of the wind.

life expectancy The average duration of life beyond any age of persons who have attained that age, calculated from a life table.

light rail Urban rail systems that operate, wholly or in part, on streets; a modern name for streetcars.

long lot A long, thin property with narrow river frontage; characteristic of French colonial settlements in the United States and Canada.

M

malapportionment An imbalance in the number of voters in various constituencies, whereby one person's vote becomes, in effect, worth more than another person's vote.

maquiladora A foreign-owned plant in Mexico that produces goods for immediate re-export and that pays lower import and export customs taxes than other manufacturing plants in the same industry.

Meech Lake Accord A 1987 proposed addition to the Canadian constitution that acknowledged Québec's distinctiveness; failed in 1990.

megalopolis An extensive, heavily populated urban complex with contained open, nonurban land, created through the spread and merging of separate metropolitan areas; (cap.) the name applied to the continuous functionally urban area of the northeastern seaboard of the United States from Maine to Virginia.

metes and bounds system A system of land surveying that uses natural and people-made features, such as streams and monuments, to mark the boundaries of properties.

Métis A person of mixed French and Canadian Indian ancestry.

metropolitan area A large functional entity, perhaps containing several urbanized areas, discontinuously built up but operating as a coherent economic whole.

Metropolitan Statistical Area (MSA) An MSA consists of a large core city (usually with at least 50,000 people) plus the surrounding core county. It can include additional contiguous counties if they are highly integrated, economically and socially, with the core county.

middle-class suburbs The communities outside the central city typically inhabited by child-raising families of average income living in comfortable single-family houses.

midlatitude cyclone A low pressure system with associated warm and cold fronts; usually associated with large-scale precipitation.

migration field An area that sends major migration flows to or receives major flows from a given place.

milieus of innovation Places typified by a work culture committed to generating new knowledge, new processes, and new products.

mini-steel mill A small steel-making plant that typically relies on electric ovens to burn scrap metal, employs nonunion labor, and makes specialized steel products.

mixed forest A forest that contains both evergreen and deciduous trees.

moldboard plow The curved plate or face of a plow designed to lift and turn the soil.

multiculturalism A Canadian policy that recognizes the uniqueness of each ethnic and racial group and seeks to protect its culture.

multiple nuclei model The idea that large cities develop by peripheral spread not from one but from several nodes of growth and that, therefore, there are many origin points of the various land use types in an urban area.

multiple-use ethic The belief that publicly owned land resources should be used for a variety of purposes as long as such use does not diminish the capacity of the land for self-renewal.

N

NAFTA The North American Free Trade Agreement, passed in 1993 to promote trade in goods and services, and to encourage direct investments, among the United States, Canada, and Mexico. Eventually virtually all barriers to the free trade of goods and services, and to cross-border investments, are to be eliminated.

nation A culturally distinctive group of people occupying a particular area and bound together by a sense of unity arising from a shared ethnicity, beliefs, and customs.

nation-State A State whose territory is identical to that occupied by a particular nation.

national forest A piece of publicly owned forested land administered by the U.S. Forest Service; together, the national forests cover some 77 million hectares (191 million acres).

national monument A place of historic, scenic, or scientific interest that may be established by presidential proclamation; the 78 national monuments are part of the National Park System.

national park A relatively large area whose natural features and ecology are of great beauty, scientific interest, or recreational and educational value; the national parks are one type of area comprising the National Park System

National Park System A collection of various types of areas (national parks, monuments, preserves, and so on) distinguished for their scenic beauty, historic importance, scientific interest, or recreational assets. The system is administered by the National Park Service, a bureau of the U.S. Department of the Interior.

National Policy   A policy introduced in 1878 that sought to protect Canadian producers with high tariffs.

national preserve   A category of publicly owned land, most of it in Alaska, administered by the National Park Service and established to allow noncompatible uses, such as sports and subsistence hunting by native people, to exist in parkland without jeopardizing traditional national park values.

national resource lands   Publicly owned land administered by the U.S. Bureau of Land Management.

National Wild and Scenic Rivers System   Ribbons of land bordering on free-flowing streams that have not been dammed, channelized, or otherwise altered by humans. Protected under the U.S. Wild and Scenic Rivers Act of 1968, they are one type of area comprising the National Park System.

National Wilderness Preservation System   Federally protected undeveloped land that by the U.S. Wilderness Act of 1964 and subsequent laws is to be managed "for the use and enjoyment of the American people in such a manner as will leave them unimpaired for future use and enjoyment as wilderness."

natural resource   A physically occurring item that a population perceives to be necessary and useful to its maintenance and well-being.

nonrenewable resource   A natural resource that is not replenished or replaced by natural processes or is used at a rate that exceeds its replacement rate.

O

Ogallala aquifer   The largest underground water supply in the United States, located in the Great Plains; it is being depleted by heavy irrigation.

open-range cattle ranching   Raising cattle or sheep on natural vegetation in an area without fences.

opponent-concentration gerrymandering   A form of political manipulation where opposition voters are placed into as few districts as possible. Also known as excess-vote gerrymandering.

opponent-dispersion gerrymandering   A form of political manipulation where opposition voters are divided among as many districts as possible, weakening the effectiveness of their votes. Also known as wasted-vote gerrymandering.

orographic precipitation   Rain or snow caused when warm, moisture-laden air is forced to rise over hills or mountains in its path and is thereby cooled.

ozone   A gas molecule consisting of three atoms of oxygen (O_3) formed when diatomic oxygen (O_2) is exposed to ultraviolet radiation. As a damaging component of photochemical smog formed at the earth's surface, it is a faintly blue, poisonous agent with a pungent odor.

P

Parti québécois   A separatist political party in the province of Québec; formed in 1968.

particular representation   An approach to governance under which a person is elected to office as the representative of a specific geographic area, such as a Congressional district or a city ward. Also called areal representation.

patriotes   French Canadians who sought to liberate Québec from British control during the Rebellion of 1837–1838.

PCBs   Polychlorinated biphenyls, compounds containing chlorine that can be biologically magnified in the food chain.

peak land value intersection   The most accessible and costly parcel of land in the central business district and, therefore, in the entire urbanized area.

periphery   The outermost part of a region, one that is tributary to the core.

permafrost   Permanently frozen subsoil.

photochemical smog   A form of polluted air produced by the interaction of hydrocarbons and oxides of nitrogen in the presence of sunlight.

plateau   An elevated, relatively flat area.

political economy approach   An explanation that focuses on how the economic and social policies of governments change the distribution of industrial and commercial activities.

political fragmentation   The division of metropolitan areas into numerous independent and often competing political jurisdictions.

pollution   The introduction into the biosphere of materials that, because of their quantity, chemical nature, or temperature, have a negative impact on the ecosystem or that cannot be readily disposed of by natural recycling processes.

polychlorinated biphenyls (PCBs)   *See* PCBs.

preservationist ethic   The belief that publicly owned land resources should be kept as ecological reserves and used only for nondestructive types of outdoor recreation.

primary activities   Those parts of the economy involved in making natural resources available for use or further processing; included are mining, agriculture, forestry, fishing and hunting, and grazing.

product cycle approach   A description of the typical changes that occur in the locations of manufacturing plants in an industry as its product matures and declines.

Q

quaternary activities   That employment concerned with research, the gathering or disseminating of information, and administration, including administration of the other economic activity levels.

Quiet Revolution   A series of social, political, and economic transformations, commencing around 1960, that made Québec into a more modern society.

R

rainshadow   The dry side of a mountain range; the side opposite from that which experiences orographic precipitation.

rang   A group of long, narrow fields in French Canada stretching out from a river or a road.

recreation   A means of refreshing or entertaining oneself after work by some pleasurable activity.

recycling   The reuse of disposed materials after they have passed through some form of treatment (such as melting down glass bottles to produce new bottles).

red tide   A reddish, brown, or yellow discoloration of sea water caused by an enormous increase in the number of certain microscopic organisms, usually algae.

region   An area of relative uniformity, such as one dominated by a particular type of farming, a type of climate, or by a kind of manufacturing.

regionalism   In political geography, identification with a particular region of a country rather than with the country as a whole.

renewable resource   A natural resource that is potentially inexhaustible because it is either constantly or periodically replenished as long as its use does not exceed its replacement rate or capacity.

representative democracy   Political decision-making where voters first elect representatives, and the latter then meet to enact legislation (in contrast to direct democracy).

right-to-work law   A state law that prohibits union-management agreements requiring a worker to join a union to obtain or hold a job.

riparian   Along the bank of a river or lake.

rural poor   Those whites and blacks living in rundown, backcountry houses outside of urban and suburban areas who have limited incomes gained from farming and service occupations.

S

St. Lawrence Seaway   An international waterway system along part of the U.S.–Canadian border that connects the Atlantic seabord with the Great Lakes; it is 3769 kilometers (2342 mi) long.

salinization   The process by which soil becomes saturated with salt, rendering the land unsuitable for agriculture; occurs when land that has poor drainage is improperly irrigated.

savanna   A grassland containing scattered trees.

secondary activities   Refers to manufacturing. Those activities which involve processing of raw materials from the primary sector, or the further processing of partially manufactured items. Can also refer to assembly of finished items into new products.

sectional normal vote   Patterns of regional voting that remain in place through, at least, a number of elections.

sectionalism   In political geography, the tendency for groups of adjacent states to vote in similar fashion on political issues.

sector model   The idea that pie-shaped wedges of different land uses radiate outward from the central business district.

sedimentary rock   Rock formed from particles of gravel, sand, silt, and clay that were eroded from already existing rocks.

segmentation   The process that gives rise to roughly homogeneous areas based on social status, family status, and ethnicity.

seigneurial system   A type of feudal organization whereby kings of France granted large tracts of land in New France to noblemen, who settled them with tenant farmers or censitaires.

service or nonbasic sector   Those economic activities of an urban unit that service the resident population.

sharecropper   A tenant farmer who gives part of the annual harvest to the landlord in place of a cash rent.

site   The local setting of a city; the immediate surroundings and their attributes.

situation   The location of a city in relation to the physical and human characteristics of other places.

small towners   People living in small towns away from urban and suburban areas that are not farm-related, such as mining, lumbering, and fishing communities.

smart buildings   Structures furnished with the latest telecommunications facilities.

soil erosion   The wearing away and removal of soil particles from exposed surfaces by agents such as moving water, wind, or ice.

South Pass   A break in the Rocky Mountains in southern Wyoming, used by many early travelers in the process of crossing the continent.

special purpose district   A form of local authority that is created, under state authorization, to serve a specific public function, such as irrigation or fire protection.

spring wheat   Varieties of wheat that are planted in spring and harvested in fall.

State   (*syn:* country) An independent political unit occupying a defined, permanently populated territory and having full sovereign control over its internal and foreign affairs.

subsistence agriculture   A type of farm economy in which most crops are grown for food, nearly exclusively for local consumption.

suburb   A functionally specialized segment of a large urban complex located outside the boundaries of the central city.

Sun Belt   The southern half of the United States, south of an east-west line starting along the North Carolina–Virginia border and extending west to the north edge of the San Francisco Bay Area. Includes Hawaii.

T

teleports   Places that provide bulk access to advanced transmission channels.

terminal moraine   A ridge of unconsolidated materials that was pushed along in front of an advancing glacier and then left at the point of its furthest advance when the glacier retreated.

tertiary activities   The service sector of the economy. Those parts of an economy concerned with the exchange and market availability of commodities; includes wholesaling, retailing, and associated transportation, government, and information services.

theme park   A type of amusement part that combines traditional amusements with a theme such as American history or country music.

thermal pollution   The introduction of heated water into the environment with consequent adverse effects on plants and animals.

Tornado Alley   The portion of North America in which tornadoes are most prevalent; generally considered to extend from eastern Texas to the Great Lakes region.

town   A nucleated settlement that contains a central business district but is smaller and less functionally complex than a city.

township and range system   A method established by the Land Ordinance of 1785 of subdividing land in the Old Northwest into a series of approximate squares.

tragedy of the commons   The observation that in the absence of collective control over the use of a resource available to all, it is to the advantage of all users to maximize their separate shares even though their collective pressures may diminish total yield or destroy the resource altogether.

Treaty of Greenville   A 1795 agreement that the Indians would accept the authority of the United States rather than that of Great Britain and renounce their claims to all but a small corner of land in the Northwest Territory.

truck crops   Intensively produced fruits and vegetables for market rather than for processing or canning.

tundra   The treeless area lying between the tree line of arctic regions and the permanently ice-covered zone.

U

unitary State   A country where the central government dictates the degree of local or regional autonomy and the nature of local government units.

U.S./Canadian Manufacturing Belt   The region of the north-central United States, southern Ontario, and southern Québec that contains a large concentration of the continent's manufacturing establishments, most of its heavy industry, and the densest transportation network.

upscale suburban   High-income areas on the edge of the city, typically containing single-family houses for those raising children and condominiums and apartments for young professionals.

upscale urban   A category that includes high-rise apartments inhabited by the well-to-do, prosperous black neighborhoods, and singles living in condominiums, garden apartments, and bungalows.

urban hierarchy   the steplike series of urban units (hamlets, villages, towns, cities, metropolises) in classes differentiated by size and function.

urban influence zones   The areas outside of a city that are still affected by it.

urban poor   The people with lowest incomes living in the least desirable parts of the city.

urbanized area   A continuously built-up urban landscape defined by building and population densities with no reference to the political boundaries of the city; it may contain a central city and many contiguous towns, suburbs, and unincorporated areas.

utilization ethic   The belief that publicly owned land resources are economic goods that should be developed to maximize profits.

V

variable costs   In economic geography, costs of production inputs that change as the level of production changes. They differ from the costs incurred by agricultural or industrial firms that are fixed and do not change as the amount of production changes.

vertical zonation   In mountainous areas, the progression of differing climatic and biotic zones as one ascends to higher and higher elevations.

vertically integrated production   A system of production in which a firm controls all phases of the manufacturing of all components of its product. Often, such firms will also control the marketing of their products. In agriculture, such a firm might unite farm production, transportation, processing, and wholesale or retail sales.

Via Rail   The government-supported corporation that operates most intercity passenger rail services in Canada.

virtual representation   A philosophy of political representation where the person elected sees himself or herself as representing the interests of the whole body politic, not just of a particular area.

visible minorities   A term used to describe nonwhites in Canada.

von Thunen, Johann Heinrich (1783–1850)   Prussian landowner who developed a model of agricultural location in a commercial economy; the model concludes that farming activities are allocated by their profit-making capabilities into concentric rings around a central market city.

voyageur Literally, a traveler; a French-Canadian boatman or guide, especially one involved in trapping and transporting furs and other supplies between remote settlements in the U.S. and Canadian northwest.

W

walking cities Cities in which most, if not all, activities are accessible to a primarily pedestrian population; that is, roughly all cities from 10,000 B.C. to A.D. 1850.

wasted votes Votes cast for a losing candidate in an election.

weather The state of the atmosphere at a given time and place.

wetland An area that is either occasionally or permanently saturated with moisture, such as a marsh or tidal flat.

wilderness As defined by the U.S. Wilderness Act of 1964, an area of undeveloped land that contains ecological, geological, or other features of scientific value or offers outstanding opportunities for solitude or primitive forms of outdoor recreation.

windward The side of a mountain or island in the direct path of the prevailing wind.

winter wheat Wheat planted in autumn for early summer harvesting.

working-class urban The separate city neighborhoods of white and black industrial workers and neighborhoods containing immigrant groups with regular, modest jobs.

Wyoming Basin A western extension of the Great Plains in southern Wyoming, which formed a part of an easy route for pioneers across the Rocky Mountain barrier.

Y

young mobiles Those living on modest incomes in small houses and apartments in college and boom towns.

Z

zero population growth (ZPG) A situation in which a population is not changing in size from year to year as a result of the combination of births, deaths, and migration.

zone in transition The area close to the central business district that is characterized by high land values but marginal existing uses; theoretically, a zone that is "in transition" to higher and better uses.

CREDITS

13.13: © Matt Bradley/Tom Stack &
13.15: © Donovan Reese/Tony Stone Images;
13.17: © Wayne Scherr/Photo Researchers;
13.19: © James Shaffer; 13.21: © Greg Vaughn;
13.23: © Mark Gibson/Visuals Unlimited;
13.25: © G. Dimijian/Photo Researchers;
13.27: © K. Scholz/H. Armstrong Roberts.

Chapter 14

Opener: © Dennis Ray/Valan Photos; **14.1:** © Kennon
Cooke/Valan Photos; **14.4:** © Valan Photos;
14.5: © Brock May/Photo Researchers; **14.8:** The
National Gallery of Canada; **14.9:** © B. Mahoney/The
Image Works; **14.10:** © Picardi/Photo Network;
14.12: © Whelan/Valan Photos; **14.16:** © Jeannie R.
Kemp/Valan Photos; **14.17a, 14.17b, 14.19:** Courtesy
of David Kaplan; **p 406(top):** © Francis Lepine/Valan
Photos; **p. 406(bottom), p. 407(top):** © Kennon
Cooke/Valan Photos; **p. 407(bottom):** © Denis
Ray/Valan Photos; **14.22:** © Kennon Cooke/Valan
Photos.

ILLUSTRATION CREDITS

Chapter 2

Fig. 2.3   After C. S. Denny, *National Atlas of the
United States,* U.S. Geological Survey. Figure from
Charles C. Plummer and David McGeary, *Physical
Geology,* 6th ed. Copyright © 1993 Wm. C. Brown
Communications, Inc., Dubuque, Iowa. All Rights
Reserved. Reprinted by permission.
Fig. 2.24(inset)   From Robert N. Wallen,
Introduction to Physical Geography. Copyright © 1992
Wm. C. Brown Communications, Inc., Dubuque, Iowa.
All Rights Reserved. Reprinted by permission.
Figs. 2.33   Data from NOAA. Figure from Michael
Bradshaw and Ruth Weaver, *Physical Geography.*
Copyright © 1993 Wm. C. Brown Communications,
Inc., Dubuque, Iowa. All Rights Reserved. Reprinted
by permission.
Fig. 2.34   Data from U.S. Navy Oceanographic
Office. Figure from Robert N. Wallen, *Introduction
to Physical Geography.* Copyright © 1992 Wm. C.
Brown Communications, Inc., Dubuque, Iowa. All
Rights Reserved. Reprinted by permission.
Fig. 2.35   From Carla Montgomery, *Physical
Geology,* 3d ed. Copyright © 1993 Wm. C. Brown
Communications, Inc., Dubuque, Iowa. All Rights
Reserved. Reprinted by permission.

Chapter 3

Fig. 3.14   From *Canada Before Confederation: A
Study in Historical Geography* by R. Cole Harris and
John Warkenton. Copyright © 1974 by Oxford
University Press, Inc. Reprinted by permission.

Chapter 4

Fig. 4.18   From John Weeks, *Population: An
Introduction to Concepts and Issues,* Updated 5th
Edition. Copyright © 1994 Wadsworth, Belmont,
Calif. Reprinted by permission.
Figs. 4.21a, 4.21b, 4.21c, 4.21d   From William P.
O'Hare, "America's Minorities—The Demographics of
Diversity," *Population Bulletin* 47, no. 4 (Washington,
D.C.: Population Reference Bureau, Inc., December
1992). Reprinted by permission.

Chapter 6

Fig. 6.11   Modified from Bernd Andreae, *Farming,
Development and Space.* Copyright © 1981 Walter de
Gruyter, Inc. Reprinted by permission.
Fig. 6.23   Data from *National Wildlife Magazine,*
26(6), October/November 1988. Figure from David A.
Castillon, *Conservation of Natural Resources: A
Resource Management Approach.* Copyright © 1992
Wm. C. Brown Communications, Inc., Dubuque, Iowa.
All Rights Reserved. Reprinted by permission.

Chapter 7

Fig. 7.10   From Truman A. Hartshorn, *Interpreting
the City.* Copyright © 1980 John Wiley & Sons, Inc.,
New York, NY. Reprinted by permission.
Fig. 7.18   From Richard Florida and Martin Kenney,
"Restructurizing in Place" in *Economic Geography,*
68:158 (1992). Copyright © 1992 Economic
Geography, Clark University, Worcester, Mass.
Reprinted by permission.
Fig. 7.20   From Richard Florida and Martin Kenney,
"Restructurizing in Place" in *Economic Geography,*
68:157 (1992). Copyright © 1992 Economic
Geography, Clark University, Worcester, Mass.
Reprinted by permission.

Chapter 8

Fig. 8.7   From Curtis W. Richards and Michael L.
Thaller, "United States Railway Traffic: An Update" in
The Professional Geographer, 30(3):251 (August
1978). Reprinted by permission of Blackwell
Publishers, Cambridge, MA.
Fig. 8.21   From Donna Cox and Robert Patterson,
National Center for Supercomputing
Applications/University of Illinois. Reprinted by
permission.

Chapter 9

Fig. 9.8   From M. P. Cozen and P. D. Phillips, "The
Nature of Metropolitan Networks" in *Modern Men
Metropolitan Systems,* edited by C. M. Christian and
R. A. Harper. Copyright © 1982 Charles E. Merrill.
Reprinted by permission.

Fig. 9.9   Graphic from M. P. Cozen and P. D.
Phillips, "The Nature of Metropolitan Networks" in
Modern Metropolitan Systems, edited by C. M.
Christian and R. A. Harper, Data from *Census of
Population, 1970,* U.S. Census Bureau and calculations
by authors. Copyright © 1982 Charles E. Merrill.
Reprinted by permission.
Fig. 9.16, 9.17   From C. D. Harris and E. L. Ullman,
"The Nature of Cities" in *The Annals of The American
Academy of Political and Social Science.* Copyright
© 1945 The Annals of The American Academy of
Political and Social Science. Reprinted by permission.

Chapter 10

Fig. 10.2   Redrawn from R. A. Murdie, "Factorial
Ecology of Metropolitan Toronto." Research Paper
116, *Department of Geography Research Series,*
University of Chicago, 1969. Reprinted by permission.

Chapter 12

Fig. 12.9   Map reprinted by permission of
International Joint Commission.
Fig. 12.21   From Carla Montgomery, *Physical
Geology,* 3d ed. Copyright © 1993 Wm. C. Brown
Communications, Inc., Dubuque, Iowa. All Rights
Reserved. Reprinted by permission.
Fig. 12.27   From William P. Cunningham and
Barbara Woodworth Saigo, *Environmental Science,* 2d
ed. Copyright © 1992 Wm. C. Brown
Communications, Inc., Dubuque, Iowa. All Rights
Reserved. Reprinted by permission.

Chapter 14

Fig. 14.7   From C. C. Harris and S. Warkenton,
Canada Before Confederation. Copyright © 1991
Carleton University Press, Ottawa. Reprinted by
permission.
Fig. 14.15   From D. G. G. Kerr, *A Historical Atlas of
Canada,* 3d ed., 1975. Published by Nelson Canada, A
Division of International Thomson Limited. Reprinted
by permission.
Fig. 14.20   Figure 2.22 "Main Street, Canada" from
The North American City, 4th ed. by Maurice Yeates.
Copyright © 1990 by Harper & Row, Publishers, Inc.
Reprinted by permission of HarperCollins Publishers,
Inc.

INDEX

Note: Page numbers followed by letters *f* and *t* refer to figures and tables, respectively.

A

Absaroka-Beartooth Wilderness Area, 307*f*
Acadia, 383, 387
Acid rain, 324, 330–332
    aquatic effects of, 332
    areas affected by, 332*f*
    formation and effects of, 331*f*
    material effects of, 332
    terrestrial effects of, 332, 333*f*
Advertising, and homogenization of
        U.S., 278–279
African Americans
    in Canada, 398*t*
    infant mortality rate for, 98–99
    in Megalopolis, 351
    migration history of, 71–72, 72*f*, 274–275
    prosperous urban families, 262
    rural poor, 270
    urban poor, 262–263
    in U.S. counties, 106*f*
    working class, 263
Age, and recreation, 286
Agglomeration, 182
    economies of, 88–89
Agriculture, 143–163
    American Indian, 53, 53*f*
    colonial, 146
    depression in, 147
    mechanization of, 145–146, 147*f*, 151–152
    modernization and productivity, 148*f*-149*f*,
        148–151
    new landscape of, 155–163
    19th-century, 145–146
    physical limitations on, 144
    plantation, 57–58
        westward expansion of, 65–66, 66*f*
    regional specialization in, 75, 146
    specialization and commercialization
        of, 152–154
    subsistence, 58, 145
    20th-century, 146–155
    von Thünen's model of, 155*f*, 155–156, 156*f*
    and water pollution, 321–323
    water use in, 318
Airlines, 212–217
    deregulation of, 212, 214
        effect on route networks, 215*f*
    hub-and-spoke pattern, 214–216
    leading U.S. routes, 213*f*
    regional patterns, 216–217

Air masses
    source regions for, in North America, 39*f*
    tropical and polar, meeting of, 34
Air pollution, 327–333
    acid rain, 324, 330–332
    controlling, 332–333
    factors affecting, 328
    photochemical smog, 329–330
    sources of, 327, 327*f*
Airports
    leading U.S., 214*t*
    location of, 216
Alaska, 373*f*
    air transportation in, 216
    climate of, 40
    cultural characteristics of, 372–374
    national preserves in, 294, 296*f*
    purchase of, 60
    soils and vegetation of, 45–46
Alaska Highway, 206–207, 208*f*
Alaska Range, 26
Alberta, 379*f*
    economic indicators for, 404*t*
    population growth in, 84, 86*f*
    recreational resources in, 308–310
Allophones, 391
Alluvium
    in Central Valley of California, 28–29
    definition of, 21
    in Mississippi alluvial valley, 20–21
Alpine (mountain) glaciers, 16
Amenity locations, 90
    population shift to, 90–92
        economic factors for, 92–95
America, use of term, 3
American Indians
    agriculture of, 53, 53*f*
    in Canada, 380*f*, 380–381, 390–391
    location of major native groups, 52*f*
    population of U.S. counties, 107*f*, 108
    status before 1607, 52–54
Amtrak, 210
Amusement parks, 290
Anaheim, California, 371, 372*f*
Anglo-America, use of term, 3
Anglo-conformity, 396, 399
Anglophones, 391, 400
Animal wastes, and water pollution, 323, 324*f*
Annexation, 122
Appalachia, 348, 360, 360*f*
Appalachian Mountains, 22
    Mohawk Gap in, 55, 62
Appalachian Plateau, 22
Architecture, urban, 228, 228*f*

Arctic north. *See* North
Areal (particular) representation, 124
Areal specialization
    in agriculture, 75, 146
    transportation and, 202–203
Arizona. *See also specific cities*
    Grand Canyon, 23, 24*f*, 284*f*, 294, 295
    population growth in, 84
Asians
    immigration to Canada, 399, 401*f*
    immigration to U.S., 72–73, 104
    population growth, predictions for, 108
    population of U.S. counties, 107*f*
Atlanta, Georgia
    Hartsfield International Airport, 214*f*
    metropolitan area, population of, 239*t*
    population growth in, 90
Atlantic Provinces, Canadian, 23, 118
    recreational resources in, 307
Atwood, Margaret, 378
Austin, Texas, state capitol building in, 119*f*
Automobile, dominant role of, 203, 205
Automobile industry, locations of, 188, 188*f*
Avant-garde, 261–262
    meeting place of, 262*f*

B

Baby boom, 80
Baltimore, Lord, 57
Baltimore, Maryland
    industrialization and, 61–62
    waterfront of, 255, 255*f*
Banff National Park, 310, 310*f*
Basic sector, 235, 288
Bay Area Rapid Transit system, San Francisco,
        212, 212*f*, 306
Belt-ways, 205
Bermuda high pressure system, 39*f*
Bethlehem Steel Corporation, 196
Biocides, 322–323
Birth rate(s)
    in Canada, areal differences in, 97, 97*f*
    crude, 110
    French Canadian, 388, 392
    Mormon, 111, 364
    religion and, 111
    for rural population, 97
    for urban population, 97–98
    in U.S., areal differences in, 95–97, 96*f*
Black Hills, 16
    Grasslands, 45*f*
Blacks. *See* African Americans
Blizzards, 42
Blue bloods, 266

Index

Open-range cattle ranching, 68–70
Opponent-concentration gerrymandering, 127
Opponent-dispersion gerrymandering, 127
Oregon Country, settlement of, 60, 66
Organized labor, 190, 275
Orlando, Florida, 278
Orographic precipitation, 33, 35f, 37
  in Hawaiian Islands, 37
  on West Coast, 39
ORVs. See Off-road vehicles
Ottawa, Ontario, 308
  Parliament Hill, 411f
Ozarks, 16
Ozone, 329
  hazards associated with, 330
Ozone pollution, 330, 330f

P

Pacific Ocean, effect on climate, 39–40
Pacific Plate, 25
Parks
  amusement, 290
  national. See National parks
  urban, 288–289, 297
Particular (areal) representation, 124
Parti québécois, 396
Partriotes, French-Canadian, 395
Passenger-kilometers, 203
PCBs. See Polychlorinated biphenyls
Peak land value intersection (PLVI), 243–244
Penn, William, 58
Pennsylvania. See also specific cities
  settlement of, 58
  soils and vegetation of, 45, 45f
Percy, George, 2
Performing arts, 291
Periphery regions, 74–76, 75f
Petroleum
  foreign imports of, 192f
  geographic distribution of, 170f
  pipelines, 220–222, 221f
pH factor, 331
Philadelphia, Pennsylvania
  colonial, 58, 59f
  industrialization and, 61–62
  manufacturing in, 179
  metropolitan area, population of, 239t
  residential patterns in, 270–272, 271f
  site of, 230, 231f
Phoenix, Arizona, 366
  metropolitan area, population of, 240t
Photochemical smog, 329–330
Physiography, 4–5
Pike, Zebulon, 68
Pilgrims, 58
Pipelines, oil and natural gas, 220–222, 221f
Pittsburgh, Pennsylvania
  in Midwest economic region, 62, 64
  population loss in, 92
  site and situation of, 231
  steel industry in, 74, 74f
Planned community, 265f
Planned industrial park, 199f
Plantation agriculture, 57–58
  westward expansion of, 65–66, 66f

Plateau(s)
  Appalachian, 22
  definition of, 22
  Western, 23–24
PLVI. See Peak land value intersection
Plymouth, Massachusetts, 52
Plymouth Colony, 58
PMSAs. See Primary Metropolitan Statistical
      Areas
Point sources, of water pollution, 321
Political economy approach, to enterprise
      location, 187–190
Political fragmentation, 255
Political geography, 115–139
  definition of, 116
  future directions for, 133–137
Political patterns, in cities, 254–255
Pollution. See Environmental pollution
Polychlorinated biphenyls (PCBs), 323
Poor
  housing projects for, 265
  rural, 270
  urban, 262–263, 280
Population. See also Population density;
      Population growth; Population shift
  Canadian, 391–392
    in metropolitan areas, 241t
  of North America, 7–8
  U.S.
    ethnic diversity of, 105–108
    future patterns of, 108–110
    in metropolitan areas, 238t–240t
    rural, 89–90
    urban, 8, 88–89
  world, leading countries, 8t
Population density, 7–8, 84–88, 87f
  economic factors for, 92–95
  of ten most populous countries, 8t
  world, 9f
Population geography, 80
Population growth
  areas with greatest, 90–92, 91f
  in Canada, 80, 81t, 84, 85f, 86f
  colonial, 55, 58–59
  measuring, 110
  and recreation, 285
  regional patterns of, 80–84
  since 1950, 80–84, 81t
  in Sun Belt, 82f, 84, 90–92
    developments responsible for, 92–95
  in U.S., 80–84, 81t
    by state, 82f, 83f
  zero, 80
Population policy, 84
Population shift
  to amenity locations, 90–92
    economic factors for, 92–95
  in U.S., 81, 81f
Port Royal, Nova Scotia, 55
Ports, 220
  leading U.S., 221f
Prairie Grain Belt culture region, 361–363, 362f
Prairies, Canadian. See Canadian Prairies
Precipitation, 32–33
  causes of, 33

convectional, 33, 36, 37, 37f
cyclonic (frontal), 33, 36f
hail, 40f, 40–41
in humid continental climate, 34
in humid subtropical climate, 36
orographic, 33, 35f, 37, 39
in steppe climates, 37
on West Coast, 39, 40
Preservationist ethic, 298
Preserves, national, 294, 296f
Presidential elections, geography of, 116f,
    132–133
Primary economic activities/industries,
    93, 142, 143
Primary Metropolitan Statistical Areas
    (PMSAs), 89, 230
Product cycle approach, to enterprise
    location, 187
Providence, Rhode Island, 61
  population of, 92
Public administration, U.S. cities specializing
    in, 237f
Pueblo natives, 55
Puerto Rico, immigrants from, 72
Puget Sound Lowland, 29
Puritans, 58

Q

Quaternary sector, 93
Québec, 379f. See also specific cities
  dam construction in, 320
  economic indicators for, 404t
  English speakers in, 394
  immigrants in, 400–401
  manufacturing in, 180
  19th-century, 395–396
  political demands of, 409
  recreational resources in, 307–308
  regional culture of, 406
  20th-century, 396
Québec City, 136f, 307–308
  Château Frontenac, 309f
  founding of, 55
Quiet Revolution, 396

R

Race and ethnicity. See also Ethnic diversity
  city distributions by, 253–254
  redistricting on basis of, 129–131
  and residential location, 259–260, 260f
Radio, 222
Radioactive waste, 342–343
  storage of, 316, 316f, 345f
  storage sites in U.S., 344f
Rail cars, double-stacked containers on, 209, 209f
Railroads, 209–212, 210f
  and manufacturing development, 179
Rainshadow, 39–40
Raleigh, Sir Walter, 142
Range livestock farming, 160–162, 161f
  geographical regions for, 157f
Rang settlements, 385, 386f
Rate of natural increase, 110
Real estate, U.S. cities specializing in, 237f
Reapportionment revolution, 128–131